AF794419

Nitrogen and Phosphorus Nutrition of Cattle

Reducing the Environmental Impact of Cattle Operations

Nitrogen and Phosphorus Nutrition of Cattle

Reducing the Environmental Impact of Cattle Operations

Edited by

Ernst Pfeffer

Institut für Tierernährung der Universität Bonn, Endenicher Allee 15, D-53115 Bonn, Germany

and

Alexander N. Hristov

Department of Animal & Veterinary Science, University of Idaho, PO Box 44-2330, Moscow, ID 83844-2330, USA

CABI Publishing

CABI Publishing is a division of CAB International

CABI Publishing
CAB International
Wallingford
Oxfordshire OX10 8DE
UK

Tel: +44 (0)1491 832111
Fax: +44 (0)1491 833508
E-mail: cabi@cabi.org
Web site: www.cabi-publishing.org

CABI Publishing
875 Massachusetts Avenue
7th Floor
Cambridge, MA 02139
USA

Tel: +1 617 395 4056
Fax: +1 617 354 6875
E-mail: cabi-nao@cabi.org

©CAB International 2005. All rights reserved. No part of this publication may be reproduced in any form or by any means, electronically, mechanically, by photocopying, recording or otherwise, without the prior permission of the copyright owners.

A catalogue record for this book is available from the British Library, London, UK.

A catalogue record for this book is available from the Library of Congress, Washington, DC, USA.

Library of Congress Cataloging-in-Publication Data
Nitrogen and phosphorus nutrition in cattle / edited by Alexander A. Hristov and Ernst Pfeffer.
p. cm.
Includes bibliographical references (p.).
ISBN 0-85199-013-4 (alk. paper)
1. Cattle--Feeding and feeds. 2. Nitrogen in animal nutrition. 3. Phosphorus in animal nutrition. I. Hristov, Alexander A. II. Pfeffer, Ernst. III. Title.
SF203.N58 2005
636.2′0852--dc22
2004022637

ISBN 0 85199 013 4

Typeset by SPI Publisher Services, Pondicherry, India
Printed and bound in the UK by Biddles Ltd, King's Lynn

Contents

Contributors

Dr Bannink, Wageningen University and Research Centre, Institute for Animal Science and Health, PO Box 65, 8200 AB Lelystad, The Netherlands.

Dr Beede, Michigan State University, Department of Animal Science, 2265K Anthony Hall, East Lansing, MI 48824-1225, USA.

Dr Ferguson, University of Pennsylvania, Department of Clinical Studies, New Boldon Center, 382 West Street Road, Kennett Square, PA 19348, USA.

Dr Firkins, Ohio State University, Department of Animal Sciences, College of Food, Agriculture and Environmental Science, Columbus, OH 43210, USA.

Dr Hristov, University of Idaho, Department of Animal and Veterinary Science, PO Box 44-2330, Moscow, ID 83844-2330, USA.

Dr Huhtanen, MTT Agrifood Research Centre, Animal Production Research, FIN-31600, Jokioinen, Finland.

Dr Hunt, University of Idaho, Department of Animal and Veterinary Science, PO Box 44-2330, Moscow, ID 83844-2330, USA.

Dr Hvelplund, Institute of Agricultural Sciences, Department of Animal Nutrition and Physiology, PO Box 50, DK-8830 Tjele, Denmark.

Dr Jouany, Institut National de la Recherche Agronomique, Centre de Clermond-Ferrand – Theix, F-63122 Saint Genes Champanelle, France.

Dr Kincaid, Washington State University, Animal Sciences Department, Pullman, WA 99164-6310, USA.

Dr Kohn, University of Maryland, Department of Animal and Avian Sciences, College Park, MD 20742, USA.

Dr Newbold, Rowett Research Institute, Bucksburn, Aberdeen, AB21 9SB, UK.

Professor Pfeffer, Institut für Tierernährung der Universität Bonn, Endenicher Allee 15, D-53115 Bonn, Germany.

Dr Reynolds, College of Food, Agriculture and Environmental Sciences, Wooster, OH 44691, USA.

Dr Rodehutscord, Martin-Luther-Universitat Halle-Wittenberg, Institut fur Ernahrungswissenschaften, D-06108 Halle (Saale), Germany.

Dr Schröder, Plant Research International, Wageningen University and Research Centre, PO Box 16, 6700 AA Wageningen, The Netherlands.

Dr Schwab, University of New Hampshire, Department of Animal and Nutrition Sciences, Ritzman Lab, 22 Colovos Road, Durham, NH 03824, USA.

Dr Sklan, Hebrew University, Faculty of Agriculture, PO Box 12, Rehovot 76-100, Israel.
Dr Valk, Animal Sciences Group, Edelhertweg 15, PO Box 65, NL 8233 AB Lelystad, The Netherlands.
Dr Walker, Rowett Research Institute, Bucksburn, Aberdeen, AB21 9SB, UK.
Dr Wallace, Rowett Research Institute, Bucksburn, Aberdeen, AB21 9SB, UK.

Preface

Animals depend on regular supply of a number of nutrients serving different functions in their metabolism. These nutrients have to be provided by feeds ingested by the animals. Normally, nutrients yielding metabolizable energy are responsible for most of the feed cost. For this reason it appeared logical for a long time to aim at maximum efficient utilization of feed energy as the target of calculating rations for farm animals, while more or less generous 'safety margins' were recommended with respect to less expensive nutrients by advisors in all countries until recently.

This purely economical approach of optimizing rations did not take into consideration the fate of that part of ingested nutrients which is not transferred into the animal products. Only towards the end of the 20th century was it generally recognized that animal units may be the cause of dramatic local or regional surpluses of nutrients creating serious impacts on soil, water and air.

Limiting nitrate in drinking water to lowered concentrations after changed legislation appeared especially critical from groundwater found in regions with high stocking densities of farm animals and it was estimated that dairy cows were responsible for more than half of the ammonia emitted into the air, consequently causing accumulations of nitrogenous compounds in natural precipitation. Even after removal of phosphates from detergents intensive growth of algae was observed in lakes and streams and this was interpreted to a great proportion as a consequence of phosphate enrichment in particulate matter transferred from fields into surface water due to erosion. Again, the highest phosphate concentrations of soils were found in regions with very high stocking densities.

Animal nutritionists increasingly realized that this situation is to be seen as a challenge to their scientific discipline. Avoiding nutrient deficiencies by allowing unnecessary safety additions may ignore the ecological demand that production of food for humans has to be sustainable.

A great number of studies dealing with details of sustainable animal production has been carried out and published and any attempt to survey the present state of the art has to be restricted with respect to species as well as nutrients. This book, therefore, is restricted to nitrogen and phosphorus in cattle, from basic biological facts to practical feeding and farm management.

The editors are grateful to all authors for their respective contributions and to CABI for publishing this book. In September 2004 we received the sad news of the death of David Sklan, he will be remembered as a respected scientist and a dear colleague.

Ernst Pfeffer and Alex Hristov

Bonn, Germany, and Moscow, Idaho, October 2004.

Abbreviations

Chapter 1:

AFO	Animal feeding operation
CAA	Clean Air Act
CAFO	Concentrated animal feeding operation
CWA	Clean Water Act
DM	Dry matter
ELG	Effluent limitations guidelines
EPA	Environmental Protection Agency
NMP	Nutrient management plan
NPDES	Nutrient pollution discharge elimination system
NPN	Non-protein nitrogen
PM_x	Particulate matter (equivalent diameters less than $\times$ μm)
TMR	Total mixed ration
VAPS	Voluntary alternative performance standards
VOC	Volatile organic compounds

Chapter 2:

AA	Amino acids
AAT	Amino acids absorbed from the small intestine
ADG	Average daily gain
ADIN	Acid detergent insoluble nitrogen
ATP	Adenosine tri-phosphate
BW	Body weight
CNCPS	Cornell Net Carbohydrate and Protein System
CP	Crude protein
dCHO	Intake of digestible carbohydrates
DIM	Days in milk
DIP	Digestible intake protein
DK	Danish system of protein evaluation
DM	Dry matter

DMI	Dry matter intake
DOM	Digestible organic matter
DUP	Digestible undegraded protein
DVE	Darm Verteerbar Eiwit
EAA	Essential amino acids
ECM	Energy corrected milk
ECP	Endogenous crude protein
EDP	Effective protein degradability
EE	Ether extract
EDP	Effective degradability of protein
EPD	Effective protein degradability
EQSBW	Equivalent shrunk body weight
ERDP	Effective rumen degradable protein
FIN	Finnish system of protein evaluation
FME	Fermentable metabolizable energy
FOM	Fermentable organic matter
GER	German system of protein evaluation
His	Histidine
INRA	Institut Nationale de la Recherche Agronomique
L	Leeding of feeding (multiple of maintenance)
Leu	Leucine
Lys	Lysine
MCP	Microbial crude protein
Met	Methionine
MP	Metabolizable protein
MPY	Milk protein yield
MSPE	Mean squared prediction error
MUN	Milk urea nitrogen
NDF	Neutral detergent fibre
NAN	Non-ammonia nitrogen
NE	Net energy
NPN	Non-protein nitrogen
NRC	National Research Council
NSC	Non-structural carbohydrates
nXP	Utilizable crude protein
OM	Organic matter
PBV	Protein balance value in the rumen
PDIA	Truly digestible rumen undegraded protein
PDIE	Protein value, when energy is limiting microbial growth
PDIN	Protein value, when nitrogen is limiting microbial growth
QDP	Quickly degraded protein
RDP	Rumen degradable feed protein
RE	Retained energy
RMSE	Root mean square error
RUP	Rumen undegradable feed protein
SDP	Slowly degraded protein
TDN	Total digestible nutrients
Thr	Threonine
TP	Tissue protein
VAL	Valine
VFA	Volatile fatty acids
WG	Weight gain

Chapter 3:

AA	Amino acids
Ala	Alanine
Arg	Arginine
ATP	Adenosine tri-phosphate
BAC	Bacterial artificial chromosome
CFB	Cytophaga-flexibacter-bacteroides
CP	Crude protein
CPCR	Competitive polymerase chain reaction
DCCD	Dicyclohexylcarbodiimide
DIC	Diphenyliodonium chloride
DM	Dry matter
DNA	Deoxy ribonucleic acid
DPP	Dipeptide peptidase
EDTA	Ethylene diamine tetraacetic acid
GDH	Glutamate dehydrogenase
GIT	Gastro-intestinal tract
Gly	Glycine
GM	Genetically modified
HAP	Ammonia hyperproducing bacteria
Leu	Leucine
LPNA	Leucine p-nitroanilide
Lys	Lysine
LysAlaMNA	Lysine alanine 4-methoxy-2-nitroanilide
mRNA	Messenger ribonucleic acids
NAD	Nicotinamide adenosine dinucleotide
NADP	Nicotinamide adenosine dinucleotide phosphate
NSAAPPPNA	N-Succinyl alanine alanine phenylalanine proline p-nitroanilide
PCR	Polymerase chain reaction
Pro	Proline
RDP	Rumen degradable protein
RDNA	Ribosomal deoxy ribonucleic acid
RNA	Ribonucleic acids
scFA	Short chain fatty acids
SDS-PAGE	Sodium dodecyl sulphate polyacrylamide gel electrophoresis
TCA	Tri-carboxylic acid

Chapter 4:

ATP	Adenosine tri-phosphate
BCFA	Branched chain fatty acids
BW	Body weight
CHO	Carbohydrates
CP	Crude protein
CT	Condensed tannins
DM	Dry matter
DMI	Dry matter intake
EO	Essential oils
ESBM	Expeller soybean meal
FA	Fatty acid

GLU	Corn dextrose
HMEC	High moisture ear maize
HT	Hydrolysable tannins
MN	Microbial nitrogen
MPS	Microbial protein synthesis
MUN	Milk urea nitrogen
NAN	Non-ammonia nitrogen
NDF	Neutral detergent fibre
NE	Net energy
NFC	Non-fibre-carbohydrates
NPN	Non-protein nitrogen
NSC	Non-structural carbohydrates
OM	Organic matter
PUN	Plasma urea nitrogen
RDP	Ruminally degradable dietary protein
RUP	Ruminally undegradable protein
RUSITEC	Rumen simulation technique
SSBM	Solvent soybean meal
STA	Corn starch
TNC	Total non-structural carbohydrates
VFA	Volatile fatty acids
WSC	Water soluble carbohydrates

Chapter 5:

ATP	Adenosine tri-phosphate
BUN	Blood urea nitrogen
CP	Crude protein
DM	Dry matter
DMI	Dry matter intake
MRNA	Messenger ribonucleic acid
NAD	Niacin adenosine dinucleotide
NAN	Non-ammonia nitrogen
NANMN	Non-ammonia non-microbial nitrogen
NEL	Net energy for lactation
NPN	Non-protein nitrogen
PDV	Portal drained viscera
RDP	Rumen degradable protein
RNA	Ribonucleic acids
RUP	Rumen undegradable protein
TDN	Total digestible nutrients

Chapter 6:

ADG	Average daily gain
ATP	Adenosine tri-phosphate
FTU	Unit of phytase activity
P_i	Inorganic phosphate

Chapter 7:

CP	Crude protein
DipM	Disintegrations per minute
DM	Dry matter
DMI	Dry matter intake
P_i	Inorganic phosphate
PTH	Parathyroid hormone
SA	Specific radioactivity

Chapter 8:

ATP	Adenosine tri-phosphate
cAMP	Cytosolic adenosine monophosphate
CL	Corpora lutea
CP	Crude protein
CR	Conception rate
DIPR	Difference between requirement for and dietary supply of rumen degradable protein
DM	Dry matter
DMI	Dry matter intake
DNA	Deoxy ribonucleic acids
LH	Luteinizing hormone
LR	Likelihood Ratio
MP	Metabolizable protein
MUN	Milk urea nitrogen
P_i	Inorganic phosphate
RDN	Rumen degradable protein
RUP	Rumen undegradable protein
SPC	Services per conception
TDN	Total digestible nutrients

Chapter 9:

A	Milk and meat
AN	Additional nitrogen requirement
ANU	Additional nitrogen requirement per unit milk and/or meat
CP	Crude protein
CF	Nutrients of crops appearing as feed
DM	Dry matter
EX	Fraction of harvested nutrients being exported
F	Feed and bedding
FP	Transfer of nutrients from feed to product (efficiency of nutrient utilization)
I	Nutrient input
IM	Fraction of nutrients in feed and bedding material being imported
IMN	Permitted feed nitrogen import per hectare
IMNU	Permitted feed nitrogen imported per unit milk and/or meat
IMP	Permitted feed phosphorus import per hectare

IMPU	Permitted feed phosphorus imported per unit milk and/or meat
L	Nutrient losses
M	Loss in faeces, urine and worn bedding
MACN	Maximum attainable crop nitrogen per hectare
MAON	Maximum attainable nitrogen output per hectare
MP	Metabolizable protein
MS	Transfer of manure nutrients to soil
O	Nutrient output
RDP	Rumen degradable protein
SC	Transfer of nutrients from soil to harvested crops

1 Interactions between Cattle and the Environment: a General Introduction

E. Pfeffer[1] and A.N. Hristov[2]
[1]*Institut für Tierernährung der Universität Bonn, Bonn, Germany*
[2]*Department of Animal and Veterinary Science, University of Idaho, Moscow, Idaho, USA*

1.1 Role of Animals in Man's Search for Food

At the beginning of human civilization, hunting animals was the predominant way to find food for man in most parts of the world. Domestication of animals was a remarkable step to secure food when, as a consequence of the growing density of human population, natural resources limited the potential quantity of food to be found just by hunting.

Developing pastoral systems were characterized by large areas producing little or no crops that could be consumed directly by man. Most of the vegetation growing on these areas could be utilized only as feed for the herds, mostly consisting of ruminants. Regular bleeding of animals and using the blood as food, from time to time slaughtering individual animals from the flock and finally allowing the offspring to drink only a part of the milk produced by their dams, in order to use the remaining milk as food for human consumption, were phases of developing more intensive forms of animal husbandry. Each of these phases ranging from nomadic systems to intensive grassland management can still be found in some regions of the world. The major function of animals in these systems is to extract nutrients from vast areas and concentrate them into food for man. In this phase excreta of the animals usually raise hardly any interest in herdsmen.

In order to increase the amount of food harvested per unit of area, land was ploughed and crop production was started in areas where climate and access to water allowed this. Density of human population usually is much higher in these crop-producing than in pastoral systems, i.e. land often is limiting the potential amount of food produced. Animals in such systems have the function to increase yields per unit of area and this is achieved by using them as draught animals and by using their excreta as fertilizer on the fields. The old German expression of 'pasture as the mother of arable land' illustrates this situation: draught animals and animals grazing on extensive rangeland during the daytime were flocked or kept in stalls overnight; excreta voided during

©CAB International 2005. *Nitrogen and Phosphorus Nutrition of Cattle* (eds E. Pfeffer and A.N. Hristov)

the night were conserved and used to increase the concentration of plant nutrients in the soil of tilled fields. The author of the first German textbook of agricultural science expressed his opinion about the function of animals in farms as follows:

> Die Tiere sind bloß wie Maschinen anzusehen, welche . . . die Fütterung zum . . . bei weitem größern Theil . . . in Mist . . . verwandeln (The animals are to be regarded just like machines which to by far the greater part convert feed into manure) (Thaer, 1809, p. 257).

Although plant nutrients were not yet identified, it was recognized that without returning excreta of animals as manure fertility of the fields could not be sustained. Today, in most areas farmers and extension workers no longer regard manure as the only source of plant nutrients, but 'cut and carry' systems in some areas seem to still follow this line. As long as farmers do not purchase fertilizer or feeds they are in danger of having negative nutrient balances in their fields, and for this reason excreta of animals are regarded as a saving box for plant nutrients which have to be returned to the land from which they were originally extracted and transferred into plant material.

Up to a certain degree, therefore, 'horizontal movement of nutrients' can be an intended effect of animal husbandry by which animals carry nutrients from wide areas into folds or stalls, where their excreta are regarded as a major product of high value.

More than a 100 years after Albrecht Thaer, Theodor Brinkmann, professor of farm management in Bonn, tried to determine the value of the various production factors for the farmer. Although he no longer regarded excreta as the main animal product, he pointed out that purchased concentrate feeds not only promoted milk and meat production directly but also imported plant nutrients into the farm. The monetary value of these plant nutrients had to be taken into account; he critically added that this, however, was valid only as long as the respective plant nutrients were truly missing in the farm because otherwise purchased feeds would only increase existing surpluses (Brinkmann, 1922, p. 109). This latter situation of excessive presence of nutrients has developed towards the end of the 20th century in wide regions of Europe and North America with the consequence of negative ecological effects. A first attempt to create a comprehensive international overview on emission of ammonia was made more than 10 years ago (Klaassen, 1992) and feeding strategies to decrease potentials for nitrogen (N) and phosphorus (P) pollution have gained increasing relevance (CAST, 2002). This book intends to summarize scientific aspects related to nitrogen and phosphorus supply and use by cattle and resulting impacts on sustainability of agriculture.

The restriction to N and P appears justified at present as these nutrients have been found to play a predominant role in the fertility of soils and in impacts on the environment, but other elements will have to be taken into consideration as well in the near future.

1.2 Historical Highlights in Research Concerning N and P as Nutrients

Of the more than 100 elements found in the periodic table today, only a dozen were known 350 years ago, among them carbon, sulphur, iron, copper, silver and gold. The term 'element' was not used in today's meaning and alchemists were convinced that they could, by experimentation, find the 'philosopher's stone' by which they could turn worthless materials into gold. One of these alchemists was Henning Brand in Hamburg who in 1669 heated concentrated urine without admitting air and found a snow-white substance, which immediately burned out when exposed to air, thereby illuminating the dark room (Childs, 2003; Van der Krogt, 2003d). This property of giving light was the base for naming of the substance discovered by Brand, from the Greek words φωσ [phos] = light; and φερω [phero] = to carry, to bring. Phosphorus thereby was the first element to be identified in modern times. About 100 years after Brand's discovery, the Swedish chemists Gahn and Scheele found calcium phosphate to be a major constituent of bone (McDowell, 1992). Today it is common knowledge that P is involved in practically all metabolic processes as phosphate ($H_2PO_4^-/HPO_4^{2-}$) or as phosphate-containing organic compounds.

About a century after the finding of P, the identification of three gases substantially promoted the scientific understanding of nature (Van der Krogt, 2003a,b,c):

1. In 1766, Henry Cavendish reported to the Royal Society in England about 'inflammable air from the metals'.

2. In 1772, Daniel Rutherford in Scotland showed that air in which animals had breathed (even after removal of the exhaled 'fixed air' – carbon dioxide) was no longer able to burn a candle, he named this entity '*aer malignus*' or noxious air.

3. In 1774, Joseph Priestly obtained a colourless gas by heating red mercuric oxide in which a candle would burn 'with a remarkable flame' (Carl Wilhelm Scheele in Sweden had discovered the same gas in 1766, but his publication was delayed until 1777, due to neglect by his publisher).

Antoine Lavoisier (1743–1794) suggested names for these gases derived from Greek. They include the syllable 'gène' from γεινομαι (geinomai) = to engender, bring forth.

As combustion of the 'inflammable air' always produced water, it was characterized by the word υδωρ (hydro) = water, hydrogène (H) in French and hydrogen in English. The German name Wasserstoff means the identical (Wasser = water; Stoff = material).

The major property of the gas causing the 'remarkable flame' was thought to be the formation of acids. Therefore, the word οξυσ (oxys) = acid became characteristic for oxygène (O) in French, oxygen in English and Sauerstoff in German (sauer = acid, sour).

Referring to the gas discovered by Daniel Rutherford, Lavoisier pointed out:

> nous l'avons donc nommé azote, de l'α privatif des Grecs, et de ζωη, vie, ainsi la partie non respirable de l'air sera le gaz azotique (we, therefore, named it azote, from the Greek alpha privativum and from ζωη, life, thus the not respirable part of the air will be the azotique gas).

Following the same thought, the gas was named Stickstoff in German, derived from the verb ersticken = to suffocate. In 1790, Jean Antoine Chaptal proposed the name nitrogène. The Greek word νιτρον [nitron] was used for saltpetre (potassium nitrate), thus the name nitrogène means 'making soda/saltpetre' (Van der Krogt, 2003b). The latter name was adopted in English as nitrogen.

With carbon and sulphur known for a long time and the three elements nitrogen, oxygen and hydrogen discovered before the end of the 18th century, interest increased in the quantitative analyses of elements in various organic materials at the beginning of the 19th century. Mulder (1838) carried out a large series of analyses in what he called the 'most important substances in the animal kingdom' – fibrin, albumin and gelatine. Regularly, he found that these substances contained more than 50% carbon, about 22% oxygen, between 15.5% and 16% nitrogen, about 7% hydrogen, and less than 1% phosphorus and sulphur. He stated:

> La matière organique, étant un principe général de toutes les parties constituantes du corps animal, et se trouvant, comme nous verrons tantot, dans le règne végétal, pourrait se nommer *Protéine* de πρωτειοσ primarius (the organic matter, being a general principle of all parts forming the animal body and to be found, as we shall soon see, in the plant kingdom as well, may be named *Protein* from *proteios* [Greek] = *primarius* [Latin]).

Thus, the name protein was meant to indicate that organic compounds containing nitrogen are by no means adverse to life (azotique) but, on the contrary, are of primary importance and play a predominant role in biological processes.

This thought was immediately taken up by Justus von Liebig who is often referred to as 'father of agricultural chemistry'. Liebig (1840, p. 64) wrote:

> In dem humusreichsten Boden kann die Entwicklung der Vegetabilien nicht gedacht werden ohne das Hinzutreten von Stickstoff, oder einer stickstoffhaltigen Materie (In soil, even richest in humus, it is impossible to imagine development of plants without the presence of nitrogen or nitrogen containing material).

He then continues to explain that there is no reason for believing that N from the air can participate in processes of animals or plants and that, on the other hand, he had found strong correlations between the amount of ammonia taken up through the roots and the amount of gluten formed in grains. Further, he observed that the presence of P was essential for the transformation of N from ammonia into protein formed by plants.

Liebig's conviction that there were only three proteins and that these were transferred without any change from plants as food into animal tissues (Liebig, 1843) was challenged by the work of Voit (1872) who found considerable differences in N balances of dogs fed varying proportions of meat and gelatine. Thomas (1909) balanced N in his own body over periods in which he ingested a constant N-free basal diet of starch and sugar either alone or supplemented by different vegetable or animal products as sole sources of protein. From the results, he concluded that clear differences exist in the 'biological value' of the protein in different foods. Mitchell (1924), taking up the basic idea of Thomas (1909), defined the 'biological value' of a diet component fed to rats as the percentage of absorbed N equivalent to the sum of metabolic faecal N, endogenous urinary N and retained N. A more complete review of the history of research and understanding of protein metabolism is given by Munro (1964).

Amino acids were identified in the period between 1806 and 1935 (Meister, 1965). Once the biological function of these components of all natural proteins had been discovered, analyses of indispensable amino acids became more meaningful than the biological value of complete proteins. In non-ruminant nutrition nowadays, free amino acids are frequently used for upgrading natural proteins and requirements, as well as recommendations for supply, and are increasingly based on amino acids absorbed prior to the caecum, i.e. from the small intestine.

Towards the end of the 19th century, fundamental differences between non-ruminants and ruminants with regard to utilization of N became obvious. Zuntz (1891), at the end of a review dealing with digestion of cellulose, addressed the finding that asparagine as the sole source of dietary nitrogen is worthless in dogs but has positive effects in ruminants. He proposed the hypothesis that nitrogen of asparagine and comparable amides might be incorporated into microbial protein, which then could be digested by ruminants. This is seen as the starting point of research into non-protein nitrogen (NPN) use in ruminants (Bergner, 1986).

More than 50 years after Zuntz's hypothesis, Loosli *et al.* (1949) presented concentrations of the ten essential amino acids in rumen material, faeces and urine of three sheep and two goats fed diets containing urea as the sole source of dietary N; the results were clear evidence of massive amino acid synthesis in the rumen. Lambs fed this diet gained about 100 g daily. Microbial synthesis of all amino acids was fully confirmed in rumen-fistulated calves by Duncan *et al.* (1953). Long-term feeding experiments in Finland finally proved that cows fed purified rations with urea and ammonium salts as the sole sources of N could not only survive but reproduce and produce moderate milk yields with normal composition over repeated lactations (Virtanen, 1966).

The potential of microorganisms to utilize NPN is not restricted to urea as a feed additive – it is also relevant for urea synthesized in the liver of their host animal. Simonnet *et al.* (1957) found in anaesthetized sheep that urea accumulated in the isolated forestomach filled with saline and concluded the existence of a cycle by which urea present in the blood was returned into the digestive tract. Schmidt-Nielsen *et al.* (1957) showed in a camel on very low N intake that not only quantities of urea in the urine were minimized but also that intravenously infused urea was retained in the body. From measuring urea clearance rates and glomerular filtration rates, these authors concluded that fractions of the filtered urea excreted were about 40% during normal N intake but only 1–2% during extremely low N intake. One way for blood urea to enter the rumen is via saliva, but there is also a direct transfer through the mucosa of the rumen wall, which has been reviewed by Houpt (1970). The role played by bacteria adhering to the rumen wall in the transfer of urea N from the blood into the rumen was reviewed by Cheng and Costerton (1980).

Rapidly growing knowledge about factors influencing the quantity of amino acids flowing to the duodenum of cattle led to the consequence that digestible crude protein could no longer be regarded as an adequate basis for describing requirements and supply of N in ruminants, and alternative systems were proposed (Roy *et al.*, 1977; Satter and Roffler, 1977; Vérité *et al.*, 1979; Madsen and Hvelplund, 1984; Rohr *et al.*, 1986). The present state of the art with respect to N requirement and systems of feed evaluation is reviewed in Chapter 2 of this book. Chapters 3 and 4 summarize the present knowledge about N metabolism in ruminal microorganisms and discuss potential strategies for improving the efficiency of N utilization by manipulation of microbial metabolism.

1.3 Resources of N and Phosphate as Plant Nutrients

Only very low concentrations of N are found in rocks from which soil originates. Fixation of N_2 from the air can be achieved by some microorganisms, free-living or in symbiosis with higher plants. Among the latter, legumes are of particular importance in agriculture. When a certain concentration of organic matter has accumulated in the soil, primarily through microbial fixation of N_2, organically bound N can be mobilized again into low-molecular-weight compounds like amino acids, ammonia and nitrate, which are taken up by plant roots. Nitrogen may be lost from soil by diffusion of nitrate into groundwater or by volatilization of ammonia.

Rocks are the major reservoir of phosphates. When soil is formed from rocks, orthophosphate is formed from apatites. Phosphorus in the soil is present on the surface of various adsorbents as precipitates with several inorganic cations or as organically bound phosphate. The central pool through which these separate pools communicate is the small amount of ionized orthophosphate in the soil solution. Plants and soil organisms take up ionized phosphate. Phosphorus may be lost by diffusion of phosphate into the groundwater or by erosion of adsorbing particles into surface water.

Insufficient replacement of nutrients extracted by plants from the soil of fields was a major reason for low crop yields with the consequence of increasing poverty and famines at regular intervals in Europe over long periods. In the 19th century, acidulating bones with the aim of increasing the solubility of phosphate was attempted empirically in several places and finally the industrial production of superphosphate, predominantly from bones, was developed. Considerable quantities of plant nutrients were transported from South America to Europe in the form of Chile nitre (mainly sodium nitrate) mined in the Atacama desert and of guano, excreta of birds on the Peruvian islands, rich in salts of nitric acid and phosphoric acid.

Phosphate ores were first mined in relatively small amounts in the 1840s in England, France and Spain and later in other countries; today most of the phosphate fertilizer and phosphate chemicals are produced from phosphate rock (Beaton, 2003). Table 1.1 shows today's important areas of phosphate mining. Phosphate-containing ore bodies are finite, non-renewable resources. Reserves are defined as deposits that may potentially be feasible at some time in the future. Reserve base is that part of an identified resource that meets specified minimum production practices. Reserve and reserve base at present cost less than \$36/t and \$90/t, respectively. At current production levels, the world's reserve and reserve base are estimated to last for less than 100 years and about 340 years, respectively (Roberts and Stewart, 2002).

The most important step towards overcoming the shortage of plant nutrients was taken in 1909

Table 1.1. World phosphate rock production, reserves and reserve base. (From Roberts and Stewart, 2002.)

Country	Production 1997–2001 (thousand t/year)	Reserves (million t)	Reserve life (years)	Reserve base (million t)	Reserve base life (years)
Morocco/Western Sahara	25,346	6,281	248	23,142	913
Tunisia	8,697	110	13	661	76
Senegal	1,860	55	30	176	95
Togo	1,917	33	17	66	34
South Africa	3,152	1,653	524	2,755	874
USA	44,851	1,102	25	4,408	98
Brazil	4,875	364	75	408	84
Jordan	6,350	992	156	1,873	295
Israel	4,487	198	44	882	196
Syria	1,955	110	56	882	451
China	24,134	1,102	46	11,020	457
Russia	11,020	220	20	1,102	100
Other countries	12,364	1,322	110	4,408	357
Total (rounded)	151,000	13,224	88	51,794	343

when Fritz Haber informed the directors of Badische Anilin und Soda Fabrik (BASF) that his search for combining nitrogen and hydrogen to ammonia had functioned successfully in the laboratory. Carl Bosch then found ways of making the principle work under industrial conditions. By application of the Haber–Bosch process, about 4000 t of ammonia were produced in 1913, and today the global output of ammonia is estimated at about 130 million t/year (Smil, 1999). Due to this invention, the 'not respirable air' discovered by Daniel Rutherford became the infinite raw material for production of nitrogen fertilizer.

1.4 Elementary Balances in Animal Production

Chemical elements can be neither produced nor destroyed in the animal's metabolism. They can only be transferred from one form into another and a very great part of research in animal nutrition is simply based on balancing elements. This is demonstrated in Table 1.2 for five elements in a dairy cow weighing 650 kg, assumed to produce 30 kg of milk daily. Further it is assumed that body mass and composition are constant. In order to cover the requirements of energy and all nutrients for maintenance and production, this cow is assumed to consume 50 kg of a total mixed ration (TMR) containing 40% dry matter (DM) plus 80 l of water per day.

A more detailed investigation may disclose that this cow daily excretes 40 kg of faeces containing 15% DM and 30 l of urine and that microbial fermentation in her digestive tract causes a daily emission of 500 l methane (CH_4). Finally, her daily consumption of oxygen from inspired air may amount to 6000 l and a corresponding volume of carbon dioxide (CO_2) may be expired daily. When elements are analysed in dietary DM, drinking water, milk and all excreta, then daily movements of the analysed elements into and out of the animal's body can be calculated, as shown in Table 1.2 for carbon, hydrogen, oxygen, N and P.

The efficiency by which the consumed elements are turned into compounds of milk in this example is 7% for oxygen, 23% and 25% for carbon and hydrogen and about 30% for N and P, respectively. Only in recent years, potential impacts on the environment of that unutilized part of the ingested elements has found scientific interest.

Expiration of CO_2 is not a net contribution to the greenhouse effect (global warming) because carbon contained in the feed must have been captured from CO_2 in the atmosphere in the preceding period of vegetation. Expired CO_2 is thus recycled into the atmospheric pool and is ready for again getting captured for photosynthesis according to the equation:

$$6CO_2 + 6H_2O \rightarrow C_6H_{12}O_6 + 6O_2 \quad (1)$$

Carbon contained in faeces and urine will finally be oxidized to CO_2 when exposed to aerobic conditions and the same should happen to methane, and thus the cycle of carbon between

Table 1.2. Approximate balance of five elements in dairy cows producing 30 kg of milk daily and fed according to common recommendations (g/day)[a].

Element	Carbon	Hydrogen	Oxygen	Nitrogen	Phosphorus
Input:					
Dietary dry matter	9000	1200	8500	550	90
Respiration (O_2)			8500		
Output:					
Milk	2100	300	1200	170	27
Methane	300	100			
Faeces	2600	300	2600	170	62
Urine	400	100	400	210	1
Respiration (CO_2)	3600		9600		
Metabolic water		400	3200		

[a]Constant body mass and composition assumed; for further assumptions see text.

atmospheric carbon dioxide and organic matter is completed. Methane and its oxidation products, especially carbon monoxide, have great importance for the chemistry of the atmosphere (Crutzen, 1995), but this point will not be followed in this book.

Oxidation of hydrogen to water in the metabolic chain of reactions is the principle for providing the organism with metabolizable energy. Water formed in this way does not have any impact on the environment.

Nitrogen is excreted in the urine mostly as urea. When contaminated with faeces, this urea may readily be hydrolysed by microbial urease according to the equation:

$$OC(NH_2)_2 + H_2O \rightarrow CO_2 + 2NH_3 \qquad (2)$$

When excreta are applied to the soil, ammonia is formed and may be taken up by plants through their roots, either directly or after conversion to nitrate. If excreted N accumulates in concentrations exceeding the capacity of plants, considerable emissions of ammonia into the air and nitrate into groundwater may occur. Both phenomena are regarded as having impact on the environment.

When cattle are grazing on pasture, enrichment of N will result in those spots where the animals urinate and enrichment of P will be found where they defecate. Thus, a certain degree of horizontal movement of nutrients will be found within the grazed paddocks.

Principally, the same phenomenon has to be registered on a much larger scale as a consequence of transporting great quantities of concentrate feeds, regardless of whether grains or by-products of the food industry, from the site of their production into areas of high animal density.

1.5 Environmental Regulations in the USA and the European Union

Although progress has been made (Børsting *et al.*, 2003), N and P are routinely overfed to ruminants, which, in combination with the continuous trend to concentrate animal units in intensive animal systems, leads to nutrient surpluses at farm and system levels (Jonker *et al.*, 2002; Ondersteijn *et al.*, 2002; Dou *et al.*, 2003). Compared to crops, production of nutrients from farm animals, particularly ruminants, is an inherently inefficient process (Domburg *et al.*, 2000; Ondersteijn *et al.*, 2002). The efficiency of utilization of dietary nutrients for milk or meat production is a simple formula:

$$\text{Efficiency} = \frac{\text{Nutrient in usable products}}{\text{Nutrient intake}} \qquad (3)$$

A reduction of the denominator or an increase of the numerator will enhance efficiency, i.e. less N input and/or greater milk N output will result in an increased efficiency of conversion of dietary N into milk N, for example. Crude protein content and composition of the diet can have a profound effect on N losses and ammonia release from manure (Swensson, 2003) and must be publicized by nutrition consultants and extension professionals as an immediately available tool for reduction of N losses from cattle operations. Alternatively, N (and P) from animal waste may be converted into value-added products, thus reducing nutrient loads to soil and atmosphere (Cowling and Galloway, 2001). Management practices, however, often have minimal impact on milk N efficiency (Jonker *et al.*, 2002), although when backed by legislative actions, farm management is critical in controlling nutrient pollution from dairy operations (Ondersteijn *et al.*, 2003). Similar conclusions can be drawn at whole-farm and agricultural system levels (De Vries *et al.*, 2001).

Concentration of livestock in large feeding operations has been associated with concerns regarding water and air quality and nuisance issues such as odour. In the USA, the Environmental Protection Agency (EPA) is the government body responsible for implementing environmental regulations, including regulations applicable to animal feeding operations (for details, see Meyer and Mullinax, 1999; Meyer, 2000; and Powers, 2003; most recent revisions can be found at the EPA web site, http://www.epa.gov/npdes/caforule; Federal Register, Vol. 68, No. 29, 12 February 2003).

In retrospect, the EPA rules regulating animal feeding operations (AFO) stemmed from the 1972 Federal Clean Water Act (CWA, Section 502) classifying beef feedlots as point sources of pollution. In 1974 effluent guidelines for feedlots were established and in 1976 regulations were issued defining Concentrated Animal Feeding Operations (CAFO) requiring National Pollutant Discharge Elimination System (NPDES) (Sweeten and Miner, 2003). Under the current regulations, AFO are required to have an NPDES permit if the animals are fed or housed in a confined area for more than 45 days in any 12-month period and crops, vegetation, forage growth or postharvest residues are not sustained in the normal growing season over any portion of the lot or facility. Animal operations are grouped into large (≥1000 beef cattle or dairy heifers, or ≥700 mature dairy cattle), medium (300 to 999 beef cattle or dairy heifers, or 200 to 699 mature dairy cattle) and small (<300 beef cattle or dairy heifers, or <200 mature dairy cattle). In most situations, large AFO are defined as CAFO and are required to have NPDES. Medium and small AFO can be classified as CAFO if animals are in direct contact with surface water running through the confinement area or the operation discharges into US waters through a manmade ditch, flushing system or other devices, or the permitting authority determines the facility is a significant contributor of pollutants and designates it as a CAFO (Koelsch, 2003). Historically, medium and small AFO have been designated CAFO status only following an on-site inspection. By definition pasture systems are not regulated by CAFO rules.

The process of obtaining an NPDES permit involves the development and implementation of a Nutrient Management Plan (NMP) by the CAFO. Federal regulations require dairy operators to have NMP in place by 31 December 2006. States may have additional requirements. Effluent Limitations Guidelines (ELG) for dairy CAFO imply no discharge of manure, litter or process wastewater from the production area, except in cases when rainfall causes the discharge and the production area is designed, operated and maintained to contain all of the manure, litter and process wastewater plus runoff from a 25-year, 24-h rainfall event (Wright, 2003). Under the new regulations, ELG for large CAFO require that manure, litter and processed wastewater be applied to agricultural fields using rates and methods that: (i) 'ensure appropriate agricultural utilization of nutrients'; and (ii) 'minimize P and N transport from the field to surface waters' (Davis, 2003). Large CAFO are required to evaluate the potential for N and P loss on all fields receiving manure. Manure applications may be limited or eliminated on fields having a high potential for P loss (determined using a risk assessment method). Based on the assessment for risk of nutrient loss, manure is applied based on P or N requirements. Medium and small CAFO are required to apply manure ensuring appropriate agricultural utilization of the waste nutrients (Sheffield and Paschold, 2003). In many situations, application of manure, based on N, overdoses P in soil; manure N:P ratios are significantly lower compared to N:P ratios in plants (Heathwaite *et al.*, 2000). Ammonia N volatilization from manure further concentrates P and contributes to P accumulation in soil.

Through the Voluntary Alternative Performance Standards (VAPS) the new EPA regulations provided an alternative to the traditional waste management systems under the ELG. Examples of alternative approaches are as follows (Sweeten *et al.*, 2003):

- reduction in nutrient excretion and/or dietary nutrient requirements through nutrition;
- grass filters, buffer strips, infiltration areas and vegetative systems reducing solid, nutrient and hydraulic loading;
- air quality process-based models to improve emission estimates from manure holding facilities;
- constructed wetlands following pre-treatment to allow release of wastewater to receiving water seasonally or continually;
- hybrid aerobic or anaerobic treatment systems shifting emissions to N_2 gas rather than ammonia;
- improving the cost effectiveness of systems (anaerobic digestion and thermal conversion) to recover energy and reduce atmospheric emissions from agricultural waste;
- cost-effective methods for recovery of marketable by-products (N and P);
- accelerating the recovery of value-added reuse of waste materials.

The contribution of ruminants to global ammonia emissions is the largest of all farm animal species and animals are the main contributors to overall ammonia N emissions from agriculture (Bouwman *et al.*, 1997). The contribution of farm animals to global or US ammonia emissions is estimated to be 48% and 50%, respectively (NRC, 2003). The contribution to N_2O, NO or CH_4 emissions is estimated at 33% and 25%, 1% (both) and 19% and 18%, respectively (NRC, 2003). The role of agriculture in greenhouse gas emission is also significant (Tamminga, 2003). Odour and human health concerns have driven regulations related to air quality impact of animal operations in the USA. With the 1990 Clean Air Act (CAA) amendments, the EPA was required to establish standards for pollutants considered harmful to human health. Standards were established for CO, NO_2, O_3, Pb and SO_2 as well as PM_{10} particulate matter (airborne particles with aerodynamic equivalent diameters less than 10 μm) (Powers, 2003). Particulate matter of 2.5 μm ($PM_{2.5}$) was proposed as pollutant with a 1997 amendment to the CAA, but a federal court blocked this addition in a 1999 ruling (Powers, 2003). The adoption of more stringent policies by the EPA is expected with the next revision of the CAA. The following is a brief overview of the important air pollutants originating from farm animal systems (NRC, 2003):

- Ammonia is produced through microbial hydrolysis of urinary urea in manure. Emitted in the atmosphere, ammonia can be converted to ammonium aerosol and removed by dry or wet deposition. Once removed from the atmosphere, ammonia or ammonium contributes to ecosystem fertilization, acidification, eutrophication and can impact visibility, soil acidity, forest productivity, terrestrial ecosystem biodiversity, stream acidity and coastal productivity (Galloway and Cowling, 2002). Ammonia also contributes indirectly to $PM_{2.5}$ through formation of ammonium salts.
- Nitrous oxide is formed through microbial nitrification and denitrification and contributes to tropospheric warming and stratospheric ozone depletion.
- Direct emission of nitric oxide from animal manure appears to be of minor importance, but fertilizer N applied to soil can be emitted as nitric oxide. Nitric oxide and nitrogen dioxide (referred to as NO_x) are rapidly interconverted in the atmosphere and removed through wet and dry deposition. NO_x is an important precursor in ozone production and aerosol nitrate is a contributor to $PM_{2.5}$ and N deposition (as HNO_3).
- Methane is produced through anaerobic fermentation of organic matter in the rumen. It is an important greenhouse gas contributing to global warming.
- Volatile organic compounds (VOC) from animal operations include organic sulphides, disulphides, C_4 to C_7 aldehydes, trimethylamine, C_4 amines, quinoline, demethylpyrazine, short-chain organic acids and aromatic compounds, and can have various environmental effects.
- Hydrogen sulphide is formed through anaerobic reduction of sulphate in water and decomposition of sulphur-containing organic matter in manure. In the atmosphere, hydrogen sulphide is oxidized to sulphur dioxide and removed by dry or wet (as aerosol sulphate) deposition. On a global scale, it appears that hydrogen sulphide emissions from farm animal systems have relatively minor ecological effects.
- PM_{10} and $PM_{2.5}$ particulate matter directly or indirectly originate from animal operations through animal activities, housing fans, air incorporation of mineral and organic material from soil, manure and water droplets and conversion to aerosols of ammonia, nitric oxide and hydrogen sulphide. Both particle types can cause health effects through deposition in airways and can affect visibility.
- Odour from animal operations, although difficult to quantify, has a significant societal, primarily local, impact and will likely be an important target in future environmental regulations.

Comparable regulations exist in most states of the European Union, which aim at protection of the environment against impacts of intensive animal production. These regulations differ in details not only between different members of the EU, but also between different regions within individual states. Depending on the respective authorities, different means for achieving the goal are considered adequate:

- limiting the number of animals kept per unit of available land;
- limiting the quantity of feed that may be purchased from external sources;
- forcing farmers to compare import and export of nutrients into their farm.

Numbers of animals and available land are easy to find out, but stocking density does not provide very reliable information about the degree of emission from a farm. Comparison of nutrient fluxes, on the other hand, is rather complicated, but gives a valid description of the degree of sustainability, if based on correct primary recordings. These recordings must include quantities and nutrient concentration of purchased fertilizer and feeds as major routes of nutrient import as compared to quantities and nutrient concentration of marketed goods of plant and animal origin. Knowledge of nutrient fluxes may provide strategies for improving nutrient efficiency and for combining profitability with sustainability of producing food.

This book intends to present the state of the art of supplying dairy cows properly with N and P without causing unwanted emissions of these elements.

References

Beaton, J. (2003) Fertilizer use – a historical perspective. Phosphorus (phosphate, P_2O_5). Available at: http://www.agcentral.com/imcdemo/01History/01-09.htm

Bergner, H. (1986) *Nathan Zuntz* and 100 years non-protein-nitrogen research in Berlin. *Archives of Animal Nutrition* 36, 127–130.

Børsting, C.F., Kristensen, T., Misciattelli, L., Hvelplund, T. and Weisbjerg, M.R. (2003) Reducing nitrogen surplus from dairy farms. Effects of feeding and management. *Livestock Production Science* 83, 165–178.

Bouwman, A.F., Lee, D.S., Asman, W.A.H., Dentener, F.J., Van Der Hoek, K.W. and Olivier, J.G.J. (1997) A global high-resolution emission inventory for ammonia. *Global Biochemical Cycles* 11, 561–587.

Brinkmann, T. (1922) Die Oekonomik des landwirtschaftlichen Betriebes. In: Weber, M. (ed.) *Grundriß der Sozialökonomik, 3. Wirtschaft und Gesellschaft.* Verlag J.C.B. Mohr, Thübingen, pp. 27–124.

CAST [Council for Agricultural Science and Technology] (2002) Animal diet modification to decrease the potential for nitrogen and phosphorus pollution. *Issue Paper* No. 21. E-mail: cast @ cast-science.org

Cheng, K.-J. and Costerton, J.W. (1980) Adherent rumen bacteria – their role in the digestion of plant material, urea and epithelial cells. In: Ruckebusch, Y. and Thivend, P. (eds) *Digestive Physiology and Metabolism in Ruminants.* MTP Press Ltd, Lancaster, UK, pp. 227–250.

Childs, P.E. (2003) Phosphorus: fire from urine. 1. The discovery of phosphorus. Available at: http://www.ul.ie/~childsp/CinA/Issue60/TOC55_Urine.htm

Cowling, E.B. and Galloway, J.N. (2001) Challenges and opportunities facing animal agriculture: optimizing nitrogen management in the atmosphere and biosphere of the Earth. *Journal of Animal Science* 80 (E. Suppl. 2), E157–E167.

Crutzen, P.J. (1995) The role of methane in atmospheric chemistry and climate. In: Engelhardt, W.v., Leonhardt-Marek, S., Breves, G. and Giesecke, D. (eds) *Ruminant Physiology: Digestion, Metabolism, Growth and Reproduction.* Ferdinand Enke Verlag, Stuttgart, Germany, pp. 291–315.

Davis, J. (2003) Making decisions about application rates. *Livestock and Poultry Environmental Curriculum.* Fact sheet #25. Available at: http://www.mwpshq.org

De Vries, W., Kros, H., Oenema, O., Reinds, G.J. and Erisman, J.W. (2001) Animal production impacts on nitrogen emissions to air and ground water: a Dutch case with a European perspective. American Society of Animal Science, International Animal Agriculture and Food Science Conference, Indianapolis in Indianapolis, pp. 1–50.

Domburg, P., Edwards, A.C., Sinclair, A.H. and Chalmers, N.A. (2000) Assessing nitrogen and phosphorus efficiency at farm and catchment scale using nutrient budgets. *Journal of the Science of Food and Agriculture* 80, 1946–1952.

Dou, Z., Ferguson, J.D., Fiorini, J., Toth, J.D., Alexander, S.M., Chase, L.E., Ryan, C.M., Knowlton, K.F., Kohn, R.A., Peterson, A.B., Sims, J.T. and Wu, Z. (2003) Phosphorus feeding levels and critical control points on dairy farms. *Journal of Dairy Science* 86, 3787–3795.

Duncan, C.W., Agrawala, I.P., Huffman, C.F. and Luecke, R.W. (1953) A quantitative study of rumen synthesis in the bovine on natural and purified rations. *Journal of Nutrition* 49, 41–49.

Galloway, J.N. and Cowling, E.B. (2002) Reactive nitrogen and the world: two hundred years of change. *Ambio* 31, 64–71.

Heathwaite, L., Sharpley, A. and Gburek, W. (2000) A conceptual approach for integrating phosphorus and nitrogen management at watershed scale. *Journal of Environmental Quality* 29, 158–166.

Houpt, R. (1970) Transfer of urea and ammonia to the rumen. In: Phillipson, A.T. (ed.) *Physiology of Digestion and Metabolism in the Ruminant.* Oriel Press Ltd, Newcastle upon Tyne, UK, pp. 119–131.

Jonker, J.S., Kohn, R.A. and High, J. (2002) Dairy herd management practices that impact nitrogen utilization efficiency. *Journal of Dairy Science* 85, 1218–1226.

Klaassen, G. (ed.) (1992) Ammonia emission in Europe: emission coefficients and abatement costs. Proceedings of a workshop held at the International Institute for Applied Systems Analysis (IIASA) in Laxenburg, Austria, 4–6 February 1991. CP-92-04, 288 pp.

Koelsch, R. (2003) Do I need an NPDES permit for my livestock or poultry operation? *Livestock and Poultry Environmental Curriculum.* Fact sheet #2. Available at: http://www.mwpshq.org

Liebig, J. (1840) *Die organische Chemie in ihrer Anwendung auf Agricultur und Physiologie.* Verlag von Friedrich Vieweg und Sohn, Braunschweig, 352 pp.

Liebig, J. (1843) *Die Thier-Chemie oder die organische Chemie in ihrer Anwendung auf Physiologie und Pathologie.* Verlag von Friedrich Vieweg und Sohn, Braunschweig, 344 pp.

Loosli, J.K., Williams, W.E., Thomas, W.E., Ferris, F.H. and Maynard, L.A. (1949) Synthesis of amino acids in the rumen. *Science* 110, 144–145.

Madsen, J. and Hvelplund, T. (1984) Prediction of amino acid supply to the small intestine of cows from analysis of the feed. *Canadian Journal of Animal Science (Supplement)* 86–88.

McDowell, L.R. (1992) *Minerals in Animal and Human Nutrition.* Academic Press, San Diego, California, 524 pp.

Meister, A. (1965) *Biochemistry of Amino Acids*, 2nd edn, Volume 1. Academic Press, New York and London, 119 pp.

Meyer, D. (2000) Dairying and the environment. *Journal of Dairy Science* 83, 1419–1427.

Meyer, D. and Mullinax, D.D. (1999) Livestock nutrient management concerns: regulatory and legislative overview. *Journal of Animal Science* 77 (Suppl. 2), 51–62.

Mitchell, H.H. (1924) A method of determining the biological value of protein. *Journal of Biological Chemistry* 58, 873–903.

Mulder, G.J. (1838) Sur la composition de quelques substance animales. *Bulletin des Sciences Physiques et Naturelles en Néerlande* 1, 104–119.

Munro, H.N. (1964) Historical introduction: the origin and growth of our present concepts of protein metabolism. In: Munro, H.N. and Allison, J.B. (eds) *Mammalian Protein Metabolism*, Volume I. Academic Press, New York and London, pp. 1–29.

NRC [National Research Council] (2003) *Air Emissions From Animal Feeding Operations: Current Knowledge, Future Needs.* National Academies Press, Washington, DC, pp. 50–73.

Ondersteijn, C.J.M., Beldman, A.C.G., Daatselaar, C.H.G., Giesen, G.W.J. and Huirne, R.B.M. (2002) The Dutch Mineral Accounting System and the European Nitrate Directive: implications for N and P management and farm performance. *Agriculture, Ecosystems and Environment* 92, 283–296.

Ondersteijn, C.J.M., Beldman, A.C.G., Daatselaar, C.H.G., Giesen, G.W.J. and Huirne, R.B.M. (2003) Farm structure or farm management: effective ways to reduce nutrient surpluses on dairy farms and their financial impacts. *Livestock Production Science* 84, 171–181.

Powers, W.J. (2003) Keeping science in environmental regulations: the role of the animal scientist. *Journal of Dairy Science* 86, 1045–1051.

Roberts, T.L. and Stewart, W.M. (2002) Inorganic phosphorus and potassium production and reserves. *Better Crops* 86, 6–7.

Rohr, K., Lebzien, P., Schafft, H. and Schulz, E. (1986) Prediction of duodenal flow of non-ammonia nitrogen and amino acid nitrogen in dairy cows. *Livestock Production Science* 14, 29–40.

Roy, J.H.B., Balch, C.C., Miller, E.L., Orskov, E.R. and Smith, R.H. (1977) Calculation of N-requirement for ruminants from nitrogen metabolism studies. In: *Protein Metabolism and Nutrition. Proceedings of the Second International Symposium on Protein Metabolism and Nutrition* held at 'Flevohof', the Netherlands, May 2–6, 1977. European Association of Animal Production Publication No. 22, Pudoc, pp. 126–129.

Satter, L.D. and Roffler, R.E. (1977) Calculating requirements for protein and nonprotein nitrogen by ruminants. In: *Protein Metabolism and Nutrition. Proceedings of the Second International Symposium on Protein Metabolism and Nutrition* held at 'Flevohof', the Netherlands, May 2–6, 1977. European Association of Animal Production Publication No. 22, Pudoc, pp. 133–136.

Schmidt-Nielsen, B., Schmidt-Nielsen, K., Houpt, T.R. and Jarnum, S.A. (1957) Urea excretion in the camel. *American Journal of Physiology* 188, 477–484.

Sheffield, R. and Paschold, J. (2003) What is required in a Nutrient Management Plan? *Livestock and Poultry Environmental Curriculum.* Fact sheet #20. Available at: http://www.mwpshq.org.

Simonnet, H., Le Bars, H. and Mollé, J. (1957) Le cycle de l'urée administrée par voie buccale chez les ruminants. *Comptes Rendús Hebdomadaires des Séances de l'Académie des Sciences, Paris*, 244, 943–945.

Smil, V. (1999) Detonator of the population explosion. *Nature* 400, 415.

Sweeten, J. and Miner, R. (2003) A brief history and background of the EPA CAFO rule. *Livestock and Poultry Environmental Curriculum.* Fact sheet #1. Available at: http://www. mwpshq.org

Sweeten, J., Miner, R. and Auvermann, B. (2003) Alternative treatment systems. *Livestock and Poultry Environmental Curriculum.* Fact sheet #7. Available at: http://www.mwpshq.org

Swensson, C. (2003) Relationship between content of crude protein in rations for dairy cows, N in urine and ammonia release. *Livestock Production Science* 84, 125–133.

Tamminga, S. (2003) Pollution due to nutrient losses and its control in European animal production. *Livestock Production Science* 84, 101–111.

Thaer, A. (1809) *Grundsätze der rationellen Landwirtschaft*, Erster Band. Realschulbuchhandlung, Berlin, 380 pp.

Thomas, K. (1909) Über die biologische Wertigkeit der Stickstoffsubstanzen in verschiedenen Nahrungsmitteln. *Archiv für Anatomie und Physiologie – Physiologische Abteilung des Archivs für Anatomie und Physiologie* 219–302.

Van der Krogt, P. (2003a) Elementymology & Elements Multidict: 01 Hydrogen. Available at: http://elements.vanderkrogt.net/elem/h.html

Van der Krogt, P. (2003b) Elementymology & Elements Multidict: 07 Nitrogen. Available at: http://www.vanderkrogt.net/elem/n.html

Van der Krogt, P. (2003c) Elementymology & Elements Multidict: 08 Oxygen. Available at: http://elements.vanderkrogt.net/elem/o.html

Van der Krogt, P. (2003d) Elementymology & Elements Multidict: 15 Phosphorus. Available at: http://www.vanderkrogt.net/elements/p.html

Vérité, R., Journet, M. and Jarrige, R. (1979) A new system for the protein feeding of ruminants: the PDI system. *Livestock Production Science* 6, 349–367.

Virtanen, A.I. (1966) Milk production of cows on protein-free feed. *Science* 153, 1603–1614.

Voit, C. (1872) Ueber die Bedeutung des Leimes bei der Ernährung. *Zeitschrift für Biologie* 8, 297–387.

Wright, P. (2003) CAFO requirements for dairy barnyards. *Livestock and Poultry Environmental Curriculum.* Fact sheet #11. Available at: http://www.mwpshq.org

Zuntz, N. (1891) Bemerkungen über die Verdauung und den Nährwerth der Cellulose. *Pflügers Archiv für die gesamte Physiologie des Menschen und der Tiere* 49, 477–483.

2 Nitrogen Requirements of Cattle

C.G. Schwab[1], P. Huhtanen[2], C.W. Hunt[3] and T. Hvelplund[4]
[1]*Department of Animal and Nutritional Sciences, University of New Hampshire, Durham, New Hampshire, USA*
[2]*MTT Agrifood Research Finland, Animal Production Research, Jokioinen, Finland*
[3]*Department of Animal and Veterinary Science, University of Idaho, Moscow, Idaho, USA*
[4]*Department of Animal Nutrition and Physiology, Danish Institute of Agricultural Sciences, Tjele, Denmark*

©CAB International 2005. *Nitrogen and Phosphorus Nutrition of Cattle*
(eds E. Pfeffer and A.N. Hristov)

2.1 Introduction

2.1.1 Overview

Many countries have instituted environmental legislation that has made it necessary for beef and dairy producers to quantify and adjust the nitrogen (N) balance on their farms. The legislation is designed to minimize the accumulation of manure N in the environment and to protect water and air quality. Major concerns are the release of ammonia and nitrous oxide to the atmosphere and nitrate contamination of groundwater. The challenge to the beef and dairy industries is to store and handle manure in ways that minimize N release into the environment and to increase the conversion of dietary crude protein (CP) into meat and milk proteins. Ration formulation decisions and more precise N feeding practices according to animal requirements are the initial control points for reducing the potential for N pollution. These offer opportunities to decrease both intake and excretion of N without impairing growth and milk production.

Several strategies can be used to increase the conversion of feed N into meat and milk protein and to reduce N wastage. One strategy is to feed for increased synthesis of microbial protein, which increases the opportunity to capture recycled N and the end products of protein breakdown in the rumen. Feeding for greater synthesis of microbial protein also has the benefit of improving the efficiency of use of absorbed amino acids (AA) because microbial protein has an AA profile that is thought to more closely approximate the profile required by the animal than virtually all feed proteins (NRC, 2001). A second strategy is to fine-tune and balance the supply of rumen-degraded feed protein (RDP) and rumen-undegraded feed protein (RUP) such that the requirements for both are met but not exceeded; in this case, neither portion of dietary CP is overfed and intake of N is minimized. A third strategy is to fine-tune and balance diets more precisely for essential AA (EAA). The last two approaches require accurate characterization of feedstuffs and use of metabolizable protein (MP) systems that provide guidance to combining feeds and feed supplements in ways that meet but not exceed the N requirements of ruminal fermentation and the AA requirements of the animal.

2.1.2 Microbial and animal requirements

The N requirements of rumen microorganisms are met by ammonia, AA and peptides, the end products of microbial breakdown of protein and recycled urea. The proteins that are broken down in the rumen include feed protein (i.e. RDP), microbial protein and the endogenous proteins of saliva and sloughed epithelial cells (respiratory tract, oesophagus, rumen and reticulum). Breakdown of microbial protein (i.e. intraruminal recycling of microbial protein) occurs in the rumen because of the consumption and lysis of bacteria by protozoa, bacteriophage-mediated lysis of bacteria, bacterial lysis caused by starvation and autolysis of protozoa (Morrison and Mackie, 1996). Many bacteria and all protozoa participate in rumen degradation of protein by synthesizing and using a variety of proteases, peptidases and deaminases. Bacteria are the most abundant microorganisms in the rumen and are the principal microorganisms involved in protein degradation. Forty per cent or more of isolated species exhibit proteolytic activity (Cotta and Hespell, 1984; Broderick *et al.*, 1991; Wallace, 1996). For bacteria, protein degradation is an extracellular event (Broderick *et al.*, 1991). Released oligopeptides are degraded to smaller peptides and free AA before cellular uptake occurs. Once inside the cell, peptides are hydrolysed to free AA. Intracellular free AA are either used for protein synthesis or catabolized to ammonia and carbon skeletons. Protozoa and anaerobic fungi are also involved in protein breakdown but are less active than bacteria. The ammonia and to a lesser extent the free AA and short peptides that result from protein breakdown serve to meet the N requirements of rumen microorganisms. See Chapter 3 for a discussion on N metabolism in the rumen.

AA are required nutrients for the host animal. Absorbed AA, used principally as building blocks for protein synthesis, are required for maintenance, growth, reproduction and lactation of cattle. Absorbed AA are provided by ruminally synthesized microbial protein, RUP and to a much lesser extent, by endogenous protein. In most feeding situations, microbial protein is the primary source of absorbed AA. However, that is not the case when feed intake is high and large amounts of RUP are fed.

2.1.3 Importance of meeting but not exceeding N requirements

As indicated previously, ruminants have two sets of N requirements, the N requirements of ruminal fermentation and the AA requirements of the host animal. Not meeting either set of requirements decreases animal performance and profitability. A shortage of RDP has been shown to reduce microbial digestion of carbohydrates (Mehrez *et al.*, 1977; Erdman *et al.*, 1986; Caton *et al.*, 1988; Nagadi *et al.*, 2000; Griswold *et al.*, 2003; Klevesahl *et al.*, 2003), reduce synthesis of microbial protein (Satter and Slyter, 1974; Aldrich *et al.*, 1993; Martin-Orue *et al.*, 2000; Griswold *et al.*, 2003), decrease feed intake (Mehrez and Ørskov, 1978; Wheeler *et al.*, 2002), decrease weight gains of growing cattle (Zinn *et al.*, 1994, 2003) and cow weight gains (Anderson *et al.*, 2001) and reduce milk yield (Kwan *et al.*, 1977; Canfield *et al.*, 1990). A shortage of absorbed AA by cattle, either because of decreased synthesis of microbial protein or less than required intakes of RUP, may decrease weight gains of growing cattle (Bagg *et al.*, 1985; Pirlo *et al.*, 1997; Lammers and Heinrichs, 2000), postpartum weight gains of cows (Wiley *et al.*, 1991; Patterson *et al.*, 2003), milk production (Kalscheur *et al.*, 1999) and reproductive efficiency (Wiley *et al.*, 1991; Triplett *et al.*, 1995), possibly through the effects on endocrine function (Kane *et al.*, 2002).

It goes without saying that overfeeding CP increases excretion of N in urine and faeces and increases the potential for N pollution. However, overfeeding CP can also lower animal performance. For example, several experiments have shown that overfeeding CP can reduce fertility (Canfield *et al.*, 1990; McCormick *et al.*, 1999; NRC, 2001; Rajala-Schultz *et al.*, 2001; Chapter 8 of this book). There are many theories as to why excess dietary CP decreases reproductive performance. These include: (i) decreased energy status because of the energy costs associated with urea synthesis; (ii) direct action of urea on the process of oocyte maturation; and (iii) diet-induced alterations in uterine pH (NRC, 2001; Ocon and Hansen, 2003). In theory, it may be expected that overfeeding may decrease weight gains of growing cattle and milk yield of lactating cows because of the energy costs associated with metabolic disposal of excess N. Indeed, evidence exists that demonstrates that feeding high levels of RDP may decrease milk production (NRC, 2001). The Cornell Net Carbohydrate and Protein System (CNCPS; Fox *et al.*, 1992) considers the energetic cost to excrete N (urea) in excess of bacterial and tissue needs and lowers the amount of energy available for growth or lactation accordingly. It is acknowledged, however, that in many experiments feeding excess CP did not decrease weight gains or milk production (Broderick, 2003; see also Chapter 5).

Overfeeding CP to lactating cows also increases milk urea N (MUN) and milk non-protein N (NPN) concentrations (Broderick, 2003; Nousiainen *et al.*, 2004), increases urine volume (Dinn *et al.*, 1998; Leonardi *et al.*, 2003), increases urinary N output (Nousiainen *et al.*, 2004) and may decrease milk protein content (Leonardi *et al.*, 2003). The decrease in milk protein concentrations is most common when the additional protein that is being supplied is RUP and the RUP has a poor AA balance (e.g. maize gluten meal) (Santos *et al.*, 1998). In cows fed grass silage-based diets feeding additional protein increased milk protein concentration, but this increase was mainly associated with increased MUN concentration (Huhtanen and Nousiainen, 2004).

There is also ample evidence that high levels of MUN have a negative effect on the processing quality of milk. Millet (1989) demonstrated that addition of urea to milk before ripening resulted in a more fragile curd with longer curd cutting time, higher residual lactose and higher pH than control milk, indicating incomplete acidification. Cheeses made with urea-supplemented milk always had greater openness and had no slits. Podhorsky and Cvak (1989) concluded that milk with increased urea content is difficult to process into cultured products and cheese; urea inhibited activity of yoghurt-started culture and to some extent ripened cream-started culture. Studies from Switzerland (Bachmann and Jans, 1995) and France (Martin *et al.*, 1997) demonstrated that MUN negatively affected characteristics and quality of cheese. Milk with high urea content caused lower acidification rate in the cheese mould and ripening after unmoulding and cheeses produced from such milk were significantly less firm, less pasty and less chalky (Martin *et al.*, 1997). Cheeses made with milk from cows having higher MUN content were found to be of inferior quality; compared to control milk, high-MUN milk had significantly lower curd score and shorter, firmer texture (Bachmann and Jans, 1995). In a study involving 876 herds, Pecorari *et al.* (1993) found that milk from herds having lower MUN (17.7 mmol/l) had better technological parameters: higher titratable activity, higher protein content and higher coagulation capacity. Coulon *et al.* (1998) studied the effect of the stage of lactation on cheese making properties of milk and quality of Saint-Nectaire type cheese. Although milk protein, casein and calcium and phosphorus content remained unchanged, MUN concentration increased with lactation stage: from 15.6 mmol/l during the first 4 weeks of lactation to 22.9 mmol/l during 225 to 255 days in milk (DIM). In the later lactation stage, higher MUN milk was associated with reduced firmness and increased melting, more intense and persistent taste, and significantly lower texture and taste scores of cheeses.

2.1.4 Demonstrated potential for reduced N feeding

Several studies have been conducted which indicate that more precise feeding can have substantial effects on the efficiency of use of dietary N as compared to more traditional ways of feeding. For example, Klopfenstein and Erickson (2002) reported that phase-feeding multiple diets to finishing calves and yearlings to match RDP, RUP and MP requirements according to NRC (1996) vs. feeding the industry average 13.5% CP to feedlot cattle throughout the feeding period decreased N inputs by 11% to 18% without affecting weight gains. Decreasing dietary CP decreased N excretion by 13% to 22%. Volatilization in the open-dirt feedlot pens was reduced by 15% to 33%. Using a well-managed case study farm involving 320 lactating cows, Klausner *et al.* (1998) reported that more precise feeding for energy and protein allowed for a reduction in CP content of the rations from 20.2 to 18.3%, a 34% reduction in total N excretion, and a 13% increase in milk production. Evaluation and refinement of diets in this experiment were conducted using the CNCPS as described by Fox *et al.* (1992).

The extent to which dietary N levels can be reduced in cattle diets by more precise feeding is probably still not fully appreciated because of the inadequacy of existing diet formulation and evaluation models. Nevertheless, studies have been conducted that indicate that precision feeding affords significant opportunities to decrease N intake and excretion without impairing growth and milk production. One index of efficiency of N use in the lactating dairy cow is the portion of feed N that is captured in milk. A review of 62 recently published papers indicated an average milk N efficiency of 27% (16.2% to 45.2%) (Chase, 2003). In this study, diet CP averaged 17.5% of dry matter (DM) (10.2% to 24.6%). The dietary factors most affecting milk N efficiency were dietary CP content and rumen degradability, carbohydrate source and method of grain processing, AA balance and frequency of feeding. When there has been an attempt to balance diets for RDP, RUP and AA in high producing, early lactation cows with models available to the researchers at the time the experiments were initiated, milk N efficiency values have varied between 31% and 38% (x = 34%) (Armentano *et al.*, 1993; Wu *et al.*, 1997; Dinn *et al.*, 1998; Robinson *et al.*, 1998; Leonardi *et al.*, 2003; Noftsger and St-Pierre, 2003). In these six experiments, diet CP averaged 15.8% and ranged from 14.4% to 16.9%. In four of the experiments, a higher protein-containing diet was fed and in no case was there a loss in milk protein production by feeding the lower protein, better balanced diet (Armentano *et al.*, 1993; Dinn *et al.*, 1998; Leonardi *et al.*, 2003; Noftsger and St-Pierre, 2003).

2.1.5 The need for protein models

Considerable progress has been made over the last 30 years to develop models/systems that predict protein requirements and allow for evaluation of protein adequacy of diets for cattle. These efforts continue and are essential for better definition of N requirements, for more precise feeding of protein, NPN and AA supplements, and for more accurate prediction of animal performance (weight gain, composition of weight gain, milk protein yield (MPY) and milk composition). The greatest challenge in developing more sophisticated protein systems is to increase accuracy in predicting: (i) dietary supply of RDP and RUP; (ii) extent of N recycling; (iii) requirements of rumen microorganisms for RDP; (iv) microbial protein supply/synthesis; (v) the quantity of total and individual absorbable AA provided by microbial protein and RUP; and (vi) the AA requirements of the host animal.

The purpose of this chapter is to review the current understanding of the N requirements of rumen microorganisms and the AA requirements of cattle, to explain how supply and requirements for MP and AA have been estimated, and to evaluate five different systems in their ability to predict MP requirements.

2.2 Metabolic Requirements for N

2.2.1 Nitrogen requirements of rumen microorganisms

Attempts to define the N requirements for optimum growth of the mixed rumen microbial population have been challenging. This is due largely to the complexity of ruminal N metabolism, the unique differences in N metabolism of the different strains and species of microorganisms that inhabit the rumen, the ever uncertainty of the strains and species that predominate the microbial ecosystem in any given feeding situation, and the incomplete understanding of the interrelationships among the microorganisms that exist.

Ammonia is a key metabolite in rumen N metabolism. It is required by several species and strains of bacteria, and is widely used by others. For several strains each of several cellulolytic bacterial species, such as *Ruminococcus flavefaciens*, *Ruminococcus albus*, *Bacteroides amylophilus*, *Bacteroides succinogenes*, *Butyrivibrio fibrisolvens*, *Fibrobacter succinogenes* and *Eubacterium ruminantium*, ammonia is an absolute requirement (Bryant and Robinson, 1961, 1962; Hungate, 1966). Bryant (1973) concluded that the principal cellulolytic bacteria in the rumen use ammonia as the main source of N and they are often inefficient in using pre-formed cell monomers such as AA. For some strains, ammonia may not be required but it stimulates growth rates (Bryant and Robinson, 1961). In a study involving 89 freshly isolated strains of predominant culturable ruminal bacteria, Bryant and Robinson (1962) observed that ammonia was essential for 25% of the strains (five morphological groups) and 56% (four morphological groups) grew with either ammonia or casein hydrolysate as the main source of N. It has been concluded that more than 80% of culturable rumen bacteria are capable of good or normal growth with ammonia as the sole N source (Morrison and Mackie, 1996).

AA and peptides are also key metabolites in rumen N metabolism. It has been demonstrated that several species of bacteria require AA and peptides (Abou Akkada and Blackburn, 1963; Pittman and Bryant, 1964; Hungate, 1966). It is estimated that about 20% of rumen bacteria require pre-formed AA or peptides for growth (Bryant and Robinson, 1961). Moreover, all protozoa, and presumably rumen fungi as well, require pre-formed AA or peptides for protein synthesis. Protozoa are not able to synthesize AA from ammonia (Jouany and Ushida, 1999) and thus require AA and peptides for protein synthesis (Coleman, 1979). Much less is known about the N requirements of fungi in the rumen but it has been concluded that like protozoa, their N needs are best met by AA and peptides (Morrison and Mackie, 1996).

Estimates of the contribution of ammonia (vs. pre-formed AA) to protein synthesis by the mixed rumen population have proven to be highly variable. Using $^{15}NH_3$ or [^{15}N] urea infused in the rumen or added as a single dose to label the ammonia pool has indicated that 18% to 100% of the N incorporated into microbial protein passed through the ammonia pool (Pilgrim *et al.*, 1970; Al-Rabbat *et al.*, 1971; Mathison and Milligan, 1971; Nolan and Leng, 1972; Nolan *et al.*, 1976; Salter *et al.*, 1979). In a similar fashion, and also using ^{15}N to label the ammonia pool, researchers using *in vitro* techniques have reported that 16% to 100% of the N in microbial cells were derived from ammonia (Atasoglu *et al.*, 1998, 1999, 2001).

A considerable amount of research has been conducted to determine the rumen ammonia-N concentrations that are needed to maximize microbial protein synthesis or carbohydrate digestion and to examine the stimulatory effect of pre-formed AA and peptides. Less work has been done to define the optimal ratios and concentrations of ammonia-N, AA-N and peptide-N.

2.2.1.1 Ammonia requirements for maximum synthesis of microbial protein

A variety of *in vivo* and *in vitro* methods has been used to determine the ammonia-N needs for bacterial protein production. In all cases, ammonia concentrations were varied in the 'rumen' by supplying differing amounts of urea.

Hume *et al.* (1970) fed a virtually protein-free purified diet (cellulose, starch, sucrose, polythene chips, minerals and molasses) containing 0.9%, 1.8%, 3.5% and 6.7% urea to mature sheep. Intakes of diets were restricted to approximately 80% of *ad libitum* intakes and were fed at 2-h intervals. Ruminal ammonia-N concentrations averaged 4.5, 6.2, 9.4 and 21.8 mmol/l of rumen fluid. Flows of total protein to the omasum were 33, 39, 50 and 48 g/day and protein synthesized per 100 g organic matter (OM) digested in the rumen was 9.1, 10.5, 12.8 and 13.3 for the respective diets. Results indicated that a ruminal ammonia-N concentration of 6.2 mmol/l was adequate to maximize the concentration of protein in the rumen, but 9.4 mmol/l was needed to maximize flow of protein from the rumen. In this study, because a protein-free diet was fed, measured protein would be the sum of microbial protein and endogenous protein. It is not clear why a higher rumen ammonia concentration was needed to maximize flow of protein than to maximize content of protein in rumen digesta because treatment had no effect on rumen fluid volume, or passage of digesta out of the rumen.

Using a continuous culture system, Satter and Slyter (1974) observed that a concentration of 1.4 mmol/l of ammonia-N was adequate to support maximum microbial protein production but concluded that a concentration of 3.6 mmol/l may be warranted to give a margin of safety. Their observations were similar for a protein-free purified diet (cerelose, starch, wood pulp, minerals and refined soybean oil), an all concentrate diet (maize, molasses and minerals), or a mixed diet (maize, cerelose, lucerne hay, timothy hay, molasses and minerals).

Allen and Miller (1976) examined the requirement for ammonia-N in the rumen of sheep by substituting part of the starch in a cereal-based diet (45.8% barley, 30.9% starch, 10% straw, and 10% molasses/sphagnum moss, and minerals and vitamins) with 0%, 0.8%, 1.6% and 2.4% urea to achieve dietary CP concentrations of 6.0%, 8.0%, 10.0% and 12.0%. The animals were limit fed 24 times per day. Ruminal ammonia-N concentrations averaged 8.2, 9.7, 11.4 and 15.7 mmol/l, respectively. Flow of non-ammonia N (NAN) to the abomasum increased linearly with urea supplementation (10.3, 10.6, 12.4 and 12.8 g of N/day).

Okorie *et al.* (1977) infused variable amounts of urea into the rumen of sheep fed a basal diet of starch, glucose, straw, barley, grass, molasses/peat mixture, vegetable oil and minerals and vitamins. The basal diet contained 5% CP and was fed using a continuous feeding apparatus. Passage of microbial protein to the duodenum was maximized at a rumen ammonia-N concentration of about 5 mmol/l.

Wallace (1979) observed an apparent increase in total viable bacteria ($5.3 \pm 1.8 \times 10^9$ vs. $2.8 \pm 0.7 \times 10^9$) and numbers of pectinolytic bacteria ($8.3 \pm 5.4 \times 10^7$ vs. $4.9 \pm 1.5 \times 10^6$) in the rumen of sheep when a whole barley diet was supplemented with urea to increase rumen ammonia-N concentrations from 6.1 to 13.3 mmol/l. The diet was fed continuously using automated feeders.

Slyter *et al.* (1979) altered rumen ammonia concentrations in eight steers fed an 8% CP diet (cracked maize, cerelose, lucerne hay, timothy hay molasses and minerals) by infusing variable amounts of urea into the rumen. The diet was fed four times daily. Animals were infused with eight different amounts of urea ranging from 0 to 140 g/day such that ration CP levels of 8.0%, 9.5%, 11.1%, 13.3%, 16.9%, 17.8%, 18.6% and 19.5% were achieved. The respective ammonia-N concentrations that resulted were 0.8, 0.8, 1.6, 3.2, 4.8, 10.1, 7.2 and 16.0 mmol/l. Tungstic acid precipitable N in whole rumen digesta was 1.1, 1.7, 2.6, 2.7, 2.9, 2.6, 2.5 and 2.2 g/kg. Increasing ammonia-N content beyond 1.6 mmol/l of rumen fluid resulted in no further increase in content of protein in rumen digesta.

Two experiments have examined the effects of incremental urea supplementation of a basal diet low in RDP on ruminal ammonia-N concentrations and formation of microbial protein in dairy cows. In the first experiment, Kang-Meznarich and Broderick (1981) supplemented a basal diet of 75% ground dry maize and 20% cottonseed hulls containing 8.3% CP with six levels of urea (0%, 0.4%, 0.7%, 1.1%, 1.6% and 2.3%) to create diets that contained 8.3%, 9.4%, 10.7%, 12.0%, 13.8% and 15.0% CP. The diets were pelleted and fed hourly to two non-lactating Holstein cows. Rumen ammonia-N concentrations averaged 0.9, 2.3, 6.0, 9.8, 16.2 and 20.5 mmol/l and rumen diaminopimelic acid concentrations (marker for microbial protein) averaged 1.5, 2.1, 2.8, 2.9, 2.7 and 2.1 nmol/kg DM, respectively, for the six diets. The authors concluded that a ruminal ammonia-N concentration of approximately

6.0 mmol/l was needed to maximize bacterial protein formation.

The second lactating dairy study was conducted in the senior author's laboratory (Ferguson, unpublished). The basal diet contained (DM basis) 32% processed maize silage, 16% grass silage, 4% chopped lucerne hay, 19% finely ground maize, 6% finely ground barley, 4.5% soybean hulls, 3% citrus pulp, 7% soybean meal, 1.3% high-RUP protein supplement and 4.4% fat and minerals and vitamins. Dietary treatments were 0%, 0.3%, 0.6% and 0.9% urea in diet DM. The total mixed rations were fed three times daily to lactating Holstein cows. The consumed basal diet (20.8 kg/day) contained 9.2% RDP in DM and had a predicted RDP balance of −170 g/day (NRC, 2001). Feeding increasing amounts of urea increased rumen ammonia-N concentrations (6.4, 8.4, 9.1 and 12.4 mmol/l; quadratic, $P < 0.05$), increased passage of microbial N to the small intestine (quadratic, $P < 0.01$) and increased microbial N as a percentage of NAN in duodenal digesta (quadratic, $P < 0.05$). Microbial protein synthesis was maximized with the 0.6% urea treatment, which resulted in a mean rumen ammonia-N concentration of 9.1 mmol/l. The diurnal variation of ammonia-N concentration as measured every 1.5 h of a 24-h day is depicted in Fig. 2.1. It is of interest to note that not only were ruminal ammonia-N concentrations of the cows fed the highest level of urea highest at each sampling time throughout the 24-h period, but the diurnal variation was also the highest. Rumen ammonia-N concentrations for the 0.0%, 0.3% and 0.6% urea treatments varied between 3.6 and 10.6 mmol/l throughout a 24-h period, with the exception of a few observations. However, rumen ammonia-N concentrations for the 0.9% urea treatment varied from about 6.4 to 25 mmol/l throughout a 24-h period.

Figure 2.2 shows a summary of five experiments examining the relationship between rumen ammonia concentration and rumen N balance in cows fed grass silage-based diets. There was a very strong negative relationship between rumen ammonia-N concentration and rumen N balance. Rumen N losses were more closely related to ammonia-N concentration ($R^2 = 0.85$) than to dietary CP content ($R^2 = 0.74$, figure not shown) demonstrating the effect of degradability on ammonia-N. Efficiency of microbial protein synthesis [g microbial N per kg digestible OM (DOM)] tended to decrease with increasing rumen ammonia-N concentration (data not shown). This may be interpreted as a result of lower ATP supply from RDP compared to

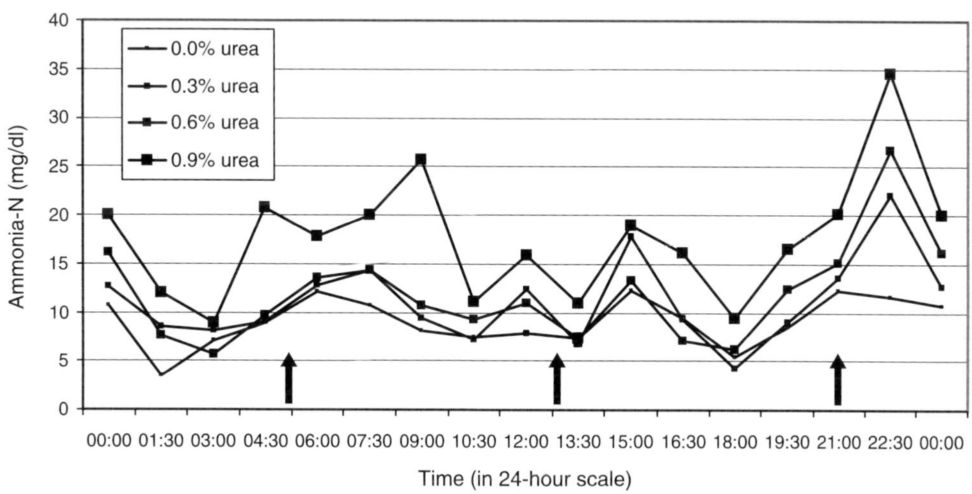

Fig. 2.1. Diurnal variation of ammonia-N concentrations in rumen fluid of lactating Holstein dairy cows fed diets containing different concentrations of urea in diet DM (0.0%, 0.3%, 0.6% and 0.9%) and fed three times daily (Ferguson, unpublished). Arrows indicate time of feeding.

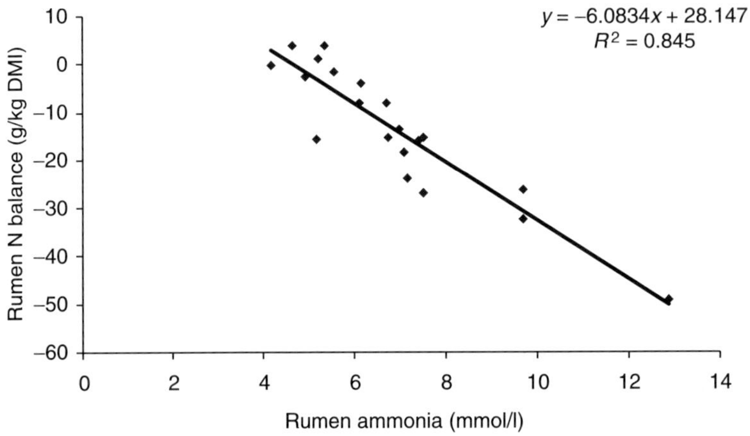

Fig. 2.2. Relationship between rumen ammonia concentration and rumen N balance. Rumen N balance was estimated by omasal sampling technique and the values are adjusted for random study effect. The data are from Ahvenjärvi *et al.* (1999, 2002a,b), Korhonen *et al.* (2002) and Korhonen (unpublished).

digestible carbohydrates and a lack of any stimulatory effects of protein supplements on microbial growth with grass silage-based diets. Madsen and Hvelplund (1988) also observed a significant relationship between mean rumen ammonia-N concentrations and calculated protein balance in the rumen. These results suggest that attempts to maximize microbial N by increasing dietary RDP content will take place at the expense of increased rumen N losses after RDP requirements are met.

In summary, available evidence indicates that rumen ammonia-N concentrations of 5 to 11 mmol/l are needed to maximize flows of microbial N from the rumen (Hume *et al.*, 1970; Allen and Miller, 1976; Okorie *et al.*, 1977; Ferguson, unpublished). These concentrations are considerably higher than the concentration of 1.4 mmol/l determined to be adequate to maximize flows of microbial protein in continuous culture (Satter and Slyter, 1974), and somewhat higher than the concentrations of 1.6 and 6.0 mmol/l that were required to maximize content of microbial protein in rumen digesta (Slyter *et al.*, 1979; Kang-Meznarich and Broderick, 1981). However, the data shown in Fig. 2.2 suggest that rumen ammonia-N concentrations higher than 5 mmol/l will result in increased N losses from the rumen.

2.2.1.2 Ammonia requirements for maximum bacterial degradative activities

Several of the experiments described in the previous section as well as others have examined the effects of changes in rumen ammonia-N concentrations on microbial activity and feed digestion in the rumen. A variety of approaches that include *in vitro, in situ* and *in vivo* techniques have been used and as in the experiments already discussed, rumen ammonia-N concentrations were varied by supplying different amounts of urea. For the above experiments in which the authors presented data relevant to this discussion, the experiments will be mentioned in the sequence discussed previously.

Hume *et al.* (1970) reported no statistically significant effects of increasing rumen ammonia-N concentrations (4.5, 6.2, 9.4 and 21.8 mmol/l) on ruminal pH, rumen fluid volume or liquid flow out of the rumen, concentration of total volatile fatty acids (VFA) in the rumen fluid, molar proportion of the individual VFA or cellulose digestion. However, there was a tendency for total VFA in the rumen fluid to increase (82.1, 87.5, 86.5 and 91.2 mmol/l) with increasing concentrations of rumen ammonia. To ensure that a relatively normal rumen microbial population was maintained *in vitro*, Satter and Slyter (1974) counted cellulolytic bacteria in one experiment where the purified diet was fed to the continuous culture fermentors. They reported numbers of 0.1, 2.4, 3.9, 1.1 and 7×10^8/g fermentor contents with increasing amounts of urea.

Wallace (1979) observed no effect of increasing rumen ammonia-N concentrations from 6.1 to 13.3 mmol/l on total VFA concentrations but did observe an increase in the degradation rates of rolled barley, wheat gluten and wheat bran with the higher concentration of rumen ammonia. Slyter *et al.* (1979) observed that a minimal ruminal ammonia-N concentration of 3.2 mmol/l was needed to maximize total VFA concentrations and the amount of N retained by the animals. Kang-Meznarich and Broderick (1981) observed an increase in the rate of DM digestion in the rumen when ammonia-N concentration was increased from 0.9 to 2.3 mmol/l, but no further increases were seen with the higher levels of rumen ammonia. Ferguson (unpublished) observed linear ($P < 0.05$) increases in total VFA concentrations and butyrate, expressed as a percentage of total VFA in rumen fluid, as ammonia-N concentrations increased from 6.4 to 12.3 mmol/l. In this experiment, a trend for a linear increase in acetate as a percentage of total VFA was also observed.

Several other experiments have been reported in which the authors examined the effect of rumen ammonia concentrations on *in situ* degradation rates of feeds. Mehrez *et al.* (1977) fed whole barley fortified with six levels of a urea solution using automated continuous feeders to maintain steady states of rumen ammonia concentrations. The ammonia-N concentration needed to maximize disappearance of barley DM from the polyester bags suspended in the rumen varied between 11 and 16 mmol/l.

Erdman *et al.* (1986) evaluated the effect of rumen ammonia-N concentrations on *in situ* digestion of ground maize, soybean meal, maize gluten feed, cottonseed meal and ground lucerne hay. The feeds were incubated in the rumen of dry Holstein cows fed a 7.4% CP diet consisting of 47.4% ground maize, 50.0% cottonseed hulls and 2.6% minerals and vitamins. The diet was fed as a total mixed ration twice daily with 10 kg fed at each feeding, and no feed was refused. Treatments consisted of continuous rumen infusion of 0, 33, 67 and 100 g/day of urea-N, which resulted in mean rumen ammonia-N concentrations of 3.0, 7.2, 12.2 and 17.8 mmol/l. Estimated effective DM degradation based on the *in situ* generated data increased in a linear fashion for maize (67.9%, 72.1%, 73.1% and 74.4%) and soybean meal (77.5%, 76.6%, 79.9% and 80.3%) whereas degradation of maize gluten feed (67.0%, 70.1%, 71.4% and 68.4%) and cottonseed meal (56.7%, 58.3%, 60.1% and 57.9%) was maximized with the third level of urea feeding. Lucerne hay DM and neutral detergent fibre (NDF) degradation were not increased with urea infusion. Erdman *et al.* (1986) concluded from this experiment and previous research that the minimum rumen ammonia concentrations required to maximize digestion depend on the fermentability of the feed and are considerably higher when digestibility is high than when digestibility is low.

The conclusion of Erdman *et al.* (1986) was supported by the work of Odle and Schaefer (1987) who demonstrated that barley is degraded at a faster rate in the rumen than maize and that a higher rumen ammonia-N concentration was needed to maximize the degradation rate of barley (8.9 mmol/l) than to maximize the degradation rate of maize (4.3 mmol/l). The experiment was conducted with steers given barley and maize diets supplemented with graded levels of an ammonium acetate solution.

It remains unclear as to what the exact ammonia-N requirements of rumen microorganisms are to maximize rumen digestion and maximize synthesis of microbial protein. There are several issues to consider. First, it is concluded from the above summary of studies that there is no 'fixed' optimum ammonia concentration. The optimum concentration appears to be dependent on diet and influenced by type of N supplements, carbohydrate fermentability, and maybe passage rates of ruminal digesta as affected by dry matter intake (DMI) and other dietary factors. Second, it appears that rumen ammonia concentrations required to maximize rumen digestion are at least as high as those required to maximize ruminal synthesis of microbial protein and that the optimal concentrations depend on the fermentability of the feed. Third, it is not only the average ammonia concentration that is important, but also the time that the concentration falls below some critical level. This is suggested by the work of Madsen and Hvelplund (1988) who observed a significant relationship between mean rumen ammonia-N concentrations and protein balance in the rumen and hours of ammonia-N concentrations below 7, 11 and 14 mmol/l rumen fluid. However, there was no significant relation to the hours between 1.4 and 3.6 mmol/l. Determining a critical ammonia concentration is difficult because diurnal variation exists, even when NPN supplements are supplied to the rumen in a continuous fashion (Erdman *et al.*, 1986; Odle and Schaeffer, 1987). Odle and Schaeffer (1987) observed a range in rumen ammonia-N concentrations between 7.8 and 10.6 mmol/l when steers were fed an hourly diet that was sprayed with ammonium acetate. Erdman *et al.* (1986) observed a range in rumen ammonia concentrations from 10.6 to 21.3 mmol/l when diets were fed twice daily and urea was infused continuously so that 100 g/day of urea was provided. And finally, defining a critical rumen ammonia-N concentration will also depend on the fermentability of the diet. A higher rumen ammonia-N concentration may be required after feeding if readily fermentable carbohydrates are available, but the required concentration may be less as the proportion of forage to concentrate in the rumen increases. Therefore, it may not be as important to maintain a certain critical ammonia-N level throughout the day as it is to better match that level with the needs of the rumen microbes as dictated by supply of fermentable carbohydrates or content of forage in the diet.

2.2.1.3 Amino acid and peptide requirements for maximum microbial growth

It could be argued that because some research has indicated that as much as 100% of the N incorporated into microbial protein passed through the rumen ammonia pool (Salter *et al.*, 1979; Atasoglu *et al.*, 1999), the mixed rumen microbial population has no dietary requirement for AA. This argument is supported by the observations of Virtanen (1966) and Oltjen *et al.* (1969) who demonstrated that cattle can lactate, reproduce and gain weight when 98% or more of the N in diets is supplied by urea. However, the latter observations can be realized, not because the mixed rumen microbial population does not contain microorganisms that have metabolic requirements for AA and possibly peptides, but because of intraruminal recycling of microbial protein, thereby eliminating the absolute need for a dietary supply of AA and peptides.

As previously discussed, AA and peptides are key metabolites in rumen N metabolism, being required nutrients for a portion of the bacterial population and all protozoa. In addition, research with many pure and mixed-batch cultures (Maeng *et al.*, 1976; Argyle and Baldwin, 1989; Cruz Soto *et al.*, 1994; Atasoglu *et al.*, 1998; Kajikawa *et al.*, 2002) and continuous cultures (Cotta and Russell, 1982; Griswold *et al.*, 1996; Carro and Miller, 1999) has indicated that pre-formed AA and peptides have stimulatory effects on bacterial growth and increase growth rates and microbial protein synthesis. This is true even when ammonia and carbohydrates exceed requirements (Maeng and Baldwin, 1976b; Cotta and Russell, 1982; Argyle and Baldwin, 1989). Also observed have been increases in fibre digestion (Merry *et al.*, 1990; McAllan, 1991; Griswold *et al.*, 1996; Carro and Miller, 1999). And finally, there may be different responses to peptides compared with AA (Argyle and Baldwin, 1989) depending on the microbial population present (Armstead and Ling, 1993; Ling and Armstead, 1995). For example, several experiments have indicated that peptide carbon was used more efficiently or at a faster rate than AA carbon (Pittman and Bryant, 1964; Pittman *et al.*, 1967; Wright, 1967; Chen *et al.*, 1987a,b; Yang, 2002). Of particular interest was the observation by Yang (2002) that in several cases, improvement in NDF digestibility was greater for dipeptide addition of valine–valine and leucine–leucine, than for the addition of the corresponding AA.

Argyle and Baldwin (1989) conducted a series of *in vitro* experiments to determine the effects of AA and peptides on microbial growth in cultures containing ammonia. They confirmed the stimulatory effect of AA and peptides on bacterial growth and concluded: (i) that peptides are more stimulatory than a complete mixture of free AA; (ii) that only a complete mixture of free AA stimulated growth whereas subgroups of AA did not stimulate growth; (iii) that the relationship between free AA and peptide concentrations and cellular growth is quadratic in nature (addition of 1 mg/l each of AA and peptides increased microbial growth over twofold whereas 10 and 100 mg/l of each increased microbial growth over ammonia threefold and fourfold, respectively); and (iv) that growth of mixed ruminal bacteria is a linear function of carbohydrate fermented and that peptides and AA 'act as multiplying factors' to microbial growth.

Using continuous culture techniques, Cotta and Russell (1982) evaluated the AA needs of five species of rumen bacteria known to be active users of AA and present in the rumen in large numbers under a variety of dietary conditions: *Selenomonas ruminantium*, *Prevotella ruminicola*, *Megasphaera elsdenii*, *Streptococcus bovis* and *B. fibrisolvens*. Peptide and AA concentrations of 0.016, 0.031, 0.062, 0.125, 0.25 and 0.50 g/l were tested. The highest concentrations of peptides and AA resulted in the highest yields of bacterial protein. Reducing concentrations below 0.062 g/l had the most dramatic effect in decreasing yield of bacterial protein.

In summary, optimal concentrations of peptides and AA for maximum synthesis of microbial protein have been difficult to define since the highest concentration of these substrates usually resulted in the highest growth rates (Cotta and Russell, 1982; Argyle and Baldwin, 1989).

2.2.1.4 Proportional need of ammonia-N, AA and peptides

Several experiments have been published in which the goal was to determine the balance and concentrations of NH_3-N, AA and peptides that are needed to optimize microbial growth. Maeng and Baldwin (1976a) reported that microbial cell and protein yield were highest in *in vitro* incubations of mixed rumen bacteria when two-thirds of the added N came from AA and one-third came from urea. No further benefit was observed by providing all supplemental N in the form of AA-N, but lower growth rates were observed when two-thirds of the added N came from urea and one-third came from AA. Similar observations were made by Russell *et al.* (1983). They determined from *in vitro* studies that microorganisms that ferment non-structural carbohydrates (NSC) derived 34% of their N from ammonia and 66% of their N from peptides or AA. This proportion was not affected by the growth rate of the microorganisms. Ling and Armstead (1995) examined uptake of AA and peptides in five species of rumen bacteria; *P. ruminicola*, *S. ruminantium*, *F. succinogenes*, *Anaerovibrio lipolytica* and *S. bovis*. When growth of the cultures was exponential, samples were mixed with ^{14}C-labelled AA or peptides. Based on uptake and metabolism data obtained on these species, and assuming that a rumen population could consist of equal proportions of these five bacterial species, the authors calculated that peptides and AA could supply up to 43% and 62% of the N requirements, respectively. Griswold *et al.* (1996) showed no apparent benefit in bacterial N yield or OM digestion in continuous culture by providing combinations of N sources compared to when the N sources were fed alone. Nitrogen forms provided were isolated soy protein, soy peptides, individual AA blended to profile soy protein and urea. All individual forms and all possible combinations were examined. In contrast to the observations of Griswold *et al.* (1996), Jones *et al.* (1998) determined that microbial growth and digestion of OM and protein were maximized at a ratio of 54% peptide-N to 46% urea-N in continuous culture involving mixed bacteria.

A preference of rumen microorganisms to use non-ammonia rather than ammonia-N for cell synthesis has been demonstrated in several studies. Hristov *et al.* (1997) designed an experiment to investigate the effect of different levels of carbohydrates and simultaneous provision of ammonia and amino N on utilization of α-amino N by mixed rumen microorganisms. Rumen inoculum obtained from a steer fed either a 50% grain diet or a 95% grain diet was incubated with five levels of carbohydrates: 0, 1, 5, 15 and 30 g/l (75% sugar and 25% starch) and five N sources (ammonia, casein-free AA, ammonia plus casein-free AA, tryptic digest of casein and ammonia plus tryptic digest of casein). The ammonia pool was labelled with $(^{15}NH_4)_2SO_4$ in order to measure incorporation of ammonia-N into microbial protein. Increasing levels of carbohydrates up to the highest level increased N depletion, increased VFA production and increased incorporation of ammonia-N into microbial protein in a linear fashion. The efficiency of N utilization was the lowest for ammonia and was improved by amino N. Further improvement was observed when ammonia-N was simultaneously provided. Ammonia treatment resulted in the highest percentage of ammonia-derived N in microbial protein (up to an average of 39%), whereas the casein-free AA and the tryptic digest of casein (peptide-bound N) reduced the percentage of ammonia-derived microbial N to 15.5% and 11.8%, respectively. When ammonia-N was provided in addition to the free AA or peptide-bound N, incorporation of ammonia-derived N increased to 23.0% and 20.1%, respectively. These data suggest that the level and efficiency of utilization of α-amino N for cell growth in the rumen is not a constant and may depend on the availability of energy and ammonia-N.

Atasoglu *et al.* (1998) examined the proportional use of ammonia by three pure cultures of predominant non-cellulolytic bacteria (*Prevotella bryantii*, *S. ruminantium* and *S. bovis*) in the presence of increasing concentrations of peptides or free AA (0, 1, 5, 10 and 30 g/l). At peptide and AA concentrations of 1 g/l, which is more similar to peptide concentrations in the rumen, 64–83% and 53–86% of total N was derived from ammonia, respectively. At the high concentrations of peptides and AA (10 and 30 g/l), 14–30% and 23–52% of total N was derived from ammonia, respectively.

In a follow-up study, Atasoglu *et al.* (2001) examined the proportional use of ammonia by three cellulolytic ruminal bacteria (*F. succinogenes*, *R. flavefaciens* and *R. albus*) in the presence of normal (1 g/l) or high (10 g/l) concentrations of peptides. Increasing the concentration of peptides in the growth media from 1 to 10 g/l decreased the amount of cell N derived from ammonia from 80% to 47%.

2.2.1.5 Summary and conclusions

Although a considerable amount of research has been conducted, it remains unclear as to what is the proportional need of ammonia, AA and short peptides and what their optimal concentrations in rumen digesta are to optimize rumen function under common feeding practices. This has been a challenge because of the complexity of rumen fermentation, the uncertainty of the predominant microorganisms that make up the microbial population in a given feeding situation, the uncertainty in a given feeding situation of the extent of ruminal recycling of microbial protein, the rate and extent to which the usable end products of protein breakdown are captured by the microorganisms and the different methods that have been used to determine requirements (i.e. *in vitro*, *in situ* or *in vivo*). The lack of culture procedures for protozoa and fungi has complicated microbiological study and limits current understanding of these microorganisms as compared to rumen bacteria.

It is necessary in the advancement of protein systems for cattle that appropriate rumen submodels be developed that predict ammonia-N, AA and peptides concentrations in the rumen. The availability of such models would permit *in vivo* determination of the optimal rumen concentrations of these N metabolites for different diets and feeding strategies and would allow optimal formulation of dietary RDP. It appears that such models will have to predict the relative size of the protozoa population. Protozoa are net exporters of ammonia, and because of the extensive recycling of protozoal N that occurs, faunated animals almost always have higher rumen ammonia-N concentrations than defaunated animals. In 16 of 17 experiments, rumen ammonia-N concentrations were significantly or numerically higher (+75%; range = 8% to 159%) in faunated compared to defaunated animals (Broudiscou and Jouany, 1995; Jouany, 1996).

2.2.2 Metabolizable protein requirements of cattle

MP is defined as the true protein that is digested post-ruminally and the released AA absorbed by the small intestine. The absorbed AA are provided by ruminally synthesized microbial protein, RUP and to a lesser extent, by endogenous protein. Microbial protein is derived from a complex mixture of microorganisms flowing out of the rumen, including bacteria associated with the fluid and particle phases, plus protozoa and fungi.

A primary function of absorbed AA is their use in the synthesis of proteins, a biosynthetic event that is vital to the maintenance, growth, reproduction and lactation of cattle. The following discussion is limited to a brief description of the MP requirements of cattle for these physiological processes and some of the challenges associated with determining the requirements.

2.2.2.1 Maintenance

It is generally assumed that the maintenance requirement includes the AA needed for the synthesis of endogenous urinary protein, the AA needed for the synthesis of scurf protein (skin, skin secretions and hair) and the AA needed for the synthesis of metabolic faecal protein when animals are fed N-free diets. Urinary endogenous protein and metabolic faecal N losses could be interpreted as two routes for the excretion of the endogenous N lost from the normal recycling of protein in the body. Metabolic faecal N losses consist of digestive enzymes, bile, desquamated epithelial cells and mucus.

It is difficult to measure urinary and faecal losses independently of each other and it is also difficult to measure scurf losses. It is difficult to separate microbial cell losses in the faeces from true metabolic losses. For these reasons, different approaches have been used to make direct measurements of these losses of protein but regardless of the method used, the losses must be divided by an assumed efficiency of conversion of MP to the net protein that is lost. The resulting value for each loss is the predicted MP requirement for that function. Efficiencies of use of MP for these losses of protein vary between 0.67 and 1.0.

Different equations are used to estimate urinary endogenous protein, scurf protein and metabolic faecal protein and different efficiencies of conversion of MP to net protein are also used. Some of the resulting equations that are used for predicting the MP requirement for endogenous urinary protein are:

$$[2.75 \times (\text{BW} - \text{conceptus weight})]^{0.50}/0.67$$

(NRC, 2001) and

$$5.9206 \times \log_{10} \text{BW} - 6.76$$

(ARC, 1965; GfE, 1986), where BW is body weight. Some of the equations to predict scurf protein are:

$$[0.2 \times (\text{BW} - \text{conceptus weight})]\text{BW}^{0.60}$$

(NRC, 2001) and

$$0.018 \times \text{BW}^{0.75}$$

(GfE, 1986). Some of the equations used for metabolic faecal N are:

$$(30 \times \text{DMI}) - 0.50\ (\text{microbial MP}/0.80) - \text{microbial MP}$$

(NRC, 2001) and

$$2.19 \times \text{DMI}$$

(GfE, 1986), where DMI is in kg.

In some protein systems, urinary endogenous protein, scurf protein and metabolic faecal protein losses are predicted from a single equation and thus, the MP requirement for those losses is calculated using a single equation. Some of the equations used are:

$$3.25\ \text{g/kg BW}^{0.75}$$

(Vérité and Peyraud, 1989; Tuori *et al.*, 2002),

$$3.0\ \text{g/kg BW}^{0.75}$$

(Madsen *et al.*, 1995) and

$$3.80\ \text{g/kg BW}^{0.75}$$

(NRC, 1996).

This approach has been used because of the difficulty in measuring urinary and faecal losses independently of each other and because it is difficult to separate microbial cell losses in the faeces from true metabolic losses. In the AFRC system (1992) MP requirements for maintenance are estimated as $(2.1875 \times \text{BW}^{0.75} + 0.1125 \times \text{BW}^{0.60})$, the first part representing endogenous N loss and the second, scurf proteins.

In the DVE/OEB system (Tamminga *et al.*, 1994), MP requirements for maintenance are restricted to endogenous losses in urine and scurf protein. The equation is (g/day):

$$(2.75 \times \text{BW}^{0.50} + 0.2 \times \text{BW}^{0.60})/0.67$$

Because the excretion of metabolic faecal protein is related to the indigestibility of DM in a feed, metabolic faecal protein losses are taken into account in the true protein digestibility of each feedstuff.

2.2.2.2 Growth

There is an obvious MP requirement for growing cattle because of the net protein accretion that occurs. Different equations are used to predict protein accretion in the different protein systems.

In the DVE/OEB system (Tamminga *et al.*, 1994), it is assumed that a direct relationship exists between energy and protein in body reserves. It is assumed that 10% of the energy in body reserves is protein and that each 6.9 MJ of energy contains about 0.7 MJ in protein. Under the assumption that there are 24 MJ/kg of protein, it is calculated that there are 29 g of protein in 6.9 MJ of tissue energy. Using an efficiency of use of digested protein of 50%, the digested protein requirement for growth for each 6.9 MJ is (g/day) 29/0.50 = 58.

The MP requirements for growth in NRC (2001) are those of heifers and steers in NRC (1996). Two equations are used, one for equivalent shrunk BW (EQSBW) less than or equal to 478 kg:

$$\frac{\text{WG} \times \{268 - [29.4 \times (\text{RE/ADG})]\}}{[83.4 - (0.114 \times \text{EQSBW})]/100}$$

and one for EQSBW greater than 478 kg:

$$\frac{\text{WG} \times \{268 - [29.4 \times (\text{RE/ADG})]\}}{0.28908}$$

where WG is weight gain, RE is retained energy and ADG is average daily gain.

In both cases, net protein accretion (i.e. the numerator in both equations) is calculated in the same way, ADG and model-predicted RE. The difference in the two equations resides in the denominator. If EQSBW is less than or equal to 478 kg, then the efficiency of use of MP for growth is variable and dependent on BW. If EQSBW is greater than 478 kg, then the efficiency of use of MP for growth is assumed to be a constant 28.9%.

2.2.2.3 Pregnancy

It is understood that the MP requirements to support pregnancy are a function of days pregnant and conceptus weight. Several different equations are used.

The equation used in the DVE/OEB system (Tamminga *et al.*, 1994) from 141 to 281 days of gestation (g/day) is: $[34.375 \times \exp^{(8.537-13.1201)} \times \exp^{(-0.00262-D)} - 0.00262 \times D]/0.50$, where D is days after conception between 141 and 281.

The equation accepted for NRC (2001) (190 to 279 days of gestation) is [(0.69 × days pregnant)−69.2)) × (calculated calf birth weight/45)]/0.33. The numerator predicts conceptus protein and is the first derivative of the quadratic regression equation of Bell *et al.* (1995). The efficiency of conversion of MP to conceptus protein is assumed to be 33%. In NRC (2001), cows more than 279 days pregnant have the same requirements as cows that are 279 days pregnant.

The equation used by AFRC (1992) to estimate the MP requirement for pregnancy (g/day) is $[1.01 \times W_c \times (TP_t \times e^{-0.00262 1t})]$, where W_c is calf birth weight, TP_t is tissue protein retention (g/day) and t is number of days from conception.

In the INRA (Vérité and Peyraud, 1989) and FIN systems (Tuori *et al.*, 2002), MP requirements for maintenance are increased by 75, 135 and 205 g/day as the cows are in 7th, 8th or 9th month of pregnancy and slightly different in the DK system where the allowances for the last 3 months are 95, 160 and 215 g/day (Madsen *et al.*, 1995).

2.2.2.4 Lactation

The MP requirement for lactation is a function of MPY and content of milk true protein. Because MPY is easily measured, the only challenge is to identify the most appropriate efficiency of use values for the protein system such that the model validates (i.e. shows no bias) across a wide range of milk yields. The MP efficiency of use values for milk protein synthesis in some current protein models are 0.64 (INRA, Vérité and Peyraud, 1989), 0.65 (NRC, 1996), 0.67 (NRC, 2001), 0.68 (AFRC, 1993) and 0.80 (GfE, 1986 (German system)).

In the DVE/OEB system (Tamminga *et al.*, 1994), a variable efficiency factor is used because of the recognition in production trials performed under Dutch conditions that the efficiency is variable and dependent on the amount of true protein digested in the small intestine and level of milk production. The equation for predicting the MP required for milk protein production (g/day) is:

$$1.396 \times \text{MPY} + 0.000195 \times \text{MPY}$$

In the Finnish system, the MP requirements for milk production are 45–47 g/kg energy-corrected milk (ECM) depending on milk yield (Tuori *et al.*, 2002). The requirements per kg ECM are lower at higher production levels.

In the Danish system, the MP requirement in early lactation when the cows are fed *ad libitum* is expressed as 90 g MP per total feed unit. If expressed in relation to milk production, a value of 37 g MP is used per kg ECM.

2.3 Evolution of Protein Systems

Considerable progress has been made over the last 30 years to develop systems that describe protein requirements and protein adequacy of diets for cattle. These efforts continue and are essential towards implementing more sophisticated strategies for balancing diets for protein. The greatest challenge in developing these enhanced protein systems is to be as accurate and precise as possible in predicting microbial protein synthesis, supply of RDP and RUP, requirements of rumen microorganisms for RDP, the digestibility and AA composition of RUP and the AA requirements of the host animal. The overall goal is to accurately predict animal responses to protein and AA supplements in terms of productive outcomes (i.e. weight gain, composition of weight gain, conceptus weight, milk production and milk composition).

2.3.1 Digestible protein systems

The early protein systems developed for ruminants described protein requirements and the protein value of feeds on the basis of CP or digestible

CP. Digestible CP was calculated as apparently digested CP and was used as a measure of AA availability to cattle. The requirement values were determined by feeding increasing levels of digestible CP and determining the point of maximum response. The digestible CP systems were shown to be relatively satisfactory for many traditional and well-balanced diets, but in many cases, diets with equivalent digestible CP content did not sustain the same level of productivity. The need for a new system of protein evaluation became apparent over 30 years ago with changes in feed technology and the growing use and economic advantage of using NPN in ruminant diets. It became clear that any new system should be capable of predicting the effects that processing (heating, grinding and chopping and pelleting) has on rumen degradability, rate of passage, site of digestion and intestinal digestion as well predicting the value of NPN in a variety of situations. Because these systems did not consider the fate of ingested protein and differentiate between, or consider independently, the N needs of rumen microorganisms and the AA requirements of the host animal, the systems were often poor predictors of AA absorption and animal performance. This led to the development of the previously mentioned MP-based systems, systems that not only consider the independent needs of rumen microorganisms and the host animal but also consider the separate protein requirements of the host animal for maintenance, growth, gestation, milk protein production and metabolic losses from the digestive tract.

2.3.2 Metabolizable protein systems

The MP concept was introduced by Burroughs *et al.* (1971) for feedlot cattle. The stated reasons for introducing the concept were: (i) 'the recent demonstration that lightweight, rapidly gaining feedlot cattle undergo AA deficiencies which can be overcome by feeding pre-formed protein and AA, but which cannot be fully overcome by feeding additional NPN such as urea'; and (ii) 'many discrepancies in the literature concerning quantitative protein requirements (total digestible) of feedlot cattle' (Burroughs *et al.*, 1971). MP was defined as 'the quantity of absorbed AA in the post-ruminal portion of the digestive tract of cattle and other ruminants'. Tentative MP and MP-AA requirements for different-sized steers and heifers fed to achieve different rates of gain were presented as were suggested MP and MP-AA values of some common feedstuffs (Burroughs *et al.*, 1971). This effort was extended to sheep and lactating cows, along with the inclusion of a new measurement, the urea fermentation potential of feeds, that attempted to quantify the amount of urea that can be useful in rations of ruminants (Burroughs *et al.*, 1974a,b, 1975).

Following the introduction of the MP system by Burroughs and co-workers, researchers in several European countries (INRA, 1978; Vérité *et al.*, 1979; ARC, 1980; Madsen, 1985; Bickel and Landis, 1987; Rohr, 1987), North America (NRC, 1985, 1989; Fox *et al.*, 1992; Russell, *et al.*, 1992; Sniffen *et al.*, 1992) and Australia (CSIRO, 1990) published new protein systems that incorporated prediction of MP flow to the small intestine and MP requirements. More recent versions of several of these protein systems have been published (AFRC, 1992, 1993; Tamminga *et al.*, 1994; Madsen *et al.*, 1995; NRC, 1996, 2001). These protein systems, although different in terminology and detail, are similar in concept and have allowed for better definition of N requirements of cattle, more exact feeding and more accurate prediction of growth and milk protein production.

2.4 Using Metabolizable Protein Systems to Meet Dietary N Requirements of Cattle

Using MP systems to balance diets for cattle represents a significant departure from using CP or digestible CP-based systems. Rather than formulating to meet a targeted ration concentration of CP or digestible CP, the emphasis is switched to meeting the N requirements of rumen microorganisms and the MP requirements of the host animal. Rations are balanced for RDP and RUP with the amount of RUP required being that needed to make up the difference between the model-predicted MP requirement and the model-predicted supply of MP from ruminally synthesized microbial protein.

As stated in Section 2.1.1, there are three strategies involving protein nutrition that can be used to lower dietary protein without compromising animal performance. One is to feed for increased synthesis of microbial protein by changing the quantity and quality of fermentable carbohydrates in the diet. This strategy increases the opportunity to capture recycled N and the end products of protein breakdown in the rumen and increases the efficiency of use of RDP. The second strategy is to not overfeed either RDP or RUP by matching supplies with requirements. And the third strategy is to increase the efficiency of use of MP by balancing diets more precisely for EAA. This strategy is aimed at decreasing the need for RUP. Current protein models/systems and current methods of feed analysis are not adequate to fully exploit each of these feeding strategies. However, as indicated in the next two sections where we review some selected published studies involving growing and lactating cattle, it becomes apparent that opportunities exist in reducing dietary CP without jeopardizing animal performance by using MP systems.

2.4.1 Beef cattle

Beef cattle consume a wide variety of forages and grains that vary widely in content of fermentable carbohydrates, RDP and RUP, and proportional content of fermentable carbohydrates and RDP. Moreover, the requirements of growing cattle for MP relative to that for metabolizable energy (ME) change with advancing age and increasing BW and are highest at birth and lowest at time of slaughter. Such variation in feedstuffs and requirements necessitates the need for protein systems that recognize the independent requirements of microbial requirements for RDP and the MP requirements of the host animal. The following discussion highlights four important issues in beef cattle nutrition: (i) defining the RDP and RUP requirements of young cattle fed high concentrate diets; (ii) defining the RDP requirements of cattle fed low quality forages; (iii) defining the RDP requirements of cattle fed grains that differ in starch digestibility; and (iv) determining the proportional amount of supplemental RDP that can be provided as urea.

2.4.1.1 RDP and RUP requirements of young cattle fed high concentrate diets

In most current feeding systems (e.g. AFRC, 1993; NRC, 1996), CP recommendations for calves under 200–250 kg BW exceed 16% of diet DM. Concentrations of CP of 17% to 18% have been recommended for rapidly growing animals (Kertz *et al.*, 1987). However, experiments have been conducted which indicate that dietary CP levels can be less than the above recommendations if diets are balanced for RDP and RUP and the diets allow for efficient synthesis of microbial protein.

In some countries it is common for growing cattle from weaning to slaughter to be fed concentrate and cereal straw for *ad libitum* consumption. Because of the uncertainty of how RDP and RUP requirements are affected with increasing BW when fed this type of diet, Devant *et al.* (2000) evaluated the effect of CP concentration and degradability on performance and N metabolism in crossbred heifers consuming barley straw and concentrate on an *ad libitum* basis. Dietary CP concentration and degradability were altered by feeding one of four concentrates that differed in content of CP (17% vs. 14% of DM) and ruminal degradability of CP (higher vs. lower). The concentrate consisted of 32% to 38% maize, 27% barley, 17% tapioca, 6% sunflower meal, either 16% or 9% soybean meal or 16% or 9% treated soybean meal (to obtain the two levels of rumen degradability) and minerals and vitamins. The barley straw contained 2.8% CP and 83.1% NDF. The heifers weighed an average of 102 kg at the start of the experiment and remained on the experiment for 16 weeks. The experiment was divided into four consecutive 4-week periods to evaluate the age (period) effect. There was no effect of CP concentration and degradability or their interaction on concentrate DM intake (4.4 kg), straw DM intake (0.3 kg/day), final BW (234 kg), average daily gains (1.2 kg/day) and feed efficiency (0.25 kg gain per kg DM intake). Ruminal ammonia-N concentrations averaged 3.8 mmol/l for the high protein, high degradability diet, 1.1 mmol/l for the high protein, low degradability diet, 1.8 mmol/l for the low protein, high degradability diet and 0.8 mmol/l for the low protein, low degradability diet. Urinary excretion of allantoin and uric acid for the four diets was 65.3 and 4.7, 66.1 and 4.9, 62.1 and 4.8 and 59.1 and 5.8 mmol/day, respectively. The lower CP diets caused less excretion of urinary N (22 vs. 41 g/day) and the lower RUP diets resulted in less excretion of faecal N (30 vs. 36 g/day). In contrast, retained N was not affected by protein concentration or degradability but retained N as a percentage of

N intake was highest with the low protein diets (46.5% vs. 38.2%). In summary, there was no evidence in this experiment that altering the protein degradability of the concentrate to increase ruminal ammonia-N to concentrations higher than 0.7 to 1.5 mmol/l or increasing RUP beyond that supplied by the mixed cereal grains and oilseed meals was needed to meet the RDP and RUP requirements of the calves.

The experiment by Devant *et al.* (2000) indicates that a dietary CP concentration considerably less than recommended amounts can be fed to young calves when high concentrate diets similar to that described are fed. Other researchers have come to similar conclusions. For example, Lana *et al.* (1997) did not find differences in ADG or feed efficiency when CP concentration of diet DM was increased from 13.5% to 16.6% when Holstein steers were fed a 90% concentrate diet from 150 to 277 kg BW. It is of interest to note that in the experiment by Devant *et al.* (2000) as well as in experiments by others (Ganev *et al.*, 1979; Loerch *et al.*, 1983) that rumen degradability of soybean meal is less in animals fed high concentrate diets than those fed diets containing more forage.

2.4.1.2 RDP requirements of cattle fed low quality forages

Cattle in many parts of the world consume low quality forages (<7% CP) because of seasonal declines in pasture quality or a reduction in quality of harvested grasses because of delayed harvest or inclement weather. While low in CP, low quality forages and roughages are valuable ruminant feed resources worldwide and considerable research has been conducted to identify protein supplementation strategies that optimize their utilization. Supplementing low quality forages and roughages with protein supplements that are high in RDP has been shown in many experiments to increase intake and digestibility of these feeds (e.g. Kartchner, 1980; Guthrie and Wagner, 1988; Stokes *et al.*, 1988; DelCurto *et al.*, 1990; Heldt *et al.*, 1999; Mathis *et al.*, 1999). Infusion experiments indicate that these responses are largely the result of the supplements providing additional RDP rather than RUP (Köster *et al.*, 1996; Olson *et al.*, 1999; Mathis *et al.*, 2000; Bandyk *et al.*, 2001).

Köster *et al.* (1996) infused incremental amounts of casein (0, 180, 360, 540 and 720 g/day) into the rumen of Angus × Hereford cows that had *ad libitum* access to water and low quality, tallgrass-prairie hay (1.9% CP, 77% NDF) which was fed twice daily. The sodium caseinate was infused twice daily immediately before feeding the hay. Intake of hay and duodenal flow of total N increased quadratically with increasing supplemental RDP, reaching maximums at the 540 g/day level of casein infusion. Ruminal fluid dilution rates, flow of microbial N to the duodenum and efficiency of microbial protein synthesis (g N/kg OM truly digested in the rumen) increased linearly with increasing casein infusion. Ruminal digestion of NDF increased with the first increment of RDP but exhibited moderate and somewhat variable responses with the higher amounts of RDP. Ruminal ammonia-N concentrations averaged 0.2, 1.3, 3.5, 5.1 and 6.8 mmol/l for the five amounts of infused casein. Using a single-slope, broken line model, the authors concluded that intake of digestible OM (DOM) was maximized when it contained 11.1% RDP. This concentration of RDP in DOM corresponded to an infused level of casein of slightly over 400 g/day and calculates to a required content of RDP in diet DM of 6.1%.

Olson *et al.* (1999) infused increasing amounts of sodium caseinate (0.03%, 0.06%, 0.09% and 0.12% of initial BW) into the rumen of Hereford × Angus steers (initial BW = 264 kg) that were given *ad libitum* access to low quality tallgrass-prairie hay (4.9% CP, 72% NDF). Intake of hay, total tract digestion of OM and NDF, and particulate and liquid passages increased linearly with supplemental RDP. Ruminal ammonia-N concentrations averaged 0.3, 0.4, 0.45 and 1.2 mmol/l. Intake of DOM was highest when RDP intake was equal to 11.6% of DOM (obtained with the highest level of casein infusion). This concentration of RDP in DOM corresponds to a required content of RDP in diet DM of 7.3%.

Mathis *et al.* (2000) conducted three infusion experiments to examine the effect of incremental amounts of supplemental RDP on utilization of low and medium quality hays in Angus × Hereford steers (initial BW = 280–315 kg). The hays were forage sorghum (4.3% CP, 60% NDF), bromegrass (5.9% CP, 65% NDF) and bermudagrass (8.2% CP, 71% NDF). Sodium caseinate was used as the source of supplemental RDP. The casein (0%, 0.041%, 0.082% and 0.124% of initial BW) was infused into the rumen once daily immediately before the once daily feeding of hay.

When the forage sorghum (4.3% CP, 60% NDF) was fed, ruminal ammonia-N concentrations averaged 3.0, 4.8, 5.8 and 7.0 mmol/l for the four levels of infused casein. Supplemental RDP increased intake of hay (64, 79, 87 and 88 g/kg $BW^{0.75}$) and total tract digestion of OM (46%, 54%, 59% and 62% of intake) and NDF (35%, 44%, 52% and 54% of intake) linearly. At the highest level of RDP supplementation, RDP constituted approximately 12.8% of total DOM (7.9% of DM intake).

When the bromegrass hay (5.9% CP, 65% NDF) was fed, ruminal ammonia-N concentrations averaged 0.4, 1.8, 3.1 and 4.1 mmol/l for the four levels of infused casein. Although the observed ruminal ammonia-N concentration for the basal diet was lower in this experiment than in the experiment where the forage sorghum was fed (0.4 vs. 3.0 mmol/l), there was no effect of casein infusion on intake of hay or digestion of OM and NDF. However, there was a linear increase in intake of DOM (62, 69, 73 and 71 g/kg $BW^{0.75}$). Using the treatment group in which the maximum intake of DOM was observed, total RDP intake constituted about 9.8% of total DOM intake (6.0% of DM intake). Hay intake and total tract digestion of OM and NDF averaged 112 g/kg $BW^{0.75}$, 60% of intake and 54% of intake.

When the bermudagrass hay (8.2% CP, 71% NDF) was fed, ruminal ammonia-N concentrations averaged 5.8, 9.9, 13.9 and 16.8 mmol/l for the four infusion levels of casein. None of the measures of forage utilization were affected by RDP. Hay intake and total tract digestion of OM and NDF averaged 89 g/kg $BW^{0.75}$, 63% of intake and 64% of intake, respectively. The RUP supplied by the hay was estimated to be 8.2% of total tract DOM (5.2% of DM intake).

Bandyk *et al.* (2001) compared the effects of ruminal vs. post-ruminal administration of 400 g/day of casein when Angus × Hereford steers were fed a low quality, tallgrass-prairie hay (3.4% CP, 77% NDF). The steers weighed 563 kg at the start of the experiment. Casein was administered once daily before feeding. Both methods of casein administration improved forage utilization (hay intake, OM and NDF digestion), but increases in hay intake and total DOM were greater when casein was infused into the rumen than when it was infused into the abomasum. Ruminal ammonia-N concentrations were increased by both methods of casein administration, but the increase was higher with ruminal infusion than with post-ruminal infusion. The ruminal ammonia-N concentrations averaged 0.6, 1.3 and 4.1 mmol/l for the control, post-ruminal and ruminal infusion treatments. This experiment clearly indicates that a portion of digestible RUP recycles back to the rumen.

The results of these experiments indicate variability in response of cattle to RDP supplementation when consuming low to medium quality forages. The results also suggest that RDP supplementation is probably not necessary if the forages contain more than 6% to 7% CP. It has been concluded previously that there is little or no benefit to protein supplementation of forages when the forage contains more than approximately 7% CP (Mathis *et al.*, 2000). And finally, the need for supplemental RDP may be related more to forage CP content than to rumen ammonia-N concentrations.

When RDP intake is adequate to support normal rumen function, there appears to be little or no value to beef cows to supplement low quality forages with RDP (Hunter and Magner, 1988; Olson *et al.*, 1999; Sletmoen-Olson *et al.*, 2000).

2.4.1.3 RDP requirements of feedlot cattle fed grains that differ in rumen fermentable starch

Grains are the principal source of fermentable carbohydrates in diets for feedlot cattle. Starch is the major energy component of grains with values usually ranging between 55% and 80% depending on species (wheat > maize and sorghum > barley and oats), effects of variety, location, year, climatic conditions and agronomic practices. Average concentrations of starch as reported in several publications were 77% for wheat, 72% for maize and sorghum and 57% to 58% for barley and oats (Huntington, 1997). Because of the high content of starch in cereal grains, differences in rate and extent of fermentation of the constituent starch as influenced by grain type (barley and wheat usually are fermented more rapidly than maize and sorghum) and especially by method of processing (grinding, dry-rolled, tempering before rolling, steam-rolled, high-moisture and steam flaking) can have profound effects on ruminal starch digestion (Feng *et al.*, 1995; Huntington, 1997; Barajas and Zinn, 1998; Zinn *et al.*, 1998; Cooper *et al.*, 2002b) and synthesis of microbial protein

(Feng *et al.*, 1995; Barajas and Zinn, 1998; Zinn *et al.*, 1998; Cooper *et al.*, 2002b).

There are limited studies that have evaluated the effects of starch availability and grain processing on RDP requirements of feedlot cattle. Although the quantity of fermentable carbohydrate in the diet is considered to be the primary 'driver' of microbial protein synthesis in the rumen, a source of nitrogenous compounds from RDP and recycled N are needed for protein synthesis and it can be expected that the requirement for RDP in diet DM is at least somewhat proportional to the quantity of fermentable carbohydrate in diet DM. A deficiency in RDP may not only decrease bacterial protein flow from the rumen (Martin-Orue *et al.*, 2000; Griswold *et al.*, 2003) but also decrease energy yield from carbohydrate fermentation (Russell *et al.*, 1992). The following discussion reviews some experiments that have attempted to demonstrate that the different rates of starch digestion in grains as affected by grain type and grain processing will affect dietary requirements of feedlot cattle for RDP.

Several studies have shown that barley has a faster rate of starch digestion than maize (Herrera-Saladana *et al.*, 1990; Campling, 1991; Feng *et al.*, 1995) and that replacing maize with barley in diets fed to growing cattle (Spicer *et al.*, 1986; Feng *et al.*, 1995) and lactating dairy cows (McCarthy *et al.*, 1989) increased microbial protein synthesis in the rumen. Because of these results, Kennington *et al.* (2003) hypothesized that the more highly degraded starch in barley would increase the requirement for RDP in barley-based diets as compared to maize-based diets. Crossbred steers (381 kg) were assigned to a 2 × 3 factorial arrangement of treatments to determine the effect of grain type (barley and maize) and content of CP in diet DM (11.5%, 12.8% and 14.0%) using soybean meal and urea as protein supplements. The diets contained 5% lucerne hay, 7% maize silage and 88% concentrate. The respective RDP values (% of DM) were 8.0, 9.0 and 10.1 for the barley diets and 7.2, 8.3 and 9.4 for the maize diets. Grains were tempered by adding 9% units of water and coarsely rolled. As expected, the barley contained less starch than the maize (48% vs. 63%), but the barley had a faster rate of starch (19.1% vs. 6.1%/h) and DM (10.3% vs. 5.7%/h) degradation in the rumen. As RDP increased, average daily gains increased linearly (1.58, 1.64 and 1.73 kg/day) and there was a trend for increased gain:feed ratios (0.165, 0.169 and 0.171 kg weight gain/kg DM intake). There was no grain by RDP interaction for growth performance variables, indicating that type of grain did not affect the RDP requirement under the conditions of this experiment.

Cooper *et al.* (2002a) conducted three trials to determine the effect of maize processing (high-moisture, steam-flaked and dry-rolled) on the RDP requirement of crossbred feedlot cattle fed high concentrate diets. The high-moisture maize was harvested at approximately 71% DM, rolled and ensiled. The steam-flaked maize was processed to a flake density of 0.37 kg/l. The mean particle size of the three processed maize diets at the time of feeding was 722, 2278 and 2850 μm, respectively. All diets contained 82% maize, 5% each of lucerne, cottonseed hulls and dry supplement, and 3% molasses. Urea was the source of RDP supplementation.

In Trial 1, the steers (379 kg) were fed high-moisture maize diets containing 0.0%, 0.4%, 0.8% and 1.2% urea of diet DM to provide RDP levels of 7.0%, 8.2%, 9.3% and 10.5% of DM. Average daily gains increased linearly as RDP increased (1.70, 1.72, 1.82 and 1.85 kg/day). Non-linear analysis predicted maximal feed efficiency at 10.2% RDP.

In Trial 2, the steers (355 kg) were fed steam-flaked maize diets containing 0.0%, 0.4%, 0.8%, 1.2%, 1.6% or 2.0% urea (DM basis) to provide RDP levels of 4.7%, 5.8%, 7.0%, 8.2%, 9.3% and 10.5% of DM. Average daily gains responded quadratically (1.44, 1.74, 2.00, 2.00, 2.02 and 2.04 kg/day) as RDP increased. Non-linear analysis predicted maximal feed efficiency at 7.1% dietary RDP.

In Trial 3, the three types of processed maize (dry-rolled, high-moisture and steam-flaked) were evaluated simultaneously. The average weight of the steers was 278 kg. Dietary urea concentrations were 0.0%, 0.5%, 1.0% and 2.0% of DM to provide dietary RDP values of 4.8%, 6.3%, 7.8%, 9.2% and 10.7% for the dry-rolled maize diets, 6.7%, 8.1%, 9.6%, 11.1% and 12.5% for the high-moisture maize diets and 4.7%, 6.1%, 7.6%, 9.0% and 10.5% for the steam-flaked maize diets.

Non-linear analysis could not predict an RDP requirement for the dry-rolled maize diets because feed efficiency was not improved beyond the first increment of dietary RDP. This suggests that the RDP requirement was met at 6.3% of DM. This requirement value for RDP is consistent with the findings of Milton *et al.* (1997a,b) and Shain *et al.* (1998) who concluded that the required content of RDP in dry-rolled maize finishing diets is between 6.4% and 7.2% of diet DM. Level 1 of NRC (1996) predicts that the RDP requirement for a typical finishing diet containing dry-rolled maize is approximately 6.8%.

The RDP requirement for maximal feed efficiency of the high-moisture maize diets was predicted by non-linear analysis to be 10.0% of diet DM (Cooper *et al.*, 2002a). This value is similar to that obtained in Trial 1 (10.2%).

The RDP requirement for maximal feed efficiency of the steam-flaked maize diets was predicted to be 9.5% of DM. This value was higher than the predicted value of 7.1% obtained in Trial 2. The authors suggested that reasons for this difference were not clear but did highlight the fact that there were a few differences between Trials 3 and 2 such as initial BW of steers (278 vs. 355 kg) and grain adaptation procedure. In Trial 3, the steers were abruptly switched to the grain diet but were transitioned slowly from restricted feeding (1.8% of BW) to *ad libitum* feeding which took approximately 21 days to accomplish. In Trial 2, the steers were transitioned over 21 days from a 40% lucerne hay diet to the experimental diets.

In a companion study to Trial 3, Cooper *et al.* (2002b) fed the 2.0% urea-containing maize diets (dry-rolled, high-moisture and steam-flaked) to study the effect of maize processing on nutrient digestion. This study provided contradictory results as to whether the RDP requirement for maximal feed efficiency of the steam-flaked maize diet was 9.5% or 7.1% of DM. Apparent ruminal starch digestibility values for the respective three maize diets were 76.2%, 91.7% and 89.6%. Regression of these values on the RDP requirements estimated for the three diets in Trial 3 provided a perfect relationship ($R^2 = 1.0$), suggesting that the RDP requirement for the steam-flaked diet was closer to the value of 9.5% predicted in Trial 3 than the value of 7.1% observed in Trial 2. However, there was no difference in bacterial N flows to the duodenum in this study between the dry-rolled and steam-flaked diets. This would lead to the conclusion that the RDP requirement for a steam-flaked maize diet is not greater than that for a dry-rolled maize diet and that the RDP requirement for a steam-flaked maize diet is closer to 7.1% than 9.5%.

More research is needed to better understand the interaction between starch availability in feed grains and RDP requirements of feedlot cattle. However, MP systems provide a basis on which to quantify these interrelationships.

2.4.1.4 Urea as an RDP supplement

Urea is a widely used RDP supplement in growing cattle diets because of its low cost relative to that of other protein supplements. Galyean (1996) reported the results of a survey of six consulting nutritionists, who were responsible for feeding 3.6 million finishing beef cattle in the USA, and found that none of the consultants formulated for escape protein and that all used urea with amounts ranging from 0.5% to 1.5% of diet DM. Ration CP varied from 12.5% to 14.4% of diet DM with urea usage being the lowest when higher protein grains such as wheat or barley were fed or when the diets contained high-moisture maize in which more of the N is in soluble forms. Grains were typically processed, most commonly by steam flaking. Dry-rolled and high-moisture grains were always fed in combination with each other or in combination with steam-flaked grain. The results of this survey were consistent with an earlier survey as reported by Galyean (1996) that involved 12 consulting nutritionists in which it was found that ration CP ranged from 12.5% to 13.8% with urea levels ranging from 0.8% to 1.5%.

Two fundamental questions involving the use of urea as a protein supplement in the above types of diets are: (i) is the efficiency of microbial capture of the released ammonia in the rumen from urea similar or lower than the efficiency of capture of the combined ammonia, free AA and short peptides that result when natural protein sources are fed? and (ii) what is the optimum ratio of urea to natural proteins in beef cattle diets?

When cattle are fed low quality forages, several experiments indicate that significant amounts of urea (up to 1.9% to 2.5% of diet DM) can replace true protein supplements such as soybean meal without reducing microbial N production or efficiency of microbial N synthesis (Kropp *et al.*, 1977a,b; Petersen *et al.*, 1985). However, urea was inferior to soybean meal for supporting digestion of OM in Kropp *et al.* (1977a,b) but not in Petersen *et al.* (1985). Köster *et al.* (2002) observed that urea could replace between 20% and 40% of the RDP in a 30% protein supplement without affecting supplement palatability, forage intake or OM digestion provided that there was sufficient RDP in the diet to maximize DOM intake.

Conclusions have varied as to the optimum inclusion rate of urea in finishing cattle diets. When cattle (332 kg) were fed a dry-rolled maize diet (90% concentrate) that contained 10% prairie hay, the optimal level of dietary urea was 0.9% of diet DM for average daily gain and gain/feed ratio. Ration CP was 10.9% (Milton *et al.*, 1997a,b). The optimal level of urea in diet DM for ruminal OM and starch digestion was 0.5%. When similar sized cattle were fed dry-rolled maize diets that contained 10% lucerne hay instead of 10% prairie hay, the optimal level of urea for average daily gain and gain:feed ratio was 0.5% of DM (Milton *et al.*, 1997a,b). In both experiments all supplemental N was provided as urea. Shain *et al.* (1998) determined that 0.88% urea of diet DM was beneficial to feed efficiency and rate of gain when steers were fed a dry-rolled maize diet (90% concentrate). An NRC (1996) evaluation of the diets determined that all diets exceeded the steer's requirements for MP, but the diets without urea supplementation were deficient in RDP. So, a response to the supplemental urea was probably due to an increased flow of microbial protein to the small intestine. The NRC predicted that the 0.88% urea diet was still slightly deficient in digestible intake protein (DIP) (−61 g/day); however, the animal response to this treatment indicates that the rumen microorganisms were provided with adequate RDP. Nitrogen recycling to the rumen may have compensated for the slight deficiency in DIP as predicted by the model.

Duff *et al.* (2003) observed that replacing soybean meal with isonitrogenous amounts of urea up to 1.0% of diet DM in steam-flaked, maize-based receiving diets (70–75% concentrate) had no adverse effects on weight gains or feed efficiency. Zinn *et al.* (2003) determined that ADG was optimized by urea supplementation by dietary inclusion of 0.8% urea when steers were fed a steam-flaked barley-based finishing diet. Improvements in ADG were due to treatment effects on DMI.

2.4.2 Lactating dairy cows

Like growing cattle, lactating dairy cows consume a wide variety of forages and grains that vary widely in content of fermentable carbohydrates, RDP and RUP, and proportional content of fermentable carbohydrates and RDP. However, unlike growing cattle, many lactating cows have traditionally been fed large amounts of high N-containing grass and legume silages. In addition, also more than in growing cattle, there has been a tendency to overfeed protein supplements, especially high RUP-containing protein supplements. The net result has been feeding 18% to 20% CP diets that contain excesses of both RDP and RUP. Yet, many experiments have now been reported that indicate that more closely matching model-predicted RDP and RUP supplies with model-predicted RDP and RUP requirements results in diets that approximate 16% CP with little or no loss in milk and MPY when fed to high-yielding dairy cows (Armentano *et al.*, 1993; Dinn *et al.*, 1998; Broderick, 2003; Leonardi *et al.*, 2003; Noftsger and St-Pierre, 2003). This is particularly true when some attempt is made to optimize the concentrations of lysine and methionine in MP (e.g. Noftsger and St-Pierre, 2003).

2.5 Evaluation of Metabolizable Protein Systems for Lactating Dairy Cows

Although widely used with demonstrated benefit in improving the efficiency of use of dietary N, it is acknowledged that existing MP systems are still in their infancy and that further development and refinement are needed. It is important that the models be constantly evaluated to identify their shortcomings and strengths with the goal of developing a universal and common MP system that will work equally well across diverse diets and feeds of variable composition. Currently, each model is different in how MP requirements and MP flows are determined. Accurate predictions of MP requirements and supplies are fundamental to predicting the animals' requirements for RDP and RUP. Therefore, for the purposes of this chapter, we considered it important to do a comparative evaluation of some of the more commonly used MP systems as designed for lactating cows.

Supply and requirements of MP were estimated according to six different systems: AFRC (1992, 1993), INRA (Vérité and Peyraud, 1989), the German system (GER; GfE, 1986), NRC (2001) and two Scandinavian systems, the Danish (DK) and Finnish (FIN) systems

(Madsen, 1985; Madsen *et al.*, 1995, 2003; Tuori *et al.*, 1998; Møller *et al.*, 2000; Hvelplund *et al.*, 2003). The evaluation was based on determined relationships between predicted MP supply and observed MPY, and between estimated MP requirement and predicted supply of MP.

2.5.1 Description of systems

2.5.1.1 AFRC (1992)

Metabolizable protein in the AFRC system is defined as: MP (g/day) = 0.6375 × MCP + digestible RUP, where MCP means microbial CP and RUP means rumen undegraded protein. Effective protein degradability (EPD) in the rumen is estimated as: $EPD = a + (b \times c)/(c + k_p)$ where a, b and c are degradation constants (Ørskov and McDonald, 1979). The passage rate is affected by the level of feeding (L, multiple of maintenance) as follows: $k_p = -0.024 + 0.179 \times [1 - e^{(-0.278L)}]$. The k_p value increases from 0.019/h at the maintenance level to 0.104/h at the feeding level of 4.5 times L, respectively.

Rumen degradable CP is defined as the sum of the CP that is quickly degraded in the rumen (QDP) and that which is slowly degraded in the rumen (SDP). It is assumed that the efficiency of capture of QDP by rumen microorganisms is 0.8 and that the efficiency of capture of SDP is 1.0. QDP is described as $a \times$ CP (g/kg DM) and SDP is described as $(b \times c)/(c + k_p) \times$ CP (g/kg DM). Effective rumen degradable protein (eRDP) is defined as: eRDP (g/kg DM) = 0.8 × QDP + SDP. Undegraded Dietary Protein (UDP) is defined as: UDP (g/kg DM) = CP − (QDP + SDP). Digestible undegraded protein (DUP) is calculated from acid detergent insoluble N (ADIN) as: DUP (g/kg DM) = 0.9 × (UDP − 6.25× ADIN).

The efficiency of microbial protein synthesis is related to the feeding level as follows: $Y\ (\text{g MCP/MJ FME}) = 7.0 + 6.0 \times [1 - e^{(-0.35L)}]$, where FME represents the fermentable ME and L the feeding level as a multiple of maintenance. FME is discounted for ME from fat and silage fermentation products, which do not provide energy for rumen bacteria. MCP supply is calculated from the supply of FME in the diet using the following equations:

$$\text{MCP (g/day)} = \text{FME (MJ/day)} \times \text{Y(g MCP/MJ FME)} \quad \text{when eRDP} \geqslant \text{MCP}$$

$$\text{MCP (g/day)} = \text{eRDP (g/day)} \quad \text{when eRDP supply} \leqslant \text{eRDP requirement.}$$

2.5.1.2 INRA (1989)

The French PDI (corresponds to MP) system calculates two protein values for each feed or diet, a protein value when RDP is limiting for microbial growth (PDIN) and a protein value when energy is limiting for microbial growth (PDIE). The protein values are calculated as: PDIN = PDIA + PDIMN and PDIE = PDIA + PDIME, where PDIA represents the truly digestible RUP, PDIMN the digestible microbial true protein based on RDP supply and PDIME the digestible microbial true protein based on energy available in the rumen. The lower of the two values for feeds (PDIN and PDIE) is the real value of the feed when it is fed alone. The higher of the two values is the potential value that can be obtained if the feed is fed with a suitable complementary feed. When calculating the PDI value of the diet, the PDIN and PDIE values of the different feeds are summed separately to achieve PDIN and PDIE values for the diet and the actual PDI value of the diet is the lower of the two sums.

Microbial CP (g/day) is estimated from fermentable OM (FOM) as 145 × FOM (kg/day). FOM corresponds to DOM minus those digestible fractions which are of low or no value as an energy source for rumen microbes: FOM = DOM − EE − fermentation products − DUP, where DUP refers to the undegradable dietary protein remaining in the nylon bag. Microbial CP based on the supply of RDP is calculated as CP × [1 − 1.11× (1 − deg)] × 0.9, where deg is theoretical degradability *in sacco*. The PDI system assumes that RDP is captured by rumen microorganisms for microbial protein synthesis with an efficiency of 0.90.

Truly digestible RUP (PDIA) is computed using the equation: PDIA = CP× 1.11 ×(1 − deg)× 1.0 ×dsi, where dsi is the true digestibility of DUP in the small intestine.

2.5.1.3 Dairy NRC (2001)

Ruminally synthesized MCP is calculated from intake of calculated total digestible nutrients

(TDN) or from RDP, whichever is the most limiting. A mean RDP:MCP ratio of 1.18 is used to define RDP requirements of rumen bacteria (i.e. it is assumed that RDP is captured for microbial protein synthesis with a constant efficiency of 0.85). The equations for calculating MCP (g/day) are 130 × kg TDN (when RDP intake equals or exceeds 1.18 × TDN-predicted microbial CP) and 0.85 × RDP intake, when RDP intake is less than 1.18 × TDN-predicted microbial CP yield. Intake of TDN is discounted for the effects of feeding level, which depends on diet TDN concentration. Ruminally synthesized MCP is assumed to contain 80% true protein and 80% of the true protein is assumed to be digested in the small intestine.

The equations for calculating RDP and RUP values (% of CP) are: RDP $= A + B\,[k_d/(k_d + k_p)]$ and RUP $= B[k_p/(k_d + k_p)] + C$. Fraction A is the fraction of CP that is assumed to be completely degraded in the rumen, fraction B is the fraction of CP that is potentially degradable and fraction C is the fraction of CP which cannot be degraded. In addition to the need for the three CP fractions and the digestion rate (k_d) of fraction B, use of the above equations also requires an estimate of passage rate (k_p) of each feed. Three equations were developed and adopted, one for wet forages [$k_p = 3.054 + 0.164 \times$ (DMI, %BW)], one for dry forages [$k_p = 3.362 + 0.479 \times$ (DMI, %BW) $- 0.007 \times$ (% concentrate in diet DM) $- 0.017 \times$ (% NDF in forage DM)] and one for concentrates [$k_p = 2.904 + 1.375 \times$ (DMI, %BW) $- 0.020 \times$ (% concentrate in diet DM)]. The three N fractions and the k_d for fraction B were determined by *in sacco* experiments.

Variable digestibility of RUP is recognized in NRC (2001). Digestibility coefficients were derived from mobile bag and *in vitro* studies.

The model also includes endogenous protein as a source of MP (2.1 g N/kg DMI). True protein content in endogenous CP (ECP) is assumed to be 0.50 and the true protein of ECP is assumed to have a digestibility of 0.80.

In summary, total MP supply in NRC (2001) is calculated as: MP = 0.64 × MCP + digestible RUP + 0.4 × ECP.

2.5.1.4 Scandinavian systems (DK and FIN)

In the Scandinavian feed protein evaluation system (Madsen *et al.*, 1995), the protein value of the diet is expressed as AA absorbed from the small intestine (AAT) and protein balance value in the rumen (PBV). AAT corresponds to MP in other systems and PBV describes the balance between the dietary supply of RDP and microbial requirements for RDP.

Although the Scandinavian systems are similar, there are some minor differences in calculating the AAT values of the feedstuffs. In this comparison, the AAT and PBV values were estimated according to the Danish and Finnish systems. The modifications made in the Finnish system are described in detail by Tuori *et al.* (1998). The main difference is in the estimation of MCP, which in the Danish system is defined as: MCP (g/day) = 179 × dCHO (kg/day), where dCHO is intake of digestible carbohydrates. In the Finnish system RDP is included as an energy substrate for microbial growth [MCP (g/day) = 179 (g/kg) × dCHO (kg/day) + RDP (kg/day)]. The coefficient is the same (179), which results in higher AAT values and lower PBV values than in the Danish system. In practice this does not produce major problems because the higher AAT intakes in the Finnish system are taken into account in feeding recommendations, and PBV is allowed to be −20 g/kg DM before the supply of RDP is considered to be limiting.

An additional difference between the two systems is the lower k_p values used in calculating the effective degradability of protein (EDP) values in the Finnish system. The passage rates in the Finnish system are based on simple first-order k_p values derived from kinetic parameters using the two compartmental rumen model described by Allen and Mertens (1988). Therefore, compared to most of the other systems, the k_p values (0.02 for forages and 0.03–0.04 for concentrates) are lower, and consequently, the contribution of MCP to MP supply is higher. In the Finnish system a constant value (0.82) is used for digestibility of DUP, whereas variable values based on the mobile bag technique are used in the Danish system (Hvelplund *et al.*, 1992, 2003).

In both systems, AAT requirements are expressed in terms of ECM instead of MPY. In the Danish system the requirements are corrected for the feed efficiency factor (Strudsholm *et al.*, 1999), which is a function of energy intake and production potential. This empirical correction factor takes into account both reduced diet digestibility and possible negative associative effects in digestion at higher feeding levels. In the model comparisons carried out, the AAT requirements were also calculated on the basis of MPY. Requirements for MPY were estimated as AAT requirement (g/kg ECM)/protein content of ECM (g/kg). Calculated efficiencies of MP utilization were 0.81 and 0.64 for the DK and FIN systems, respectively.

2.5.1.5 German system (GER)

The German system is based on the utilizable CP flow at the duodenum. Utilizable CP (nXP) can be estimated from 12 alternative equations. In the present comparison of the systems, Equation 1a of those twelve equations was used: [nXP = 8.76 × ME (MJ/day) + 0.36 × CP (g/day)]. Equation 1a was used because all data needed were available. Urea is excluded from CP in calculating nXP. Utilizable CP was converted to MP as follows: MP = nXP × 0.73 × 0.85. The coefficients of 0.73 and 0.85 represent the proportion of AA-N in duodenal NAN and the efficiency of absorption of AA-N, respectively. The German system is very simple compared to the other systems, since it does not require estimates of ruminal feed protein degradability or digestibility of RUP. There is no predicted estimate of MCP. MP derived from ME can be assumed to be MCP and that from CP from undegraded CP.

2.5.2 Materials and methods

2.5.2.1 Experimental approach

Data from eight Finnish production experiments conducted with lactating dairy cows were used to evaluate the models. The experiments included 72 dietary treatments. All studies were conducted using changeover designs with 3 to 4 week experimental periods. The studies were selected so that a wide range of diets and production levels could be evaluated. The treatments included the most common strategies to manipulate the supply of MP (level and type of protein supplementation, heat-treatment of protein supplement, proportion of concentrate in the diet, silage digestibility, silage CP content as manipulated by N fertilization, extent of *in-silo* fermentation and replacement of grass silage with whole crop barley silage). Grass silage was fed *ad libitum* in each study. The origin of the data used in the evaluation is shown in Table 2.1.

There were considerable ranges in DMI and proportion of concentrate and CP concentration in diet DM (Table 2.2). As a result of these wide ranges in DMI and concentrate and CP concentrations in diet DM, milk production varied considerably. Standard errors of means of the production parameters were small (coefficient of variation was <5.0%). The average supplies of ME and MP (AAT) were 105.1% and 101.8%, respectively, of the Finnish requirements (Tuori *et al.*, 2002), indicating that protein probably was slightly more limiting than energy.

To estimate rumen microbial CP synthesis, intakes of ME (GER) and FOM (Vérité and Peyraud, 1989) were estimated using *in vivo* or *in vitro* cellulase digestibility measurements for forages and analysed chemical composition and digestibility coefficients (Tuori *et al.*, 2002) were used for the concentrate ingredients. Discounts for digestible fat (34 MJ/kg) and silage fermentation acids

Table 2.1. Studies and experimental factors used for the evaluation of protein models.

Reference	Factor I	Factor II	Factor III
Heikkilä *et al.* (1998)	Silage fermentation	Protein supplementation	
Jaakkola *et al.* (unpublished)	Grass vs. whole-crop barley silage	Protein supplementation	
Rinne *et al.* (1999a)	Silage digestibility	Forage:concentrate ratio	Protein supplementation
Rinne *et al.* (1999b)	Protein supplementation	Protein source	
Saarisalo *et al.* (2002)	Forage:concentrate ratio	Protein supplementation	
Sairanen *et al.* (unpublished)	Forage:concentrate ratio	Protein supplementation	
Shingfield *et al.* (2001)	Grass N fertilization	Protein source	
Shingfield *et al.* (2003)	Protein supplementation	Protein source	

Table 2.2. Feed intake, crude protein and predicted metabolizable protein concentrations of the diet, and the relevant production data from the studies used for the evaluation of protein models ($N = 72$).

	Mean	SD	Min.	Max.
DMI (kg/day)				
Forage	11.81	1.25	8.81	14.13
Concentrate	7.9	1.96	2.81	12.86
Total	19.71	2.37	12.9	22.12
Concentrate in diet (g/kg DM)	395	70	216	584
Crude protein (g/kg DM)	152	18	112	192
MP (g/kg DM)				
AFRC	86.5	10.92	65.4	111.3
DK	79.6	4.54	70	91.0
FIN	90.2	4.85	78	103.5
GER	94.5	5.57	80.9	104.7
INRA	88.5	7.33	70	107.7
NRC	86.9	7.11	70.8	106.6
Production				
Milk (kg/day)	26.6	4.88	13	34.3
ECM (kg/day)	29.1	4.82	15.6	38.1
Protein (g/day)	875	144.5	463	1133
Protein (g/kg)	33.3	1.6	30.3	37.8

(15 MJ/kg) were based in most cases on measured intakes in cows using AFRC (1992). In some cases, when the digestibility of ether extract was not measured, the coefficients were derived from tabular values (Tuori *et al.*, 2002). Discounts in FOM intake (INRA) were based on analysed concentrations of ether extract and silage fermentation products. In NRC (2001), MCP was estimated both from TDN estimated using the NRC equations or from a combination of *in vitro* cellulase digestibility measurements for forages and analysed chemical compositional data and published tabular digestibility coefficients for concentrate ingredients (Tuori *et al.*, 2002). Because the two sets of TDN values were strongly correlated ($R^2 = 0.987$) and the mean bias in TDN intake between the two estimates was small (0.42 kg/day), estimated MP supplies were similar and only those obtained with the NRC (2001) TDN equations are presented. In the DK and FIN systems, MCP synthesis was estimated from dCHO (DK) or from dCHO + RDP (FIN). Digestible CHO was estimated using the same approach as used for ME, or FOM.

Ruminal degradability of feed protein was estimated using the parameter values described in each respective system. Feeding level effects were taken into account in the AFRC (1992) and NRC (2001) systems. The values for digestibility of RUP were taken from the respective system, or calculated from ADIN (AFRC, 1992). If ADIN values were not available in AFRC (1992), they were taken from NRC (2001) feed tables. The discounts needed for bag DUP in the INRA model were based on the INRA (Vérité and Peyraud, 1989) coefficients.

To evaluate the effects of limited RDP supply on MP supply, MP supply was also estimated assuming that RDP did not limit MP supply (AFRC, 1992; NRC, 2001). In the INRA (Vérité and Peyraud, 1989) system, using PDIE as PDI assumes that RDP was not limiting MCP synthesis.

2.5.2.2 *Statistical analysis*

Relationships between MP supply and MPY in each system were estimated using the MIXED procedures of SAS (Littel *et al.*, 1996). The model was: MPY = Exp + MP + e, where Exp is a random effect of experiment, MP is linear effect of MP and e is an error term. Relationships between predicted and observed MPY were also estimated by a simple linear regression model. Root mean square errors (RMSE) and adjusted R^2 were used to compare the fit of the different models. For the mixed model R^2 and RSME values were estimated from the values adjusted for random study effect.

The relationship between calculated MP requirement and estimated MP supply was estimated by a linear regression analysis. The mean squared prediction error (MSPE) was calculated as $[(3\ (\text{supply} - \text{requirement})^2)]/n$, where $n = 72$. The MSPE was decomposed into error due to the overall bias, error due to deviation of the slope from unity and error due to variation around the regression line (Bibby and Toutenburg, 1977).

2.5.3 Results

2.5.3.1 Estimating MP supply

There were some numerical differences in the average predicted MP concentrations of the diets (Table 2.2). The average MP concentration was the lowest in the DK system. This is partly because the proportions of AA-N in microbial N (0.70 vs. 0.75–0.80) and RUP (0.65–0.85 vs. 1.00) are lower in the Scandinavian systems as compared to the other systems. The average MP concentration was the highest for the GER system, probably because the calculations are based on duodenal CP flow and because the equation that was used was derived from flow data that includes flow of endogenous N. The average MP concentration was also high for the FIN system, mainly because the coefficient for the efficiency of MCP synthesis is higher than in other systems.

There were considerable differences in the standard deviation and range of MP concentrations between the systems. Variation was highest in AFRC followed by NRC and INRA, whereas the variation was markedly lower for the Scandinavian and German systems. The highest variability in AFRC can be attributed to a number of factors. First, the k_p values are strongly related to feeding level and are considerably higher than in any other systems. This increases the contribution of RUP to the total MP supply, and also increases differences between the diets because of the wide range in DMI among the studies. Secondly, the strong association between the efficiency of MCP synthesis and DMI widens the range in dietary MP concentrations. And thirdly, the RDP requirements were not always met, which resulted in discounted MP intakes for these diets and further increased the range in MP concentrations.

The smaller ranges in predicted MP concentrations for the Scandinavian systems appeared to have resulted because of one or more reasons. First, the systems consistently predict higher contributions of MCP to MP supply. This is especially true for the FIN system. Secondly, estimates of ruminal protein degradability or efficiency of MCP synthesis are not considered to be affected by differences in DMI. And finally, neither system imposes a restriction on synthesis of MCP because of limited supplies of RDP. The lower variation in predicted MP concentrations in diet DM for the GER system is the result of the simplicity of the model and its failure to consider many of the factors that can affect flow of MP to the small intestine.

2.5.3.2 Relationship between predicted MP supply and milk protein yield

The observed linear relationships between predicted MP supply and MPY for the different models are presented in Table 2.3. All models predicted variation in MPY relatively well as indicated by R^2 values ranging from 0.83 to 0.95. Interestingly, the R^2 values were higher and the MSPE were lower in the AFRC, INRA and NRC systems when MP supply was always predicted from available energy and not from RDP in the cases where RDP was predicted to be limiting with the model. This may suggest that the RDP requirements are overestimated in these systems, or that the RDP deficiencies were covered by recycling of urea-N into the rumen. In the FIN system, prediction of MPY was improved marginally when the AAT supply was discounted for PBV values below −20 g/kg DM. In the DK system, the corresponding value was −8 g/kg DM. The GER system predicted MPY accurately, even though it uses a constant degradability for all dietary protein except urea. These results indicate that our methods to estimate feed protein degradability by the nylon bag technique are either not very accurate or precise in describing the variation in ruminal protein degradability that exists. Another possible explanation for this observation is that efficiency of MCP synthesis is reduced with decreased ruminal protein degradability.

The degree of curvilinearity between MP supply and MPY varied among the systems (data not shown), being the strongest in the AFRC and NRC systems (quadratic regression coefficients

Table 2.3. Linear relationships between predicted supply of MP and milk protein yield (Y = A + BX).

		A	SE	*P*-value	B	SE	RSME	R^2
AFRC		231	35.5		0.375	0.020	60.0	0.828
AFRC	eRDP[a]	146	26.1	<0.001	0.398	0.014	39.4	0.919
DK		−50	31.9	0.124	0.588	0.020	40.0	0.924
FIN		−48	25.4	0.062	0.517	0.014	32.3	0.950
FIN	PBV-20[b]	−43	24.4	0.081	0.515	0.014	31.2	0.953
GER		−72	27.2	0.010	0.507	0.014	33.6	0.946
INRA		69	35.0	0.054	0.460	0.020	49.0	0.885
INRA	PDIE[c]	5	29.3		0.488	0.016	39.0	0.927
NRC2001		−64	45.3	0.159	0.560	0.027	54.0	0.860
NRC2001	RDP[d]	−137	44.2	0.003	0.597	0.026	49.6	0.882

[a]eRDP, no discounts were made for limited supply of RDP.
[b]PBV-20, discounts in MP supply were made for PBV concentrations below −20 g/kg DM.
[c]PDIE, no discounts were made for limited supply of RDP.
[d]RDP, no discounts were made for limited supply of RDP.

−0.00019 and −0.00027). The quadratic effect was non-significant in the GER and FIN systems. In addition to the diminishing production responses with increased MP intake, the increased MP concentrations with increased DMI explain partly the strong quadratic effects. In the AFRC system, the strong discounts in MP supply for a deficiency of RDP would further widen the range in MP supply and increase the quadratic effect. The stronger curvilinear effect in the DK system compared to the FIN system may have resulted because of the higher relative protein values assigned to concentrates as compared to the values assigned to forages in the DK system.

The relationships between MP supply and MPY using a mixed model with a random study effect are presented in Table 2.4. The slopes represent the marginal responses in MPY to increasing supplies of MP within experiments. The smaller slopes compared to those obtained with the simple regression model probably indicate diminishing responses when additional MP is fed to the same cows, whereas the slope of a simple regression model also represents variation within animals and production levels between the studies. Most noticeable by the observed low RMSE (below 20 g/day) and the high R^2 (0.98) values is the fact that the Scandinavian systems were the

Table 2.4. Linear relationships between predicted supply of MP and milk protein yield (Y = A + BX + exp).

		A	SE	B	SE	RMSE[a]	R^2 model[a]
AFRC		395	42.9	0.280	0.020	31.5	0.907
AFRC	eRDP[b]	188	38.0	0.375	0.019	26.4	0.961
DK		48	35.4	0.526	0.020	18.6	0.978
FIN		24	31.6	0.478	0.016	16.9	0.983
FIN	PBV-20[c]	52	30.4	0.463	0.015	16.5	0.983
GER		−9	36.0	0.475	0.018	19.3	0.979
INRA		236	43.2	0.364	0.022	27.3	0.964
INRA	PDIE[d]	76	38.5	0.448	0.020	22.2	0.976
NRC 2001		218	48.5	0.391	0.024	28.0	0.918
NRC 2001	RDP[e]	144	55.8	0.432	0.030	31.2	0.908

[a]Milk protein yield adjusted for a random study effect.
[b]eRDP, no discounts were made for limited supply of RDP.
[c]PBV-20, discounts in MP supply were made for PBV concentrations below −20 g/kg DM.
[d]PDIE, no discounts were made for limited supply of RDP.
[e]RDP, no discounts were made for limited supply of RDP.

most accurate in predicting observed differences in MPY between the diets within experiments. The mixed model analysis also indicates that MPY responses were predicted more accurately in AFRC, INRA and NRC by assuming that RDP did not limit microbial CP synthesis. This again shows either a limitation in estimating degradability of dietary CP or predicting RDP requirements for microbial protein synthesis.

The marginal responses to increased MP supply within the experiments used in this evaluation ranged from 0.28 to 0.53. These values are higher than the value of 0.20 in a similar model evaluation by AFRC (1992) and Webster (1992). This indicates that MP supply in the present data set was more limiting than in the AFRC (1992) data set. Moreover, the within experimental milk yield response to increased PDI supply (INRA system) was 0.9 kg/100 g PDI. Based on an analysis of 17 INRA experiments (Vérité and Peyraud, 1989), this observation also indicates that the average MP supply was a limiting factor in this data set. These observations lend support to the fact the data set was probably good for comparing the protein systems.

The better performance of the Scandinavian systems may have resulted because the systems were developed using production data from diets similar to those in the test data set. This may indicate that developing a universal factorial system for predicting the MP value of diets and MPY responses to additional MP across a wide variety of diets may be difficult, if not impossible. Developing dynamic mechanistic models may be a better approach to improve our protein evaluation systems. In addition, the ability of the dynamic mechanistic models to deal with different diets and interactions between dietary components is markedly better than that of the current systems. For example, most of the current systems do not take into account the reduced energy supply for rumen microbes that results because of the negative effects that NSC have on ruminal cell wall digestion and site of starch and NDF digestion. Only ruminal digestion provides energy for synthesis of MCP.

2.5.3.3 Relationship between MP supply and requirements

Predicted MP supplies and requirements and their relationships to each other are presented in Table 2.5. The greatest discrepancies between predicted

Table 2.5. Calculated supply and requirement of metabolizable protein (MP), the difference between predicted MP supply and requirement, parameters of the regression equation: MP requirement (g/day) = intercept + slope × MP supply (g/day), and the mean square prediction error (RMSE) between supply and requirement.

System	Modification	MP supply	MP requirement	Bias	Intercept	Slope	R^2	RMSE
AFRC		1717	1484	232	551	0.544	0.834	296
AFRC	eRDP[a]	1831	1484	347	428	0.577	0.925	383
DK		1573	1432	141	247	0.753	0.901	165
DK	Feed eff.[b]	1573	1550	22	−69	1.030	0.903	83
DK	Feed eff.[b] + MPY req.[c]	1573	1548	25	−89	1.041	0.925	75
FIN		1784	1750	35	274	0.827	0.908	93
FIN	MPY req.[c]	1784	1749	36	237	0.847	0.955	75
GER	a1[d]	1866	1627	239	−50	0.899	0.931	251
INRA		1752	1749	4	426	0.754	0.892	106
INRA	PDIE[e]	1780	1749	31	323	0.801	0.934	89
NRC (2001)		1678	1945	−267	189	1.047	0.843	289
NRC (2001)	RDP[f]	1693	1945	−252	39	1.126	0.878	271

[a]eRDP, assumed that eRDP never limited synthesis of microbial protein.
[b]Feed eff., corrected for reduced feed efficiency with increased feeding level.
[c]MPY req., production requirements were estimated from milk protein yield instead of ECM yield.
[d]a1, equation used to estimate MP supply according to the GER system.
[e]PDIE, assumes that RDP was not limiting MP supply in any case.
[f]RDP, assumes that RDP was not limiting MP supply in any case.

MP supplies and requirements occurred for the AFRC, GER and NRC systems.

The AFRC system predicted an average supply of MP that greatly exceeded the average predicted requirement, even when the discount for deficient supply of RDP was used. The slope between supplies and requirements was only slightly above 0.50, indicating that predicted supplies increased more than the requirements. This is probably the result of the previously mentioned strong effect that DMI has on passage rates of undigested feed protein and predicted supplies of MP in the AFRC system. As expected, excluding the discount for deficient supplies of eRDP increased the bias between requirement and supply, but the relationship between predicted supplies and requirement became better as evidenced by the higher R^2 value (0.925 vs. 0.834). Although the total variance between MP supply and requirement increased when it was assumed that eRDP was not limiting rumen MCP synthesis, the reduced slope and random variance suggest that the deficient supplies of eRDP influenced total MP supply less than what was predicted by the model. The mean oversupply of MP by the AFRC system would allow for about 5 kg/day more milk.

The DK system predicted an average AAT supply of MP that exceeded the average MP requirement by 141 g/day when the requirement was not corrected for reduced 'feed efficiency' with increased DMI but when the requirement was corrected for reduced feed efficiency, the corrected MP requirement was very close to the predicted MP supply. The correction factor empirically takes into account reduced diet digestibility and possible negative associative effects on digestion at higher feeding levels. The production requirements in the DK and FIN systems are based on ECM yields. Converting the MP requirements based on MPY improved the relationship between MP supply and requirement in the DK system. Theoretically, MP requirements based on MPY can be better justified than MP requirements based on ECM yield, since increasing ECM yield by increasing milk fat content does not increase the MP requirement.

In the FIN system, MP requirements based on protein yield rather than ECM yield also improved the relationship between MP supply and requirements. Most of the variance resulted from random variation (Table 2.6), and partly from the slope being marginally below 1.00.

Table 2.6. Distribution of variance between MP supply and requirement between variance resulting from bias, slope and random variation.

		Variance				Proportion		
System	Modification	Bias	Slope	Random	Total	Bias	Slope	Random
AFRC		54,034	25,661	7,245	86,940	0.622	0.295	0.083
AFRC	eRDP[a]	120,347	21,734	3,280	145,361	0.828	0.150	0.023
DK		19,913	3,404	3,471	26,788	0.743	0.127	0.130
DK	Feed eff.[b]	500	50	6,342	6,892	0.073	0.007	0.920
DK	Feed eff.[b] + MPY req.[c]	616	93	4,890	5,599	0.110	0.017	0.873
FIN		1,209	2,214	5,127	8,550	0.141	0.259	0.600
FIN	MPY req.[c]	1,276	1,735	2,529	5,541	0.230	0.313	0.456
GER	a1[d]	56,926	784	4,608	62,318	0.913	0.013	0.074
INRA		13	5,275	6,006	11,294	0.001	0.467	0.532
INRA	PDIE[e]	981	3,220	3,671	7,872	0.125	0.409	0.466
NRC (2001)		71,451	127	11,723	83,301	0.858	0.002	0.141
NRC (2001)	RDP[f]	63,824	825	9,117	73,766	0.865	0.011	0.124

[a]eRDP, assumed that eRDP never limited synthesis of microbial protein.
[b]Feed eff., corrected for reduced feed efficiency with increased feeding level.
[c]MPY req., production requirements were estimated from milk protein yield instead of ECM yield.
[d]a1, equation used to estimate MP supply according to the GER system.
[e]PDIE, assumes that RDP was not limiting MP supply in any case.
[f]RDP, assumes that RDP was not limiting MP supply in any case.

The INRA system predicted an average MP supply and requirement that were very similar (bias = +4 g/day) and as indicated by the high R^2 values, the relationship between supplies and requirements was good. When it was assumed that a deficient supply of RDP never limited microbial CP synthesis, there was a small increase in the bias between supply and requirement (+31 g/day), but the slope and random errors were decreased. Predicted MP supply was clearly higher than requirement in the GER system, but the slope and random errors were small, suggesting that the system was able to predict differences between the diets in MP supply very accurately.

In contrast to the AFRC system, the NRC system predicted an average supply of MP that was considerably less than the average predicted requirement. However, the cows produced about 5 kg more milk than what was allowed by the NRC predicted MP supply. The bias and random errors decreased when it was assumed that RDP did not limit ruminal MCP synthesis, but the variance resulting from the slope error increased. High values for the slope in the NRC system may result from too large an increase in MP requirements for maintenance with increasing DMI. It is possible that the NRC system, which is based on markedly different diets from those used in the studies of the present data set, underestimates the MP supply of the diets used in this comparison. For example, it is very likely that the proportion of OM digested in the rumen is higher for the diets used in this comparison than for those used to derive the NRC equation for MCP synthesis. Considerably more maize starch escapes rumen fermentation than barley or oats starch, which decreases the energy supply for rumen microbes. This also emphasizes the importance of dynamic mechanistic models, by which different rates of digestion of carbohydrate fractions can better be taken into account in estimating rumen MCP synthesis.

Studying the relationships between predicted MP supplies and requirements and some dietary parameters may reveal some reasons why the current protein systems predict or fail to predict MP supply precisely and accurately. High negative intercepts and high positive slopes both for forage and concentrate DMI suggest that generally the supply of MP with increasing DMI is overestimated in the AFRC system. This is probably the result of too strong an effect of DMI on k_p, and consequently on the efficiency of rumen MCP synthesis and on ruminal protein degradability. In comparison, feeding level has no influence on the concentration of FME in the diet. The effect of dietary CP concentration was also strongly overestimated, partly because of high RDP requirements and reduced synthesis of microbial CP, and partly due to estimates of k_p that are too high and cause overestimation of UDP supply. This effect was considerably smaller when it was assumed that RDP was not limiting MCP synthesis.

In the Scandinavian systems, the negative intercepts and slopes for DMI were much smaller than with the other systems, suggesting that these systems better predict the effects of feeding level on MP supply. However, the coefficients were considerably higher for concentrate DMI compared with forage DMI, suggesting that these systems overestimate the MP values of concentrates compared to forages. This may partly result from the lower concentration of AA-N in forage UDP compared with concentrate UDP. Feed AA-N flow estimated from the AA profile of omasal digesta, rumen microbes and feed samples suggest that the supply of RUP from grass silage is higher and that from barley smaller than estimated by the AAT-PBV systems.

Similar coefficients for both concentrate and forage DMI in the INRA and NRC systems suggest that these systems correctly estimate the relative MP values of forages and concentrates. However, both systems overestimate the effect of dietary CP concentration on MP supply, especially when MP from microbial protein is discounted for deficient supply of RDP.

2.5.3.4 Effect of intake and diet composition on predicted MPY

In an attempt to further evaluate the AFRC, DK, FIN, GER, INRA and NRC systems it was of interest to determine how the six systems differ regarding the effect of DMI, forage to concentrate ratios and dietary CP concentration on predicted MPY responses. To conduct this evaluation of the models, the same test diets were evaluated with all models. The diets consisted of grass silage and concentrate. The concentrate was comprised of an energy (40% barley, 40% oats and 20% molasses sugarbeet pulp) and protein (60% rapeseed meal and 40% soybean meal) supplement. The digestibility coefficients obtained from the Finnish feed tables (Tuori *et al.*, 2002) were used in all systems to estimate microbial protein synthesis. TDN was estimated from DOM by adding 1.25 × fat (= ether extract − 1; for silage fatty acid content 25 g/kg DM was used). For the AFRC and INRA systems, in which discounts are made for silage fermentation products in estimating MCP synthesis, lactic and acetic acid concentrations of 50 and 20 g/kg DM were used.

Dietary MP concentrations were estimated for DMI of 12, 15, 18, 21 and 24 kg/day, and for diets containing 20%, 30%, 40%, 50%, 60% and 70% concentrate (DM basis). These model evaluations were conducted by keeping dietary CP constant (160 g/kg DM) in both silage and concentrate. The increases in dietary CP concentration (141, 150, 159, 169, 178, 187 and 196 g/kg DM) were achieved by increasing the proportion of protein supplement in the concentrate from 0% to 30% using 5% intervals. These calculations were made for two forage:concentrate ratios (60:40 and 40:60) and for DMI of 12, 15, 18, 21 and 24 kg/day, respectively.

MP-allowable MPY was calculated as (MP supply − MP used for maintenance)/efficiency of MP utilization. An average BW of 550 kg was used in all calculations. Estimated marginal responses to increased DMI, proportion of concentrate in diet DM and dietary CP concentration were compared to regression coefficients derived from a large data set ($n = 306$) of similar diets (see Nousiainen *et al.*, 2004).

There were considerable differences among the systems in predicted MPY in response to changes in DMI, both at each level of DMI and also in marginal responses to increased DMI (Fig. 2.3). The AFRC system predicted the highest and NRC predicted the lowest MPY at each level of DMI. The observed differences between DK, FIN and INRA were relatively small. At least two reasons could be suggested to explain why NRC predicted the lowest MP-allowable MPY. First, the coefficient in the equation for predicting MCP synthesis may be too low for the diets used in this evaluation of the models. The data used to derive the NRC equation were derived from a high proportion of studies that involved maize-based diets. In contrast to barley and oats, more of the starch in maize is digested post-ruminally, and therefore, a greater proportion of the starch does not provide energy for MCP synthesis. Second, it is possible that the TDN discount factor has too strong an effect with diets based on grass silage, barley and oats. With maize-based diets, both digestibility of starch or cell solubles and NDF decrease with increasing DMI (Tyrrell and Moe, 1974), whereas with barley and oats-based concentrates, ruminal starch digestibility was not influenced by increasing DMI (Volden, 1999). And third, the maintenance requirement increased in the NRC system with increased DMI decreasing the increment in

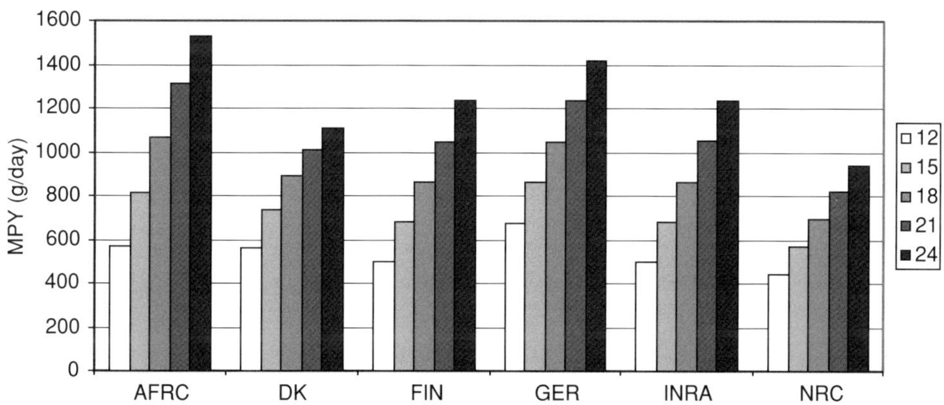

Fig. 2.3. Effect of DM intake (DMI) on MP-allowable milk protein yield (MPY) estimated according to different systems when DMI = 12, 15, 18, 21 and 24 kg/day.

MP available for milk production with increased feed intake.

Predicted MPY at 21 kg DMI were 1193, 975, 1018, 1193, 1006 and 754 g/day for the AFRC, DK, FIN, GER, INRA and NRC systems, respectively (Fig. 2.3). The corresponding MPY estimated with a mixed model regression (random study effect) from the large data set of Nousiainen *et al.* (2004) for a 550 kg cow consuming 21 kg DM/day of a diet containing 60% forage (DM basis) and 160 g CP/kg DM (i.e. the same used in this model evaluation) was 989 g/day. The DK, FIN and INRA systems predicted MPY with a relatively small bias, whereas AFRC and GER strongly overestimated and NRC strongly underestimated MP-allowable MPY. The overestimation of MPY by AFRC would have been even greater but at the two highest levels of DMI, RDP limited MCP synthesis.

Marginal MPY responses to increased DMI were 70, 45, 60, 60, 59 and 38 g/kg for AFRC, DK, FIN, GER, INRA and NRC, respectively. The corresponding regression coefficient derived from the data set was 46 (simple regression) or 38 g/kg DM (mixed model; within study response), when the effects of dietary CP concentration, BW and proportion of concentrate were excluded. It is possible that in a data set of actual production trials, marginal responses would be smaller compared to those based on calculated MP-allowable responses. This may be due to the diminishing returns at higher feed intake levels and nutrient partitioning towards body tissues. In spite of this, AFRC most likely overestimates MP-allowable milk protein potential. Dietary MP concentrations increased considerably with increased DMI even though RDP became limiting at 18 kg/day DM (from 88.8 to 96.3 g/kg DM). In AFRC, MP concentration increases with increasing DMI because both efficiency of MCP synthesis and UDP supply are positively related to feeding level, whereas no discounts are made for decreasing digestibility or potential negative associative effects, such as reduced cell wall digestibility with increased concentrate supplementation.

In contrast, dietary MP concentration estimated by the NRC system decreased slightly with increasing DMI, at least for the diet used here. This is because the TDN discount factor had a stronger negative effect on the supply of TDN for MCP synthesis than the positive effect of increased passage on RUP supply. These relative changes may depend on diet composition. The relationship between DMI and maintenance requirement is another point in which NRC is different from the other systems, except for the German system. The MP requirement for maintenance increases by approximately 30 g for each kg increase in DMI, thereby requiring that dietary MP concentrations increase with increasing DMI to meet MP requirements.

In the DK system, the relationship between DMI and MP-allowable MPY is curvilinear due to the effects of the feed efficiency factor. Predicted MPY responses decreased from 60 g/kg increase in DM at DMI of 12–15 kg/day to 31 g/kg increase in DMI at DMI of 21–24 kg/day. At the highest DMI, predicted milk protein responses may be too small. The feed efficiency factor is based on an empirical relationship between estimated and observed output, and probably takes into account both feeding level effects and negative associative effects. In the present analysis, only feeding level effects are considered.

The effects of increasing concentrate in diet DM on MP-allowable MPY are shown in Fig. 2.4. The data are based on DMI of 18 and 21 kg/day. Predicted responses to increased concentrate feeding were similar for DK, FIN, INRA and NRC systems (32.1, 30.8, 33.1 and 31.6 g of milk protein per 10% increase in the proportion of concentrate, respectively). The value was considerably lower (16.8 and 15.3) for the AFRC and GER systems. However, in the AFRC system this was mainly because RDP limited MCP synthesis almost immediately when the proportion of concentrate in the diet was increased. When RDP was not limiting at the lower DMI, the corresponding MPY responses varied between 27 and 45 g per 10% increase in concentrate. Lower predicted response to increased concentrate feeding in the GER compared to the other systems was mainly because the system uses constant protein degradability for forages and concentrates.

Except for the GER system, in each system, the predicted responses in MPY were considerably higher than the values of 20.4 and 12.8 g per 10% increase in concentrate proportion when estimated using simple and mixed model regression (provided that RDP was not limiting like it was in AFRC in this model evaluation). The data ($n = 80$) were derived from studies designed to investigate the effects of concentrate level using the data of Nousiainen *et al.* (2004). The effects of DMI and CP

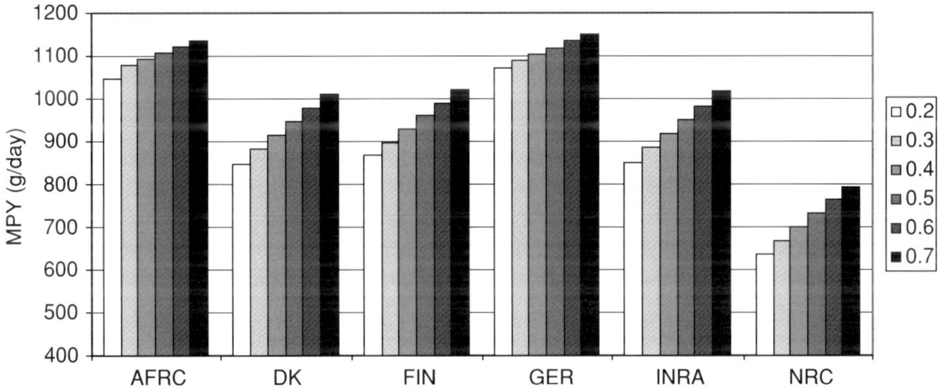

Fig. 2.4. Effect of proportion of concentrate on MP allowable milk protein yield (MPY) estimated according to different systems when proportion of concentrate in diet DM = 0.2, 0.3, 0.4, 0.5, 0.6 and 0.7.

concentration were excluded from the variation to make the conditions comparable to the present analysis. The discrepancy may be related partly to the negative associative effects in digestion, with the models overestimating the supply of energy for rumen microbes with an increasing proportion of concentrate. However, the NRC system, which takes into account a greater depression in digestion of highly digestible diets with increasing DMI, predicted similar responses to the other systems. It is also possible that rumen degradability of some concentrate ingredients is higher than the default values in different systems. When estimated from the AA profile of omasal digesta, ruminal protein degradability of barley was markedly higher (0.90) than the values of 0.70–0.80 in the feed tables (Huhtanen *et al.*, unpublished). One possible factor here could also be lower pH and reduced efficiency of MCP synthesis with higher concentrate diets (Strobel and Russell, 1986).

Increasing dietary CP concentration by replacing an energy supplement with rapeseed meal and soybean meal increased MP-allowable MPY in all systems, but predicted responses varied markedly (Fig. 2.5). Predicted responses in MPY to additional dietary CP were clearly highest using the AFRC system. At a lower level of protein supplementation, MP-allowable MPY was increased by 6.2 g/day per 1 g/kg increase in dietary CP. At the high level of protein supplementation, the corresponding value was 3.9. At the lower levels of protein supplementation, the MP supply in AFRC was increased because of an increased synthesis of microbial CP, the result of an elimination of an RDP deficiency, as well as an increased supply of UDP. In the other systems, the increases in MP-allowable MPY were more similar and consistent over the wide range of dietary CP concentrations that were used (3.0, 2.4, 3.1, 3.8 and 4.2 g/day per 1 g/kg increase in dietary CP for DK, FIN, GER, INRA and NRC systems, respectively).

The corresponding response in a data set from studies investigating responses to increased protein concentration in concentrate supplement was 2.6 per 1 g/kg CP when estimated with a mixed model (random study effect) from protein supplementation studies in the data set of Nousiainen *et al.* (2004). However, in the data, protein supplementation was associated with a significant increase in silage DM intake, and part of the MPY responses could be attributed to that. Excluding the effect of silage DM intake in the analysis decreased MPY response to 2.0 g per 1 g/kg CP. Taking into account decreasing marginal responses at the higher levels of protein supplementation, MPY responses predicted by the other systems, except AFRC, seem to be realistic, but perhaps too high for the NRC system. The higher predicted response in NRC compared to FIN is

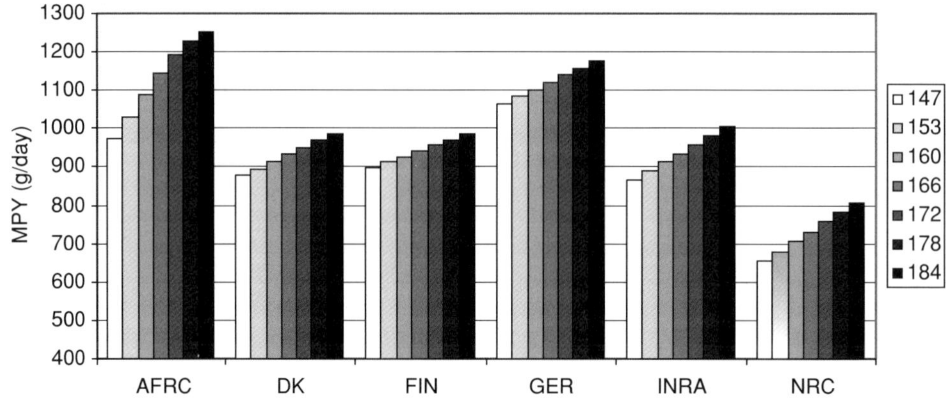

Fig. 2.5. Effect of increasing dietary CP concentration obtained by replacing an energy supplement with a mixture of rapeseed meal and soybean meal, on MP allowable milk protein yields (MPY) when CP concentration = 147, 153, 160, 166, 172, 178 and 184 g/kg DM.

most likely related to the higher passage rates in NRC leading to a greater increase in the contribution of RUP to MP supply with increasing proportions of protein supplement in the diet. In the AFRC system, the passage rates used to estimate ruminal protein degradability are unrealistically high. The passage rate at 18 kg DMI is about 0.093/h (11 h retention time), whereas Huhtanen and Hristov (2001) estimated a 39-h compartmental retention time using internally labelled lucerne ADF-^{15}N as a kinetic marker.

2.5.3.5 Comparison of three systems using US data

Of interest to us was to compare the FIN, GER and NRC systems using data from US production experiments where the diets were considerably different from the grass silage-based diets used in the aforementioned evaluation of the six different models. The NRC system was used because the data were already available. The FIN system was used in the analysis of the US data set because the data in the other comparison were from FIN studies and the INRA and DK systems predicted similar MPY responses in FIN diets. The GER system was used in the comparison because of its simplicity in model structure and feed inputs, and it performed well with FIN diets.

For this exercise, we selected 13 experiments published in the *Journal of Dairy Science* involving 59 dietary treatments (Barney *et al.*, 1981; Alhadhrami and Huber, 1992; Broderick, 1992; Grings *et al.*, 1992; Holter *et al.*, 1992; Cadorniga and Satter, 1993; Schingoethe *et al.*, 1996, 1999; Armentano *et al.*, 1997; Kalscheur *et al.*, 1999; Mowrey *et al.*, 1999; Wu and Satter, 2000). All studies were conducted with Holstein cows, ranging from early to late lactation.

The studies were selected to get diets that differed in number and types of forages, number and types of energy feeds, number and types of protein supplements, percent forage in diet DM and ranges in DMI and milk yield. Fourteen diets contained lucerne silage as the sole forage, nine contained maize silage as the sole forage and six contained lucerne hay as the sole forage. Four diets contained a mixture of lucerne hay and lucerne silage, and 26 diets contained a mixture of maize silage with lucerne silage, lucerne hay or grass silage. With respect to energy feeds, 46 diets contained maize, 19 contained barley, 18 contained maize distillers grains, ten contained whole cottonseed, eight contained maize gluten feed, five contained wheat middlings, five contained soybean hulls, four contained beet pulp, three contained oats and two contained brewers grains. With respect to protein supplements, 39 diets contained soybean meal, 24 contained maize gluten meal, six contained roasted or extruded soybeans, six contained cottonseed meal, six contained fishmeal, six contained urea, four contained protected soybean meal and three contained sunflower. The diets varied in number of forages (1–3), energy feeds (1–5) and protein supplements (1–3). Lastly, studies were selected to achieve a range in DMI (12.8–26.5 kg/day; mean = 21.2), milk yield (13.5–42.0 kg/day; mean = 29.5) and percent forage in diet DM (40–98). Ranges and (means) for diet RDP balance, RUP balance and MP balance were −487 to +1559 g/day (+439), −536 to +1039 g/day (+149) and −396 to +885 g/day (+129) using NRC.

The NRC model was used as designed for predicting MP flows and MPY. The AAT supply in FIN was calculated using the NRC coefficients to estimate the supply of digestible carbohydrates and RDP for MCP synthesis. Other coefficients were from the Finnish feed tables (e.g. efficiency of MCP synthesis, passage rates, digestibility of MCP and RUP). AAT supply was also calculated by discounting the MP values for RDP deficiency when PBV was below −20 g/kg DM. MP supply was also estimated from ME and CP intakes according to the German system. ME was estimated according to NRC (2001) at the maintenance feeding level.

The resulting MP supplies generally predicted MPY responses poorly compared to when the Finnish data set was used (compare data in Table 2.7 to data in Table 2.3). This is probably because the Finnish data set consisted of more uniform diets (all grass silage based) and studies were conducted in the same institute. The variation in animals, environmental conditions and experimental techniques was greater in the US data set than in the Finnish experiments. When estimated using simple regression, differences between the systems in prediction accuracy were relatively small. Including a quadratic term in the model improved predicted MPY for all systems, indicating decreasing marginal responses at the higher levels of MP (Table 2.7). The lack of significant differences between the systems in predicting MPY supports the conclusion made with the Finnish data set, which is that the prediction is biased because of the nature of the diets used.

Including experiment as a random factor in the regression model improved predictions considerably (Table 2.8), indicating that a large proportion of the variation originated from variation among experiments. The differences between the systems still remained small, as observed for the simple regression model. Predicted MP supply accounted for 80–93% of the variation in MPY adjusted for the random study effect. The slope between MP supply and MPY was markedly higher for the FIN system as compared to the NRC system. It is interesting to note that the GER system, which assumes a constant rumen degradability of feed proteins and a constant intestinal digestibility of RUP, performed at least as well as the NRC and FIN systems. As noted previously, both rumen

Table 2.7. Linear relationships between predicted supply of MP and milk protein yield ($Y = A + BX$, or $Y = A + BX + CX^2$) for the US diets.

	A	B	C	*P*-value	RMSE[a]	R^2 model[a]
Linear						
NRC	153	0.324			111.8	0.648
FIN[b]	5	0.423			105.9	0.678
FIN-PBV[c]	0	0.427			101.7	0.704
GER	115	0.354			104.7	0.685
Quadratic						
NRC	−793	1.242	−0.00021	0.000	98.6	0.721
FIN[b]	−743	1.232	−0.00021	0.051	102.3	0.700
FIN-PBV[c]	−894	1.396	−0.00025	0.015	96.4	0.734
GER	−905	1.394	−0.00025	0.003	97.5	0.727

[a]Milk protein yield adjusted for a random study effect.
[b]Digestible CHO from NRC (2001), RDP from NRC (2001) degradation parameters and FIN k_p.
[c]AAT intake discounted for PBV below −20 g/kg DM.

Table 2.8. Linear relationships between predicted supply of MP and milk protein yield (Y = exp +A + BX or Y = exp + A + BX + CX^2) for the US diets.

	A	B	C	*P*-value	RMSE[a]	R^2 model[a]
Linear						
NRC	425	0.194			44.7	0.801
FIN[b]	199	0.323			41.3	0.891
FIN-PBV[c]	202	0.323			42.2	0.888
GER	48	0.503			39.5	0.928
Quadratic						
NRC	−266	0.817	−0.00013	0.004	46.1	0.861
FIN[b]	−385	0.937	−0.00016	0.070	41.6	0.897
FIN-PBV[c]	−416	0.971	−0.00016	0.057	43.1	0.900
GER	−629	1.113	−0.00019	0.035	40.6	0.936

[a]Milk protein yield adjusted for a random study effect.
[b]Digestible CHO from NRC (2001), RDP from NRC (2001) degradation parameters and FIN k_p.
[c]AAT intake discounted for PBV below −20 g/kg DM.

protein degradability and RUP digestibility vary among feeds in NRC and rumen protein degradability varies in the FIN system. The excellent performance of the GER system agrees with Tuori *et al.* (1998) who observed that using a constant rumen degradability (0.80) for all feeds predicted differences in the protein value of the diet better than using EPD values determined by ruminal *in situ* incubation. It is unclear why ignoring differences in ruminal degradability and RUP digestibility of feeds had no effect on the predictions. The *in situ* technique has problems, but research indicates that it does rank feeds fairly well with respect to rumen degradability and intestinal digestibility. It is probably a combined effect of errors in estimating RUP and RUP digestibility and predicting MCP. Errors may counterbalance each other (i.e. it is possible that increased RUP supply decreases the efficiency of MCP synthesis). Also, and probably more important, current methods probably overestimate differences in degradability; e.g. if the measured values for feeds A and B are 60 and 80 and the true values are 65 and 75, then using a constant value 70 can result in a better prediction of MPY than using values of 60 and 80 even though the feeds were ranked correctly.

The relationships between estimated MP supply and calculated requirements were relatively similar for all systems when using the US data. However, MP supply exceeded MP requirements in 44 of the 59 cases when using NRC and MP supply exceeded requirements in all cases when using the other systems. The bias between supply and requirements was considerably smaller for the NRC systems. The slopes when requirements were regressed on supplies were below 1.00 for all systems, indicating that the supply increased more than the requirement at higher DMI. It is also interesting to note that in the US data set, there was an average oversupply of MP of 129 g/day according to NRC but an average undersupply of MP of 267 g/day in the Finnish data. When using the FIN system, the MP supply and requirement values were relatively similar when using the Finnish data set (bias 35 g/day) but when using the US data, there was an oversupply of 285 g/day. A large difference between the data sets in MP balance estimated either by NRC or FIN systems questions the validity of using one factorial MP system in all circumstances, at least as currently designed.

Analysing the US and Finnish data together showed poor relationships between MP supply and MPY with R^2 values of 0.42 and 0.64, respectively, for the NRC and FIN systems. Including a quadratic effect in the simple regression model improved the prediction more for the NRC system ($R^2 = 0.55$) than for the FIN system ($R^2 = 0.72$); however, the RMSE (87–125 g/day) are unacceptable for practical prediction purposes. Using a mixed regression model to allow investigation of the relationship between MP supply and MPY within experiment showed much more acceptable errors (40 and 30 g/day for NRC and FIN, respectively). Again, introducing a quadratic term in the model improved prediction accuracy for NRC (RMSE = 35) but not for FIN. Slightly better prediction of the FIN system resulted mainly from the Finnish data, whereas there were no marked differences between the two systems when the comparison was based on the US data.

Distribution of variance observed in the differences between MP supplies and requirements was smaller for the NRC system than for the different variations of the FIN system for the combined data set. This was entirely the result of a smaller bias variation. Random variation was smaller for all versions of the FIN system as compared to the NRC system, indicating that the differences among diets in their MP content were predicted more accurately with the FIN system. This is consistent with the slightly higher R^2 values and the smaller RMSE in the mixed model regression analysis.

2.5.4 Limitations of existing models for predicting N requirements of cattle

It is well recognized that both rumen degradability of feed protein and intestinal digestibility of UDP differ among feeds. However, including these sources of variation in feed protein evaluation models does not seem to have improved the accuracy of MP predictions based on MPY responses. As stated previously, MPY was predicted more accurately with the German system, where constant values for protein degradability and intestinal digestibility of UDP are assumed (Equation 1a), than with the more complicated systems. In agreement with this comparison, Tuori *et al.* (1998) observed less variance between MP supplies and requirements when a constant protein degradability value was used than when variable degradability values based on *in situ* incubation were used. The analysis was based on 157 treatment means. Compared to the other systems, the FIN system, in which the contribution of UDP to total MP supply is smaller and less variable and a constant digestibility for UDP (0.82) is used, was more accurate in predicting MPY responses than systems using variable digestibility coefficients for UDP with the Finnish diets. These findings suggest that our current methods for estimating ruminal protein degradability and intestinal digestibility of UDP are not accurate enough to determine differences in the two important parameters of our current protein evaluation systems.

2.5.4.1 Estimating ruminal degradability of feed proteins

The first point of concern is the use of the nylon bag technique to estimate ruminal degradability of feed proteins. The limitations of the method have been discussed in detail (Michalet-Doreau and Ould-Bah, 1992; Nozière and Michalet-Doreau, 2000). Bacterial contamination of undegraded feed residues leads to considerable and variable underestimations of CP degradability. A second problem is the rapidly degradable fraction (*a*-fraction), which includes not only rapidly degraded sources of NPN but also soluble protein that is not instantaneously degraded and small undegraded particles that are washed out of the bags without degradation. Recent studies (Choi *et al.*, 2002; Volden *et al.*, 2002) have clearly demonstrated that variable portions [33–79 g N/kg N (Choi *et al.*, 2002), 74–122 g N/kg soluble N (Volden *et al.*, 2002)] of the feed N can escape from the rumen in the liquid phase as non-ammonia, non-microbial N. Peptide-N is quantitatively the most important amino N fraction flowing out of the rumen in the liquid phase (Choi, 2002). This problem is partly taken into account in the Dutch protein evaluation system by assuming that 5% of soluble N in ensiled feeds is washed out from the rumen in the liquid phase (Tamminga *et al.*, 1994). Hvelplund and Weisbjerg (2000) presented a correction equation for particle loss from nylon bags, but their approach may also present some problems because it assumes the same degradation characteristics for the escaped particles as for the original feed sample. Microbial colonization within the bag also has been demonstrated to be different from that of the surrounding rumen digesta, particularly for the cellulolytic population. Meyer and Mackie (1986) reported lower numbers of cellulolytic bacteria within the bags than in the surrounding digesta. Later studies by Huhtanen and Khalili (1992) and Nozière and Michalet-Doreau (1996) demonstrated markedly lower particle-associated fibrolytic enzyme activities within the bags than in the surrounding digesta. Enzyme activities in a bag appear to be a function of both pore size and open surface area (Huhtanen *et al.*, 1998).

Despite many serious attempts to standardize the *in sacco* method for estimating ruminal protein degradability, ring tests have shown considerable variation among laboratories. For example, ruminal protein degradability of soybean meal varied from 40% to 80% in a European ring test (Madsen and Hvelplund, 1994). The results of this and other ring tests suggest that low repeatability and lack of reproducibility are serious limitations of the method.

Probably the most serious problem in our current systems for estimating ruminal protein degradability is the kinetic models that are used to estimate degradability from degradation kinetic parameters and passage rates. The model suggested by Ørskov and McDonald (1979), which is now used in most of the modern protein evaluation systems [$\text{RDP} = A + B \times k_d(k_d + k_p)$], assumes that the rumen is a single compartment system and that the probability of particles escaping the system is independent of factors such as particle size, functional specific gravity, age, etc. However, studies involving duodenal digesta sampling have clearly demonstrated selective retention of externally (Pond *et al.*, 1988; Ellis *et al.*, 1994) and internally (Huhtanen and Hristov, 2001) labelled forage particles in the rumen. Because of the smaller particle size of concentrates as compared to forages, it could be assumed that concentrate particles are not selectively retained in the rumen and follow the same passage kinetics as liquid phase markers. However, the findings of Huhtanen *et al.* (1993) clearly showed using duodenal sampling that concentrate particles also were selectively retained in the reticulo-rumen. This was indicated by the ascending phase of the excretion curve of Yb-labelled concentrate particles. Estimating the k_p values only from the descending phase of the marker excretion curve will seriously underestimate retention time in the rumen, and consequently, overestimate the supply of RUP. Using the reciprocal of total mean rumen retention time will partly solve the problem. A more correct approach would be to use either a two-compartment rumen model that includes selective retention or to estimate simple first-order k_p as suggested by Allen and Mertens (1988): $k_p = (k_r \times k_e)/(k_d + k_r + k_e)$, where k_r, k_e and k_d are the rate of release from the non-escapable compartment to the escapable compartment, the rate of escape from the escapable compartment and the rate of digestion. The simple first-order passage is not only a function of passage kinetic parameters but also of each fraction's digestion rate.

Mean rumen retention times of forage particles in the reticulo-rumen compartments have been shown to vary between 35 and 45 h when determined by external markers (Beauchemin and Buchanan-Smith, 1989), internal markers (Huhtanen and Hristov, 2001) or by rumen evacuation using indigestible NDF as a marker (Stensig and Robinson, 1997; Rinne *et al.*, 2002). These values are markedly longer than the reciprocal of passage rates adopted in most of the protein evaluation systems. For example, INRA uses 0.06/h for all feeds, whereas AFRC (1992) and NRC (2001) use model calculated k_p values related to feeding level. At DMI of 20 kg/day, the AFRC and NRC predict forage k_p of 0.096 and 0.056/h, respectively. The AFRC passage rate seems physiologically unrealistic.

Despite the several problems and weaknesses of the nylon bag method in estimating ruminal protein degradability, the enthusiasm for using the method has not been dampened. This is probably partly because of the simplicity of the method. Using experimental resources for further testing of the methodological aspects of the technique is also of concern because despite these efforts, the RDP values determined by the technique do not seem to improve the accuracy of the MPY predictions compared to using a constant degradability value for feeds.

A more detailed analysis of the NRC (2001) system suggests that most of the inaccuracies in predicting MP supply in terms of MPY responses are associated with predicting the contribution of RUP to MP. The amount of MP derived from MCP alone predicted MPY responses at least as accurately as from total MP supply, when the data were analysed with a simple (RMSE 47.8 vs. 49.6 g/day) or with a mixed regression model (30.5 vs. 31.2 g/day). When both microbial MP and feed MP were used as independent variables in the model, accuracy of MPY prediction was improved considerably as indicated by RSME values of 34.7 and 21.8 g/day for the simple and mixed models, respectively. The most interesting observation was that the coefficient for microbial MP was markedly higher than for feed MP (0.90 vs. 0.31 with simple model). The difference in the slopes suggests that the relative contribution and range in the supply of feed MP was probably overestimated by the NRC (2001) system. This could be a result, at least in part, of the higher passage rates used in NRC (2001). Also, the differences between MP supplies and requirements were strongly related to dietary concentration of digestible RUP (slope = 13.1 g MP per 1 g/kg DM increase in digestible RUP). However, production responses to feed MP may be expected to be smaller than those from microbial MP, because increases in microbial MP flows are always associated with increased ME intake.

The importance of considering differences in model-default values for UDP digestibility was examined in the AFRC and NRC systems by using total UDP or digestible UDP as independent variables together with microbial MP for predicting MPY. A simple bivariate, mixed regression model was used. There were no differences in prediction accuracy for either the NRC system (RMSE = 34.7 vs. 35.4) or the AFRC system (RMSE = 58.8 vs. 58.5) when digestible UDP vs. total UDP was used. These findings suggest that very little can be achieved in feed protein evaluation models by introducing variable digestibility coefficients based on our current analytical methods, particularly if the coefficients that are used are model-default values.

2.5.4.2 Estimating MP requirements

As reviewed earlier in the chapter, the models differ considerably in predicting MP requirements. This is particularly true for maintenance. The AFRC, DK, FIN and INRA systems predict the MP requirement solely from BW. In these systems, the daily MP requirement for maintenance of a 550 kg cow will be satisfied by 2.6, 4.1, 3.9 and 3.9 kg of DM from a 60:40 forage:concentrate diet with the concentrate containing 20% of a protein supplement. In contrast to these systems, the MP requirement for maintenance in the NRC system is affected by DMI. This is illustrated by the fact that the MP requirement for maintenance is increased from 5.1 to 9.3 kg/day when intake of the above-described diet is increased from 12 to 24 kg/day.

A more detailed analysis of factors influencing the difference between predicted MP supplies and MP requirements was performed for the NRC and FIN systems. For this exercise, the differences between supply and requirement (residuals) were regressed on actual MPY (Fig. 2.6) and DMI (Fig. 2.7). It might appear that if the low DMI/low MPY data points from Figs 2.6 and 2.7 were removed that the slope bias would not exist. However, while excluding the low DMI diets reduced the slope bias from 42 to 27 g/kg DMI, the slope bias was still apparent. The systems differed markedly in their prediction of MP supplies and requirements with increased feed intake and MPY. As noted in Figs 2.6 and 2.7, the residuals were positively related to DMI and MPY for the FIN

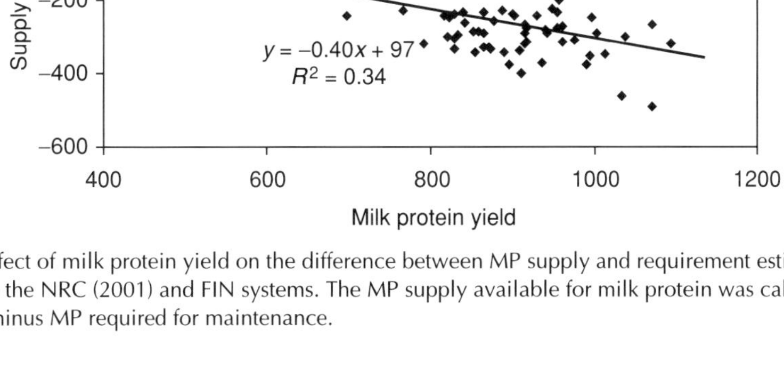

Fig. 2.6. Effect of milk protein yield on the difference between MP supply and requirement estimated according to the NRC (2001) and FIN systems. The MP supply available for milk protein was calculated as MP supply minus MP required for maintenance.

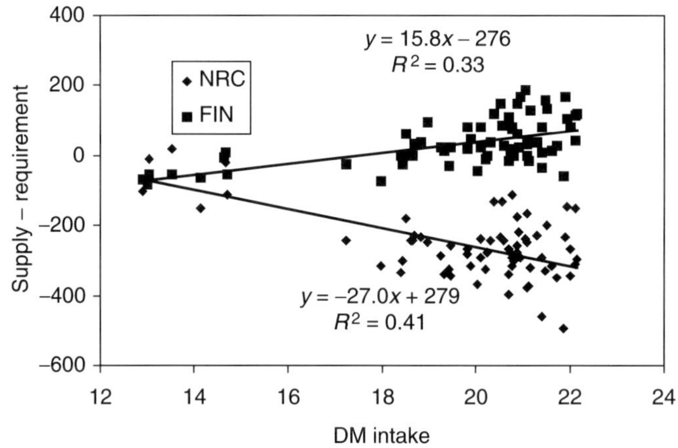

Fig. 2.7. Effect of DM intake on the difference between MP supply and requirement estimated according to the NRC (2001) and FIN systems. The MP supply available for milk protein was calculated as MP supply minus MP required for maintenance.

system and negatively related for the NRC system. Some possible reasons for these slope biases are: (i) the maintenance requirement of cattle is related to DMI; (ii) the efficiency of milk protein synthesis is related to protein yield; or (iii) MP supply is not correctly estimated with increasing DMI. The effects of DMI and MPY on the difference between predicted MP supplies and requirements were similar in the INRA system to those observed for the FIN system; i.e. the differences between supplies and requirements increased with increasing DMI and MPY.

When considering the effects of DMI on MP supply, it is not likely that the different slopes observed in Fig. 2.7 for the NRC and FIN systems can result from the relatively small differences between the two systems in the effect that DMI has on dietary MP concentration. In the FIN system, DMI has no effect on calculated MP values of diets whereas in the NRC system, increasing DMI can either slightly decrease or slightly increase dietary MP concentrations, depending on the relative effects that increasing DMI has on TDN discounts and RUP supplies.

As noted previously, the MP requirement for maintenance in NRC is strongly related to DMI (about 30 g MP/kg increase in DMI), whereas in the FIN system the maintenance requirement is only a function of the BW of the cow. However, the data presented in Figs 2.6 and 2.7 indicate that neither system predicts the MP requirement of cows for MPY correctly, particularly if it is assumed that the efficiency of use of MP for milk protein synthesis is constant across different levels of production or feed intake. As noted in Figs 2.5 and 2.6, the MP supply available for milk protein synthesis was calculated as total MP supply minus the MP required for maintenance. If the MP requirement for maintenance is calculated correctly, and the efficiency of use of MP for milk protein synthesis is constant across different levels of production and feed intake, then regression of the residuals (supply−requirements) on MPY should yield a slope of zero. However, this did not happen for either of the systems. The NRC system indicated an increased efficiency of use of MP with increasing MPY (Fig. 2.6) whereas the FIN indicated a decrease in efficiency of use of MP. The effects were highly significant ($P < 0.001$) with both systems. The same relationships were observed for increasing DMI (Fig. 2.7). In the Dutch protein evaluation system (Tamminga *et al.*, 1994), the efficiency of MP for MPY is dependent on the production level. Calculated efficiencies of MP utilization are 0.661, 0.636 and 0.613 when daily MPY is 600, 900 and 1200 g, respectively. In the Dutch system, endogenous losses of digestion are taken into account in calculating the protein value of each feedstuff. These losses are calculated as 0.075 × indigestible DM (g/kg DM). Assuming that the mean digestibility of DM in the diets used in the present data set is about 70%, ECP losses are calculated to be about 25 g/kg DMI. These losses are slightly less than the 30 g/kg DMI estimated by the NRC system. In the GER system, a value of 2.19 g N/kg DMI (13.7 g CP) is used which corresponds to 18.3 g CP/kg DMI with the efficiency of 0.75. The AFRC, INRA, DK and FIN systems do not express faecal endogenous (metabolic) losses separately and maintenance requirements are related to metabolic live weight. In these systems, maintenance requirement related to live weight is higher than endogenous urinary N requirement in NRC, GER or Dutch systems.

The observed different effects of DMI (and also MPY) on the difference between MP supply and requirement suggest that the contribution of endogenous and metabolic faecal CP to the maintenance requirement is probably too high in NRC, but ignoring it completely such as in the INRA and FIN systems is also not a correct approach. Using Excel Solver to estimate the coefficients for the different components of the maintenance requirement reduced the variance between supply and requirement, mainly because of reduced slopes and less bias. The following coefficients were estimated: 1.00 g/kg $BW^{0.75}$ for endogenous urinary N, 18.0 g/kg DMI for metabolic faecal N and 0.67 for the efficiency of MP utilization in milk protein synthesis. Testing these parameters using a large data set ($n = 306$) resulted in a higher R^2 value and the slope was considerably closer to 1.00 than our current requirements. The mean endogenous urinary N requirement was 118 g MP/day, which is very close to the corresponding requirements in the Dutch (112), German (98) and NRC (111) systems. The MP requirement of 18.0 g/kg DMI for metabolic faecal N is similar to the GER system, but lower than the values adopted by NRC (2001) or in the Dutch system (Tamminga *et al.*, 1994). The efficiency of MP utilization of 0.67 for milk protein synthesis corresponds well with the values adopted in the AFRC (0.68), INRA (0.64), NRC (0.67) and Dutch (0.61–0.66 depending on MPY) systems.

2.5.4.3 Estimating RDP requirements

There are considerable differences in the calculated RDP requirements between the systems. In the present data, RDP requirements were met on average at dietary CP concentrations of 159, 134, 149, 139 and 133 g/kg DM for AFRC, DK, FIN, INRA and NRC systems, respectively. The practical recommendation of −20 g/kg DM in the FIN systems for PBV corresponds to 123 g CP/kg DM. Microbial protein flow was not increased in cows fed grass silage and barley-based diets (about 130 g CP/kg DM) in response to supplemental protein (Ahvenjärvi *et al.*, 1999; Korhonen *et al.*, 2002) when flows were measured using omasal sampling techniques and triple-marker systems. The following equation between rumen dietary CP concentration and rumen CP balance [= 6.25 × omasal NAN flow − CP intake (g/kg DMI)] was estimated with a mixed model regression from data derived using the omasal sampling technique: CP balance (g/kg DMI) = 104.3 (± 22.6) − 0.80 (± 0.15)× CP (g/kg DM) ($n = 20$, $R^2 = 0.85$, RMSE = 6.9). According to this equation, zero rumen N balance was achieved at a dietary CP concentration of 131 g/kg DM. This value is slightly higher than the corresponding value of 123 at PBV−20 g/kg DM, and suggests that the RDP requirement can partly be recovered by recycling of urea-N into the rumen. The estimated rumen CP balance according to the previous equation was −6.4 g/kg DMI at a dietary CP concentration of 123 g/kg DM, below which MCP synthesis appears to be compromised. The contribution of endogenous N to NAN flows can be assumed to be smaller for omasal sampling than for duodenal sampling, and therefore, RDP requirements based on data derived from duodenal sampling may be overestimated.

The high RDP requirements in the AFRC system are related to the increased efficiency of rumen MCP synthesis and decreased ruminal protein degradability at high feeding levels, since there is no discount for reduced diet digestibility or possible negative associative effects in the model for high feeding levels. Passage rates of 0.10/h or higher for forages seem unrealistically high, even at high feeding levels, when compared to estimates derived from rumen evacuation data or from duodenal sampling using marker techniques with appropriate kinetic models (Pond *et al.*, 1988; Ellis *et al.*, 1994; Huhtanen and Hristov, 2001; Rinne *et al.*, 2002). Using the passage rate of 0.10/h to estimate ruminal NDF digestibility implies that only 0.33 of potentially digestible NDF will be digested in the rumen if the NDF digestion rate is 0.05/h.

2.6 Amino Acid Requirements of Cattle

Intestinally absorbed AA are required nutrients for the host animal. Tissues of cattle, like those of other animals, cannot synthesize the carbon chain of certain AA. The AA that need to be absorbed, and thus are considered essential, are the same in ruminants as they are for other mammals. The primary function of absorbed AA is their use in the synthesis of proteins, a biosynthetic event that is vital to the maintenance, growth, reproduction and lactation of cattle. Based on work with swine and poultry, it probably can be assumed that an optimum AA profile exists for each of these physiological functions. If these were known for cattle, and if the absolute amounts and the profile of absorbed AA could be optimized, it is unclear what effect this would have on increasing the efficiency of use of MP for maintenance and productive functions. However, even if AA nutrition was optimized, it is understood, because of the dynamics and inefficiencies of whole body and tissue turnover, that there will still be large and unavoidable losses of AA as a result of catabolism (Lobley, 2003). All AA can serve as immediate sources of metabolic energy when oxidized to CO_2 and H_2O.

In addition to their role in protein synthesis, AA are also used as substrates for other metabolic pathways. For example, AA other than leucine serve as precursors for gluconeogenesis and all can be converted to fatty acids. AA also contribute N to the many NPN compounds synthesized in the body. And finally, it is understood that many AA are also involved in signalling pathways and the integration of metabolism, including N metabolism. Regarding the latter point, available research indicates, for example, that AA signalling stimulates protein synthesis and inhibits proteolysis, increases cell volume (which in turn has been shown to increase synthesis of protein, glycogen and lipid), increases production of insulin (because of increased B-cell production) which increases AA uptake by tissues for biosynthetic reactions, as well as other positive outcomes of metabolism. Because of the functions that AA are involved in beyond those of protein synthesis, it is questionable as to whether or not the ideal profile of AA in MP can be established accurately from knowledge of the AA composition of synthesized proteins (i.e. lean tissue and milk protein).

2.6.1 Response of cattle to supplemental AA

Many experiments have shown that the efficiency of use of MP for protein accretion in growing cattle or yield of milk protein in lactating dairy cows is often less than optimum because of the proportionality of absorbed AA. The experimental approaches that have been used to administer selected individual or combinations of AA have included infusion studies (abomasal, duodenal and intravenous) and feeding the AA in ruminally protected form, as well as taking advantage of the reflex closure of the reticular groove in young calves. Responses to the administered AA have included changes in N retention, growth rates and yield of milk protein.

The following discussion provides a brief summary of the responses of growing cattle and lactating dairy cows to selected AA.

2.6.1.1 Growing cattle

Methionine (Met), lysine (Lys), histidine (His), threonine (Thr), leucine (Leu) and valine (Val) have all been implicated in one or more experiments as being in less than optimum concentrations in MP for N retention or growth of growing cattle. Methionine has been shown in several experiments to be the first limiting AA when ruminally synthesized microbial protein was the predominant source of MP (Richardson and Hatfield, 1978; Campbell *et al.*, 1997; Greenwood and Titgemeyer, 2000). The AA that have been shown to be limiting after Met, when most of the absorbed AA are provided by microbial protein are Lys, His, Leu, Val and Thr (Richardson and Hatfield, 1978; Greenwood and Titgemeyer, 2000; Löest *et al.*, 2001).

Titgemeyer and Merchen (1990) observed a 17% increase in N retention with abomasally infused Met when 310 kg steers gaining 0.9 kg/day were fed a semi-purified diet based on ammoniated maize cobs, maize starch, molasses and urea; a small amount of casein was included in the diet to provide ruminal microorganisms with a supply of AA and peptides. Lusby (1994) observed a 9%

increase in weight gains of light-weight calves grazing native pasture when the diet was supplemented with 5 g/day of Smartamine M™, a rumen-protected Met product. A summary of four studies by Kunkle and Hopkins (1999) indicated that supplementing 2.6 to 5.0 g/day of a rumen-protected Met product (Smartamine M™) increased weight gains from 0.07 to 0.15 kg/day in growing cattle fed medium quality forages supplemented with molasses-based supplements.

When significant amounts of RUP are fed, the sequence of Met and Lys limitation is determined by their relative concentrations in RUP. As reviewed in NRC (2001), Met has been identified as first limiting for young post-weaned calves and growing cattle when most of the supplemental RUP was provided by soybean products or animal-derived proteins. In contrast, Lys has been identified as first limiting for young post-weaned calves and growing cattle when maize and feeds of maize origin provided most or all of the RUP. Also, as reviewed in NRC (2001), responses of growing cattle to improved supplies of Lys and Met in MP include variable increases in weight gains and feed efficiency and variable decreases in urinary N excretion. Williams *et al.* (1999) observed that Lys and Met were the first limiting AA for growing cattle fed maize/maize silage-based diets but not for heifers grazing tall fescue.

2.6.1.2 *Lactating dairy cows*

Met, Lys and His have been identified most often as the most limiting AA for lactating dairy cows. Similar to the observations with growing cattle, Met is typically first limiting when most of the RUP is provided by soybean protein, animal-derived proteins or a combination of the two, Lys is first limiting when maize and feeds of maize origin provide most or all of RUP, and Met and Lys have been identified as co-limiting AA for milk protein production when cows were fed maize silage-based diets containing complementary feed proteins (NRC, 2001). Production responses of lactating dairy cows to increased supplies of Lys and Met in MP include variable increases in content and yield of milk protein and milk yield (NRC, 2001).

Histidine has been shown to be more limiting than Lys or Met when cows are fed grass silage-based diets (Kim *et al.*, 1999, 2000, 2001a,b; Vanhatalo *et al.*, 1999; Korhonen *et al.*, 2000; Huhtanen *et al.*, 2002). In all cases, the diets were devoid of maize and contained barley and oats as the supplemental energy feeds. The diets were fed with or without feather meal as the sole source of supplemental RUP. Abomasal infusions of Lys and Met had no effect on MPY in cows fed grass silage-based diets (Varvikko *et al.*, 1999) and these AA did not produce any further response in addition to His when infused either alone or in a combination (Vanhatalo *et al.*, 1999). These findings make an important contribution to our understanding of AA requirements and highlight the impact that diet composition has on the sequence of AA limitation.

Data presented in Table 2.9 provide at least two indications as to why His might have been more limiting in these studies. First, His may be more limiting in ruminally synthesized bacteria than either Lys or Met for milk protein synthesis. As noted in Table 2.9, His is 2.0% of CP in rumen bacteria and His is 2.7% of CP in milk. In contrast, concentrations of Lys and Met are both very similar in rumen bacteria and milk (7.9% and 7.6%, and 2.6% and 2.7%, respectively). This is mentioned because in these experiments, it is expected that bacterial protein constituted a larger percentage of total MP than in cows fed maize-based diets. The CP of grass silage, barley and oats contains considerably less RUP than maize silage and ground maize (NRC, 2001). A smaller contribution of RUP to MP means that the AA composition of RUP has less of an effect on the AA composition of total MP than feeds that are less degradable and have a higher content of RUP in CP. And second, the His content of these feeds is low. The His content of barley and oats is lower than the His content of maize (2.3–2.4% vs. 3.1% of CP). The His content of feather meal is considerably lower than the His content of other protein supplements (1.2% vs. 2.0–2.8% of CP). The high content of His in blood (6.4% of CP) is noteworthy and may give blood meal an additional advantage over other protein supplements, particularly when higher grass silage, lower maize diets are fed. However, the global use of blood meal is limited because many countries have banned the feeding of it to ruminant animals.

2.6.2 Ideal profile of essential AA in MP

Based on the above discussion, it seems important that the ideal concentrations of the most limiting

Table 2.9. A comparison of the essential amino acid composition of body lean tissue, milk and ruminal bacteria with that of some common feeds[a].

Item	Arg	His	Ile	Leu	Lys	Met	Phe	Thr	Trp	Val
	(% of CP)									
Lean tissue	6.6	**2.5**[b]	2.8	6.7	**6.4**	**2.0**	3.5	3.9	0.6	4.0
Milk	3.4	**2.7**	5.8	9.2	**7.6**	**2.7**	4.8	3.7	1.5	5.9
Bacteria	5.1	**2.0**	5.7	8.1	**7.9**	**2.6**	5.1	5.8	–	6.2
Lucerne silage	3.9	**1.7**	3.9	6.4	**4.4**	**1.4**	4.2	3.8	0.9	5.0
Maize silage	2.0	**1.8**	3.3	8.6	**2.5**	**1.5**	3.8	3.2	0.4	4.5
Grass silage	3.1	**1.7**	3.6	6.1	**3.3**	**1.2**	4.4	3.3	1.1	4.9
Barley	5.1	**2.3**	3.5	7.0	**3.6**	**1.7**	5.1	3.4	1.2	4.9
Maize	4.6	**3.1**	3.3	11.2	**2.8**	**2.1**	4.6	3.6	0.7	4.0
Oats	6.8	**2.4**	3.8	7.3	**4.2**	**2.9**	5.2	3.5	1.2	5.2
Wheat	4.7	**2.4**	3.3	6.6	**2.8**	**1.6**	4.6	2.9	1.2	4.2
Brewers grains	5.8	**2.0**	3.9	7.9	**4.1**	**1.7**	4.6	3.6	1.0	4.8
Rape meal	7.0	**2.8**	3.8	6.8	**5.6**	**1.9**	4.1	4.4	1.5	4.7
Maize DDG w/sol	4.1	**2.5**	3.7	9.6	**2.2**	**1.8**	4.9	3.4	0.9	4.7
Maize gluten meal	3.2	**2.1**	4.1	16.8	**1.7**	**2.4**	6.4	3.4	0.5	4.6
Cottonseed meal	11.1	**2.8**	3.1	5.9	**4.1**	**1.6**	5.3	3.2	1.2	4.2
Soybean meal	7.3	**2.8**	4.6	7.8	**6.3**	**1.4**	5.3	4.0	1.3	4.6
Sunflower meal	8.2	**2.6**	4.1	6.4	**3.6**	**2.3**	4.6	3.7	1.2	5.0
Blood meal	4.4	**6.4**	1.3	12.8	**9.0**	**1.2**	6.9	4.3	1.6	8.7
Feather meal	6.9	**1.2**	4.9	8.5	**2.6**	**0.8**	4.9	4.7	0.7	7.5
Fishmeal	5.8	**2.8**	4.1	7.2	**7.7**	**2.8**	4.0	4.2	1.1	4.8
Meat meal	7.1	**2.1**	3.0	6.3	**5.4**	**1.4**	3.6	3.4	0.7	4.4

[a]Amino acid values for lean tissue, milk and ruminal bacteria are from O'Connor *et al.* (1993) and amino acid values for feeds are from NRC (2001).
[b]The values for His, Lys and Met are in bold as they are first limiting most frequently in lactating dairy cows.

AA be determined for cattle. If these concentrations were known and diets could be formulated to achieve these ideal concentrations, then the efficiency of use of MP for growth and milk protein production could be maximized. Progress has been made in determining the ideal concentrations of Lys and Met in MP for lactating dairy cows. The NRC (2001) publication contains dose–response plots that relate measured milk protein content and yield responses to changes in predicted percentages of Lys and Met in MP.

The breakpoint estimates for the required concentrations of Lys and Met in MP for maximal content of milk protein were 7.2% and 2.4%, respectively (3.0:1.0 ratio; Fig. 2.8). The breakpoint estimates for the required concentrations of Lys and Met in MP for maximal yield of milk protein were 7.1% and 2.4% (plots not shown). Examination of the dose–response plots indicates little or no expected loss in content or yield of milk protein when Lys and Met in MP are 6.9% and 2.3%, respectively. The senior author's field experience indicates no advantage of exceeding these concentrations when using NRC (2001) for diet evaluation. Because these concentrations are often difficult to achieve, particularly in high producing cows fed maize-based diets, the 'practical recommendations' for percentages of Lys and Met in MP are considered to be 6.6 and 2.2, respectively. As more information is obtained for His, similar breakpoint estimates can be calculated to determine the required concentrations of His in MP for maximal yield of milk protein.

2.6.3 Is there a need for AA sub-models?

Because the profile of absorbed AA generally appears to be less than optimum, there appears to be little doubt that AA requirements need to be defined in terms of individual AA and not as MP. Because of that, model level 2 in NRC (1996) was the first attempt by NRC to accomplish this for growing cattle. The model, using factorial

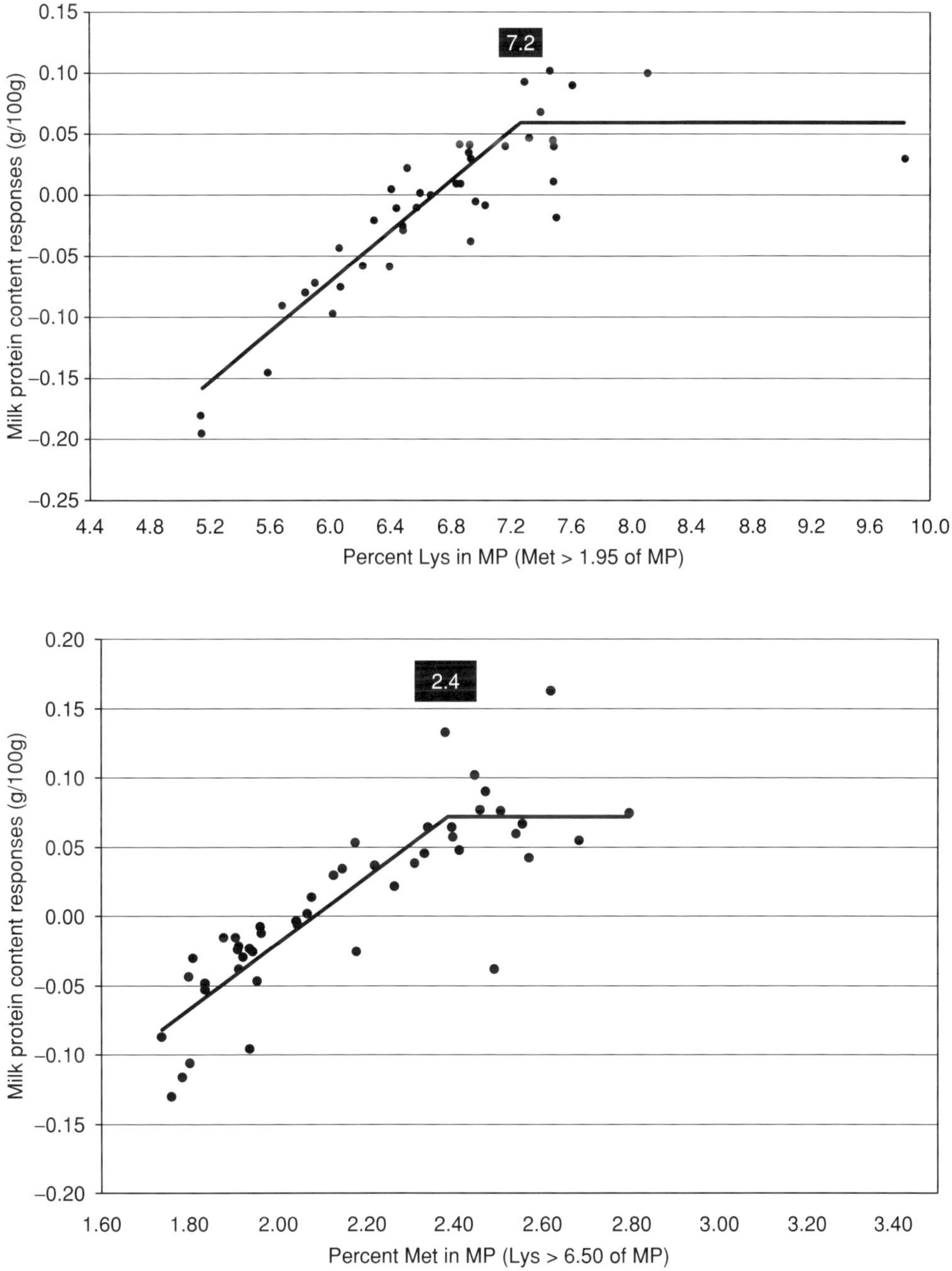

Fig. 2.8. Milk protein content responses as a function of percent Lys and Met in MP. For the Lys plot, the regression analysis was limited to data where Met were predicted to be 1.95% or more of MP. For the Met plot, the regression analysis was limited to data where Met were predicted to be 6.50% or more of MP (NRC, 2001).

approach, predicts both the quantity and proportion of metabolizable EAA provided by the diet and required by the animal. To advance research on AA requirements and to allow for implementation of the results, the NRC (2001) committee decided to extend the MP model to one that would most accurately predict the profile of EAA in duodenal protein and flows of metabolizable EAA to the small intestine. A multivariate regression approach is used. However, it was the opinion of the 2001 NRC committee that knowledge was too limited, both for model construction and model evaluation, to put forth a model that 'quantifies' AA requirements for dairy cattle. However, an alternate and first step to that approach is to begin to define the ideal content of EAA in MP. This requires establishing dose–response relationships between changes in concentrations of EAA in MP (at least those considered to be the most limiting) and animal responses. Because the model predicts concentrations of EAA in MP, and because several studies have evaluated milk protein responses to changes in concentrations of Lys and Met in duodenal protein, the prerequisites were in place to use the model to define the requirements for Lys and Met in MP for lactating cows (Fig. 2.8). The approach that was used was that described by Rulquin *et al.* (1993).

Current diet evaluation models that predict passage of MP-AA to the small intestine (e.g. NRC, 1996, 2001; CNCPS, 2000) in their present form are not as useful as they could be in predicting the effect that changes in supplies of MP-Lys and MP-Met have on milk and milk component production. Until such systems are in place, it will remain difficult to predict the effect that changes in protein and AA supplementation strategies have on predicting growth and MPY responses to changes in AA supply.

In an attempt to determine if MPY can be predicted more accurately from predicted supplies of MP-Lys and MP-Met than from MP (the sum of absorbed AA), Schwab *et al.* (2003) used the NRC (2001) model in conjunction with published experiments to examine the relationships between predicted supplies of MP, MP-Met and MP-Lys and yields of milk and milk protein. This effort has been extended for this chapter.

Over 300 diets from experiments published in the *Journal of Dairy Science* were entered into the NRC (2001) model. In most of these experiments the objective was to compare the effects of feeding different protein supplements on milk production and milk composition, and in some cases, passage of N fractions to the small intestine. Relevant data from the *Summary and Duodenal Amino Acid Supply Reports* were recorded.

In order to generate plots of measured yields of milk and milk protein vs. predicted supplies of MP, data were restricted to diets in which Net Energy (NE)-allowable milk was higher than MP-allowable milk, and actual milk yield was between −6 kg and +6 kg of MP-allowable milk. The former restriction was imposed to help ensure that MP was more limiting than NE. The latter restriction was imposed to avoid the use of experiments in which factors other than MP or NE limited lactation performance or situations where excessive protein mobilization may have been occurring.

To generate plots of measured yields of milk and milk protein vs. predicted supplies of MP-Lys and MP-Met, data were restricted to diets in which MP balance was within −250 and +100 g/day of zero balance. This was done with the hope of further ensuring that Lys and Met were limiting. For the Met plots, the restriction that the Lys:Met ratio in MP had to be greater than 3.0:1.0 to make more certain that Met was more limiting than Lys was imposed. For the Lys plots, it was necessary to add the restriction that the ratio of Lys to Met in MP had to be less than 3.0:1.0 to ensure that Lys was more limiting in MP than Met. However, only in a few cases was the ratio of Lys to Met in MP less than 3.0:1.0. Therefore, to provide an adequate number of data points from which to get some idea of the relationship between yields of milk and milk protein vs. predicted supplies of Lys, diets yielding predicted Lys:Met ratios up to 3.25:1.0 were used.

The resulting plots are presented in Fig. 2.9. There are at least three observations that are worthy of mention. First, in all cases (for MP, MP-Met and MP-Lys), it appears that protein yields can be predicted more accurately than milk yields. This would be expected because of the changes in milk protein percentages that often occur with changes in protein nutrition. Second, as expected, predicting yields of milk and milk protein from intestinal supplies of the most limiting AA is more precise than predicting yields from MP supply. Third, while the current data are too limited and not adequate for this

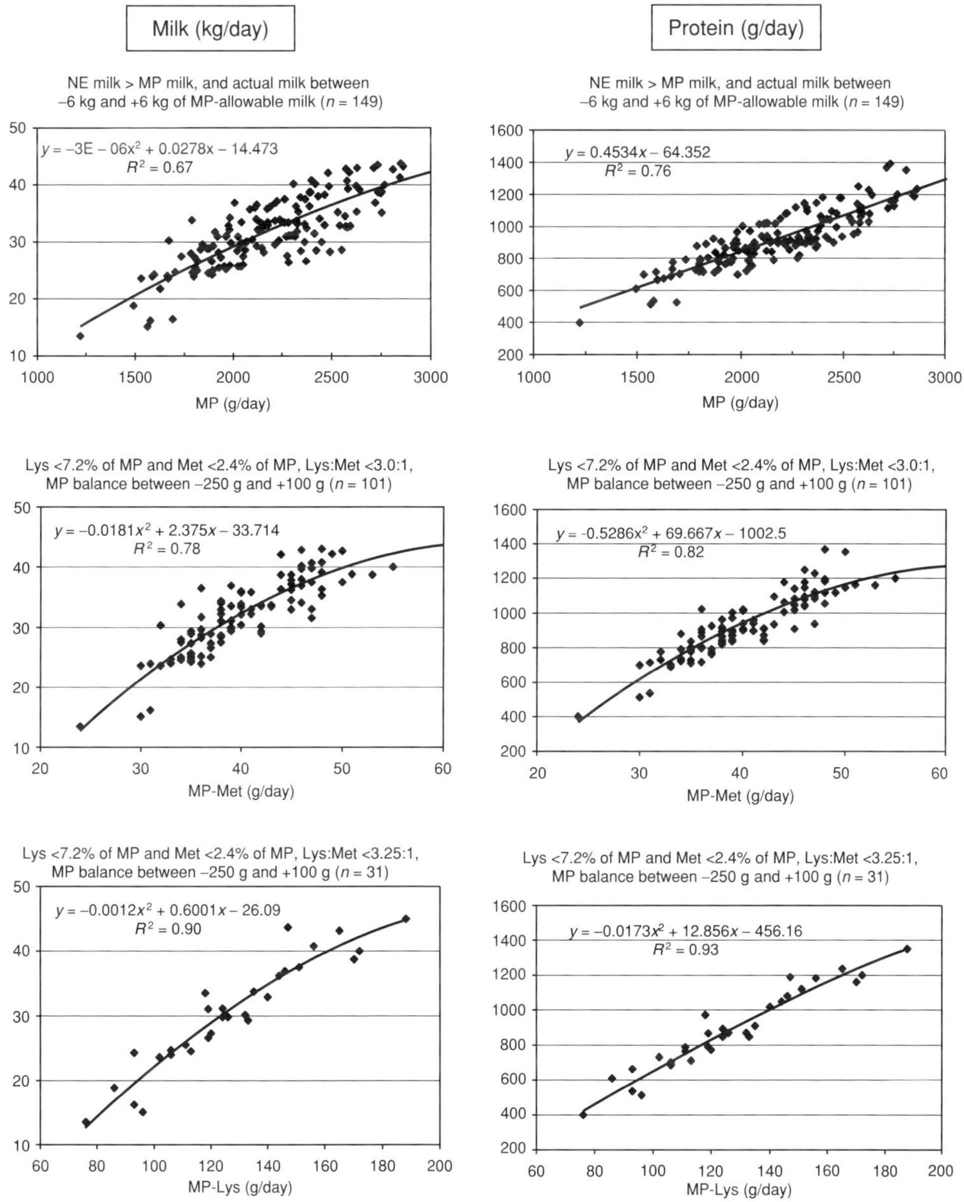

Fig. 2.9. Plots of measured milk and milk protein yields vs. NRC (2001) predicted flows of metabolizable protein (MP) and MP-Lys and MP-Met. Data were selected from a database involving 321 diets fed to Holstein cows without AA supplementation (restrictions used for selecting data are indicated above each of the plots).

exercise, it appears that a very strong relationship exists between milk and MPY and predicted MP-Lys supplies. This should probably be expected given the fact that Lys, unlike Met, has only one function in the body, i.e. protein synthesis.

2.6.4 Opportunities for reduced N losses with more precise balancing of diets for AA

There are two lines of evidence with lactating dairy cows that suggest that balancing diets for AA will increase efficiency of use of MP for milk production. The first line of evidence is that provided in Figs 16.4 and 16.5 in NRC (2001) which shows that the difference between MP-allowable milk and actual milk increases as the concentration of Lys decreases from 6.5% of MP and as the concentration of Met decreases from 1.9% of MP. As stated in NRC (2001), 'This suggests that although supply of total MP was adequate (according to the model) in many of these experiments, the balance of absorbable may have been incorrect and limiting milk production'. The experiments used for this evaluation of the protein portion of the NRC (2001) were from continuous lactation trials published in the *Journal of Dairy Science* using only conventional feedstuffs and no ruminally protected AA.

The second line of evidence is that provided by individual experiments where, by selective use of protein and ruminally protected AA supplements, higher predicted concentrations of Lys and Met in MP are achieved than could otherwise be achieved without a deliberate attempt to optimize the balance of AA in MP. For example, Noftsger and St-Pierre (2003) increased the efficiency of conversion of feed N to milk N from 31.7% to 35.0% in early lactation cows when the predicted Met concentration in MP was increased from 1.8% to 2.1%; Lys was held constant at 6.8% of MP. While this observed improvement in efficiency of use of dietary N by lactating cows is greater than typically observed in the studies that have described MPY responses to post-ruminal Lys and Met supplementation, it is reasonable to expect that an improved profile of AA in MP will increase efficiency of use of MP for milk production.

2.7 Conclusions

Reducing wastage of N by beef and dairy cattle requires close matching of feed N supply to the requirements of rumen microorganisms and the AA requirements of the animal. Many new protein systems have been developed in the last 15 years to provide ways by which to better define protein requirements and to match the requirements with a corresponding supply. These systems represent a significant step forward in achieving the goal of more precise feeding for N but as we have tried to highlight in this chapter, significant shortcomings still exist. Each protein system discussed is constructed differently and uses different feed inputs, different equations for predicting passage of digestible or MP from microbial protein and RUP, and different equations for predicting requirements for MP. And only a few protein systems make any attempt to predict passage of individual digestible EAA or consider requirements of EAA. Clearly, a better understanding is needed of the balance and quantities of NH_3-N, AA and peptides required for optimum rumen function and yield of microbial protein, a better understanding is needed of the AA requirements of the host animal, and more precise analysis and characterization of the N constituents in feedstuffs are needed to more accurately predict passage of digestible microbial protein, digestible RUP and digestible endogenous protein to the small intestine.

References

Abou Akkada, A.R. and Blackburn, T.H. (1963) Some observations on the nitrogen metabolism of rumen proteolytic bacteria. *Journal of General Microbiology* 31, 461–469.

AFRC [Agricultural and Food Research Council – Technical Committee on Responses to Nutrients] (1992) Nutritive Requirements of Ruminant Animals: Protein, Report No. 9. *Nutrition Abstracts and Reviews, Series B* 62, 787–835.

AFRC [Agricultural and Food Research Council] (1993) *Energy and Protein Requirements of Ruminants.* CAB International, Wallingford, UK, 159 pp.

Ahvenjärvi, S., Vanhatalo, A., Huhtanen, P. and Varvikko, T. (1999) Effects of supplementation of a grass silage and barley diet with urea, rapeseed meal and heat-moisture-treated rapeseed cake on omasal digesta flow and milk production in lactating dairy cows. *Acta Agriculturae Scandinavica, Section A Animal Science* 49, 179–189.

Ahvenjärvi, S., Joki-Tokola, E., Vanhatalo, A., Jaakkola, S. and Huhtanen, P. (2002a) Effects of gradual replacement of grass silage with whole-crop barley silage on ruminal metabolism, digestibility and milk production in dairy cows. In: McGechie, L. and Thomas, C. (eds) *Conference Proceedings of the XIII*

International Silage Conference, September 11–13, 2002, Auchincruive, Scotland, pp. 320–321.
Ahvenjärvi, S., Vanhatalo, A., and Huhtanen, P. (2002b) Supplementing barley or rapeseed meal to dairy cows fed grass-red clover silage. I. Rumen degradability and microbial flow. *Journal of Animal Science* 80, 2176–2187.
Aldrich, J.M., Muller, L.D., Varga, G.A. and Griel, L.C., (1993) Non-structural carbohydrate and protein effects on rumen fermentation, nutrient flow, and performance of dairy cows. *Journal of Dairy Science* 76, 1091–1105.
Alhadhrami, G. and Huber, J.T. (1992) Effects of alfalfa hay of varying fiber fed as 35 or 50% of diet on lactation and nutrient utilization by dairy cows. *Journal of Dairy Science* 75, 3091–3099.
Allen, M.S. and Mertens, D.R. (1988) Evaluating constraints of fiber digestion by rumen microbes. *Journal of Nutrition* 118, 261–270.
Allen, S.A. and Miller, E.L. (1976) Determination of nitrogen requirement for microbial growth from the effect of urea supplementation of a low N diet on abomasal N flow and N recycling in wethers and lambs. *British Journal of Nutrition* 36, 353–368.
Al-Rabbat, M.F., Baldwin, R.L. and Weir, W.C. (1971) Microbial growth dependence on ammonia nitrogen in the bovine rumen: a quantitative study. *Journal of Dairy Science* 54, 1162–1172.
Anderson, L.P., Paterson, J.A., Ansotegui, R.P., Cecava, M. and Schmutz, W. (2001) The effects of degradable and undegradable intake protein on the performance of lactating first-calf heifers. *Journal of Animal Science* 79, 2224–2232.
ARC [Agricultural Research Council] (1965) *The Nutrient Requirements of Farm Livestock No. 2, Ruminants.* HMSO, London, 264 pp.
ARC [Agricultural Research Council] (1980) *The Nutrient Requirements of Ruminant Livestock. Technical Review.* CAB, Farnham Royal, UK, 351 pp.
Argyle, J.L. and Baldwin, R.L. (1989) Effects of amino acids and peptides on rumen microbial growth yields. *Journal of Dairy Science* 72, 2017–2027.
Armentano, L.E., Bertics, S.J. and Riesterer, J. (1993) Lack of response to addition of degradable protein to a low protein diet fed to midlactation dairy cows. *Journal of Dairy Science* 76, 3755–3762.
Armentano, L.E., Bertics, S.J. and Ducharme, G.A. (1997) Response of lactating cows to methionine or methionine plus lysine added to high protein diets based on alfalfa and heated soybeans. *Journal of Dairy Science* 80, 1194–1199.
Armstead, I.P. and Ling, J.R. (1993) Variations in the uptake and metabolism of peptides and amino acids by mixed ruminal bacteria in vitro. *Applied and Environmental Microbiology* 59, 3360–3366.
Atasoglu, C., Valdes, C., Walker, N.D., Newbold, C.J. and Wallace, R.J. (1998) De novo synthesis of amino acids by the ruminal bacteria *Prevotella bryantii* $B_1$4, *Selenomonas ruminantium* HD4, and *Streptococcus bovis* ES1. *Applied and Environmental Microbiology* 64, 2836–2843.
Atasoglu, C., Valdes, C., Newbold, C.J. and Wallace, R.J. (1999) Influence of peptides and amino acids on fermentation rate and *de novo* synthesis of amino acids by mixed microorganisms from the sheep rumen. *British Journal of Nutrition* 81, 307–314.
Atasoglu, C., Newbold, C.J. and Wallace, R.J. (2001) Incorporation of [^{15}N]ammonia by the cellulolytic ruminal bacteria *Fibrobacter succinogenes* BL2, *Ruminococcus albus* SY3, and *Ruminococcus flavefaciens* 17. *Applied and Environmental Microbiology* 67, 2819–2822.
Bachmann, H.P. and Jans, F. (1995) Fütterung kann Käsequalität beeinflussen. *Schweizerische Milchzeitung* 121, 3.
Bagg, J.G., Grieve, D.G., Burton, J.H. and Stone, J.B. (1985) Effect of protein on growth of Holstein heifer calves from 2 to 10 months of age. *Journal of Dairy Science* 68, 2929–2939.
Bandyk, C.A., Cochran, R.C., Wickersham, T.A., Titgemeyer, E.C., Farmer, C.G. and Higgins, J.J. (2001) Effect of ruminal vs. postruminal administration of degradable protein on utilization of low-quality forage by beef steers. *Journal of Animal Science* 79, 225–231.
Barajas, R. and Zinn, R.A. (1998) The feeding value of dry-rolled and steam-flaked corn in finishing diets for feedlot cattle: influence of protein supplementation. *Journal of Animal Science* 76, 1744–1752.
Barney, D.J., Grieve, D.G., MacLeod, G.K. and Young, L.G. (1981) Response of cows to dietary crude protein during midlactation. *Journal of Dairy Science* 64, 655–661.
Beauchemin, K.A. and Buchanan-Smith, J.G. (1989) Evaluation of markers, sampling sites and models for estimating rates of passage of silage or hay in dairy cows. *Animal Feed Science and Technology* 27, 59–75.
Bell, A.W., Slepetis, R. and Ehrhardt, R.A. (1995) Growth and accretion of energy and protein in the gravid uterus during late pregnancy in Holstein cows. *Journal of Dairy Science* 78, 1954–1961.
Bibby, J. and Toutenburg, H. (1977) *Prediction and Improved Estimation in Linear Models.* Cp. 1.5.4. Wiley, London, 188 pp.
Bickel, H. and Landis, J. (1987) Present situation of protein evaluation for ruminants in Switzerland. In: Jarrige, R. and Alderman, G. (eds) *Feed Evaluation and Protein Requirement Systems for Ruminants.* Proceedings of a CEC seminar held in Brussels, June 25–27, 1986. CEC L 2935, Luxembourg, pp. 41–47.
Broderick, G.A. (1992) Relative value of fishmeal versus solvent soybean meal for lactating dairy cows fed alfalfa silage as sole forage. *Journal of Dairy Science* 75, 174–183.

Broderick, G.A. (2003) Effects of varying dietary protein and energy levels on the production of lactating dairy cows. *Journal of Dairy Science* 86, 1370–1381.

Broderick, G.A., Wallace, R.J. and Ørskov, E.R. (1991) Control of rate and extent of protein degradation. In: Tsuda, T., Sasaki, Y. and Kawashima, R. (eds) *Physiological Aspects of Digestion and Metabolism in Ruminants.* Academic Press, San Diego, California, pp. 541–592.

Broudiscou, L. and Jouany, J.P. (1995) Reassessing the manipulation of protein synthesis by rumen microbes. *Reproduction, Nutrition, Development* 35, 517–535.

Bryant, M.P. (1973) Nutritional requirements of the predominant rumen cellulolytic bacteria. *Federation Proceedings* 32, 1809–1813.

Bryant, M.P. and Robinson, I.M. (1961) Studies on the nitrogen requirements of some ruminal cellulolytic bacteria. *Applied Microbiology* 9, 96–103.

Bryant, M.P. and Robinson, I.M. (1962) Some nutritional characteristics of predominant culturable ruminal bacteria. *Journal of Bacteriology* 84, 605–614.

Burroughs, W., Trenkle, A.H. and Vetter, R.L. (1971) Some new concepts of protein nutrition of feedlot cattle (metabolizable protein and metabolizable amino acids). *Veterinary Medicine and Small Animal Clinician* 66, 238–247.

Burroughs, W., Nelson, D.K. and Mertens, D.R. (1974a) Evaluation of protein nutrition by metabolizable protein and urea fermentation potential. *Journal of Dairy Science* 58, 611–619.

Burroughs, W., Trenkle, A.H. and Vetter, R.L. (1974b) A system of protein evaluation for cattle and sheep involving metabolizable protein (amino acids) and urea fermentation potential of feedstuffs. *Veterinary Medicine and Small Animal Clinician* 69, 713–722.

Burroughs, W., Nelson, D.K. and Mertens, D.R. (1975) Protein physiology and its application in the lactating cow: the metabolizable protein feeding standard. *Journal of Animal Science* 41, 933–944.

Cadorniga, C. and Satter, L.D. (1993) Protein versus energy supplementation of high alfalfa silage diets for early lactation cows. *Journal of Dairy Science* 76, 1972–1977.

Campbell, C.G., Titgemeyer, E.C. and St-Jean, G. (1997) Sulfur amino acid utilization by growing steers. *Journal of Animal Science* 75, 230–238.

Campling, R.C. (1991) Processing cereal grains for cattle – a review. *Livestock Production Science* 28, 223–234.

Canfield, R.W., Sniffen, C.J. and Butler, W.R. (1990) Effects of excess degradable protein on postpartum reproduction and energy balance in dairy cattle. *Journal of Dairy Science* 73, 2342–2349.

Carro, M.D. and Miller, E.L. (1999) Effect of supplementing a fibre basal diet with different nitrogen forms on ruminal fermentation and microbial growth in an in vitro semi-continuous culture system (RUSITEC). *British Journal of Nutrition* 82, 149–157.

Caton, J.S., Freeman, A.S. and Galyean, M.L. (1988) Influence of protein supplementation on forage intake, in situ forage disappearance, ruminal fermentation and digesta passage rates in steers grazing dormant blue grama rangeland. *Journal of Animal Science* 66, 2262–2271.

Chase, L.E. (2003) Nitrogen utilization in dairy cows – what are the limits of efficiency? *Proceedings of Cornell Nutrition Conference,* Cornell University, Ithaca, New York, pp. 233–244.

Chen, G., Russell, J.B. and Sniffen, C.J. (1987a) A procedure for measuring peptides in rumen fluid and evidence that peptide uptake can be a rate-limiting step in ruminal protein degradation. *Journal of Dairy Science* 70, 1211–1219.

Chen, G., Strobel, H.J., Russell, J.B. and Sniffen, C.J. (1987b) The effect of hydrophobicity on utilization of peptides by ruminal bacteria in vitro. *Applied Environmental Microbiology* 53, 2021–2025.

Choi, C.W. (2002) Assessment of the flow of soluble dietary non-ammonia nitrogen escaping degradation in the rumen of dairy cows fed grass silage based diets. *Helsingin yliopiston kotieläintieteen laitoksen julkaisuja* 66, 60 p. + encl. Diss. Helsingin yliopisto, 2002.

Choi, C.W., Ahvenjärvi, S.,Vanhatalo, A., Toivonen, V. and Huhtanen, P. (2002) Quantitation of the flow of soluble non-ammonia nitrogen entering the omasal canal of dairy cows fed grass silage based diets. *Animal Feed Science and Technology* 96, 203–220.

CNCPS [Cornell University Nutrient Management Planning System] (2000) *The Net Carbohydrate and Protein System for Evaluating Herd Nutrition and Nutrient Excretion.* CNCPS version 4.0, November 3. Model Documentation.

Coleman, G.S. (1979) Rumen ciliate protozoa. In: Levandowsky, M. and Hunter, S.H. (eds) *Biochemistry and Physiology of Protozoa,* Vol. 2. Academic Press, New York, pp. 381–408.

Cooper, R.J., Milton, C.T., Klopfenstein, T.J. and Jordan, D.J. (2002a) Effect of corn processing on degradable intake protein requirement of finishing cattle. *Journal of Animal Science* 80, 242–247.

Cooper, R.J., Milton, C.T., Klopfenstein, T.J., Scott, T.L., Wilson, C.B. and Mass, R.A. (2002b) Effect of corn processing on starch digestion and bacterial crude protein flow in finishing cattle. *Journal of Animal Science* 80, 797–804.

Cotta, M.A. and Hespell, R.B. (1984) Protein and amino acid metabolism of rumen bacteria. In: Milligan, L.P., Grovum,W.L. and Dobson, A. (eds) *Control of Digestion and Metabolism in Ruminants.* Prentice-Hall, Englewood Cliffs, New Jersey, pp. 122–136.

Cotta, M.A. and Russell, J.B. (1982) Effect of peptides and amino acids on efficiency of rumen bacterial protein synthesis in continuous culture. *Journal of Dairy Science* 65, 226–234.

Coulon, J.-B., Verdier, I., Pradel, P. and Almena, M. (1998) Effects of lactation stage on the cheese making properties of milk and the quality of Saint-Nectaire-type cheese. *Journal of Dairy Research* 65, 295–305.

Cruz Soto, R., Muhammed, S.A., Newbold, C.J., Stewart, C.S. and Wallace, R.J. (1994) Influence of peptides, amino acids and urea on microbial activity in the rumen of sheep receiving grass hay and on the growth of rumen bacteria in vitro. *Animal Feed Science and Technology* 49, 151–161.

CSIRO [Commonwealth Scientific and Industrial Research Organization] (1990) In: Corbett, J.L. (ed.) *Feeding Standards for Australian Livestock: Ruminants.* Standing Committee on Agriculture, Ruminants Sub-Committee. CSIRO Publishers, Melbourne, Victoria, Australia.

DelCurto, T., Cochran, R.C., Harmon, D.L., Beharka, A.A., Jacques, K.A., Towne, G. and Vanzant, E.S. (1990) Supplementation of dormant tallgrass-prairie forage. I. Influence of varying supplemental protein and(or) energy levels on forage utilization characteristics of beef steers in confinement. *Journal of Animal Science* 68, 515–531.

Devant, M., Ferret, A., Gasa, J., Calsamiglia, S. and Casals, R. (2000) Effects of protein concentration and degradability on performance, ruminal fermentation, and nitrogen metabolism in rapidly growing heifers fed high-concentrate diets from 100 to 230 kg body weight. *Journal of Animal Science* 78, 1667–1676.

Dinn, N.E., Shelford, J.A. and Fisher, L.J. (1998) Use of the Cornell Net Carbohydrate and Protein System and rumen-protected lysine and methionine to reduce nitrogen excretion from lactating dairy cows. *Journal of Dairy Science* 81, 229–237.

Duff, G.C., Malcolm-Callis, K.J., Galyean, M.L. and Walker, D.A. (2003) Effects of dietary urea concentration on performance and health of receiving cattle and performance and carcass characteristics of finishing cattle. *Canadian Journal of Animal Science* 83, 569–575.

Ellis, W.C., Matis, J.H., Hill, T.M. and Murphy, M.R. (1994) Methodology for estimating digestion and passage kinetics of forages. In: Fahey, G.C., Collins, M., Mertens, D.R. and Moser, L.E. (eds) *Forage Quality, Evaluation and Utilization.* American Society of Agronomy, Madison, Wisconsin, pp. 682–756.

Erdman, R.A., Proctor, G.H. and Vandersall, J.H. (1986) Effect of rumen ammonia concentration on in situ rate and extent of digestion of feedstuffs. *Journal of Dairy Science* 69, 2312–2320.

Feng, P., Hunt, C.W., Pritchard, G.T. and Parish, S.M. (1995) Effect of barley variety and dietary barley content on digestive function in beef steers fed grass hay-based diets. *Journal of Animal Science* 73, 3476–3484.

Fox, D.G., Sniffen, C.J., O'Connor, J.D., Russell, J.B. and Van Soest, P.J. (1992) A net carbohydrate and protein system for evaluating cattle diets. III. Cattle requirements and diet adequacy. *Journal of Animal Science* 70, 3578–3596.

Galyean, M.L. (1996) Protein levels in beef cattle finishing diets: industry application, university research, and systems results. *Journal of Animal Science* 74, 2860–2870.

Ganev, G., Ørskov, E.R. and Smart, R. (1979) The effect of roughage or concentrate feeding and rumen retention time on total degradation of protein in the rumen. *Journal of Agricultural Science, Cambridge* 93, 651–656.

GfE [Ausschuss für Bedarfsnormen der Gesellschaft für Ernährungsphysiologie] (1986) *Energie und Nährstoffbedarf landwirtschaftlicher Nutztiere, Nr. 3. Milchkühe und Aufzuchtrinder.* DLG-Verlag, Fankfurt am Main, Germany, 92 pp.

Greenwood, R.H. and Titgemeyer, E.C. (2000) Limiting amino acids for growing Holstein steers limit-fed soybean hull-based diets. *Journal of Animal Science* 78, 1997–2004.

Grings, E.E., Roffler, R.E. and Deitelhoff, D.P. (1992) Responses of dairy cows to additions of distillers dried grains with solubles in alfalfa-based diets. *Journal of Dairy Science* 75, 1946–1953.

Griswold, K.E., Hoover, W.H., Miller, T.K. and Thayne, W.V. (1996) Effect of form of nitrogen on growth of ruminal microbes in continuous culture. *Journal of Animal Science* 74, 483–491.

Griswold, K.E., Apgar, G.A., Bouton, J. and Firkins, J.L. (2003) Effects of urea infusion and ruminal degradable protein concentration on microbial growth, digestibility, and fermentation in continuous culture. *Journal of Animal Science* 81, 329–336.

Guthrie, M.J. and Wagner, D.G. (1988) Influence of protein or grain supplementation and increasing levels of soybean meal on intake, utilization and passage rate of prairie hay in beef steers and heifers. *Journal of Animal Science* 66, 1529–1537.

Heikkilä, T., Toivonen, V. and Huhtanen, P. (1998) Effects of and interactions between the extent of silage fermentation and protein supplementation in lactating dairy cows. *Agricultural and Food Science in Finland* 7, 329–343.

Heldt, J.S., Cochran, R.C., Mathis, C.P., Woods, B.C., Olson, K.C., Titgemeyer, E.C., Nagaraja, T.G., Vanzant, E.S. and Johnson, D.E. (1999) Effects of level and source of carbohydrate and level of degradable intake protein on intake and digestion of low-quality tallgrass-prairie hay by beef steers. *Journal of Animal Science* 77, 2846–2854.

Herrera-Saladana, R.E., Huber, J.T. and Poore, M.H. (1990) Dry matter, crude protein, and starch degradability of five cereal grains. *Journal of Dairy Science* 73, 2386–2393.

Holter, J.B., Hayes, H.H., Urban, W.E., Ramsey, S. and Rideout, H. (1992) Response of Holstein cows to corn gluten meal used to increase undegradable protein in early or later lactation. *Journal of Dairy Science* 75, 1495–1506.

Hristov, A.N., McAllister, T.A. and Cheng, K.J. (1997) Effect of carbohydrate level and ammonia availability on utilization of α-amino nitrogen by mixed ruminal microorganisms in vitro. *Proceedings, Western Section, American Society of Animal Science* 48, 186–190.

Huhtanen, P. and Hristov, A.N. (2001) Estimating passage kinetics using fibre-bound ^{15}N as an internal marker. *Animal Feed Science and Technology* 94, 29–41.

Huhtanen, P. and Khalili, H. (1992) The effect of sucrose supplements on particle-associated carboxymethylcellulase (*EC* 3.2.1.4) and xylanase (*EC* 3.2.1.8) activities in cattle given grass-silage-based diets. *British Journal of Nutrition* 67, 245–255.

Huhtanen, P. and Nousiainen, J. (2004). Dietary factors influencing milk protein content of cows fed grass silage-based diets. *Journal of Dairy Science* 87 (Suppl. 1), 466.

Huhtanen, P., Jaakkola, S., Kukkonen, U. and Nousiainen, J. (1993) Influence of passage model in estimating ruminal protein degradability. *Journal of Animal Science* 71 (Suppl. 1), 263.

Huhtanen, P., Vanhatalo, A. and Varvikko, T. (1998) Enzyme activities of rumen particles and feed samples incubated in situ with differing types of cloth. *British Journal of Nutrition* 79, 161–168.

Huhtanen, P., Vanhatalo, V. and Varvikko, T. (2002) Effects of abomasal infusions of histidine, glucose, and leucine on milk production and plasma metabolites of dairy cows fed grass silage diets. *Journal of Dairy Science* 85, 204–216.

Hume, I.D., Moir, R.J. and Somers, M. (1970) Synthesis of microbial protein in the rumen. I. Influence of the level of nitrogen intake. *Australian Journal of Agricultural Research* 21, 283–296.

Hungate, R.E. (1966) *The Rumen and its Microbes.* Academic Press, New York, 533 pp.

Hunter, R.A. and Magner, T. (1988) The effect of supplements of formaldehyde-treated casein on the partitioning of nutrients between cow and calf in lactating *Bos indicus* × *Bos taurus* heifers fed a roughage diet. *Australian Journal of Agricultural Research* 39, 1151–1162.

Huntington, G.B. (1997) Starch utilization by ruminants: from basics to the bunk. *Journal of Animal Science* 75, 852–867.

Hvelplund, T. and Weisbjerg, M.R. (2000) In situ techniques for the estimation of protein degradability and postrumen availability. In: Givens, D.I., Owen, E., Axford, R.F.E. and Omed, H.M. (eds) *Forage Evaluation in Ruminant Nutrition.* CAB International, Wallingford, UK, pp. 233–258.

Hvelplund, T., Weisbjerg, M.R. and Andersen, L.S. (1992) Estimation of the true digestibility of rumen undegraded protein in the small intestine of ruminants by the mobile bag technique. *Acta Agriculturae Scandinavica, Section A, Animal Science* 42, 34–39.

Hvelplund, T., Madsen, J. and Weisbjerg, M.R. (2003) Proteinvurdering. In: Hvelplund, T. and Nørgaard, P. (eds) *Kvægets ernæring og fysiologi.* Bind 1 – *Næringsstofomsætning og fodervurdering.* DJF rapport, Husdyrbrug nr. 53, Forskningscenter Foulum, Denmark, pp. 583–602.

INRA [Institut Nationale de la Recherche Agronomique] (1978) Jarrige, R. (ed.) *Alimentation des Ruminants.* Institut National de la Recherche Agronomique, Nouzilly, France, 597 pp.

INRA [Institut National de la Recherche Agronomique] (1989) *Ruminant Nutrition: Recommended Allowances and Feed Tables.* John Libbey Eurotext, Paris, 389 pp.

Jones, D.F., Hoover, W.H. and Miller Webster, T.K. (1998) Effects of concentrations of peptides on microbial metabolism in continuous culture. *Journal of Animal Science* 76, 611–616.

Jouany, J.-P. (1996) Effect of rumen protozoa on nitrogen utilization by ruminants. *Journal of Nutrition* 126, 1335S–1346S.

Jouany, J.P. and Ushida, K. (1999) The role of protozoa in feed digestion. Review. *Asian-Australian Journal of Animal Science* 12, 113–128.

Kajikawa, H., Mitsumori, M. and Ohmomo, S. (2002) Stimulatory and inhibitory effects of protein amino acids on growth rate and efficiency of mixed ruminal bacteria. *Journal of Dairy Science* 85, 2015–2022.

Kalscheur, K.F., Vandersall, J.H., Erdman, R.A., Kohn, R.A. and Russek-Cohen, E. (1999) Effects of dietary crude protein concentration and degradability on milk production responses of early, mid, and late lactation diary cows. *Journal of Dairy Science* 82, 545–554.

Kane, K.K., Creighton, K.W., Petersen, M.K., Hallford, D.M., Remmenga, M.D. and Hawkins, D.E. (2002) Effects of varying levels of undegradable intake protein on endocrine and metabolic function of young post-partum beef cows. *Theriogenology* 57, 2179–2191.

Kang-Meznarich, J.H. and Broderick, G.A. (1981) Effects of incremental urea supplementation on ruminal ammonia concentration and bacterial protein formation. *Journal of Animal Science* 51, 422–431.

Kartchner, R. J. (1980) Effects of protein and energy supplementation of cows grazing native winter range forage on intake and digestibility. *Journal of Animal Science* 51, 432–438.

Kennington, L.R., Hunt, C.W., Hinman, D.D., Szasz, J.I. and Sorensen, S.J. (2002) Effect of added degradable intake protein on performance of cattle fed corn or barley-based finishing diets. *Proceedings, Western Section, American Society of Animal Science* 53, 511–514.

Kertz, A.F., Prewitt, L.R. and Ballam, J.M. (1987) Increased weight gain and effects on growth parameters of Holstein heifer calves from 3 to 12 months of age. *Journal of Dairy Science* 70, 1612–1622.

Kim, C.-H., Choung, J.J. and Chamberlain, D.G. (1999) Determination of the first limiting amino acid for milk production in dairy cows consuming a diet of grass silage and a cereal-based supplement containing feather meal. *Journal of the Science of Food and Agriculture* 79, 1703–1708.

Kim, C.-H., Choung, J.J. and Chamberlain, D.G. (2000) Variability in the ranking of the three most-limiting amino acids for milk protein production in dairy cows consuming grass silage and a cereal-based supplement containing feather meal. *Journal of the Science of Food and Agriculture* 80, 1386–1392.

Kim, C.-H., Kim, T.G., Choung, J.J. and Chamberlain, D.G. (2001a) Effects of intravenous infusion of amino acids and glucose on the yield and concentration of milk protein in dairy cows. *Journal of Dairy Research* 68, 27–34.

Kim, C.-H., Choung, J.J. and Chamberlain, D.G. (2001b) Estimates of the efficiency of transfer of L-histidine from blood to milk when it is the first-limiting amino acid for secretion of milk protein in the dairy cow. *Journal of the Science of Food and Agriculture* 81, 1150–1155.

Klausner, S.D., Fox, D.G., Rasmussen, C.N., Pitt, R.E., Tylutki, T.P., Wright, P.E., Chase, L.E. and Stone, W.C. (1998) Improving dairy farm sustainability. I. An approach to animal and crop nutrient management planning. *Journal of Production Agriculture* 11, 225–233.

Klevesahl, E.A., Cochran, R.C., Titgemeyer, E.C., Wickersham, T.A., Farmer, C.G., Arroquy, J.I. and Johnson, D.E. (2003) Effect of a wide range in the ratio of supplemental rumen degradable protein to starch on utilization of low-quality, grass hay by beef steers. *Animal Feed Science and Technology* 105, 5–20.

Klopfenstein, T.J. and Erickson, G.E. (2002) Effects of manipulating protein and phosphorus nutrition of feedlot cattle on nutrient management and the environment. *Journal of Animal Science* 80 (E. Suppl. 2), E106–E114.

Korhonen, M., Vanhatalo, A., Varvikko, T. and Huhtanen, P. (2000) Responses to graded postruminal doses of histidine in dairy cows fed grass silage diets. *Journal of Dairy Science* 83, 2596–2608.

Korhonen, M., Vanhatalo, A. and Huhtanen, P. (2002) Effect of protein source on amino acid supply, milk production and metabolism of plasma nutrients in dairy cows fed grass silage. *Journal of Dairy Science* 85, 3336–3351.

Köster, H.H., Cochran, R.C., Titgemeyer, E.C., Vanzant, E.S., Abdelgadir, I. and St-Jean, G. (1996) Effect of increasing degradable intake protein on intake and digestion of low-quality, tallgrass-prairie forage by beef cows. *Journal of Animal Science* 74, 2473–2481.

Köster, H.H., Woods, B.C., Cochran, R.C., Vanzant, E.S., Titgemeyer, E.C., Grieger, D.M., Olson, K.C. and Stokka, G. (2002) Effect of increasing proprotion of supplemental N from urea in prepartum supplements on range beef cow performance and on forage intake and digestability by steers fed low-quality forage. *Journal of Animal Science* 80, 1652–1662.

Kropp, J.R., Johnson, R.R., Males, J.R. and Owens, F.N. (1977a) Microbial protein synthesis with low quality roughage rations: isonitrogenous substitution of urea for soybean meal. *Journal of Animal Science* 45, 837–843.

Kropp, J.R., Johnson, R.R., Males, J.R. and Owens, F.N. (1977b) Microbial protein synthesis with low quality roughage rations: level and source of nitrogen. *Journal of Animal Science* 45, 844–854.

Kunkle, W.E. and Hopkins, D.I. (1999) Is methionine the first limiting amino acid for growing cattle fed forages? *Proceedings of the 10th Annual Florida Ruminant Nutrition Symposium*, Florida, pp. 19–29.

Kwan, K., Coppock, C.E., Lake, G.B. Fettman, M.J., Chase, L.E. and McDowell, R.E. (1977) Use of urea by early postpartum Holstein cows. *Journal of Dairy Science* 60, 1706–1724.

Lammers, B.P. and Heinrichs, A.J. (2000) The response of altering the ratio of dietary protein to energy on growth, feed efficiency, and mammary development in rapidly growing prepubertal heifers. *Journal of Dairy Science* 83, 977–983.

Lana, R.P., Fox, D.G., Russell, J.B. and Perry, T.C. (1997) Influence of monensin on Holstein steers fed high-concentrate diets containing soybean meal or urea. *Journal of Animal Science* 75, 2571–2579.

Leonardi, C., Stevenson, M. and Armentano, L.E. (2003) Effect of two levels of crude protein and methionine supplementation on performance of dairy cows. *Journal of Dairy Science* 86, 4033–4042.

Ling, J.R. and Armstead, I.P. (1995) The in vitro uptake and metabolism of peptides and amino acids by five species of rumen bacteria. *Journal of Applied Bacteriology* 78, 116–124.

Littel, R.C., Milliken, G.A., Stroup, W.W., Wolfinger, R.D. (1996) *SAS® System for Mixed Models*. SAS Institute Inc, Cary, North Carolina, p. 633.

Lobley, G.E. (2003) Protein turnover – what does it mean for animal production? *Canadian Journal of Animal Science* 83, 327–340.

Loerch, S.C., Berger, L.L., Ganiola, D. and Fahey, G.C. (1983) Effects of dietary protein source and energy

level on in situ nitrogen disappearance of various protein sources. *Journal of Animal Science* 56, 206–216.
Löest, C.A., Titgemeyer, E.C., Lambert, B.D. and Trater, A.M. (2001) Branched-chain amino acids for growing cattle limit-fed soybean hull-based diets. *Journal of Animal Science* 79, 2747–2753.
Lusby, K.S. (1994) Performance of beef calves supplemented with protein or energy with or without Smartamine-M. *1994 Oklahoma State University Animal Science Research Report* 173–178.
Madsen, J. (1985) The basis for the proposed Nordic Protein Evaluation System for ruminants. The AAT-PBV system. *Acta Agriculturae Scandinavica* (Suppl. 25), 9–29.
Madsen, J. and Hvelplund, T. (1988) The influence of different protein supply and feeding level on pH, ammonia concentration and microbial protein synthesis in the rumen of cows. *Acta Agriculturae Scandinavica* 38, 115–125.
Madsen, J. and Hvelplund, T. (1994) Prediction of in situ protein degradability in the rumen. Results of a European ringtest. *Livestock Production Science* 39, 201–212.
Madsen, J., Hvelplund, T., Weisbjerg, M.R., Bertilsson, J., Olsson, I., Spörndly, R., Harstad, O.M., Volden, H., Tuori, M., Varvikko, T., Huhtanen, P. and Olafsson, B.L. (1995) The AAT/PBV protein evaluation system for ruminants. A revision. *Norwegian Journal of Agricultural Science* (Suppl. 19), 37 pp.
Madsen, J., Misciattelli, L., Kristensen, V.F. and Hvelplund, T. (2003) Proteinforsyning til malkekøer. In: Strudsholm, F. and Sejrsen, K. (eds) *Kvægets ernæring og fysiologi*. Bind 2 – *Fodring og produktion*. DJF rapport, Husdyrbrug nr. 54, Forskningscenter Foulum, pp. 113–132.
Maeng, W.J. and Baldwin, R.L. (1976a) Factors influencing rumen microbial growth rates and yields: effects of urea and amino acids over time. *Journal of Dairy Science* 59, 643–647.
Maeng, W.J. and Baldwin, R.L. (1976b) Factors influencing rumen microbial growth rates and yields: effect of amino acid additions to a purified diet with nitrogen from urea. *Journal of Dairy Science* 59, 648–655.
Maeng, W.J., Van Nevel, C.J., Baldwin, R.L. and Morris, J.G. (1976) Rumen microbial growth rates and yields: effect of amino acids and protein. *Journal of Dairy Science* 59, 68–79.
Martin, B., Coulon, J.B., Chamba, J.F. and Bugaud C. (1997) Effect of milk urea content on characteristics of matured Redlochon cheeses. *Lait* 77, 505–514.
Martín-Orue, S.M., Balcells, J., Vicente, F. and Castrillo, C. (2000) Influence of dietary rumen-degradable protein supply on rumen characteristics and carbohydrate fermentation in beef cattle offered high-grain diets. *Animal Feed Science and Technology* 88, 59–77.
Mathis, C.P., Cochran, R.C., Stokka, G.L., Heldt, J.S., Woods, B.C. and Olson, K.C. (1999) Impacts of increasing amounts of supplemental soybean meal on intake and digestion by beef steers and performance by beef cows consuming low-quality tallgrass-prairie forage. *Journal of Animal Science* 77, 3156–3162.
Mathis, C.P., Cochran, R.C., Heldt, J.S., Woods, B.C., Abdelgadir I.E.O., Olson, K.C., Titgemeyer, E.C. and Vanzant, E.S. (2000) Effects of supplemental degradable intake protein on utilization of medium- to low-quality forages. *Journal of Animal Science* 78, 224–232.
Mathison, G.W. and Milligan, L.P. (1971) Nitrogen metabolism in sheep. *British Journal of Nutrition* 25, 351–366.
McAllan, A.B. (1991) Carbohydrate and nitrogen metabolism in the forestomach of steers given untreated or ammonia treated barley straw diets supplemented with urea or urea plus fishmeal. *Animal Feed Science and Technology* 33, 195–208.
McCarthy, R.D., Klusmeyer, T.H., Vicini, J.L., Clark, J.H. and Nelson, D.R. (1989) Effects of source of protein and carbohydrate on ruminal fermentation and passage of nutrients to the small intestine of lactating cows. *Journal of Dairy Science* 72, 2002–2016.
McCormick, M.E., French, D.D., Brown, T.F., Cuomo, G.J., Chapa, A.M., Fernandez, J.M., Beatty, J.F. and Blouin, D.C. (1999) Crude protein and rumen undegradable protein effects on reproduction and lactation performance of Holstein cows. *Journal of Dairy Science* 82, 2697–2708.
Mehrez, A.Z. (1976) Assessment of the nitrogen requirement for rumen fermentation in sheep. PhD thesis, Aberdeen University, UK, 255 pp.
Mehrez, A.Z. and Ørskov, E.R. (1978) Protein degradation and optimum urea concentration in cereal-based diets for sheep. *British Journal of Nutrition* 40, 337–345.
Mehrez, A.Z., Ørskov, E.R. and McDonald, I. (1977) Rates of rumen fermentation in relation to ammonia concentration. *British Journal of Nutrition* 38, 437–443.
Merry, R.J., McAllan, A.B. and Smith, R.H. (1990) In vitro continuous culture studies on the effect of nitrogen source on rumen microbial growth and fibre digestion. *Animal Feed Science and Technology* 31, 55–64.
Meyer, J.H.F. and Mackie, R.I. (1986) Microbiological evaluation of the intraruminal in sacculus digestion technique. *Applied and Environmental Microbiology* 51, 622–629.
Michalet-Doreau, B. and Ould-Bah, M.Y. (1992) In vitro and in sacco methods for the estimation of dietary nitrogen degradability in the rumen: a review. *Animal Feed Science and Technology* 40, 57–86.

Millet, J. (1989) Incidence de la tenaur en urée du lait et de la standardisation du pH du lait sur la qualité d'une pate pressée cuite: 'Comté'. *Process* 1039, 56–62.

Milton, C.T., Brandt, R.T. and Titgemeyer, E.C. (1997a) Urea in dry-rolled corn diets: finishing steer performance, nutrient digestion, and microbial protein production. *Journal of Animal Science* 75, 1415–1424.

Milton, C.T., Brandt, R.T. and Titgemeyer, E.C. (1997b) Effects of dietary nitrogen source and concentration in high-grain diets on finishing steer performance and nutrient digestion. *Journal of Animal Science* 75, 2813–2823.

Møller, J., Thøgersen, R., Kjedsen, A.M., Weisbjerg, M.R., Søegaard, K, Hvelplund, T. and Børsting, C.F. (2000) Fodermiddeltabelle. Sammansætning og foderværdi af fodermidler til kvæg. *Rapport nr. 91*. Landbrugets Rådgivningscenter.

Morrison, M. and Mackie, R.I. (1996) Nitrogen metabolism by ruminal microorganisms: current understanding and future perspectives. *Australian Journal of Agricultural Research* 47, 227–246.

Mowrey, A., Ellersieck, M.R. and Spain, J.N. (1999) Effect of fibrous by-products on production and ruminal fermentation in lactating dairy cows. *Journal of Dairy Science* 82, 2709–2715.

Nagadi, S., Herrero, M. and Jessop, N.S. (2000) The effect of fermentable nitrogen availability on in vitro gas production and degradability of NDF. *Animal Feed Science and Technology* 87, 241–251.

Noftsger, S. and St-Pierre, N.R. (2003) Supplementation of methionine and selection of highly digestible rumen undegradable protein to improve nitrogen efficiency for milk production. *Journal of Dairy Science* 86, 958–969.

Nolan, J.V. and Leng, R.A. (1972) Dynamic aspects of ammonia and urea metabolism in sheep. *British Journal of Nutrition* 27, 177–194.

Nolan, J.V., Norton, B.W. and Leng, R.A. (1976) Further studies of the dynamics of nitrogen metabolism in sheep. *British Journal of Nutrition* 35, 127–147.

Nousiainen, J., Shingfield, K.J. and Huhtanen, P. (2004) Evaluation of milk urea concentration as a diagnostic of protein feeding. *Journal of Dairy Science* 87, 386–398.

Nozière, P. and Michalet-Doreau, B. (1996) Validation of in sacco method: influence of sampling site, nylon bag or rumen contents, on fibrolytic activity of solid-associated microorganisms. *Animal Feed Science and Technology* 57, 203–210.

Nozière, P. and Michalet-Doreau, B. (2000) *In sacco* methods. In: D'Mello, J.P.F. (ed.) *Farm Animal Metabolism*. CAB International, Wallingford, UK, p. 233.

NRC [National Research Council] (1985) *Ruminant Nitrogen Usage*. National Academy Press, Washington, DC, 138 pp.

NRC [National Research Council] (1989) *Nutrient Requirements of Dairy Cattle*, 6th rev. edn. National Academy Press, Washington, DC, 157 pp.

NRC [National Research Council] (1996) *Nutrient Requirements of Beef Cattle*, 7th rev. edn. National Academy Press, Washington, DC, 242 pp.

NRC [National Research Council] (2001) *Nutrient Requirements of Dairy Cattle*, 7th rev. edn. National Academy Press, Washington, DC, 381 pp.

O'Connor, J.D., Sniffen, C.J., Fox, D.G. and Chalupa, W. (1993) A net carbohydrate and protein system for evaluating cattle diets. IV. Predicting amino acid adequacy. *Journal of Animal Science* 71, 1298–1311.

Ocon, O.M. and Hansen, P.J. (2003) Disruption of bovine oocytes and preimplantation embryos by urea and acidic pH. *Journal of Dairy Science* 86, 1194–1200.

Odle, J. and Schaefer, D.M. (1987) Influence of rumen ammonia concentration on the rumen degradation rates of barley and maize. *British Journal of Nutrition* 57, 127–138.

Okorie, A.U., Buttery, P.J. and Lewis, D. (1977) Ammonia concentration and protein synthesis in the rumen. *Proceedings of the Nutrition Society* 36, 38A (abstract).

Olson, K.C., Cochran, R.C., Jones, T.J., Vanzant, E.S., Titgemeyer, E.C. and Johnson, D.E. (1999) Effects of ruminal administration of supplemental degradable intake protein and starch on utilization of low-quality warm-season grass hay by beef steers. *Journal of Animal Science* 77, 1016–1025.

Oltjen, R.R., Bond, J. and Richardson, G.V. (1969) Growth and reproductive performance of bulls and heifers fed purified and natural diets. I. Performance from 14 to 189 days of age. *Journal of Animal Science* 28, 717–722.

Ørskov, E.R. and McDonald, I. (1979) The estimate of protein degradability in the rumen from incubation measurements weighted according to rate of passage. *Journal of Agricultural Sciences, Cambridge* 92, 499–503.

Patterson, H.H., Klopfenstein, T.J., Adams, D.C. and Musgrave, J.A. (2003) Supplementation to meet metabolizable protein requirements of primiparous beef heifers. I. Performance, forage intake, and nutrient balance. *Journal of Animal Science* 81, 800–811.

Pecorari, M., Mariani, M.S., Calzolari, M.G. and Tedeschi, G. (1993) Urea content in milk: variation and relationships with technological parameters. *Scienza e Tecnica Lattiero-Casearia* 44, 144–154.

Petersen, M.K., Clanton, D.C. and Britton, R. (1985) Influence of protein degradability in range supplements on abomasal nitrogen flow, nitrogen balance and nutrient digestibility. *Journal of Animal Science* 60, 1324–1329.

Pilgrim, A.F., Gray, F.V., Weller, R.A. and Belling, C.B. (1970) Synthesis of microbial protein from ammonia in the sheep's rumen and the proportion of dietary nitrogen converted into microbial nitrogen. *British Journal of Nutrition* 24, 589–598.

Pirlo, G., Capelletti, M. and Marchetto, G. (1997) Effects of energy and protein allowances in the diets of prepubertal heifers on growth and milk production. *Journal of Dairy Science* 80, 730–739.

Pittman, K.A. and Bryant, M.P. (1964) Peptides and other nitrogen sources for growth of *Bacteroides ruminicola*. *Journal of Bacteriology* 88, 401–410.

Pittman, K.A., Lakshmanan, S. and Bryant, M.P. (1967) Oligopeptide uptake by *Bacteroides ruminicola*. *Journal of Bacteriology* 93, 1499–1508.

Podhorsky, M. and Cvak, Z. (1989) The influence of non-protein nitrogen on hygiene and processing properties of milk. *Prumysl Potravin* 40, 83–84.

Pond, K.R., Ellis, W.C., Matis, J.H., Ferreiro, H.M. and Sutton, J.D. (1988) Compartmental models for estimating attributes of digesta flow in cattle. *British Journal of Nutrition* 60, 571–595.

Rajala-Schultz, P.J., Saville, W.J.A., Frazer, G.S. and Wittum, T.E. (2001) Association between milk urea nitrogen and fertility in Ohio dairy cows. *Journal of Dairy Science* 84, 482–489.

Richardson, C.R. and Hatfield, E.E. (1978) The limiting amino acids in growing cattle. *Journal of Animal Science* 46, 740–745.

Rinne, M., Jaakkola, S., Kaustell, K., Heikkilä, T. and Huhtanen, P. (1999a) Silage harvested at different stages of grass growth vs. concentrate foods as energy and protein sources in milk production. *Animal Science* 69, 251–263.

Rinne, M., Jaakkola, S., Varvikko, T. and Huhtanen, P. (1999b) Effects of the type and amount of rapeseed feed on milk production. *Acta Agriculturae Scandinavica, Section A Animal Science* 49, 137–148.

Rinne, M., Huhtanen, P. and Jaakkola, S. (2002) Digestive processes of dairy cows fed silages harvested at four stages of maturity. *Journal of Animal Science* 80, 1986–1998.

Robinson, P.H., Chalupa, W., Sniffen, C.J., Julien, W.E., Sato, H., Watanabe, K., Fujieda, T. and Suzuki, H. (1998) Ruminally protected lysine or lysine and methionine for lactating dairy cows fed a ration designed to meet requirements for microbial and postruminal protein. *Journal of Dairy Science* 81, 1364–1373.

Rohr, K. (1987) Present situation of the modern protein systems: Germany. In: Jarrige, R. and Alderman, G. (eds) *Feed Evaluation and Protein Requirement Systems for Ruminants*. Proceedings of a CEC seminar held in Brussels, June 25–27, 1986. CEC L 2935, Luxembourg, pp. 3–10.

Rulquin, H., Pisulewski, P.M., Vérité, R. and Guinard, J. (1993) Milk production and composition as a function of postruminal lysine and methionine supply: a nutrient-response approach. *Livestock Production Science* 37, 69–90.

Russell, J.B., Sniffen, C.J. and Van Soest, P.J. (1983) Effect of carbohydrate limitation on degradation and utilization of casein by mixed ruminal bacteria. *Journal of Dairy Science* 66, 763–775.

Russell, J.B., O'Connor, J.D., Fox, D.G., Van Soest, P.J. and Sniffen, C.J. (1992) A net carbohydrate and protein system for evaluating cattle diets: I. Ruminal fermentation. *Journal of Animal Science* 70, 3551–3561.

Saarisalo, E., Jaakkola, S. and Huhtanen, P. (2002) Effects of supplementing grass silage with protein on production of primiparous cows in late lactation. In: Durand, J.-L., Emile, J.-C., Huyghe, C. and Lemaire, G. (eds) *Multifunctional Grasslands: Quality Forages, Animal Products and Landscapes. Proceedings of the 19th General Meeting of the European Grassland Federation.* La Rochelle, France, pp. 594–595.

Salter, D.N., Daneshvar, K. and Smith, R.H. (1979) The origin of nitrogen incorporated into compounds in the rumen bacteria of steers given protein- and urea-containing diets. *British Journal of Nutrition* 41, 197–209.

Santos, F.A.P, Santos, J.E.P., Theurer, C.B. and Huber, J.T. (1998) Effects of rumen-undegradable protein on dairy cow performance: a 12-year literature review. *Journal of Dairy Science* 81, 3182–3213.

Satter, L.D. and Slyter, L.L. (1974) Effect of ammonia concentration on rumen microbial protein production in vitro. *British Journal of Nutrition* 32, 199–208.

Schingoethe, D.J., Brouk, M.J., Lightfield, K.D. and Baer, R.J. (1996) Lactational responses of dairy cows fed unsaturated fat from extruded soybeans or sunflower seeds. *Journal of Dairy Science* 79, 1244–1249.

Schingoethe, D.J., Brouk, M.J. and Birkelo, C.P. (1999) Milk production and composition from cows fed wet corn distillers grains. *Journal of Dairy Science* 82, 574–580.

Schwab, C.G., Ordway, R.S. and Whitehouse, N.L. (2003) The latest on amino acid feeding. In: *Proceedings of the Southwest Nutrition and Management Conference*, Phoenix, Arizona, pp. 27–41.

Shain, D.H., Stock, R.A., Klopfenstein, T.J. and Herold, D.W. (1998) Effect of degradable intake protein level on finishing cattle performance and ruminal metabolism. *Journal of Animal Science* 76, 242–248.

Shingfield, K.J., Jaakkola, S. and Huhtanen, P. (2001) Effects of level of nitrogen fertilizer application and various nitrogenous supplements on milk production and nitrogen utilization of dairy cows given grass silage-based diets. *Animal Science* 73, 541–554.

Shingfield, K.J., Vanhatalo, A. and Huhtanen, P. (2003) Comparison of heat-treated rapeseed expeller and solvent-extracted soya-bean meal supplements for dairy cows given grass silage-based diets. *Animal Science* 77, 305–317.

Sletmoen-Olsen, K.E., Caton, J.S., Olson, K.C. and Reynolds, L.P. (2000) Undegraded intake protein supplementation: I. Effects on forage utilization and performance of periparturient beef cows fed low-quality hay. *Journal of Animal Science* 78, 449–455.

Slyter, L.L., Satter, L.D. and Dinius, D.A. (1979) Effect of ruminal ammonia concentration on nitrogen utilization by steers. *Journal of Animal Science* 48, 906–912.

Sniffen, C.J., O'Connor, J.D., Van Soest, P.J., Fox, D.G. and Russell, J.B. (1992) A net carbohydrate and protein system for evaluating cattle diets: II. Carbohydrate and protein availability. *Journal of Animal Science* 70, 3562–3577.

Spicer, L.A., Theurer, C.B., Sowe, J. and Noon, T.H. (1986) Ruminal and post-ruminal utilization of nitrogen and starch from sorghum grain-, corn- and barley-based diets by beef steers. *Journal of Animal Science* 62, 521–530.

Stensig, T. and Robinson, P.H. (1997) Digestion and passage kinetics of forage fiber in dairy cows as affected by fiber-free concentrate in the diet. *Journal of Dairy Science* 80, 1339–1352.

Strobel, H.J. and Russell, J.B. (1986) Effect of pH and energy spilling on bacterial protein synthesis by carbohydrate-limited cultures of mixed rumen bacteria. *Journal of Dairy Science* 69, 2941–2947.

Stokes, S.R., Goetsch, A.L., Jones, A.L. and Landis, K.M. (1988) Feed intake and digestion by beef cows fed prairie hay with different levels of soybean meal and receiving postruminal administration of antibiotics. *Journal of Animal Science* 66, 1778–1789.

Strudsholm, F., Aaes, O., Madsen, J., Kristensen, V.F., Andersen, H.R., Hvelplund, T. and Østergaard, S. (1999) *Danske fodernormer till kvaeg.* Landskontoret for Kvaeg Rapport nr. 84. Landbrugetsukets Rådgivningscenter, 47 pp.

Tamminga, S., Van Straalen, W.M., Subnel, A.P.J., Meijer, R.G.M., Steg, A., Wever, C.J.G. and Blok, M.C. (1994) The Dutch protein evaluation system: the DVE/OEB-system. *Livestock Production Science* 40, 139–155.

Titgemeyer, E.C. and Merchen, N.R. (1990) Sulfur-containing amino acid requirement of rapidly growing steers. *Journal of Animal Science* 68, 2075–2083.

Triplett, B.L., Neuendorff, D.A. and Randel, R.D. (1995) Influence of undegraded intake protein supplementation on milk production, weight gain, and reproductive performance in postpartum Brahman cows. *Journal of Animal Science* 73, 3223–3229.

Tuori, M., Kaustell, K.V. and Huhtanen, P. (1998) Comparison of the protein evaluation systems of feeds for dairy cows. *Livestock Production Science* 55, 3–46.

Tuori, M., Kaustell, K, Valaja, J., Aimonen, E., Saarisalo, E. and Huhtanen, P. (2002) *Rehutaulukot ja ruokintasuositukset [Feed Tables and Feeding Recommendations]*, 3rd edn. Yliopistopaino Helsinki, 88 pp.

Tyrrell, H.F. and Moe, P.W. (1974) Net energy value of a corn and a barley ration for lactation. *Journal of Dairy Science* 57, 451–458.

Vanhatalo, A, Huhtanen, P., Toivonen, V. and Varvikko, T. (1999) Response of dairy cows fed grass silage diets to abomasal infusions of histidine alone or in combinations with methionine and lysine. *Journal of Dairy Science* 82, 2674–2685.

Varvikko, T., Vanhatalo, A., Jalava, T. and Huhtanen, P. (1999) Lactation and metabolic responses to graded abomasal doses of methionine and lysine in cows fed grass silage diets. *Journal of Dairy Science* 82, 2659–2673.

Vérité, R. and Peyraud, J.-L. (1989) Protein: the PDI system. In: Jarrige, R. (ed.) *Ruminant Nutrition. Recommended Allowances and Feed Tables.* INRA, Nouzilly, France, pp. 33–48.

Vérité, R., Journet, M. and Jarrige, R. (1979) A new system for the protein feeding of ruminants: the PDI system. *Livestock Production Science* 6, 349–367.

Virtanen, A.I. (1966) Milk production of cows on protein-free feed. *Science* 153, 1603–1614.

Volden, H. (1999) Effects of level of feeding and ruminally undegraded protein on ruminal bacterial protein synthesis, escape of dietary protein, intestinal amino acid profile and performance of dairy cows. *Journal of Animal Science* 77, 1905–1918.

Volden, H., Mydland, L.T. and Olaisen, V. (2002) Apparent ruminal degradation and rumen escape of soluble nitrogen fractions in grass and grass silage administered intraruminally to lactating dairy cows. *Journal of Animal Science* 80, 2704–2716.

Wallace, R.J. (1979) Effect of ammonia concentration on the composition, hydrolytic activity and nitrogen metabolism of the microbial flora of the rumen. *Journal of Applied Bacteriology* 47, 443–455.

Wallace, R.J. (1996) Ruminal microbial metabolism of peptides and amino acids. *Journal of Nutrition* 126, 1326S–1334S.

Webster, A.J.F. (1992) The metabolizable protein system for ruminants. In: Garnsworthy, P.C., Haresign, E. and Cole, D.J.A. (eds) *Recent Advances in Animal Nutrition 1992.* Butterworth-Heinemann, Oxford, UK, pp. 93–110.

Wheeler, J.S., Lalman, D.L., Horn, G.W., Redmon, L.A. and Lents, C.A. (2002) Effects of supplementation on intake, digestion, and performance of beef cattle consuming fertilized, stockpiled bermudagrass forage. *Journal of Animal Science* 80, 780–789.

Wiley, J.S., Petersen, M.K., Ansotegui, R.P. and Bellows, R.A. (1991) Production from first-calf beef heifers fed a maintenance or low level of prepartum nutrition and ruminally undegradable or degradable protein postpartum. *Journal of Animal Science* 69, 4279–4293.

Williams, J.E., Newell, S.A., Hess, B.W. and Scholljegerdes, E. (1999) Influence of rumen-protected methionine and lysine on growing cattle fed forage and corn-based diets. *Journal of Production Agriculture* 12, 696–701.

Wright, D.E. (1967) Metabolism of peptides by rumen microorganisms. *Applied Microbiology* 15, 547–550.

Wu, Z., Fisher, R.J., Polan, C.E. and Schwab, C.G. (1997) Lactational performance of cows fed low or high ruminally undegradable protein prepartum and supplemental methionine and lysine postpartum. *Journal of Dairy Science* 80, 722–729.

Wu, Z. and Satter, L.D. (2000) Milk production during the complete lactation of dairy cows fed diets containing different amounts of protein. *Journal of Dairy Science* 83, 1042–1051.

Yang, C.-M.J. (2002) Response of forage fiber degradation by ruminal microorganisms to branched-chain volatile fatty acids, amino acids, and dipeptides. *Journal of Dairy Science* 85, 1183–1190.

Zinn, R.A., Borquez, J.L. and Plascencia, A. (1994) Influence of levels of supplemental urea on characteristics of digestion and growth performance of feedlot steers fed a fat-supplemented high-energy finishing diet. *Professional Animal Scientist* 10, 5–10.

Zinn, R.A., Alvarez, E.G., Montano, M.F., Plascencia, A. and Ramirez, J.E. (1998) Influence of tempering on the feeding value of rolled corn in finishing diets for feedlot cattle. *Journal of Animal Science* 76, 2239–2246.

Zinn, R.A., Barrajas, R., Montano, M. and Ware, R.A. (2003) Influence of dietary urea level on digestive function and growth performance of cattle fed steam-flaked barley-based finishing diets. *Journal of Animal Science* 81, 2383–2389.

3 Nitrogen Metabolism in the Rumen

N.D. Walker, C.J. Newbold and R.J. Wallace
Rowett Research Institute, Aberdeen, UK

3.1 Introduction

3.1.1 Overview

Inefficient nitrogen (N) retention by ruminants has been a major concern for animal physiologists, nutritionists and microbiologists for many years. The main source of protein available to the ruminant is derived from the microbial population itself. Undegraded food and microbial protein passes from the rumen into the small intestine where it is further degraded and absorbed. One of the principal areas of research by ruminant nutritionists is the study of protein metabolism with the aim of maximizing protein utilization and increasing the flow of N from the rumen to the small intestine. As we also enter a more environmentally aware era and the impact of intensive farming practices on the environment are becoming apparent, pressure is also mounting to ensure

©CAB International 2005. *Nitrogen and Phosphorus Nutrition of Cattle* (eds E. Pfeffer and A.N. Hristov)

that the amount of N excreted is also reduced. This chapter describes the microbiological and biochemical background of N metabolism in the rumen.

3.1.2 Protein breakdown in the rumen by indigenous microorganisms

Dietary protein breakdown in the rumen is a complex process, which involves many different microorganisms that provide the necessary enzymes to hydrolyse peptide bonds. Protein is hydrolysed, releasing oligopeptides, which are then broken in turn into smaller peptides and finally amino acids (AA), which are deaminated to form ammonia (NH_3). Ammonia can freely diffuse across the rumen wall and is eventually excreted as urea, leading to a loss of N from the system (Leng and Nolan, 1984). Protein breakdown provides these microorganisms with peptides and AA necessary for growth but, especially in intensive production conditions where dietary protein is abundant, it often occurs in excess, contributing to inefficient N retention and utilization of dietary N, and leading to excessive NH_3 production (Leng and Nolan, 1984). High NH_3 production can, in turn, cause problems with environmental pollution and the removal of large amounts of nitrogenous waste. As a result, much effort has been directed at studying the different microorganisms and characterizing the various steps involved in this breakdown process with a view to decreasing wasteful peptide breakdown, thus increasing the flow of N from the rumen to the small intestine and maximizing N utilization.

3.1.3 Ecological importance of nitrogen waste and pollution

Over recent years, concerns have arisen over the increase in the contribution of agriculture, via intensive farming practices, to environmental pollution. Intensive farming practices not only result in the production of large amounts of slurry waste, a mixture of faeces and urine which contains high amounts of water, N, phosphorus (P) and potassium (K), but also contribute to the emission of methane and carbon dioxide, which contribute to the greenhouse effect. Associated with large quantities of effluent are problems with its disposal, odours, formation of NH_3 gas, nitrous oxide and nuisances such as flies, which can also present a health hazard. Attempts are being made to develop strategies to achieve whole farm nutrient balance, ensuring sustainability and reducing environmental pollution by the recycling of organic waste in the form of manure (Van Horn *et al.*, 1996). However, the correct balance of the elements N, P and K in manure must be achieved, otherwise problems can be encountered in either a net loss of nutrients, resulting in depletion of the farm's soils, or an excess, leading to problems of pollution (Van Horn *et al.*, 1996).

Of particular concern is the impact that ruminants have on global N emissions, producing high levels of NH_3 (Jarvis, 1994) and nitrates (Smith and Frost, 2000). The ruminant is relatively inefficient with regard to its N retention. A high proportion of the ruminant's daily N intake is excreted as urine and faeces, and may account for as much as 70% of the daily N consumption (Tamminga and Verstegen, 1996). Dairy cattle are classified as major N polluters in animal husbandry (Castillo *et al.*, 2001b) and it has been estimated that a single 650 kg dairy cow can excrete 116 kg N/year (Smith and Frost, 2000). Of this amount excreted, a high proportion (12%) could be lost by NH_3 volatilization (Lockyer and Whitehead, 1990). In some instances, the increased feeding of protein supplements to improve productivity actually results in lower efficiency of crude protein (CP) utilization, with N being consumed in excess of nutritional requirements. Tomlinson *et al.* (1996) observed a 77% increase in N excretion in lactating dairy cows when CP concentration increased from 120 to 180 g/kg dry matter (DM) with no significant effect on milk N secretion. Therefore, increasing the protein supplement was actually detrimental rather than beneficial in terms of N loss from the animal to the environment. No effect on animal performance or duodenal flow of microbial protein was observed in rapidly growing heifers fed high concentrate diets when CP in the concentrate was reduced to 14% (Devant *et al.*, 2000) and further studies concluded that different protein supplements may have an impact upon the N loss, AA profile and total AA flow reaching the duodenum (Devant *et al.*, 2001).

Strategies are being developed which will maximize N retention and minimize N excretion, both by manipulating the microbial population and by manipulating the nutritional qualities of the diet. Energy and protein sources have been shown to have a significant effect on N excretion in dairy cows (Castillo *et al.*, 2001a,b). Generally, in terms of environmental pollution, urinary N has a greater impact than faecal N. Lowering supplemental protein concentration and feeding diets with low rumen degradable protein (RDP) sources decreased N output in urine (Castillo *et al.*, 2001b). On low-CP concentration diets, a high efficiency of N utilization is achieved, with faeces being the main route of N excretion. As protein supplementation increases, an increase in urinary N occurs until more than 70% of N in excess of animal requirements is excreted in urine, contributing to NH_3 emissions and pollution (Castillo *et al.*, 2001b). A model has recently been developed which will allow the evaluation of different dietary regimes and their effect on the amount of N excretion and on the form in which it is excreted (Kebreab *et al.*, 2002). This model will allow strategies to be developed, which can maximize N utilization and minimize N excretion, and allow accurate measurement of the form in which N is excreted.

3.1.4 Urea recycling

The quality and composition of the diet can have a significant effect upon the rate of protein degradation and efficiency of N and nutrient utilization in the rumen, affecting the microbial population and the amount of urea, which is recycled. This ability to recycle urea helps to augment diets low in N. On low N diets, ruminants are able to transfer urea from the blood to the gastrointestinal (GI) tract where it can supplement the N supply of the ruminal microorganisms and thereby supply the host animal with AA derived from the microbial protein. It has been estimated that the ruminant can recycle up to a maximum of 6 or 24 g N/day for sheep and cattle, respectively (Houpt, 1970). Physiological changes associated with the consumption of diets low in N by ruminants include reduced plasma filtration by the kidney (Leng *et al.*, 1985), increased urea reabsorption from the initial inner collecting ducts of the kidney (Isozaki *et al.*, 1994) and an increased rate of urea clearance into the GI tract (Ford and Milligan, 1970; Kennedy and Milligan, 1980). It has been suggested that the increase in the transfer of urea into the GI tract occurs due to the presence of urea transporters lining the wall of the GI tract, which are differentially expressed in response to changes in dietary N content (Ritzhaupt *et al.*, 1997, 1998).

However, on diets high in N, the efficiency of N utilization is decreased, leading to an overall loss of N from the system. Generally, as protein supplementation increases, the efficiency decreases, with excessive amounts of N being excreted in the form of urinary N (Castillo *et al.*, 2001b). Thus, diets which are low in N and which implement the capacity to recycle N may prove to be a more efficient way to increase N utilization and decrease N excretion.

3.2 The Role of Ruminal Microbes in N Metabolism

Representatives of the ruminal bacteria (Wallace *et al.*, 1997a,b), protozoa (Forsberg *et al.*, 1984; Lockwood *et al.*, 1988) and anaerobic fungi (Wallace and Joblin, 1985) have all been implicated in the breakdown of soluble protein. The ruminal bacteria play the most significant role in protein breakdown; the bacterial fraction exhibits 6 to 10 times higher specific proteinase activity than the protozoal fraction (Brock *et al.*, 1982). Perhaps more unexpectedly, recent work has demonstrated that plant proteinases may also contribute to the breakdown of their own cell protein (see Section 3.2.1.5; Zhu *et al.*, 1999; Wallace *et al.*, 2001).

3.2.1 The catabolic cascade of proteolysis

The first step in the proteolytic cascade is the breakdown of dietary protein to oligopeptides. This step involves a highly variable population of many different proteolytic microorganisms (Falconer and Wallace, 1998), which can interact with each other to breakdown protein in a synergistic manner (Wallace, 1985). The diversity is reflected in the wide spectrum of different protease activities that are observed in rumen contents. Furthermore, inhibitor studies have demonstrated that the majority of proteinases present in rumen contents and mixed ruminal bacteria are cysteine proteases, although serine, aspartate and metalloproteinases are also present (Brock *et al.*, 1982; Kopecny and Wallace, 1982; Prins *et al.*, 1983; Attwood and Reilly, 1996). In addition to this highly variable proteolytic population, dietary factors due to proteinases present in the plant material (Section 3.2.1.5) can lead to differences in the proteolytic activity of the rumen fluid. This high degree of variability is an important point, which has to be considered when evaluating possible mechanisms, which would target ruminal proteinases as a means of decreasing protein breakdown in the rumen. It has been suggested that in addition to targeting the microbial population, plant proteinases may also be targeted, with the alteration of the proteinase activity of grass by breeding or genetic engineering, leading to enhanced protein metabolism in the rumen (Wallace *et al.*, 2001).

3.2.1.1 Bacterial proteolysis

Many ruminal bacteria from many different groups and genera have been shown to possess protease activity (Wallace and Brammall, 1985; Attwood and Reilly, 1995; Wallace *et al.*, 1997a,b). Only a few of the main cellulolytic organisms, *Fibrobacter succinogenes*, *Ruminococcus flavefaciens* and *Ruminococcus albus*, do not appear to possess protease activity or participate in the proteolytic cascade (Wallace *et al.*, 1997a). The most extensively studied ruminal proteolytic bacteria are *Ruminobacter amylophilus*, *Butyrivibrio fibrisolvens*, *Prevotella* spp. and *Streptococcus bovis*. These bacteria play a role in the degradation of soluble protein, acting as primary degraders (Wallace, 1985) and as such can influence the rate of soluble protein breakdown and subsequent loss of N from the rumen in the form of NH_3.

R. amylophilus was one of the first ruminal bacteria from which a proteinase was isolated (Lesk and Blackburn, 1971; Blackburn and Hullah, 1974). Its major proteolytic activity is a cell-associated serine proteinase (Wallace and Brammall, 1985), which has a broad pH optimum (Blackburn, 1968; Lesk and Blackburn, 1971). Although during growth this enzyme is cell-associated, during stationary phases it is released into the medium upon autolysis (Lesk and Blackburn, 1971). Both the cell-associated and the soluble form have been shown to be active against trypsin substrates and are inhibited by trypsin substrate analogues (Lesk and Blackburn, 1971). Some aminopeptidase activity has also been observed in this organism (Blackburn, 1968). Because *R. amylophilus* uses NH_3 as its principal source of N and derives only a small proportion of its cell N from protein, peptides and AA (Hobson *et al.*, 1968; Hullah and Blackburn, 1971), it has been suggested that this highly amylolytic organism produces proteinases in order to degrade structural proteins of cereal grains to allow access to starch granules (Cotta and Hespell, 1986). A similar function has also been assigned to the proteinases of another highly amylolytic organism, *S. bovis* (Griswold *et al.*, 1999a), which, when compared with the other proteolytic organisms, was only recently recognized as an important organism involved in protein breakdown (Russell *et al.*, 1981; Wallace and Brammall, 1985; Attwood and Reilly, 1995). Like *R. amylophilus*, *S. bovis* is able to use NH_3 as a source of N for growth and poorly converts exogenous soluble protein to cell protein (Russell *et al.*, 1981). Therefore, it is logical to assume that this organism produces extracellular proteolytic enzymes for reasons other than providing N for growth. Because *S. bovis* can proliferate in the rumen when animals are fed cereal grain diets which are high in starch and soluble sugars, and also in grazing animals which consume large quantities of fresh herbage high in soluble protein (Hazlewood *et al.*, 1983; Nugent *et al.*, 1983; Attwood and Reilly, 1995), it is regarded as a predominant organism which can have a significant effect upon ruminal proteolysis. Because of the phenotypic similarity observed between the proteolytic activities of several strains of *S. bovis* and its numerical abundance under different dietary regimes, it has been suggested that this organism is a prime candidate to target in ruminal proteolysis control strategies (Griswold *et al.*, 1999a).

Proteinase activity in *S. bovis* is predominantly cell-bound (Wallace and Brammall, 1985; Attwood and Reilly, 1996) and consists of a mixture of serine and cysteine proteases which exhibit high leucine aminopeptidase activity (Russell and Robinson, 1984; Wallace and Brammall, 1985; Attwood and Reilly, 1996). In addition to cell-bound activity, extracellular proteolytic activity has also been observed. Griswold *et al.* (1999a) characterized the extracellular proteolytic activity of several different strains of *S. bovis* and found a high molecular weight serine protease present in the culture medium which displayed a high degree of uniformity and phenotypic similarity between different strains. Expression of this serine proteinase was constitutive, but was found to be influenced by the composition and N source of the growth medium; casein gave the highest proteolytic activity, which was decreased upon the addition of exogenous carbohydrates and peptides to the growth medium. The addition of NH_3 and AA to the growth medium has also been shown to affect the total proteolytic activity exhibited by *S. bovis* (Sales *et al.*, 2000; Sales-Duval *et al.*, 2002), affecting the expression of both the cell-bound and extracellular proteinases. The addition of NH_3 and AA significantly decreases the cell-bound proteolytic activity, and increasing AA content also leads to a minor decrease in the extracellular proteolytic activity. It was concluded that this decrease in proteolytic activity was induced by changes in the endopeptidasic activities as a result of the simultaneous uptake of NH_3 and small peptides (Sales-Duval *et al.*, 2002).

Different strains of *B. fibrisolvens* display highly variable rates and types of proteolytic activity (Cotta and Hespell, 1986; Attwood and Reilly, 1996). These strain differences may reflect the phenotypic and phylogenetic diversity of this group of bacteria (Forster *et al.*, 1996; Willems *et al.*, 1996; Kopecny *et al.*, 2001). Serine, cysteine and metalloproteinase activities have been measured in several different isolates of *B. fibrisolvens* (Wallace and Brammall, 1985; Attwood and Reilly, 1996), with activity against several different synthetic proteinase substrates, including the chymotrypsin substrate *N*-succinyl alanine alanine phenylalanine proline *p*-nitroanilide (NSAA-PPPNA) and the leucine aminopeptidase substrate [leucine *p*-nitroanilide (LPNA)], being observed (Attwood and Reilly, 1996). Most highly proteolytic strains exhibit serine proteinase activity (Wallace and Brammall, 1985; Cotta and Hespell, 1986; Strydom *et al.*, 1986; Attwood and Reilly, 1996). The majority of proteolytic activity measured in these strains of *B. fibrisolvens* is extracellular rather than cell-associated (Wallace and Brammall, 1985; Cotta and Hespell, 1986), although some strains have been isolated which display a cell-associated proteolytic activity (Attwood and Reilly, 1996). Generally, strains which have high proteolytic activity release their proteinases into the medium (Wallace and Brammall, 1985; Cotta and Hespell, 1986; Falconer and Wallace, 1998), whereas those with low activity release only a very small proportion into the medium (Attwood and Reilly, 1996). Proteolytic activity in *B. fibrisolvens* is constitutive, but is stimulated upon the addition of either exogenous NH_3 or AA (Cotta and Hespell, 1986; Sales *et al.*, 2000). Further work has concluded that this increase in proteolytic activity was due to a better balance in the expression of serine, cysteine and metalloproteases by this organism (Sales-Duval *et al.*, 2002). Even though this organism under certain dietary conditions can be numerically abundant and can be enriched when more resistant types of protein such as albumin are present in the diet (Wallace *et al.*, 1987), because it exhibits such a diverse proteolytic profile and diversity at the genus level, it would not be a suitable candidate for targeting as a means to decrease protein breakdown in the rumen.

The *Prevotella* spp. form one of the predominant groups of proteolytic organisms (Wallace *et al.*, 1997a). Genetically diverse (Avgustin *et al.*, 1994, 1997), they are numerically abundant on both all roughage and mixed roughage–concentrate diets (Wallace *et al.*, 1997b). In some instances, the *Prevotella* spp. can comprise more than 60% of the bacterial population (Van Gylswyk, 1990). Early studies indicated that *Prevotella* spp. possessed a wide variety of cell-associated proteinases (Hazlewood *et al.*, 1981; Wallace and Brammall, 1985; Attwood and Reilly, 1996), which are released into the extracellular medium during the stationary phase of growth (Lesk and Blackburn, 1971; Hazlewood *et al.*, 1981). The proteolytic profile of *Prevotella* spp. most closely resembles that of whole rumen contents, being affected by trypsin-like inhibitors, metallo-, cysteine and serine protease inhibitors (Hazlewood and Edwards, 1981; Wallace and Brammall, 1985), reflecting the high degree of diversity within these species. These organisms have recently been reclassified into four different subgroups on the basis of their 16S ribosomal deoxy ribonucleic acid (rDNA) data and phenotypic diversity (Avgustin *et al.*, 1994, 1997) – *Prevotella bryantii* (type strain B_14), *Prevotella ruminicola* (type strain 23), *Prevotella brevis* (type strain GA33) and *Prevotella albensis* (type strain M384). Subsequent studies have shown that within the group of *Prevotella* spp. there is a large variation in the type, size and number of different proteinases present in each subgroup (Griswold *et al.*, 1999b). *P. albensis* produced low molecular weight metalloproteinases, whereas *P. bryantii* produced a single high molecular weight metalloprotease. *P. ruminicola* produced one cysteine protease and two metalloproteases which differed in size to those found in *P. albensis* and *P. bryantii. P. brevis* produced a wide variety of different cysteine, serine and metalloproteases. In addition to very diverse proteolytic profiles, these organisms exhibited extreme variation in their growth rate and total proteolytic activity with different N sources (Griswold *et al.*, 1999b), leading to the suggestion that each played a different role in the proteolytic cascade in the rumen, with *P. bryantii* being involved in both protein and peptide breakdown whereas *P. brevis* was primarily involved in peptide breakdown (Griswold *et al.*, 1999b). Kirk *et al.* (2000) demonstrated that the growth medium had an effect on the expression of proteolytic activity of *P. bryantii* with an induction of proteolytic activity in response to a decrease in N availability. The N source of the growth medium also affects the proteolytic activity of *P. albensis*, with the addition of exogenous NH_3 or AA leading to a decrease in proteolytic activity (Sales *et al.*, 2000; Sales-Duval, 2002). The inhibition of proteolysis by free AA was believed to be linked to the decrease in endo- and exopeptidases and the specialization of cell-associated endopeptidases, and the decrease in proteolysis associated with increasing NH_3 supplementation was believed to be due to decreasing aminopeptidase activity which coincided with increased dipeptidyl peptidase (DPP) activity (Sales-Duval *et al.*, 2002). Because the *Prevotella* exhibit such a large phenotypic and genetic diversity, it has been concluded that targeting protein metabolism at the level of proteinase activity by this organism would prove to be problematic.

Although *Prevotella* spp., *B. fibrisolvens* and *S. bovis* are believed to be the major organisms involved in protein breakdown due to their presence at relatively high numbers in the rumen (Stewart *et al.*, 1997), several other organisms have been isolated which also display proteolytic activity but which have been dismissed as playing a significant role in the proteolytic cascade due to their small population size. Included in this group are several *Eubacterium* strains which have mainly cell-associated serine and metallo-type proteases which show high activity against the chymotrypsin substrate NSAAPPPNA (Wallace and Brammall, 1985; Attwood and Reilly, 1995, 1996), and a *Clostridium* sp. which exhibited predominantly serine proteinase activity but also cysteine and metalloprotease activity (Attwood and Reilly, 1995; Attwood *et al.*, 1996). In addition *Lachnospira*, *Selenomonas*, *Succinivibrio* and *Fusobacterium* species have all been isolated which displayed proteolytic activity (Wallace and Brammall, 1985; Wallace *et al.*, 1997b).

The rate of protein breakdown by two of the major proteolytic organisms *S. bovis* and *Prevotella* is regulated by N source and availability. Regulation of proteolytic activity by the amount and availability of N source has already been extensively studied in the dairy starter organisms *Lactococci*, and it has been shown that proteolysis in these organisms was repressed by the end products of casein breakdown (Meijer *et al.*, 1996). This has also been found to be the case for several other rumen and colonic bacteria, where NH_3 inhibits proteolysis (Cotta and Hespell, 1986; Gibson and MacFarlane, 1988). By being able to regulate the expression of proteins involved in N metabolism in rapid response to environmental stimuli means these organisms are placed at a competitive advantage over other organisms in the same ecosystem.

Because the breakdown of dietary protein involves such a complex and diverse array of different enzymes and organisms, this first step in the proteolytic cascade would be a difficult step to target as a means of decreasing protein breakdown in the rumen. The proteolytic population is also highly variable and is influenced by dietary factors and can even vary between animals on the same diet (Falconer and Wallace, 1998). This high degree of variability is an important point, which has to be considered when evaluating possible mechanisms for decreasing protein breakdown. As a result, more effort has been concentrated on the following steps in the breakdown pathway, the breakdown of oligopeptides to dipeptides, which involves only a few different microorganisms, and the subsequent hydrolysis of dipeptides and AA, the last of which is carried out in part by a specialist microbial population.

3.2.1.2 *Bacterial breakdown of oligopeptides*

Only a few species of microorganisms are involved in the breakdown of oligopeptides to dipeptides, with *Prevotella* spp. playing a principal role in this catabolic process (Fig. 3.1; Wallace and McKain, 1991). Evidence to support this group as being the main organisms involved in oligopeptide breakdown is provided by the manner in which they break down these small peptides. *Prevotella* spp. exhibit an intracellular DPP activity, which cleaves a dipeptide from the N-terminus

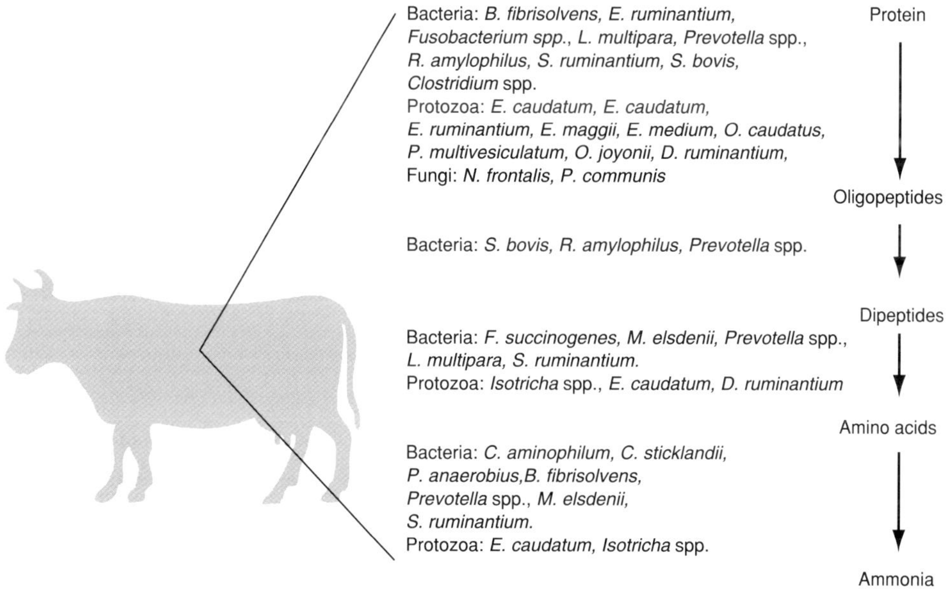

Fig. 3.1. Microorganisms involved in the catabolic sequence from protein to ammonia in the rumen.

of the oligopeptide in a sequential manner until only dipeptides and AA remain (Wallace and McKain, 1991). This is the main mechanism by which oligopeptides are seen to be broken down by mixed rumen contents (Wallace and McKain, 1989; Depardon *et al.*, 1995, 1996) and a survey of laboratory cultures (Wallace and McKain, 1991) and selective isolations from rumen fluid (McKain *et al.*, 1992) indicated that the only predominant ruminal species to possess this DPP activity were *Prevotella* spp. It should be noted that the other two organisms involved in the breakdown of oligopeptides (Fig. 3.1), *S. bovis* and *R. amylophilus*, like the majority of bacteria, exhibit a different type of peptidase activity. These two organisms cleave off a single AA from the N-terminus of the peptide by an enzyme termed aminopeptidase (E.C. 3.4.11; Wallace and McKain, 1991). Aminopeptidase activity is not predominant in rumen contents (Wallace and McKain, 1989) even though it is the main mechanism by which most bacteria break down peptides. Therefore, it was concluded that these two organisms do not play as significant a role as *Prevotella* spp. in the breakdown of oligopeptides, and would only come into effect under dietary conditions in which they were the predominant bacterial species present and the *Prevotella* spp. had been eliminated. Due to the limited number of organisms involved, this part of the proteolytic pathway would seem a likely step to target as a means of decreasing wasteful protein breakdown.

As mentioned previously, the *Prevotella* spp. are phenotypically and genetically diverse. However, all exhibit DPP activity, although the relative activities against different substrates differed markedly between the different subgroups, which may reflect different ecological niches within the rumen ecosystem (Avgustin *et al.*, 1997). Because *P. albensis* exhibited the highest DPP type I activity, reflecting what is observed in mixed rumen contents, we have focused on oligopeptide breakdown by this organism.

OLIGOPEPTIDE BREAKDOWN IN *P. ALBENSIS* M384. Ion exchange chromatography (IEC) of sonicated cell extracts of *P. albensis* M384 indicated that there are at least four different DPP activities present in this organism (Wallace *et al.*, 1997a). Although these enzymes cleave oligopeptides in the same manner, they have different substrate specificities and characteristics. The majority of bacteria display a different type of peptidase activity to what is observed in *P. albensis*. Instead of cleaving a dipeptide from the N-terminus, they remove a single AA by an amino-acyl aminopeptidase. Only a few microorganisms display DPP activity (Chan *et al.*, 1985; Atkinson *et al.*, 1995; Ogasawara *et al.*, 1996a,b, 1997). The majority of these only possess DPP type IV that recognizes peptides of the form X-Pro-X-X and cleaves the bond between the proline (Pro) and the third AA (Abiko *et al.*, 1985; Atlan *et al.*, 1990; Booth *et al.*, 1990a,b; Lloyd and Pritchard, 1991; Nardi *et al.*, 1991; Meyer-Barton *et al.*, 1993). Only mammalian systems have also been shown to possess four different types of DPP activity (McDonald and Barrett, 1986). These have similar substrate specificities but differ slightly in their characteristics when compared with those found in *P. albensis*.

P. albensis DPP type I is similar to the mammalian DPP-I (cathepsin C, E.C. 3.4.14.1), in terms of its substrate and inhibitor specificity. Wallace *et al.* (1997a) demonstrated that it was inhibited by iodoacetate and ethylene diamine tetraacetic acid (EDTA), suggesting that sulphhydryl groups and metal ions were important for catalysis, but was also affected, to some extent, by serine protease inhibitors. However, unlike its mammalian counterpart, *P. albensis* type I was oxygen-sensitive and readily hydrolysed substrates containing the basic AA arginine (Arg) in the N-terminal position (Wallace *et al.*, 1997a), which the mammalian DPP-I does not (McDonald *et al.*, 1969a,b; McGuire *et al.*, 1992). There is a cysteine protease, gingipain, in another oral bacterium, *Porphyromonas gingivalis*, that has similar substrate specificities and inhibition profile to the DPP-I of *P. albensis* (Chen *et al.*, 1992; Madeira *et al.*, 1997). However, this enzyme is an extracellular endopeptidase that is different from the intracellular exopeptidase activity of *P. albensis*. Both these organisms belong to the *Cytophaga-Flexibacter-Bacteroides* (CFB) phylum, cluster within the same group (Ramsak *et al.*, 2000) and are closely related. Therefore, certain similarities would be expected between the two organisms, and information from one may give insight to peptide breakdown mechanisms in the other.

P. albensis type II, similar to mammalian DPP-II (E.C. 3.4.14.2), hydrolysed lysine alanine 4-methoxy-2-nitroanilide (LysAlaMNA) but was unaffected by the mammalian DPP-II serine protease inhibitors and, like DPP-I, was only affected by iodoacetate and EDTA, indicating a dependence on metal ions for its activity (Wallace *et al.*, 1997a). This would indicate that *P. albensis* DPP-II is a metalloprotease instead of a serine protease like its mammalian counterpart. Although mammalian DPP-II has broad specificity, it also has a preference for both alanine (Ala) and proline (Pro) in the P1 position and for lysine (Lys) at the terminal P2 position (McDonald *et al.*, 1969a,b; McDonald, 1998), whereas the *P. albensis* type II did not tolerate Pro in the P1 position. Ogasawara *et al.* (1996b) have also identified another DPP in *Pseudomonas* that is able to hydrolyse LysAlaMNA and a variety of other substrates including GlyPhepNA and Ala_2pNA, indicating broad specificity. This would appear to be different to the type II in *P. albensis*, which had separate activity against LysAlaMNA and Ala_2pNA (Wallace *et al.*, 1997a). The hydrolysis of GlyPhepNA was not tested in *P. albensis*. The enzyme from the *Pseudomonas* spp. was also a serine peptidase, unlike that of *P. albensis*. Again this highlights not only differences between mammalian and microorganism systems, but also between individual species of bacteria.

The third type of DPP, type III, does not have a mammalian counterpart and was specific for the hydrolysis of alanine-containing peptides, Ala_4, Ala_5 and the synthetic substrates Ala_2pNA and ValAlapNA, and was strongly affected by serine protease inhibitors (Wallace *et al.*, 1997a). This enzyme appeared to be very specific and only recognized substrates that had an Ala in the P1 position. Mammalian DPP-III (E.C. 3.4.14.4) is also very specific and only hydrolyses dipeptidyl arylamides of the form Arg-Arg-X-X, although some oligopeptides containing Ala, Lys and leucine (Leu) are slightly susceptible to breakdown (Lee and Snyder, 1982). Like the *P. albensis* type III, it is also a serine protease but there the similarities end. *P. albensis* type III would appear to be unique to this organism (Wallace *et al.*, 1997a) and as a result would be a suitable candidate to target for inhibition studies because an inhibitor specific for this enzyme would only affect this enzyme. Because of this enzyme's preference for alanine-containing peptides, it has been referred to as Ala-DPP. This enzyme has been partially purified (Kim *et al.*, 2001).

The final type, which has been identified so far, is DPP type IV, which is very similar to its mammalian counterpart DPP-IV (E.C. 3.4.14.5). This enzyme has recently been cloned from *P. albensis*, sequenced and expressed in *Escherichia coli* (Walker *et al.*, 2003). As expected, this enzyme hydrolysed the test substrates GlyPropNA and GlyProMNA and was strongly inhibited by serine protease inhibitors and diprotin A (Ile-Pro-Ile), which is believed to act as a competitive inhibitor. This enzyme, like all of the other DPP-IVs identified to date, was a serine protease and had a preference for X-Pro-X-X peptides, but could also hydrolyse X-Ala-X-X peptides. Thus, DPP-IV would appear to be the most highly conserved DPP, with several different bacteria (Atlan *et al.*, 1990; Mineyama and Saito, 1991; Nardi *et al.*, 1991), yeast (Bordallo *et al.*, 1984) and fungi (Tachi *et al.*, 1992) already being described with similar substrate specificity, catalytic characteristics and inhibitor specificity to the mammalian DPP-IV, and with similar sequence identity.

A better understanding and characterization of the DPP present in *P. albensis* will allow for their possible manipulation. Potential specific inhibitors of the four different DPPs present in *P. albensis* have already been identified which are structural analogues of dipeptides (Wallace *et al.*, 2001, 2003). It is hoped that it will be possible to reduce peptide breakdown by this organism using these as potential novel feed additives.

3.2.1.3 Breakdown of small peptides to amino acids

The next step in the breakdown of dietary protein is the breakdown of dipeptides and tripeptides to AA. In contrast to the previous step, that only involved three different species of bacteria, several different bacteria and protozoa exhibit dipeptidase activity (Fig. 3.1). The main bacterial species are *Prevotella* spp., *Megasphaera elsdenii*, *F. succinogenes* and *Lachnospira multipara*. All were inhibited by 1,10-phenanthroline, a chelator of divalent metal ions, indicating that all of these dipeptidases were metalloenzymes (Wallace *et al.*, 1996). The ciliate protozoa also exhibit a high metal-dependent dipeptidase activity (Wallace *et al.*, 1990a, 1996) and, therefore, like the majority of dipeptidases are found in other systems (Lazdunski, 1989). Those found in the rumen are metallopeptidases. It has been demonstrated that in the mixed microbial population the protozoa are mainly responsible for dipeptide breakdown (Wallace *et al.*, 1996). However, even in a defaunated animal that has had all the protozoa removed, there is no obvious difference in dipeptidase activity, indicating that the bacteria take over the niche previously occupied by the protozoa. Thus, defaunating agents would have little effect on the reduction of dipeptide breakdown. Metal chelators are also very non-specific in the enzymes and cell processes they target and as a result more specific dipeptidase inhibitors would have to be designed before this step could be affected (Wallace and McKain, 1996).

In addition to the breakdown of protein and oligopeptides the *Prevotella* play a significant role in the breakdown of di- and tripeptides. The final step in the breakdown of protein to AA is the cleavage of di- and tripeptides. Tripeptidase activity would appear to be not as important as dipeptidase activity because tripeptides are only formed as the carbon (C)-terminal product formed from the breakdown of odd-numbered peptides (Wallace and McKain, 1991; Wallace *et al.*, 1993a, 1995). These tripeptides are then broken down further by a single AA being cleaved off the N-terminus, resulting in the formation of a free AA and a dipeptide, which is broken down in turn by the dipeptidase (Wallace *et al.*, 1995). Thus the dipeptidase plays an essential role in the formation of free AA. Both the dipeptidase and the tripeptidase have been shown to be non-specific in the type of peptides they break down with similar rates of hydrolysis obtained for a wide variety of substrates (Broderick *et al.*, 1988; Wallace *et al.*, 1995).

Anion exchange chromatography of sonicated cell extracts of *P. albensis* has demonstrated that di- and tripeptidase activities are quite distinct and separate from DPP activity in this organism, occurring in different fraction peaks (Wallace *et al.*, 1995). However, like DPP activity, di- and tripeptidase activities are located in the cytoplasm, and it has been suggested that transport limits the rate of their hydrolysis (Wallace *et al.*, 1995). Protonophores, ionophores or dicyclohexylcarbodiimide (DCCD), an ATPase inhibitor, had no effect on the hydrolysis of the dipeptide, Ala_2, and the tripeptide, Ala_3, in whole cells. Only 1,10-phenanthroline and EDTA inhibited the di- and tripeptidases, consistent with them being metalloproteases (Wallace *et al.*, 1995). Banding patterns on native gels visualized using activity staining showed that separate single di- and tripeptidase activities were observed which hydrolysed several different substrates (Wallace *et al.*, 1995). GlyPro activity could not be measured by the activity staining technique and, therefore, it was not determined whether *P. albensis*, like many other organisms, possessed a dipeptidase activity that was specific for X-Pro dipeptides. It would, however, be likely that this organism would possess a specialized enzyme for the hydrolysis of proline-containing dipeptides that would be readily formed by the DPP-IV enzyme from the hydrolysis of X-Pro oligopeptides. Rf values obtained for the tripeptidase were the same for the four different *Prevotella* type strains but differences were observed for the dipeptidase (Wallace *et al.*, 1995). Gel permeation chromatography (GPC) indicated that the M_r of the dipeptidase of *P. albensis* M384 was 115,000 and the M_r of the tripeptidase was 112,500. This would place them at a similar size as already identified in other di- and tripeptidases in other organisms. A peptidase-deficient strain of *E. coli* was successfully complemented with DNA from *P. albensis* and sequence analysis of the clone identified an open reading frame (ORF) which shared identity to PepD, a broad specificity dipeptidase found in several prokaryotes (Walker *et al.*, unpublished results).

3.2.1.4 Ammonia formation from amino acids

The final step in the breakdown of dietary protein is the breakdown of AA to NH_3. There is very little free AA present in rumen fluid, even 1 h after feeding (Leibholz, 1969) and the little amount present is found intracellularly (Wallace, 1979). Any free AA is rapidly deaminated, and the C-skeletons are metabolized to give a variety of short chain fatty acids (scFA) (Blackburn, 1965; Allison, 1970). The amount of free AA present in rumen fluid varies between different diets, lucerne hay giving the greatest accumulation (Leibholz, 1969). Rapidly degradable protein in the diet also leads to a greater accumulation of free AA (Broderick and Wallace, 1988). These differences in diet may reflect differences in the microbial population. The breakdown of AA involves several different species of bacteria and protozoa (Fig. 3.1).

For many years, it was assumed that NH_3 formation was carried out by some of the most numerous species of ruminal bacteria that had been identified to produce NH_3 weakly from protein or protein hydrolysates (Bladen *et al.*, 1961). Organisms, which have already been shown to play a role in other steps in the proteolytic cascade, also play a role in the breakdown of AA. *Prevotella* spp., *B. fibrisolvens*, *S. ruminantium*, *S. bovis* and *M. elsdenii* all deaminate AA, some exhibiting a selective preference for certain AA, others breaking down a substantial proportion of all AA present in the growth medium (Scheifinger *et al.*, 1976). Different strains of *M. elsdenii* display different rates of deaminase activity (Rychlik *et al.*, 2002). All strains tested were resistant to monensin and although they could not utilize branched chain AA as an energy source for growth, they were able to rapidly deaminate these AA (Rychlik *et al.*, 2002).

However, Russell and his colleagues (Chen and Russell, 1988, 1989, 1990; Russell *et al.*, 1988, 1991) calculated that these bacteria, though numerous, did not have sufficient activity to account for observed *in vitro* rates of NH_3 production by the mixed population in their cattle. Selective enrichments were carried out with Trypticase as sole source of energy, and bacteria were isolated which grew on Trypticase as sole source of energy. They were much less numerous than the others, comprising about 1% of the population, but they had a specific activity of NH_3 production from Trypticase which was an order of magnitude greater than that of the other species. These bacteria, unlike the more numerous species, were Gram-positive and highly sensitive to the feedlot ionophore, monensin. Ruminal NH_3 concentrations are lower when ruminants receive this ionophore, so it was deduced that the new species must be significant NH_3 producers *in vivo*. The species isolated were identified as *Peptostreptococcus anaerobius*, *Clostridium sticklandii* and *Clostridium aminophilum* (Paster *et al.*, 1993). They were non-saccharolytic but able to grow rapidly on Trypticase and were different from the most numerous ruminal species. However, even at a low population size they had deaminative activity sufficient to contribute significantly to NH_3 production by the mixed population *in vivo*.

In subsequent studies, similar bacteria – so-called 'ammonia-hyperproducing' (HAP) – were isolated in New Zealand (Attwood *et al.*, 1998), Australia (McSweeney *et al.*, 1999) and Great Britain (Eschenlauer *et al.*, 2002). In the first study, Attwood *et al.* (1998) isolated 14 morphologically different species from pasture-grazed cows, sheep and deer. The isolates were similar to the Russell *et al.* (1988) HAP species, but all were genotypically different. A greater diversity of HAP species than the original isolates was again indicated in the isolates made from goats receiving tannins-rich *Calliandra calothyrsus* (McSweeney *et al.*, 1999).

Some of the isolates were saccharolytic and/or proteolytic, suggesting that the HAP niche is not occupied only by asaccharolytic organisms, but also by organisms with wider metabolic functions. In the British study (Eschenlauer *et al.*, 2002) done with sheep, 1.4% of the total bacterial population grew on Trypticase alone, of which 93% were eliminated by monensin. Nineteen isolates were capable of growth on Trypticase, which fell into six phylogenetic groups. All were sensitive to monensin, and almost all were rapid NH_3 producers. The most abundant HAP species were most closely related to asaccharolytic ruminal and oral *Clostridium* and *Eubacterium* spp. (Fig. 3.2). Others included bacteria phylogenetically related to *Desulfomonas piger* (*pigra*) and *Acidaminococcus fermentans*.

Rates of NH_3 production in rumen fluid appear to vary greatly depending on diet and it is not always necessary to invoke the activity of the high-activity NH_3 producers to explain observed rates of NH_3 production by the mixed rumen population (Wallace, 1996). Published values for rates of NH_3 production vary considerably, because of possible dietary, species and methodological differences. The batch-culture type of incubation of mixed ruminal digesta, in which the bacteria grow exponentially, used in some studies to calculate NH_3 production gives higher apparent rates than shorter incubations with lower concentrations of substrate, which is probably more similar to the conditions found *in vivo* (Eschenlauer *et al.*, 2002).

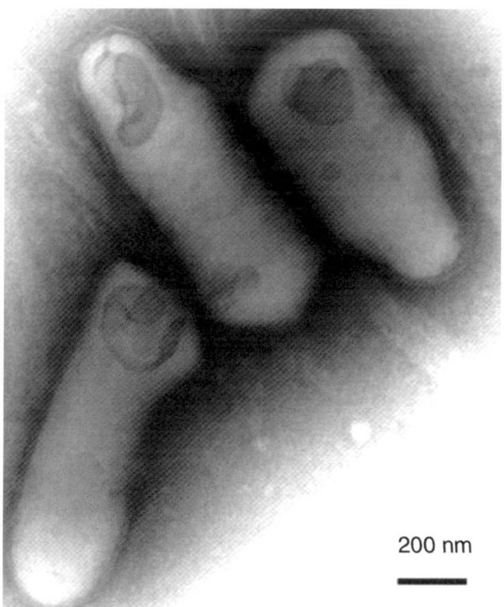

Fig. 3.2. *Eubacterium pyruvativorans*, one of the new isolates of non-saccharolytic, amino acid-fermenting bacteria isolated from the rumen (Wallace *et al.*, 2003).

Therefore, the HAP group of bacteria plays an important role in the ruminal fermentation of peptides and AA and may exist in higher numbers and more diverse species than first estimated. Dietary factors may also play a role in the numerical abundance of these bacteria; in animals fed a hay diet the population of HAP organisms is four times higher than in animals fed a grain diet (Rychlik and Russell, 2000). Evidence to support the importance of HAP organisms in the breakdown of protein comes from studies in which monensin-treated animals showed a significant decrease in ruminal deamination and the production of NH_3. It is believed that only the monensin-sensitive HAP bacteria were affected, leading to a decrease in NH_3 production (Attwood *et al.*, 1998). Sixteen S rRNA probes have demonstrated that the addition of monensin to the diet reduced the population of the initial HAP organisms isolated, *P. anaerobius* and *C. sticklandii* but not *Clostridium anaerobius* (Krause and Russell, 1996). It was concluded that because *C. aminophilum* displayed monensin resistance *in vivo* which was not observed *in vitro* batch culture, the use of monensin as a means of inhibiting the growth of all obligate AA deaminating organisms could not be entirely guaranteed (Krause and Russell, 1996). With the isolation of more diverse HAP organisms, further studies using probes based on these recently isolated HAP bacteria will have to be conducted to determine whether these are also affected by the addition of monensin to the diet. Generally, the bacteria which are recognized as low activity NH_3 producers, *M. elsdenii* and the *Prevotella* spp. are resistant or are able to adapt to growth in the presence of physiological concentrations of monensin (Newbold *et al.*, 1992; Rychlik *et al.*, 2002). The numbers of these deaminative organisms would also have to be measured in response to the addition of monensin to the diet.

In addition to sensitivity to monensin, some of these HAP organisms have also been shown to be sensitive to the bacteriocin nisin (Callaway *et al.*, 1997) and a bacteriocin produced by a ruminal strain of *B. fibrisolvens* (Rychlik and Russell, 2002).

These bacteriocins are small pore-forming peptides that cause a decrease in cell viability by causing the efflux of intracellular ions and subsequent ATP depletion. It has been suggested that these bacteriocins may be useful in controlling the population of these obligate AA fermenting bacteria (Rychlik and Russell, 2002).

The protozoa also play a significant role in deamination (see Section 3.2.5.3 and all of Section 3.3).

In the mixed population of the rumen, different AA are deaminated at different rates, and the extent of incorporation of AA vs. their deamination depends on many factors, including the availability of carbohydrate and the concentrations of AA. When presented at low concentrations (10 mg/l) of radiolabelled AA in ruminal fluid, only small quantities of leucine, tyrosine and phenylanine were labelled in microbial protein (Armstead and Ling, 1993). The other AA were catabolized. In incubations with much higher concentrations (5 g/l), concentrations of AA declined, due partly to incorporation and partly to catabolism (Atasoglu *et al.*, 2003a). Some AA were incorporated with little being degraded. These AA included isoleucine, phenylalanine, lysine and to a lesser extent leucine. The low catabolism of these acids is consistent with their C-skeleton being required, either as the AA or as precursor short and branched chain fatty acids (see Section 3.2.3.2). Others, particularly members of the glutamate family, were catabolized much more extensively, such that glutamate breakdown was almost as great as its incorporation. The minimal catabolism of these AA is not obvious from simple AA concentrations at the beginning and end of the incubation (Atasoglu *et al.*, 2003), nor was it evident from the analyses carried out by Chalupa (1976) or Broderick and Balthrop (1979), which measured only AA loss and did not distinguish between catabolism and incorporation.

3.2.1.5 *Plant proteinase activity*

Virtually all the early work on proteinase activity in the rumen made the assumption that the proteolytic enzymes were derived from the microorganisms. Early comprehensive reviews (Blackburn, 1965; Allison, 1970) did not consider that the feedstuff could contribute enzymic activity. None of the older papers deals with the digestion of plant proteins *in situ* when the diet is fresh forage. Brock *et al.* (1982) used ruminal fluid from a Holstein cow receiving lucerne hay plus maize grain supplement, and azocasein as substrate. Kopecny and Wallace (1982) used a ration consisting of grass hay and barley-based concentrate together with labelled casein as substrate. Nugent and Mangan (1981) used fraction 1 leaf protein (Rubisco) as the substrate, but it had already been extracted from plant tissues. Prins *et al.* (1983) and Falconer and Wallace (1998) also used ruminal fluid from animals fed dried rations as their starting material. In the latter study, the main proteolytic enzymes in the rumen of different animals appeared to be highly variable, so it was suggested that the proteolytic microbial population was also highly variable. Therefore, none of these studies could provide a clue as to the validity of the Theodorou *et al.* (1996) suggestion.

The first discussion of a possible contribution by plant proteinases appeared, when Theodorou *et al.* (1996) proposed that much of the rapid release of NH_3 in grazing animals might be initiated by the action of plant, rather than microbial, proteinases. They argued that grass cells contain vacuoles harbouring broad-spectrum proteinases which are known to be responsible for protein breakdown in the silo (Wetherall *et al.*, 1995); that proteins of fresh forage would not be available for microbial attack because they are trapped in plant tissues which are incompletely disintegrated; and finally that, within the microenvironment of the incompletely disintegrated plant tissue, it would be the plant proteinases themselves which would break down the plant proteins.

Zhu *et al.* (1999) set out to assess the importance of plant proteinases in four forages – ryegrass, red clover, white clover and bird's-foot trefoil – by following the breakdown pattern of plant proteins by SDS-PAGE in the presence and absence of ruminal fluid *in vitro*. The ruminal fluid was not from a grazing animal, but from a cow receiving grass silage or a sheep receiving hay. The digestion of proteins to peptides clearly occurred with both ruminal fluid in the incubation mixture and without added ruminal fluid after 24 h incubation. The authors concluded that intrinsic plant proteinases were responsible for the proteolysis. However, scFA accumulated to almost 40 mmol/l, which is about 40% of the total scFA concentration in ruminal fluid, indicating that substantial microbial fermentation had occurred. Thus, microbial growth must have occurred in the incubations without added ruminal fluid, which could have contributed in a major way to proteolysis. The inoculum was presumably epiphytic bacteria associated with the plant material.

Our own experiments (Wallace *et al.*, 2000b) aimed at evaluating the role of plant proteinases in ruminal protein breakdown in grazing animals concluded that plant proteinases made a real impact, but that its magnitude was likely to be small compared to microbial activity. The rate of release of NH_3 from homogenized grass was almost double that from homogenized, autoclaved grass, which is consistent with a significantly lower rate of hydrolysis when plant proteinases are destroyed. The proteolytic activity of the grass preparation was much lower than that of the microorganisms, indicating that the location of the plant proteinases at the same place as the proteins being broken down gives them a significance that outweighs their relatively low proteolytic activity. Thus, altering the proteinase activity of grass by breeding or genetic engineering should produce significant benefits to the protein nutrition of grazing ruminants.

3.2.2 Breakdown of non-protein nitrogen-containing materials

With only a few exceptions, the breakdown of nitrogenous compounds other than proteins by ruminal microorganisms has received little attention by researchers in recent years. Nevertheless, the importance of cheap, non-protein N as a feed ingredient remains of great significance in the developing world, while the breakdown of nucleic acids has taken on new significance as the controversy about genetically modified (GM) crops and their impact on health and the environment continues, particularly in the European Union.

3.2.2.1 Urea breakdown

Urea is converted to NH_3 and carbon dioxide (CO_2) by rumen microbial urease, which is similar to the well characterized jackbean enzyme (Wallace *et al.*, 1997b). The activity is bacterial in origin. No activity is associated with ciliate protozoa (Onodera *et al.*, 1977) or fungi (Sakurada *et al.*, 1994), and defaunation did not affect the urease activity of ruminal digesta in the study of Sahu *et al.* (2000), but urease activity was lower in defaunated calves in another experiment (Pal *et al.*, 1998). In the absence of the microbial population in germ-free lambs, urea was not hydrolysed, reaching the same concentration in the rumen as in blood (Cheng and Wallace, 1979).

The microbial ecology of ureolysis in the rumen distinguishes the bacteria of the deep ruminal digesta from those inhabiting the rumen wall (Laukova and Koniarova, 1995; Wallace *et al.*, 1997b). The former population probably hydrolyse urea entering the rumen in saliva and in the diet, while highly active facultative species, predominantly *Staphylococcus* spp., inhabit the rumen wall, where the urea originates by diffusion from blood (Wallace *et al.*, 1979). *Selenomonas ruminantium* was the most abundant strict anaerobe with urease activity in a large survey carried out by Laukova and Koniarova (1995). Urease is regulated in *S. ruminantium* by glutamine synthetase (Smith and Bryant, 1979; Smith *et al.*, 1981). Ruminal urease activity is enhanced by dietary nickel (Spears *et al.*, 1977; Spears and Hatfield, 1978), and it is a nickel-containing protein in *S. ruminantium* (Hausinger, 1986).

Urea hydrolysis by ruminal microorganisms enhances the efficiency of N retention by ruminants on a low-N diet (Leng and Nolan, 1984). Ammonia, which results from microbial activity in the digestive tract and from catabolic activity in the animal's tissues, is converted to urea. Urea in the bloodstream can then diffuse back to the rumen, and be recycled into microbial protein (Kennedy and Milligan, 1980). It is only when enhancing the N content of the diet by adding urea is considered (Roffler and Satter, 1975) that bacterial urease activity is considered excessive and the target for control. Until now, ways of exerting such control have not proved successful (see Section 3.3.2). Arguably more success was obtained by introducing modified urea or other forms of non-protein N which were degraded more slowly than urea (Schwartz, 1967; Chalupa, 1972; Merry *et al.*, 1982).

3.2.2.2 Nucleic acids

One of the earlier objectives of finding out how extensively nucleic acids are broken down in the rumen was to assess if total DNA or RNA present in rumen contents could be used as a marker of microbial biomass (Smith and McAllan, 1970). Naked DNA and RNA are hydrolysed rapidly in rumen contents (McAllan and Smith, 1973b), releasing nucleotides which are also rapidly broken down (McAllan and Smith, 1973a). Intact nucleic acids in rumen contents are, therefore, likely to be predominantly microbial, so DNA or RNA can be used as biomass markers. Which microbial species are primarily responsible for the destruction of exogenous plant nucleic acids remains unclear: Flint and Thomson (1990) investigated nuclease activity of some species of ruminal bacteria and found that the highest activities were present in *Prevotella* spp. and *F. succinogenes*. The extracellular enzymes of these species could explain the high nuclease activity of extracellular fluid from mixed rumen contents (Russell and Wilson, 1988). Genetic investigations have led to numerous observations about restriction endonucleases of ruminal bacteria (Pristas *et al.*, 1995, 2001; Miyagi *et al.*, 1998). But are these the same enzymes that degrade the bulk of plant DNA? It seems unlikely.

This type of information becomes ever more important as the GM crops debate continues, particularly in the European Union. If the consumption by ruminants of plants containing new genes leads to gene transfer to the microorganisms or to the host, some would argue that this presents a hazard to the environment or human health. It has been shown that intact genes and gene fragments persist for very short periods of time in rumen contents except in particulate plant materials where the DNA is protected in intact plant cells (Alexander *et al.*, 2002; Duggan *et al.*, 2003) which would mean that the uptake of genes by the microorganisms would be a very rare event and that uptake by the host of intact genes is very low. This argument could be used in support of the safety of GM crops as ruminant feeds. However, it has to be weighed against the knowledge that there is overwhelming evidence that horizontal gene transfer seems likely to have occurred between ruminal species (Flint and Scott, 2000) and even between organisms in different phylogenetic domains (Gilbert *et al.*, 1992).

3.2.2.3 Other nitrogenous compounds

There are other nitrogenous compounds whose metabolism has been studied in ruminal microorganisms. Some of these, such as nitrate, choline and ethanolamine, are present in the feed. Others have been synthetic compounds, which were designed to provide NH_3 at a rate of release that matched microbial requirements better than the too rapidly degraded urea. Also, the possibility that gaseous N_2 could be fixed in the rumen has been investigated. These various investigations will be described briefly here.

Nitrate is abundant in many feeds, but its interest to livestock producers has not been nutritional, rather the hazard it presents by being reduced to nitrite by ruminal microorganisms (Lewis, 1951; Holtenius, 1957). If the nitrite is not further reduced, it can be absorbed by the animal and cause nitrite poisoning. Jones (1972) investigated ways of preventing nitrite accumulation, and found that only formate provided reducing equivalents rapidly enough to prevent nitrite accumulation. It is known that *S. ruminantium* and probably other bacteria use nitrate as an electron acceptor (John *et al.*, 1974), but as far as we are aware little is known about the microbial ecology of nitrite reduction. Choline and ethanolamine are also fairly abundant in feedstuffs. Choline is particularly significant with regard to protozoal activity (see Section 3.2.5).

Under normal circumstances, the rate of N_2 fixation from atmospheric N_2 is insignificant (Moisio *et al.*, 1969; Hobson *et al.*, 1973; Jones and Thomas, 1974). Clearly if rumen fermentation could be manipulated to fix N_2 the implications would be enormous. However, attempts to achieve this by inoculating sheep with a N_2-fixing *Bacillus macerans* resulted in nutritionally insignificant N_2 fixation (Jones and Thomas, 1974).

Another manipulation that has been investigated concerns attempts to find a synthetic non-protein N compound that releases NH_3 less rapidly than urea. Compounds such as biuret, creatine, ammoniated molasses and glycosyl ureas have been investigated (Schwartz, 1967; Chalupa, 1972; Merry *et al.*, 1982) but probably the physical protection of urea in urea-molasses blocks has been the most effective means of controlling energy supply and the availability of non-protein N to ruminants. Chemical protection may only be transient because of adaptation by the rumen microflora (Nikolic *et al.*, 1980).

3.2.3 *De novo* synthesis of amino acids

One of the most important benefits of the evolution of a forestomach fermentation is that the microorganisms which result from microbial fermentation are available for gastric and intestinal digestion, whereas in hindgut fermenters, the microorganisms are voided in faeces and their nutritive value is lost. From a livestock production standpoint, this capability is very important in low-input production systems. The microbial population as a whole does not require AA: NH_3 is an adequate source of N for protein synthesis if sufficient energy is available (Leng and Nolan, 1984). Thus, urea can be recycled within the body and leads to microbial protein synthesis in the rumen. Furthermore, diets deficient in AA-N can be supplemented with NH_3-yielding non-protein N, such as urea. The ultimate demonstration of the value of this nutritional capability was made by Virtanen and his colleagues (Virtanen, 1966), who fed dairy cattle for extended periods on diets totally deficient in AA, but which contained energy and urea as N sources. Milk production was normal and the animals suffered no ill effects.

The biosynthetic events necessary for AA biosynthesis will be described in terms of how N is assimilated from NH_3 and how C-skeletons of the AA are formed. The AA requirements of ruminal microorganisms will be assessed, and compared with an important aspect of protein nutrition, namely the benefit that fermentation obtains from the availability of pre-formed AA and peptides – *benefit* rather than *requirement*.

3.2.3.1. Ammonia uptake in ruminal microorganisms

The mechanisms of NH_3 assimilation by ruminal bacteria have been described comprehensively before (Hespell, 1984; Morrison and Mackie, 1996; Wallace *et al.*, 1997b), and only a summary is given here. Some calculations have suggested that NH_3 may accumulate inside microbial cells (Russell and Strobel, 1987), indicating that NH_3 uptake may be mediated by an active carrier, but we are not aware of other work having been done on the mechanism of translocation of NH_3 into the cell. Once taken into the cells, ruminal bacteria assimilate NH_3 into AA largely via the reductive amination of glutamate by NAD-linked glutamate dehydrogenase (NAD-GDH). The evidence for this being the predominant mechanism in the mixed population comes from direct measurement of enzyme activities (Erfle *et al.*, 1977; Wallace, 1979; Lenartova *et al.*, 1987) and tracer experiments (Blake *et al.*, 1983; Atasoglu *et al.*, 1999). Other enzymes of NH_3 assimilation are also present, notably NADP-linked glutamate dehydrogenase (NADP-GDH), the glutamine synthetase–glutamate synthase (GS-GOGAT) coupled reactions, and alanine dehydrogenase (Chalupa *et al.*, 1970; Wallace, 1979). With NH_3 concentrations of 5–15 mmol/l, which occur in ruminants with adequate dietary N, NAD-GDH is the main mechanism of NH_3 assimilation: the $K_m(NH_3)$ of NAD-GDH is 20–33 mmol/l (Erfle *et al.*, 1977; Wallace, 1979) and glutamate is the most abundantly labelled AA when ^{15}N is presented (Atasoglu *et al.*, 1999). However, when NH_3 concentration falls, the higher-affinity NADP-GDH ($K_m(NH_3)$ of 2–3 mmol/l) may increase (Chalupa *et al.*, 1970; Erfle *et al.*, 1977; Wallace, 1979; Lenartova *et al.*, 1987) and, under NH_3-limiting conditions which probably seldom prevail *in vivo*, GS can also increase significantly (Erfle *et al.*, 1977). In contrast, at unusually high NH_3 concentrations, there are indications that the first AA to become labelled is alanine (Wallace, 1979; Blake *et al.*, 1983), suggesting that alanine dehydrogenase ($K_m(NH_3)$ of 70 mmol/l; Wallace, 1979) may act as an assimilatory enzyme.

Pure-culture studies carried out more recently call into question certain assumptions about NH_3-assimilatory mechanisms. The purified NADP-GDH from the cellulolytic ruminal bacterium, *R. flavefaciens*, required 0.5 mol/l KCl for optimal activity (Duncan *et al.*, 1992). A similar ionic requirement was observed with the NAD(P)-GDH of *P. bryantii* (Wen and Morrison, 1997), reflecting the slightly halophilic conditions experienced by ruminal bacteria. Therefore, previous enzyme measurements where $[K^+]$ were not high may have to be revised. Furthermore, in *P. bryantii* and *P. ruminicola*, both NAD- and NADP-linked activities were present in a single protein, and the activities with both cofactors were co-regulated in *P. bryantii*, indicating a link between the activities, at least in some *Prevotella* spp., in accord with human intestinal *Bacteroideaceae* (Wen and Morrison, 1997). Wen and Morrison (1997) suggested that, by analogy with *Bacteroides* spp., glutamine synthetase activities play a much smaller role in regulating N metabolism in ruminal *Prevotella* than in the better characterized enteric bacteria. Thus, modern molecular studies must be continued, along with flux and activity measurements, in order to amplify our understanding of NH_3 uptake by ruminal bacteria.

In rumen ciliate protozoa, an NAD^+-dependent glutamate dehydrogenase (GDH; E.C. 1.4.1.24) was cloned from the rumen ciliate *Entodinium caudatum* (Newbold *et al.*, 2000b). When expressed in *E. coli*, the enzyme had a high affinity for NH_3 and α-ketoglutarate (apparent K_m 2.33 and 0.71 mmol/l, respectively) and a low affinity for glutamate (apparent K_m 98 mmol/l). In addition, washed *E. caudatum* cells incubated in the presence of NH_3 and antibiotics had higher GDH activities and increased levels of GDH mRNA. These results suggest that the enzyme may be involved in the assimilation of NH_3 in the rumen. In fungi, the mechanisms of uptake of NH_3 appear to be NADP-linked GDH and GS, as other enzymes were not present (Dijkerman *et al.*, 1997). These activities increased in response to N limitation (Dijkerman *et al.*, 1997).

Once assimilated into glutamate and possibly alanine, the N from NH_3 then becomes rapidly distributed to other AA via aminotransferase activities. As well as the best known glutamate–oxaloacetate and glutamate–pyruvate aminotransferases widespread in bacteria, other aminotransferases are abundant in the rumen (Tsubota and Hoshino, 1969; Bhatia *et al.*, 1979, 1980). Most recently, the glutamate–phenylpyruvate aminotransferase of *Prevotella* spp. has been purified and characterized (Amin *et al.*, 2001, 2002) because of its significance in the formation of phenylalanine. To date, little or no information is available on molecular structure or genetic regulation of these vital enzymes in ruminal bacterial AA metabolism. Protozoa may also be particularly significant, because they have a higher aminotransferase activity than ruminal bacteria (Bhatia *et al.*, 1979). A strain of *Piromyces* had no apparent glutamate–pyruvate aminotransferase activity, although it did possess glutamate–oxaloacetate aminotransferase (Dijkerman *et al.*, 1997).

3.2.3.2 Amino acid synthesis by ruminal microorganisms

The mechanisms by which bacterial AA C-skeletons are synthesized have been fairly well understood for some time, but only up to a point, because detailed analysis at the molecular/genetic level is lacking. In contrast, the extent to which individual AA are actually synthesized *de novo* in the rumen and how various factors affect *de novo* synthesis has received renewed attention. This discussion will combine these considerations in dealing separately with the different classes of AA.

THE GLUTAMATE FAMILY. The glutamate family of AA comprises glutamine, proline and arginine as well as glutamate itself. In order for NH_3 to be assimilated by GDH, there has to be supply of α-oxoglutarate. This appears to be formed either by operation of the Krebs cycle reactions in the forward direction or by reductive carboxylation of succinic acid (Milligan, 1970; Sauer *et al.*, 1975). The importance of the different synthetic routes seems to vary, presumably because different individual species of bacteria employ different mechanisms for α-oxoglutarate synthesis (Wallace *et al.*, 1997b); it should be recalled that the tricarboxylic acid (TCA) cycle is not operational in anaerobic bacteria (Sauer *et al.*, 1975). *M. elsdenii* forms α-oxoglutarate by that portion of the TCA cycle enzymes working in the forward direction (Somerville and Peel, 1967; Somerville, 1968), while *Selenomonas*, *Veillonella* and *Prevotella* spp. form α-oxoglutarate by the reductive carboxylation of succinate. The C-skeleton of glutamate incorporated into microbial protein is derived slightly more from exogenous AA than C-skeletons of asparate and alanine (Atasoglu *et al.*, 2003), presumably reflecting the greater availability of oxaloacetate and pyruvate, respectively, compared to α-oxoglutarate, as intermediary metabolites. Oxaloacetate is formed readily by the carboxylation of pyruvate (Atwal and Sauer, 1974). Nevertheless, the majority of glutamate is extensively synthesized from α-oxoglutarate in the mixed population. Glutamine is formed from glutamate by glutamine synthetase, as has already been discussed. Proline and arginine seem to be anomalous AA in comparison with AA in other families, in the sense that their *de novo* synthesis is depressed in some studies to an unusual extent when pre-formed AA are available. Salter *et al.* (1979) found that, in steers receiving a straw and tapioca-based diet, most (>50%) proline-N was received from the pre-formed AA, whereas only about 20% of glutamate-N was derived from exogenous glutamate. Similarly, with sheep on a 50% grass hay, 50% concentrate diet, Atasoglu *et al.* (1999) found less than half of glutamate-N but >95% of proline-N to be derived from exogenous AA. Under these circumstances, therefore, the conversion of glutamate, via glutamyl phosphate, γ-glutamyl phosphate, glutamate γ-semialdehyde and Δ^1-pyrroline 5-carboxylate may be downregulated in order to save ATP expended during proline synthesis. A similar pattern was observed in non-cellulolytic bacteria (*P. bryantii*, *S. ruminantium* and *S. bovis* (Atasoglu *et al.*, 1998)), but not in cellulolytic bacteria (Atasoglu *et al.*, 2001) or fungi (Atasoglu and Wallace, 2002), suggesting that diet-induced population changes may influence the extent of *de novo* AA synthesis, particularly for proline. Less information is available about arginine, but arginine-N also seemed to be derived mainly from pre-formed arginine when the other AA (except proline) were formed mainly from NH_3 (Salter *et al.*, 1979). Again, downregulating the pathway of arginine synthesis may save energy and also spare the depletion of α-oxoglutarate.

THE ASPARTATE FAMILY. This family comprises aspartate, asparagine, lysine, threonine and methionine. More microbial aspartate than any other AA is synthesized *de novo*, both in terms of its utilization of NH_3 and its use of intrinsic C-skeletons (Salter *et al.*, 1979; Atasoglu *et al.*, 2003). The biosynthesis of asparagine does not appear to have been investigated – the amide group is lost during the acid hydrolysis of protein usually used to prepare free AA for analysis, so asparagine is not usually measured – but presumably it is synthesized readily by asparagine synthase. Lysine is an AA that has stimulated interest, because it is often a limiting AA in the host animal (Schwab, 1995). It is formed predominantly by the diaminopimelate pathway in the mixed population (Sauer *et al.*, 1975). Ciliate protozoa form lysine from diaminopimelate in the cell walls of the bacteria which they consume (Onodera and Kandatsu, 1973), although the quantitative significance of this mechanism of synthesis to the nutrition of the host animal is likely to be minor (Onodera *et al.*, 1991). When lysine was available exogenously, the mixed population incorporated it intact in preference to *de novo* synthesis: no *de novo* synthesis of the C-skeleton took place (Atasoglu *et al.*, 2003). Protozoal protein is particularly enriched in lysine, in comparison to bacterial protein, indicating that increasing the flux of protozoa out of the rumen may be beneficial in providing lysine to the animal (Wallace, 1994). Threonine is usually formed in bacteria directly from aspartate via homoserine (Umbarger, 1969; Patte, 1996). However, the experiments of Sauer *et al.* (1975) and Atasoglu *et al.* (2003) indicate that a substantial proportion of the C-skeleton of threonine is derived from other routes, perhaps by reductive carboxylation of propionate to α-oxobutyrate and reversal of threonine dehydratase. Isoleucine is formed from α-oxobutyrate in ruminal microorganisms as in other organisms, but the source of the α-oxobutyrate is thought to be from the reductive carboxylation of propionate, which is abundant in ruminal digesta, rather than threonine dehydratase (Sauer *et al.*, 1975), rather than from aspartate as is the case in other organisms. The labelling pattern of different C atoms in the isoleucine molecule indicated that the C-skeleton was subject to several rearrangements before incorporation into microbial protein (Atasoglu *et al.*, 2003). Methionine is also essential for the host animal, and may be the AA most limiting growth or productivity under some dietary conditions and for some purposes, e.g. wool growth (Reis *et al.*, 1990). Labelling experiments suggested that methionine was formed via aspartic semialdehyde and that CO_2 was the likely origin of the sulphhydryl methyl group (Sauer *et al.*, 1975).

THE SERINE FAMILY. The serine family consists of serine itself, plus glycine and cysteine. Serine is derived from the glycolytic intermediate, 3-phosphoglycerate, via 3-phosphohydroxypyruvate and 3-phosphoserine (Somerville, 1968; Sauer *et al.*, 1975). Serine is then formed by the exchange of a hydroxymethyl group with tetrahydrofolate, and cysteine by the uptake of hydrogen sulphide (H_2S) in exchange for water (H_2O).

THE AROMATIC FAMILY. Tyrosine, phenylalanine and tryptophan can be formed *de novo* from erythrose-4-phosphate and phosphoenolpyruvate via the chorismate acid pathway. However, most synthesis seems to occur by the reductive carboxylation of phenolic ring precursors followed by transamination (Allison, 1969; Sauer *et al.*, 1975; Atasoglu *et al.*, 2003). Thus, phenylacetic and phenylpropionic acids are important nutrients for *R. albus* (Allison, 1965; Hungate and Stack, 1982; Stack and Cotta, 1986) and bacteria of the mixed population (Amin and Onodera, 1997). Tryptophan metabolism is often not studied because it is destroyed during the acid hydrolysis of proteins. However, there is good evidence that tryptophan is formed from indole-3-acetic acid and other indole precursors (Allison and Robinson, 1967; Okuuchi *et al.*, 1993). Presumably, *de novo* synthesis occurs via the chorismate–anthranilate pathway followed by the addition of phosphoribosylpyrophosphate (Umbarger, 1969).

THE PYRUVATE FAMILY. Alanine is formed from pyruvate by transamination. Pyruvate is also the precursor of valine and leucine. Valine and leucine are formed partly via the addition of a C-2 unit to pyruvate to form acetolactate, followed by a chain of reactions which branches at 1-oxo-2-methybutyrate to valine (by transamination) and eventually to leucine by the addition of acetyl-CoA. Labelling patterns also suggest substantial synthesis of valine and particularly leucine by reduction carboxylation of isobutyrate and isovalerate, respectively (Sauer *et al.*, 1975; Atasoglu *et al.*, 2003).

HISTIDINE. The usual pathway for histidine biosynthesis is closely aligned to the biosynthesis of the purines. Both are derived from phosphoribosylpyrophosphate and ATP. While little direct information is available about the formation of histidine in ruminal microorganisms, the limited amount of labelling information from Sauer *et al.* (1975) is consistent with this route of histidine formation.

3.2.4 Optimum amino acids for rumen fermentation: benefit vs. requirement

Although NH_3 can provide all microbial cell N requirements, peptides and AA are undoubtedly beneficial to rumen fermentation because they stimulate the growth of ruminal microorganisms (Hume, 1970; Amos and Evans, 1976; Maeng and Baldwin, 1976; Maeng *et al.*, 1976; Ben-Ghedalia *et al.*, 1978; Cotta and Russell, 1982; Rooke and Armstrong, 1989; Merry *et al.*, 1990; Cruz Soto *et al.*, 1994; Chikunya *et al.*, 1996). The benefit obtained by supplying AA could occur via one of the two mechanisms: the fermentation rate of the microorganisms may be increased, leading to improved feed intake in the animal; alternatively, or perhaps additionally, the growth yield of the microorganisms might be increased, leading to a decreased flow of microbial protein from the rumen. The Cornell model for ruminal fermentation assumes a 18.7% improvement in microbial growth efficiency when pre-formed AA are used by non-cellulolytic bacteria (Russell *et al.*, 1992). Presumably the yield benefit occurs because less ATP is required for AA biosynthesis. No increase is projected for cellulolytic species, but in view of the finding that these bacteria, hitherto assumed not to incorporate AA, take up significant amounts of AA (Atasoglu *et al.*, 2001), the benefit may occur in cellulolytic as well as non-cellulolytic bacteria.

It has proved difficult to identify specific AA or groups of AA that are responsible for the benefits measured with a complete mixture. The intention here would be to find a minimal mixture of AA that might be used as a feed additive to stimulate ruminal fermentation. Maeng *et al.* (1976) and Argyle and Baldwin (1989) added single or groups of AA to ruminal fermentations *in vitro*. They found that only complete mixtures of AA gave maximum responses. Certain groups, such as aromatic AA, gave intermediate responses. However, no specific AA were identified by this supplementation approach. Labelling methods have also been inconclusive. Salter *et al.* (1979) showed that different AA are formed *de novo* to differing extents. Proline biosynthesis was affected particularly strongly when protein was added. Similar *in vitro* experiments by Atasoglu *et al.* (1999) confirmed the unusual sensitivity of proline biosynthesis to the provision of pre-formed AA, and identified glycine, valine and threonine as other AA whose biosynthesis was most sensitive to repression by added AA. Supplementation with these AA failed to replicate the stimulatory effects of a complete mixture of AA, however. The complementary approach, that of deleting individual AA from a complete mixture and observing the effects on fermentation, proved equally inconclusive (Atasoglu *et al.*, 2003a). Deletion of some AA, principally phenylalanine, leucine and serine, had some effect on fermentation rate and/or microbial growth efficiency, but selective supplementation using these AA was not effective in replicating the stimulation given by the complete mixture. It must be concluded, therefore, that the benefits of protein to rumen fermentation result from the cumulative effects of more efficient incorporation or biosynthesis of all AA – either energetically or kinetically – rather than specific requirements of certain AA by specific bacteria.

3.2.5 Role of protozoa in nitrogen metabolism in the rumen

The rumen contains a large (*ca.* 10^6/ml) and varied population of ciliated protozoa, which may make up over 50% of the microbial biomass. The ruminal protozoa are believed to play a relatively minor role in the direct breakdown of dietary protein (Section 3.2.1; Nugent and Mangan, 1981; and also Chapter 4) although they may be more important in the breakdown of insoluble particulate proteins compared to soluble substrates, in particular as noted below protozoa have a unique role in the engulfment and digestion of bacterial and fungal cells.

3.2.5.1 Protein breakdown

A number of authors have measured protein breakdown by washed protozoal populations both from the rumen and from *in vitro* incubations (Abou Akkada and Howard, 1962; Coleman, 1983; Jouany *et al.*, 1992). However, such preparations inevitably contain bacteria both as endo- and exosymbionts but also within digestive vacuoles, which complicates measurements of protozoal activity (Williams and Coleman, 1992). Pre-incubation with antibiotics can help decrease bacterial activity but may also lead to atypical ciliate activity (Coleman, 1962). Nevertheless, it is clear that both entodiniomorphid and holotrich protozoa are actively proteolytic. Indeed, we have recently cloned and expressed proteinase genes from the rumen ciliate *E. caudatum* (Newbold *et al.*, unpublished). Measuring hydrolysis of fraction 1 leaf protein by entodiniomorphid protozoa, Coleman (1983) concluded that proteolytic activity was highest in *E. caudatum* and *E. simplex* and lowest in cellulolytic species. However, it appears likely that proteolytic activity is proportional to protein solubility (Naga and El-Shazly, 1968), with entodiniomorphid protozoa contributing to the degradation of insoluble but not of soluble proteins in the rumen (Hino and Russell, 1987). As noted below, the most important example of particulate protein breakdown by protozoa is the engulfment and digestion of bacteria, but rumen protozoa also avidly engulf and subsequently digest chloroplasts (Mangan and West, 1977). In contrast to entodiniomorphid protozoa which feed primarily by engulfment, the holotrich protozoa are believed to be able to utilize both particulate and soluble proteins (Onodera and Kandatsu, 1970). It has also been suggested that they may be able to coagulate soluble proteins in the rumen thus making them available to entodiniomorphid ciliates (Onodera, 1990; Onodera and Yakiyama, 1990). Proteolytic enzymes from lysates of mixed rumen protozoa had a pH optimum of 5.8 and appeared to be mainly cysteine and to a lesser extent aspartic proteinases (Forsberg *et al.*, 1984). Coleman (1983) also found cysteine proteases to be the predominant activity in washed *E. caudatum* but here the pH optimum appeared to be below 4, although Abou Akkada and Howard (1962), working with apparently the same ciliate, reported maximal breakdown of casein at pH 6.5–7.0. Lockwood *et al.* (1988) reported that optimal activities in six different genera of rumen protozoa occurred between pH 4 and 5 and although they were predominantly cysteine proteinases, gelatin–SDS-PAGE analysis revealed multiple forms of proteinase within each ciliate and substantial differences between species (Lockwood *et al.*, 1988). Given the apparent diversity in 18S rDNA diversity and codon usage even in apparently morphologically identical rumen ciliates, it is perhaps not useful to make multiple comparisons between historical studies in which ciliates have been identified solely on morphological characteristics (Moon-van der Staay *et al.*, 2002; McEwan *et al.*, 2003).

An alternative to measuring protein degradation by washed suspensions of rumen protozoa has been the incubation of protein substrates in the rumen of faunated and defaunated animals. The use of such studies to comment on ciliate activities is obviously compromised by the changes in the bacterial population in response to defaunation (Williams and Coleman, 1992). Based on such an approach, Ushida and Jouany (1985) concluded that degradation of the insoluble protein from soybean meal increased by 11% when a mixed A-type protozoal population was inoculated into the rumen, which concurred with *in vitro* studies suggesting that the importance of protozoa decreased as protein solubility increased (Jouany, 1996). In contrast to *in vitro* studies, however, nylon bag studies suggested that larger protozoa were more important than small protozoa in the digestion of insoluble proteins (Ushida and Jouany, 1985, 1986). Measurements of non-bacterial non-ammonia nitrogen (NBNAN) flow at the duodenum of defaunated sheep and sheep sequentially reinoculated with holotrichs, large entodiniomorphid protozoa and small entodinia apparently suggested that ciliates increased the flow of dietary protein from the rumen (Ivan *et al.*, 2000a,b), presumably due to protozoal predation of proteolytic bacteria. Again it was suggested that large entodiniomorphid protozoa had a larger negative effect on NBNAN flow than small entodinia or holotrich protozoa (Ivan *et al.*, 2000a,b).

3.2.5.2 Peptide breakdown

Protozoa have been shown to contain a range of peptidase-like activities, which appear to be predominantly exopeptidase in nature (Abou Akkada and Howard, 1962; Nagasawa *et al.*, 1992). As with ruminal bacteria, they appear to be metallopeptidases (Abou Akkada and Howard, 1962; Wallace *et al.*, 1996), although protozoal peptidase activity was markedly less sensitive to 1,10-phenanthroline than mixed ruminal bacteria, possibly due to difficulties in separating protozoa from plant debris which may have bound free 1,10-phenanthroline (Wallace *et al.*, 1996). Peptidase activity was higher in the small entodinia than in holotrich or larger entodiniomorphid protozoa (Newbold *et al.*, 1989) but in all cases decreased as the peptide chain length decreased. Thus, while ciliate protozoa may contribute significantly to dipeptidase activity in the rumen, they are likely to be less important as peptide length increases (Wallace *et al.*, 1990c). Even with dipeptide breakdown, it is not obvious that removal of protozoa from the rumen would spare peptides from degradation as the resultant increased bacterial population appears to be able to occupy the same metabolic niches and overall dipeptide breakdown is not affected (Wallace *et al.*, 1987).

3.2.5.3 Amino acid breakdown

Both passive and active uptake of AA have been reported in ruminal protozoa (Williams and Coleman, 1992). Active uptake can be sensitive to both pH and salt concentrations in the rumen (Williams and Coleman, 1992). Rates of AA deamination were approximately three times higher in washed protozoa than in bacteria (Hino and Russell, 1987) and deaminase activities were higher in rumen fluid of faunated sheep compared to their defaunated compatriots. Ciliates deaminate a relatively small number of AA, with NH_3 being produced from glutamine, asparagine, citrulline, arginine and orthithine but not glutamate, asparate or histidine (Onodera *et al.*, 1983; Onodera and Goto, 1990) and it has been suggested that much of the NH_3 production when ciliates are incubated with casein results from hydrolysis of amide groups rather than AA deamination (Abou Akkada and Howard, 1962). Various AA interconversions have been reported in protozoa (Williams and Coleman, 1992), although apparently there is no interconversion of AA in *E. caudatum* (Coleman, 1967). AA are formed from and metabolized to a variety of other compounds in the ruminal protozoa, of which perhaps the most important is the formation of lysine from diaminopimelic acid in bacterial cell walls (Denholm and Ling, 1989; Onodera *et al.*, 1991).

3.2.5.4 Breakdown of bacteria

Although degradation of feed protein is one of the most obviously deleterious catabolic activities in the rumen, up to 50% of the bacterial protein formed in the rumen also may be broken down subsequently to NH_3 and thus is not available to the host (Nolan and Stachiw, 1979). This could be due to autolysis of the bacteria or lysis of the bacteria by bacteriophages or mycoplasmas. However, *in vitro* studies suggest that the engulfment and subsequent digestion of bacteria by ciliate protozoa is the most important activity regulating the turnover of bacterial protein in the rumen (Wallace and McPherson, 1987).

Table 3.1. Effect of protozoa on microbial flow from the rumen and intraruminal N recycling.

	Defaunated[a]	Faunated[a]	Defaunated[b]	Faunated[b]
Microbial protein flow from the rumen (g N/day)	13.3	8.9	17.3	10.8
NH_3 pool size (g N)	0.84	2.0	0.72	1.07
Intraruminal recycling (g N/day)	0.8	6.6		
Bacterial recycling (g N/day)			10.0	13.8

[a]Newbold *et al.* (2000a).
[b]Koenig *et al.* (2000).

Recently, we have measured the recycling of N within the rumen and in the animal (Koenig *et al.*, 2000; Newbold *et al.*, 2000a). As in previous experiments, approximately 50% of the microbial protein formed in the rumen was recycled therein, with considerable recycling via the blood (Table 3.1). Recycling dropped sharply in the absence of protozoa, although the extent of this decrease varied from 100% to 25%, suggesting that diet may have a major effect on activity of the protozoa. Consistent with the decline in intraruminal recycling, the flow of microbial protein from the rumen increased by between 35% and 50%. Bacterial breakdown by protozoa has been studied extensively *in vitro* (Williams and Coleman, 1992). The susceptibility of ruminal bacteria to breakdown by ciliate protozoa differs both between and within bacterial species (Wallace and McPherson, 1987; Williams and Coleman, 1992). Of the ciliates studied, only *E. caudatum* is apparently not selective in terms of which bacteria it ingests and digests (Coleman, 1964), with other ciliates showing selective uptake of different ruminal bacteria (Newbold and Jouany, 1997). Coleman and Hall (1972) observed that the cell wall peptidoglycan of bacteria ingested by protozoa is rapidly degraded, leaving only fragments of plasma membrane visible (Coleman and Hall, 1972). Ling (1990) argued elegantly that bacterial cell walls must represent the primary barrier to breakdown of bacteria within the protozoa. Recently, we and others have attempted to characterize and clone the bacterial cell wall-degrading enzymes from ruminal protozoa in the hope of gaining further insight into bacterial breakdown by the protozoa (Morgavi *et al.*, 1996; Newbold *et al.*, 1999; Eschenlauer *et al.*, 2000). In addition to engulfment and degradation of bacteria, ruminal protozoa also ingest and digest ruminal fungi. Again, studies are ongoing to characterize chitinase-like activities from rumen protozoa (Newbold and Hillman, 1990; Morgavi *et al.*, 1994; Komatani *et al.*, 2000).

Wallace and McPherson (1987) concluded that small entodinia were responsible for most bacterial breakdown *in vitro*, and this has recently been confirmed *in vivo* by Ivan *et al.* (2000a,b), who concluded that *Entodinium* had the most detrimental effect on duodenal N supply in a series of experiments in which ciliate-free sheep were refaunated with different ciliates (Fig. 3.3).

Recently, we have shown that, in addition to the breakdown of ruminal bacteria, protozoa may also be important in controlling the passage of

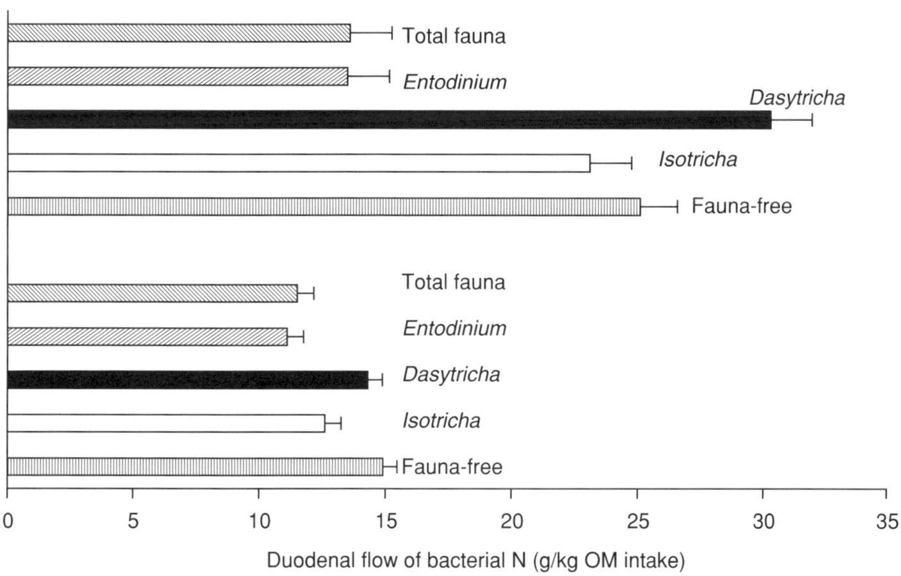

Fig. 3.3. Influence of different protozoal species on the flow of bacterial protein in the duodenum of wethers. From Ivan *et al.* (2000b).

pathogens through the rumen, including *E. coli* 0157 and *Listeria*. Indeed, passage of *Listeria* from the rumen was threefold higher in defaunated compared to faunated sheep (Shepherd *et al.*, 2000; McIntosh *et al.*, 2002). However, as with their digestion of ruminal bacteria, it is apparent that not all protozoa degrade pathogens at the same rate and it may be possible to manipulate the composition of the protozoal population to retain the pathogen barrier whilst limiting bacterial protein breakdown (Newbold *et al.*, 2001).

3.2.5.5 Control of ruminal protozoa

As noted earlier, defaunation or even partial defaunation slows the nutritionally wasteful cycle of bacterial protein breakdown and resynthesis. In a recent meta-analysis on the effects of defaunation, Eugène *et al.* (2003) concluded that defaunation stimulated duodenal N flow by over 30% and presumably as a result average daily gain increased by almost 9% and wool growth, which is more sensitive to microbial protein supply, by almost 15%. However, while many antiprotozoal agents have been tried experimentally (Williams and Coleman, 1992), none has passed into routine use because of toxicity problems, either to the rest of the rumen microbial population or to the host animal. Recently, there has been an increased interest in plant secondary metabolites and immunization for use as possible defaunating strategies (see Sections 3.3.1.6 and 3.3.2 and Chapter 4).

3.2.6 Role of rumen anaerobic fungi

Anaerobic fungi are an important component of the ruminal microflora. Research interest has focused primarily on their unique attributes in breaking down fibrous plant materials – they appear to invade tissue in a physical manner and they have exceptionally active cellulolytic enzymes (Orpin and Joblin, 1997; Lee *et al.*, 2000). They also have a role in N metabolism. Reference has already been made throughout this chapter where fungi have an involvement in N metabolism.

In terms of protein breakdown, most studies suggest that the fungi possess proteolytic activity, although the activity may have minor consequences as a part of the activity of the whole microbial population. Wallace and Joblin (1985), Asao *et al.* (1993) and Yanke *et al.* (1993) found a high metalloprotease activity to be present in fungal cultures. As this is a minority activity in the mixed rumen population (Brock *et al.*, 1982; Kopecny and Wallace 1982), and because under some conditions proteolysis may not be present (Michel *et al.*, 1993), the fungi probably play a relatively minor role in proteolysis overall. There may, however, be specific circumstances, such as the breakdown of a resistant protein–carbohydrate matrix (Yanke *et al.*, 1993), where the proteolytic activity of the fungi is significant. None of the main species of cellulolytic ruminal bacteria is significantly proteolytic. The fungi possess aminopeptidase activity (Michel *et al.*, 1993), but its significance is unknown, either to the mixed population or to the fungi themselves.

In terms of biosynthesis, NH_3 alone as N source supports growth, but AA mixtures are stimulatory (Lowe *et al.*, 1985; Orpin and Greenwood, 1986; Theodorou *et al.*, 1994); 59% of AA-N was still formed *de novo* when peptides and AA were present (Atasoglu and Wallace, 2002). Lysine was exceptional, in that its synthesis decreased much more than other AA when Trypticase or AA were added to the medium, suggesting that lysine synthesis might limit fungal growth in the rumen.

3.3 Strategies for Influencing N Metabolism

3.3.1 Effects of diet, feed additives, including ionophores, antimicrobials and natural plant compounds

3.3.1.1 Influence of diet

Dietary factors can have a significant effect upon the rate of protein breakdown, with dry forage diets leading to low rates of proteolysis. In some instances, simply by feeding a fresh herbage diet instead of dry rations, proteolysis can be increased. This is thought to occur due to an increase in the soluble protein, leading to a shift in the microbial population and a consequent enrichment of proteolytic organisms (Hazlewood *et al.*, 1983; Nugent *et al.*, 1983). Another hypothesis has suggested that this increase in proteolytic activity can be attributed to intrinsic plant proteases, which are released upon grazing and contribute towards the initial stages of proteolysis of grazed forage (Section 3.2.1.5). Feeding cereal diets instead of dry forage rations also leads to a stimulation of proteolysis, again, either due to a difference in the microbial population, with a shift towards an enrichment of proteolytic organisms which tend to also be starch degraders (Stewart *et al.*, 1997; Griswold *et al.*, 1999a) or due to cooperation and synergy between proteolytic and amylolytic organisms, leading to their subsequent enrichment (Moharrery and Das, 2001). The feeding of legumes that are high in condensed tannins can decrease protein degradation, either by altering the structure of the forage protein or by inhibiting the microbial proteinases (McSweeney *et al.*, 2001).

In addition to diet composition, temporal factors can also have an effect upon proteolysis and fermentation. Frequency of feeding can lead to an increase in proteolysis, because the total flow of soluble protein and NH_3 as well as peptides is increased upon the number of times an animal is fed (Chen *et al.*, 1987a). It has also been suggested that the synchronous supply of energy and AA-N may improve the fermentation process (Sinclair *et al.*, 1995; Chapter 4). However, fermenter studies designed to prove this have remained relatively inconclusive (Russi *et al.*, 2002), although peptides did have a small stimulatory effect upon the fermentation process.

One thing that is clear, however, is that in addition to a variation of proteolytic activity due to diet, a high degree of animal to animal variation can be observed (Falconer and Wallace, 1998; Wallace *et al.*, 2000a), a fact which must be accounted for whilst conducting feeding trials.

3.3.1.2 Influence of protein structure and chemical composition

The secondary and tertiary structures of a protein can have an effect upon its rate of degradation. For a long time it was thought that protein breakdown in the rumen was proportional to its solubility, however this has since been revised and other factors must be taken into consideration (Wallace *et al.*, 1997b). The proteins of some plants can be afforded some natural protection by being located in a polysaccharide matrix, rendering them inaccessible to proteolytic organisms (Debroas and Blanchart, 1993). Cross-linking the protein with artificial disulphide or other bridges can lead to improved resistance to proteolytic attack (Wallace, 1983; Williams *et al.*, 2002). Other chemical treatments, which affect protein structure and solubility, have also proved to be useful protection mechanisms. Heat can affect the tertiary structure and solubility of the protein, as can formaldehyde treatment (Kaufmann and Lüpping, 1982). As mentioned previously, tannins can affect the protein structure, leading to a decrease in its degradation. The protein can also be protected by coating it in an undegradable shell of either blood (Ørskov *et al.*, 1980) or heated sugar–protein mixture (Wallace and Falconer, 1992). Many other protection methods, both for protein and AA, have been investigated (Ferguson, 1975).

Recently, plant breeders have investigated the possibility of engineering plants which have proteins that are high in essential AA, but that are resistant to rumen microbial degradation whilst remaining nutritionally available to the animal (Hancock *et al.*, 1994), thus improving N availability to the host animal. Generally, these proteins have a large number of artificial disulphide bridges with a complex quaternary structure (Williams *et al.*, 2002). However, it should be noted that the stability against proteolytic attack of a given protein, once expressed in a transgenic plant, cannot be guaranteed, as it must be expressed and folded correctly to ensure that the correct quaternary structure is formed (Guenoune *et al.*, 2002).

In addition to improving N retention in the rumen by making the protein more resistant to degradation, some studies have focused on the next step in the proteolytic cascade. By either formulating methods to make peptides more resistant to degradation by rumen microbes, or targeting the organisms that are involved, the flow of N from the rumen to the small intestine can also be increased. Most peptides are broken down rapidly in ruminal contents. A pattern emerges by studying the breakdown of individual peptides (Table 3.2; Wallace *et al.*, 1990a,c, 1993a,b; Yang and Russell, 1992) that: (i) peptides with a Gly or Pro residue at or next to the N-terminal site are generally more resistant to degradation than others; (ii) acidic peptides are more resistant to degradation than basic peptides; and (iii) peptides blocked at the N-terminus are degraded slowly (see Section 3.3.1.2). The idea that hydrophobic peptides are more resistant to degradation than others (Chen *et al.*, 1987a) has not been sustained in subsequent studies (Wallace *et al.*, 1990a,c, 1993a; Williams and Cockburn, 1991;

Table 3.2. Rates of degradation of different peptides in ruminal digesta.

Dipeptide	Rate of hydrolysis[a]	Tripeptide	Rate of hydrolysis[a]	Oligopeptide	Rate of hydrolysis[a]
AlaGly	0.55	Phe_3	0.94	Ala_4	1.18
GlyAla	0.40	Ala_3-pNA	0.85	Phe_4	0.96
HisGly	0.33	Ala_3	0.76	GlyHisArgPro	0.38
Gly_2	0.33	$LeuGly_2$	0.60	ValGlySerGlu	0.32
ValAla	0.29	GluAlaGlu	0.22	$ProLeuGly_2$	0.14
LysAsp	0.25	Gly_2Arg	0.10	$GlyProGly_2$	0.07
GluLys	0.25	GlyHisLys	0.08	ValGlyAspGlu	0.06
GlyHis	0.22	Glu_3	0.07		
ValTyr	0.18	$PheGly_2$	0.06		
AspGly	0.18	Gly_2Leu	0.06		
Ala_2	0.15	$ProGly_2$	0.05		
AspLys	0.14	Gly_2Phe	0.05		
GlyPro	0.13	Gly_3	0.05		
AspPro	0.07	N-Ac-Ala_3	0.01		

[a]From Wallace *et al.* (1990a,c).

Depardon *et al.*, 1996; Debroas *et al.*, 1998). Some peptides escape degradation in the rumen for a long period, but the quantity which escapes differs according to the analysis used (Chen *et al.*, 1987b; Wallace and McKain, 1990). N-terminal modification of peptides by acetylation may protect them from degradation (Wallace *et al.*, 1993b) and it has been shown that the microbial population is unable to adapt to use acetylated peptides (Witt *et al.*, 1998). Although these acetylated peptides are resistant to microbial degradation they are still nutritionally available to the animal further down the intestinal tract (Wallace *et al.*, 1998). The method might be suitable for upgrading protein sources that have been partially degraded, such as by-products of fish or soya processing.

3.3.1.3 *Influence of ionophores*

Ionophores are compounds which catalyse the translocation of ions across membranes. The mode of action of ionophores and their effects on ruminal microorganisms were described in detail by Nagaraja *et al.* (1997). As far as we understand their action, if an ionophore reaches the lipid bilayer of a membrane, it will enable certain ions to be translocated across the membrane. The specificity of the ion(s) translocated varies with different ionophores. For example, nigericin catalyses the exchange of Na^+ and H^+ across membranes, while valinomycin catalyses K^+/H^+ exchange. A group of ionophores, but by no means all ionophores, have been found to benefit rumen fermentation, to improve ruminant nutrition, and to influence the composition of the rumen microflora. These include monensin (Na^+/H^+), salinomycin (Na^+/H^+), tetronasin (Ca^+/H^+) and lasalocid (K^+/H^+). Thus, there is no common ionic mechanism linking the ionophores which are useful in the rumen. The antimicrobial profile of the members of this group is common, however, in that Gram-positive bacteria are more sensitive and Gram-negative bacteria are less sensitive to the entire group. This suggests that it is the relative permeability of the bacterial cell envelope – Gram-negative bacteria have an outer membrane as well as the cytoplasmic membrane – that explains the selectivity of these ionophores against ruminal bacteria. Rumen ciliate protozoa are transiently sensitive to ionophores when they are introduced into the diet, but subsequently numbers recover, suggesting that the protozoa develop a mechanism for protecting themselves against the toxic effects of ionophores (Dennis *et al.*, 1986). Rumen anaerobic fungi are also sensitive to ionophores *in vitro* (Stewart and Richardson, 1989), and probably also *in vivo*.

Many benefits have been attributed to dietary ionophores in ruminants, among which improving the efficiency of N retention is an important one. The initial step of proteolysis is carried out by so many organisms that, not surprisingly, ionophores do not appear to be effective in controlling this step (Newbold *et al.*, 1990). Earlier conclusions that proteolysis was inhibited (Van Nevel and Demeyer, 1977) were misleading because the assay reflected the effect on the whole degradation sequence. Peptide breakdown is decreased in an adaptive fashion, both in ruminal fluid and in pure cultures of *Prevotella* spp. Peptides accumulate in adapted ruminal fluid in the presence of ionophores (Whetstone *et al.*, 1981; Wallace *et al.*, 1990b; Chen and Russell, 1991). In pure culture, the peptidolytic *Prevotella* spp. adapt in the presence of ionophores to decrease the permeability of their cell envelope (Newbold *et al.*, 1992; Callaway and Russell, 1999), which simultaneously protects them against the ionophores and decreases the rate of entry of peptides into the cell. The greatest effect on the proteolytic sequence is thought to be at the final step, that of deamination of AA. The HAP bacteria are typically sensitive to monensin, as described in Section 3.2.1.4, and their numbers are generally decreased in animals receiving monensin (Krause and Russell, 1996).

3.3.1.4 Influence of antibiotics

Several antibiotics are, or have been, used in ruminant nutrition, as reviewed by Nagaraja *et al.* (1997). Many of the antibiotics decrease the degradation of amino-N, leading to an improved outflow of dietary protein from the rumen (Broderick and Balthrop, 1979; Van Nevel and Demeyer, 1990). The most successful antibiotics have a spectrum of activity against different bacterial species, which is very similar to the ionophores, so one would speculate that their mode of action on N metabolism would be similar to the ionophores. Avoparcin, for example, has a similar antibacterial spectrum to monensin, affecting mainly Gram-positive bacteria (Stewart *et al.*, 1983), and also decreases the deamination of AA in ruminal fluid (Froetschel *et al.*, 1983; Jouany and Thivend, 1986). The same is true of virginiamycin (Nagaraja *et al.*, 1997). Virginiamycin and flavomycin may have another protein-sparing effect, via the suppression of *Fusobacterium necrophorum* (Nagaraja *et al.*, 1997). *F. necrophorum* is well known to be the pathogen causing liver abscesses in cattle (Tan *et al.*, 1994). The bacterium is derived from the rumen, where it invades ruminal wall tissue. Recently it was found that *F. necrophorum* is among the most sensitive of all ruminal bacteria to flavomycin (Wallace *et al.*, 2002). This observation seems to be consistent with a protein-sparing effect in the tissues (MacRae *et al.*, 1999).

3.3.1.5 Other chemicals

Many chemicals have been investigated for their effects on ruminal fermentation (Chalupa, 1980). Some of these have been investigated for their influence on protein metabolism. The chemicals, which inhibit ruminal proteolysis, are all standard proteinase inhibitors and could not be considered to be feed additives (Nagaraja *et al.*, 1997). In any case, the variability in proteinase activity in the mixed population, mentioned earlier (Section 3.2.1.1), would preclude such control.

The regulation of peptidase activity by chemical means may prove to be more straightforward than proteolysis, because of the relative importance of *Prevotella* spp. in peptide breakdown (Section 3.2.1.2). Inhibitors of DPP activity have been developed, mainly peptide analogues but also benserazide, which inhibit peptidase activity in *Prevotella*, decrease the rate of NH_3 formation in ruminal digesta, yet do not appear to be selectively bacteriocidal (Wang *et al.*, 2003). Whether the population adapts around these inhibitors to return to the same rate of peptide hydrolysis as before remains to be seen. The breakdown of dipeptides to AA is highly sensitive to metal chelators; however, as with proteolysis, none of these appeared to be a candidate as a feed additive (Wallace and McKain, 1996; Wallace *et al.*, 1996).

Amino acid deamination has proved most readily inhibited of all the steps of protein catabolism to NH_3. Once again, severe chemical inhibitors that could not be considered to be potential feed additives have been used (Broderick and Balthrop, 1979). Others seem to have much more potential. Diaryl iodonium compounds, in particular diphenyliodonium chloride (DIC), were investigated for their ability to decrease NH_3 formation by mixed ruminal microorganisms (Chalupa, 1980; Chalupa *et al.*, 1983). Another unrelated compound, an analogue of proline with the systematic name 1-[(E)-2-(2-methyl-4-nitrophenyl)diaz-1-enyl]pyrrolidine-2-carboxylic acid and coded LY29, also decreased the rate of NH_3 formation (Floret *et al.*, 1999). The mode of action of DIC and LY29 appeared to be different, however. LY29 was selectively toxic to HAP bacteria, while DIC had no discernible antimicrobial effects (Floret *et al.*, 1999). In an environmental climate where selective antimicrobial effects have connotations with antibiotic resistance, even when ill founded, DIC would appear to be a better prospect for development than LY29. It is notable that there seems to be a strong connection between inhibitors of methanogenesis and inhibitors of AA deamination. Amichloral, developed as an inhibitor of methane formation, also improved N retention (Johnson, 1974), most likely by inhibiting AA deamination (Chalupa, 1980). Conversely, the dimethyl derivative of DIC also inhibited methanogenesis (Van Nevel and Demeyer, 1992). The connection is that hydrogen accumulation affects the availability of NAD^+, which is required for the deamination of some AA (Russell and Martin, 1984; Hino and Russell, 1985).

Fats are natural materials that have many effects on ruminal fermentation. Under some circumstances, they inhibit proteolysis (Nagaraja *et al.*, 1997). The main effect is probably against ciliate protozoa, which may cause inhibition of the breakdown of particulate protein (Broudiscou *et al.*, 1990a,b; Chapter 4). The main effect is that, once protozoal predation is removed, microbial growth efficiency increases (Section 3.2.5; reviewed by Nagaraja *et al.*, 1997). Care must be taken to use the correct dose and form of fat, as fats can be very toxic to the overall fermentation (Kowalcyk *et al.*, 1977; Igwuegbu and Sutton, 1982).

Inhibition of ruminal urease activity has been an objective of research for many years. Chemical inhibition has been attempted. Most recently, Ludden *et al.* (2000a) established that *N*-(n-butyl) thiophosphoric triamide was an effective inhibitor in short-term *in vitro* incubations. However, as often happens with chemical inhibition, the population adapted to be less sensitive to the inhibitor, such that no inhibition was observed 15 days into a trial (Ludden *et al.*, 2000b).

Mineral clays have sometimes been used as feed additives, mainly as a complexing mechanism to deliver minerals to the animal. One of these, bentonite, was discovered to stimulate wool growth (Fenn and Leng, 1988) and increase the flow of dietary and bacterial protein (Ivan *et al.*, 1992), by suppressing protozoal numbers. The mechanism by which bentonite appeared to suppress protozoal activity was to interfere with the cilia, rendering the protozoa immotile (Fig. 3.4; Wallace and Newbold, 1991).

3.3.1.6 Natural products

Growth-promoting antibiotics and ionophores will be banned in Europe from the end of 2005, in an

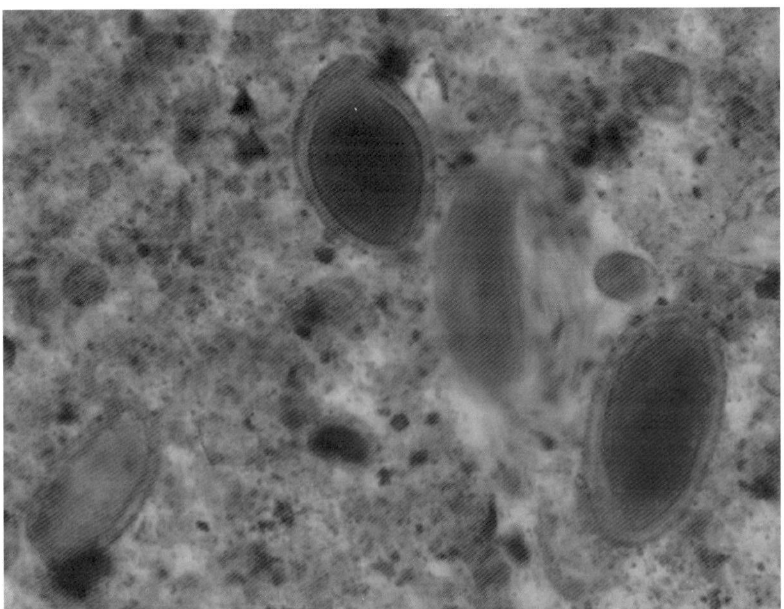

Fig. 3.4. Influence of bentonite on rumen ciliate protozoa. Note the way that there seems to be a shell at the end of the cilia, preventing ciliate motility. These protozoa are about 250 μm in length. From Wallace and Newbold (1991).

attempt to stem the dangers of antibiotic resistance arising from livestock production, and other parts of the globe may follow suit. Thus, there is increasing interest in exploiting natural products as feed additives, which bring the same benefits without the accompanying dangers to human health. Secondary metabolites in plants function both as nutrient stores and also as a mechanism for defending their structure and reproductive elements from predation by animals and insects (Haborne, 1989). Among the types of material tested, essential oils, saponins, tannins and microbial feed additives hold promise as natural feed additives for ruminants.

ESSENTIAL OILS. Essential oils are steam-volatile or organic-solvent extracts of plants, comprising monoterpenes or their aldehyde or alcohol derivatives. They have been used traditionally for their pleasant odour, flavour, or antiseptic and/or preservative properties. Essential oils were examined many years ago in ruminal bacteria, from the point of view of the oils contributing to poor palatability in some plant species (Oh *et al.*, 1967). General inhibitory activity was found across a range of plant materials, of which vinegar weed was the most potent. Oh *et al.* (1968) showed that individual oils had different effects on mixed ruminal bacteria. Monoterpene hydrocarbons were less toxic and sometimes stimulatory to microbial activity compared to the corresponding oxygenated compounds, the monoterpene alcohols and aldehydes (Oh *et al.*, 1968). The sensitivity of ruminal bacteria to essential oils of *Artemisia tridentata* (big sagebrush) was the same in captive deer as it was in wild deer, which was suggested to mean that ruminal bacteria did not adapt to essential oils (Nagy and Tengerdy, 1968). Thus, essential oils were not necessarily toxic to ruminal bacteria, and their effects might be expected to persist.

Dietary essential oils decreased rates of NH_3 production from AA in ruminal fluid taken from sheep and cattle receiving the oils, yet proteinase and peptidase activities were unchanged (McIntosh *et al.*, 2003; Chapter 4). HAP species were the most sensitive of ruminal bacteria to essential oils in pure culture (McIntosh *et al.*, 2003), and numbers of HAP bacteria were suppressed in ruminal fluid by essential oils. There are also indications that colonization of protein supplements is influenced by essential oils (Wallace *et al.*, 2002). Thus, if the key components of the essential oil mixture which influence different steps can be identified, the prospects are excellent for future development of these materials as a means of regulating protein metabolism in the rumen.

SAPONINS. Another group of natural plant compounds that has received attention as a possible means of manipulating N metabolism in the rumen are the saponins. A decrease in protozoal numbers in the rumen of sheep consuming the pericarp of *Sapindus saponaria* was reported by Diaz *et al.* (1993). A methanol extract prepared from *Sapindus rarak* fruit depressed the protozoal population in the rumen of sheep by 57% (Thalib *et al.*, 1995). Foliage from *Sesbania sesban*, a multipurpose leguminous tree from sub-Saharan Africa, inhibited protozoal activity *in vitro* and transiently depressed the number of protozoa in the rumen of sheep in the UK (Newbold *et al.*, 1997). In all cases the antiprotozoal action has been attributed to saponins in the plant material.

Saponins are found in a wide variety of different plants. Saponins are glycosides, which apparently interact with the sterols present in eukaryotic membranes but not in prokaryotic cells (Hostettmann and Marston, 1995; Cheeke, 1998; Wallace *et al.*, 2002). Consequently, they are toxic to protozoa and act as effective defaunating agents (Newbold *et al.*, 1997; Makkar *et al.*, 1998; Hristov *et al.*, 1999; Chapter 4). Defaunation can lead to improved ruminal N metabolism by increasing bacterial protein synthesis, reducing bacterial lysis through predation, and thus improving the flow of microbial protein leaving the rumen (Section 3.2.5; Koenig *et al.*, 2000). Unfortunately, however, there appears to be adaptation of the mixed microbial population to saponins or saponin-containing plants over time (Teferedegne *et al.*, 1999), thus reducing their effectiveness. Many saponins are degraded rapidly in rumen fluid *in vitro* (Makkar and Becker, 1997), although the resultant sapogenins are apparently more resistant to further degradation (Wang *et al.*, 1998). Sapogenins do not have the antiprotozoal property of the parent saponin (Teferedegne *et al.*, 1999). This observation may explain why although many tropical forages are apparently toxic to protozoa in an initial *in vitro* screen, only a few (presumably those with saponins less likely to be degraded to sapogenin) have prolonged antiprotozoal activity *in vivo* (Teferedegne, 2000). Amongst those with a prolonged activity *in vivo*, up to 14 days before ciliates return (Ivan *et al.*, 2003), *Enterolobium cyclocarpum* has been shown to increase the rate of body weight gain and wool growth in lambs (Leng *et al.*, 1992; Navas-Camacho *et al.*, 1993) presumably as a consequence of an increased supply of microbial protein. Certain fibrolytic bacteria are also detrimentally affected by the addition of saponins (Wallace *et al.*, 1994), as are anaerobic ruminal fungi, important in the initial colonization of plant material (Orpin and Joblin, 1997). These are factors that could have serious consequences on overall fermentation and the advantages and disadvantages of feeding these compounds must be considered and evaluated before use.

Other plant secondary compounds may also prove to be beneficial feed additives. Already polyphenolic compounds such as the tannins have been well characterized with respect to their effect upon ruminal N metabolism (Section 3.2.1.1; McSweeney *et al.*, 2001; Chapter 4). Alkaloids, flavonoids, glycosides, amines and non-proteic AA may also prove beneficial. It may be argued, however, that instead of concentrating on specific compounds, perhaps it would be more beneficial to look at the wider picture and test a wide range of different plant species with the aim of identifying plants which alter rumen fermentation leading to decreased bloat, lactic acidosis, methanogenesis, in addition to increased N retention. Several research consortia have already been put in place to fulfil these aims.

MICROBIAL FEED ADDITIVES. Microbial feed additives investigated in ruminants include bacteria, yeast, fungi and their extracts. To our knowledge, none has been designed with N metabolism as a target, and direct effects on microbial protein metabolism are difficult to discern. *Lactobacillus acidophilus* caused decreased proteolysis in calves (Skrivanova and Marounek, 1990) and rumen-simulating fermenters (Yoon and Stern, 2003). The effects of yeast and fungi on protein metabolism appear to be indirect. *Saccharomyces cerevisiae* addition to the diet decreased NH_3 concentrations, but this effect may result simply because the yeast stimulates growth of ruminal bacteria, which in turn use more NH_3 (Nagaraja *et al.*, 1997). Similarly, *Aspergillus oryzae* extract improves fibre digestion under some dietary conditions (Nagaraja *et al.*, 1997) which in turn would enhance the uptake of NH_3 by ruminal microorganisms. There appears to be scope for microbial feed additives targeted more towards regulating ruminal N metabolism.

3.3.2 Immunization of the animal against unwanted ruminal microorganisms

At first sight, producing a vaccine against specific ruminal microorganisms might seem an improbable way of regulating ruminal N metabolism, because the microbes are, by and large, commensals with no pathogenic effects. Experience has shown, however, that potential exists for such a strategy, provided the target organism(s) is well defined and is solely or mainly responsible for the target activity. It is presumed that the antibodies in saliva provide the antimicrobial effect.

The first report linking immunoglobulins and undesirable ruminal bacteria was that of Horacek *et al.* (1977) who suggested that *S. bovis* might be controlled, via salivary immunoglobulins, by immunization of the host animal. Shu *et al.* (1999) picked up on that idea to produce a vaccine that was successful in controlling *S. bovis* and *Lactobacillus* and thus decreasing the tendency of cattle to acidosis. Immunization against methanogenic archaea has also proved successful (Baker, 1995, 1999), although whether, as suggested above, deamination of AA might also be inhibited has not been investigated. Experiments have suggested that an antiprotozoal vaccine might prove effective (Gnanasampanthan, 1993), which in turn should lead to improved microbial protein flow (Section 3.2.5; Chapter 4). However, no publications have appeared describing vaccines against other potential targets of protein metabolism, such as *Prevotella* spp. and HAP bacteria.

Immunization has been investigated with the aim of decreasing urease activity in the rumen and thereby helping to retain as much of the urea-N in rumen microbial protein as possible. Serum antibodies from sheep immunized against jackbean urease inhibited jackbean urease but not urease from the bacteria on the rumen wall (see Section 3.2.2.1; Marini *et al.*, 2003). As a consequence, urea kinetics in the animals were unaffected. Possibly, therefore, rumen microbial urease is antigenically different to jackbean urease; alternatively, as the enzyme is predominantly intracellular, antibodies probably do not reach the bacterial cytoplasm where the enzyme is located.

3.4 Impact of the –omics Technologies on Understanding Nitrogen Metabolism in the Rumen

During the last decade the advancements of molecular techniques have helped to provide insight into, and evaluate the genetic diversity and phylogenetic relationships of the microorganisms present in the rumen ecosystem without the need for their cultivation. As with most natural ecosystems, especially those involving obligate anaerobes in gut systems, the direct microscopic count of rumen bacteria frequently exceeds the cultivable count (Ward *et al.*, 1990; Tajima *et al.*, 1999). This can be due to problems with obtaining the correct growth conditions to reproduce those encountered *in vivo* or because of the need for a syntrophic association for the growth of certain organisms (McInerney *et al.*, 1981). In addition, a bias towards the over-representation of organisms which are easily cultivated under laboratory conditions in culture-based enumeration and diversity studies can exist. As a result, it has been impossible to fulfil the basic prerequisites for ecological studies, namely population analysis and the exact enumeration and identification of specific community members with respect to their temporal and spatial organization in the rumen in response to dietary changes. However, with the recent advancement in molecular methods employed for the analysis of microbial communities using group- and species-specific 16S rRNA targeted probes (Stahl *et al.*, 1988; Krause and Russell, 1996; Wood *et al.*, 1998), enumeration of these organisms has been achieved, and further insight has been gained into the complex relationships involved in the rumen ecosystem. In addition, new organisms have been identified on the basis of the retrieval and sequencing of SSU ribosomal DNA (rDNA) from clone libraries generated in response to a specific diet (Tajima *et al.*, 1999) and in response to dietary changes which have led to changes in the microbial population (Tajima *et al.*, 2000). Such approaches will prove useful in determining the effect of potential dietary factors, which may decrease proteolysis in the rumen, upon the microbial population.

Already studies using competitive polymerase chain reaction (cPCR) with specific primers based upon 16S rDNA sequences have also been used to enumerate certain groups of proteolytic organisms (*Streptococcus*, *B. fibrisolvens*, *Eubacterium* sp. and *P. bryantii* and *Prevotella* spp.) and the total eubacterial population in ruminants fed different levels of N in their diets (Reilly *et al.*, 2002). It was concluded that the availability of N and carbohydrate in the diet had an effect upon the different species but no effect on the total eubacterial population, nor on the total number of *Prevotella* present, even though differences were observed with the different diets in the number of *P. bryantii* present, indicating an effect upon the subpopulations. The same probes were used in another study to determine the effect of condensed tannins from the forage legume *Lotus corniculatus* upon the numbers of these specific organisms (Min *et al.*, 2002) and their proteolytic activity (Jones *et al.*, 1994; Molan *et al.*, 2001). Although the condensed tannins from this legume reduced the populations of some proteolytic bacteria, total ruminal microbial protein and microbial protein outflow to the abomasum were unchanged, suggesting a species-specific effect of condensed tannins on bacteria in the rumen. Real time PCR studies have also been used to enumerate ruminal microorganisms (Tajima *et al.*, 2001; Ouwerkerk *et al.*, 2002) with specific PCR primers. However, due to the limit in the number of detection probes and PCR primers available at present for ruminal bacteria, this work has been restricted, and further work is required to design and evaluate more probes for the detection and quantification of ruminal microorganisms.

In addition to developing specific probes for the analysis and enumeration of different species and groups of organisms, further studies have been carried out in microbial ecosystems using another non-culturing technique. Termed metagenomics, this technique combines both traditional and functional genomics and allows the study of the full extent of biodiversity that can be found within a microbial ecosystem as complex as the rumen, in terms of the different genes and activities found. Collectively, the genomes of all of the microorganisms present in a system can be studied by cloning large fragments of DNA isolated directly from the environment into bacterial artificial chromosome (BAC) libraries which can be expressed and screened for a variety of different activities of interest. Sequence analysis will then provide a

Table 3.3. Genes involved in nitrogen metabolism which have been cloned from ruminal microorganisms.

Organism	Strain	Gene	Reference
Prevotella albensis	M384^T	Dipeptidyl peptidase IV	Walker *et al.* (2003)
		Dipeptidase	Walker (2001)
Prevotella bryantii	B_14^T	Glutamate dehydrogenase	Wen and Morrison (1996)
Ruminococcus flavefaciens	FD1	Glutamate dehydrogenase	Antonopoulos *et al.* (2003)
Butyrivibrio fibrisolvens	H17c	Glutamine synthetase	Goodman and Woods (1993)
Entodinium caudatum		Glutamate dehydrogenase	Newbold *et al.* (2000b)

^{T}Type strain.

basis for conducting genomic analyses, linking functional and phylogenetic information of organisms which have been previously unculturable. Proteolytic and peptidolytic activities can be easily screened for, and perhaps novel organisms which have proved previously unculturable, may be identified through this approach.

In conjunction with the above techniques, improved efficiency and speed of genomic sequencing will lead to further understanding of the complexity and microbial ecology of the rumen and the identification of genes of interest and the regulatory elements involved in their expression. To date, only a handful of genes which are involved in N metabolism have been cloned and sequenced from ruminal microorganisms (Table 3.3). Several predominant important ruminal microorganisms have had their genome sequenced. Included in this group is the type strain *P. ruminicola*, number 23, which as discussed previously is an important proteolytic organism. Through sequencing of its genome, it is hoped that genes involved in N metabolism will be identified, along with possible regulatory elements, which may allow the down-regulation of proteolytic and peptidolytic activity. Comparative genomic analysis between related organisms can identify conserved regions and subtractive hybridizations can identify differences between the genomes of related organisms. A shotgun cloning approach has also been applied to the rumen ciliate protozoa with the view to obtaining and identifying genes of interest in these organisms. Several proteases and genes which encode enzymes involved in N metabolism have been identified (Neil McEwan, 2003, Aberdeen, UK, personal communication).

The influence of N availability in the growth medium has been studied upon growth and proteolytic activity of several ruminal microorganisms (Kirk *et al.*, 2000; Sales *et al.*, 2000; Walker *et al.*, unpublished results). Like the *in vivo* studies of Reilly *et al.* (2002), these *in vitro* studies have demonstrated that N availability and concentration can affect proteolytic and peptidolytic activities. Analysis of the proteome of these organisms in response to changes in the N composition of the medium may give some indication of the regulatory mechanisms involved in these important organisms and the influence these may have on the overall N metabolism of the host animal.

Thus the combination of genomic, metagenomic and proteomic techniques will provide important influential techniques in the study and understanding of the complexity of the rumen microbial ecosystem and the role it plays in N metabolism in the host animal. By utilizing probes and markers, the influence of dietary factors upon the ruminal microbial population can be studied effectively. As it is the microbial population itself that determines the overall fermentation and metabolic activity of the host, any factor that causes a shift in the microbial population must be evaluated.

3.5 Conclusions

Extensive research has expanded the scientific understanding of the biochemical processes and ecological relationships between the different ruminal microorganisms involved in the metabolism of nitrogenous compounds in the rumen. There can be no argument that the microbial population plays a key role in determining the rate and degree of protein degradation and N utilization. However, it seems as one question is answered, and one problem is solved, more questions and problems arise. It is hoped that with the advancement in molecular techniques and an increase in the knowledge regarding the genetic elements which control the expression of enzymes of microorganisms involved in the proteolytic cascade and the different ecological interactions which occur in the ruminal ecosystem, we may finally achieve the ultimate goal of altering N retention by the host ruminant animal and thus having a positive impact on the environment by controlling N excretion.

References

Abiko, Y., Hayakawa, M., Murai, S. and Takiguehi, H. (1985) Glycylprolyl dipeptidyl aminopeptidase from *Bacteroides gingivalis*. *Journal of Dental Research* 64, 106–111.

Abou Akkada, A.R. and Howard, B.H. (1962) The biochemistry of rumen protozoa. The nitrogen metabolism of *Entodinium*. *Biochemical Journal* 82, 313–320.

Alexander, T.W., Sharma, R., Okine, E.K., Dixon, W.T., Forster, R.J., Stanford, K. and McAllister, T.A. (2002) Impact of feed processing and mixed ruminal culture on the fate of recombinant EPSP synthase and endogenous canola plant DNA. *FEMS Microbiology Letters* 214, 263–269.

Allison, M.J. (1965) Phenylalanine biosynthesis from phenylacetic acid by anaerobic bacteria from the rumen. *Biochemical and Biophysical Research Communications* 18, 30–35.

Allison, M.J. (1969) Biosynthesis of amino acids by ruminal microorganisms. *Journal of Animal Science* 29, 797–807.

Allison, M.J. (1970) Nitrogen metabolism of ruminal micro-organisms. In: Phillipson, A.T. (ed.) *Physiology of Digestion and Metabolism in the Ruminant.* Oriel Press Ltd, Newcastle, UK, pp. 456–473.

Allison, M.J. and Robinson, I.M. (1967) Tryptophan biosynthesis from indole-3-acetic acid by anaerobic bacteria from the rumen. *Biochemical Journal* 102, 36P.

Amin, M.R. and Onodera, R. (1997) Synthesis of phenylalanine and production of other related compounds from phenylpyruvic acid and phenylacetic acid by ruminal bacteria, protozoa, and their mixture in vitro. *Journal of General and Applied Microbiology* 43, 9–15.

Amin, M.R., Onodera, R., Khan, R.I., Wallace, R.J. and Newbold, C.J. (2001) Purification and properties of glutamate–phenylpyruvate aminotransferase from rumen bacterium *Prevotella albensis*. *Pakistan Journal of Biological Sciences* 4, 1377–1381.

Amin, M.R., Onodera, R., Khan, R.I., Wallace, R.J. and Newbold, C.J. (2002) Purification and properties of glutamate–phenylpyruvate aminotransferase from the ruminal bacterium, *Prevotella bryantii* $B_1$4. *Anaerobe* 8, 101–107.

Amos, H.E. and Evans, J. (1976) Supplementary protein for low quality bermudagrass diets and microbial protein synthesis. *Journal of Animal Science* 43, 861–868.

Antonopoulos, D.A., Aminov, R.I., Duncan, P.A., White, B.A. and Mackie, R.I. (2003) Characterization of the gene encoding glutamate dehydrogenase (gdhA) from the ruminal bacterium *Ruminococcus flavefaciens* FD-1. *Archives in Microbiology* 179, 184–190.

Argyle, J.L. and Baldwin, R.L. (1989) Effects of amino acids and peptides on rumen microbial growth yields. *Journal of Dairy Science* 72, 2017–2027.

Armstead, I.P. and Ling, J.R. (1993) Variations in the uptake and metabolism of peptides and amino acids by mixed ruminal bacteria in vitro. *Applied and Environmental Microbiology* 59, 3360–3366.

Asao, N., Ushida, K. and Kojima, Y. (1993) Proteolytic activity of rumen fungi belonging to the genera *Neocallimastix* and *Piromyces*. *Letters in Applied Microbiology* 16, 247–250.

Atasoglu, C. and Wallace, R.J. (2002) De novo synthesis of amino acids by the ruminal anaerobic fungi, *Piromyces communis* and *Neocallimastix frontalis*. *FEMS Microbiology Letters* 212, 243–247.

Atasoglu, C., Valdes, C., Walker, N.D., Newbold, C.J. and Wallace, R.J. (1998) *De novo* synthesis of amino acids by the ruminal bacteria *Prevotella bryantii* $B_1$4, *Selenomonas ruminantium* HD4, and *Streptococcus bovis* ES1. *Applied and Environmental Microbiology* 64, 2836–2843.

Atasoglu, C., Valdes, C., Newbold, C.J. and Wallace, R.J. (1999) Influence of peptides and amino acids on fermentation rate and *de novo* synthesis of amino acids by mixed micro-organisms from the sheep rumen. *British Journal of Nutrition* 81, 307–314.

Atasoglu, C., Newbold, C.J. and Wallace, R.J. (2001) Incorporation of $[^{15}N]$ammonia by the cellulolytic ruminal bacteria *Fibrobacter succinogenes* BL2, *Ruminococcus albus* SY3, and *Ruminococcus flavefaciens* 17. *Applied and Environmental Microbiology* 67, 2819–2822.

Atasoglu, C., Guliye, A.Y. and Wallace, R.J. (2003a) Use of a deletion approach to assess the amino acid requirements for optimum fermentation by mixed microorganisms from the sheep rumen. *Animal Science* 76, 147–153.

Atasoglu, C., Guliye, A.Y. and Wallace, R.J. (2003b) Use of stable isotopes to measure *de novo* synthesis and turnover of amino acid-C and -N in mixed micro-organisms from the sheep rumen *in vitro*. *British Journal of Nutrition* 91, 253–261.

Atkinson, P.R., Hilton, M.D., Attwood, G.T. and Lamboo y, P.K. (1995) Purification and preliminary characterization of dDAP, a novel dipeptidyl ami-

nopeptidase from *Dictyostelium discoideum*. *Biochemistry* 34, 10827–10834.
Atlan, D., Laloi, P. and Portalier, R. (1990) X-Prolyl-dipeptidyl aminopeptidase of *Lactobacillus delbrueckii* subsp. *bulgaricus*: characterisation of the enzyme and isolation of deficient mutants. *Applied and Environmental Microbiology* 56, 2174–2179.
Attwood, G.T. and Reilly, K. (1995) Identification of proteolytic rumen bacteria isolated from New Zealand cattle. *Journal of Applied Bacteriology* 79, 22–29.
Attwood, G.T. and Reilly, K. (1996) Characterisation of proteolytic activities of rumen bacterial isolates from forage-fed cattle. *Journal of Applied Bacteriology* 81, 545–552.
Attwood, G.T., Reilly, K. and Patel, B.K. (1996) *Clostridium proteoclasticum* sp. nov., a novel proteolytic bacterium from the bovine rumen. *International Journal of Systematic Bacteriology* 46, 753–758.
Attwood, G.T., Klieve, A.V., Ouwerkerk, D. and Patel, B.K. (1998) Ammonia-hyperproducing bacteria from New Zealand ruminants. *Applied and Environmental Microbiology* 64, 1796–1804.
Atwal, A.S. and Sauer, F.D. (1974) Enzymes of pyruvate and phosphoenol pyruvate carboxylation in rumen bacteria. *Canadian Journal of Animal Science* 54, 595–603.
Avgustin, G.A., Wright, F. and Flint, H.J. (1994) Genetic diversity and phylogenetic relationships among strains of *Prevotella* (formerly *Bacteroides*) *ruminicola* from the rumen. *International Journal of Systematic Bacteriology* 44, 246–255.
Avgustin, G.A., Wallace, R.J. and Flint, H.J. (1997) Phenotypic diversity among ruminal isolates of *Prevotella ruminicola*: proposal of *Prevotella brevis* sp. nov., *Prevotella bryantii* sp. nov., and *Prevotella albensis* sp. nov. and redefinition of *Prevotella ruminicola*. *International Journal of Systematic Bacteriology* 47, 284–288.
Baker, S.K. (1995) *Method for Improving Utilization of Nutrients by Ruminant or Ruminant Like Animals*. International Patent No. WO9511041.
Baker, S.K. (1999) Rumen methanogens and inhibition of methanogenesis. *Australian Journal of Agricultural Research* 50, 1293–1298.
Ben-Ghedalia, D., McMeniman, N.P. and Amstrong D.G. (1978) The effect of partially replacing urea nitrogen with protein N on N capture in the rumen of sheep fed a purified diet. *British Journal of Nutrition* 39, 37–44.
Bhatia, S.K., Pradhan, K. and Singh, R. (1979) Microbial transaminase activities and their relationship with bovine rumen metabolites. *Journal of Dairy Science* 62, 441–446.
Bhatia, S.K., Pradhan, K. and Singh, R. (1980) Ammonia anabolizing enzymes in cattle and buffalo fed varied nonprotein nitrogen and carbohydrates. *Journal of Dairy Science* 63, 1104–1108.
Blackburn, T.H. (1965) Nitrogen metabolism in the rumen. In: Dougherty, R.W., Allen, R.S., Burroughs, W., Jacobson, N.L. and McGilliard, A.D. (eds) *Physiology of Digestion in the Ruminant*. Butterworths, London, pp. 322–334.
Blackburn, T.H. (1968) The protease liberated from *Bacteroides amylophilus* strain H18 by mechanised disintegration. *Journal of General Microbiology* 53, 37–51.
Blackburn, T.H. and Hullah, W.A. (1974) The cell-bound protease of *Bacteroides amylophilus* H18. *Canadian Journal of Microbiology* 20, 435–441.
Bladen, H.A., Bryant, M.D. and Doetsch, R.N. (1961) A study of bacterial species from the rumen which produce ammonia from protein hydrolyzate. *Applied Microbiology* 9, 175–180.
Blake, J.S., Salter, D.N. and Smith, R.H. (1983) Incorporation of nitrogen into rumen bacterial fractions of steers given protein- and urea-containing diets. Ammonia assimilation into intracellular bacterial amino acids. *British Journal of Nutrition* 50, 769–782.
Booth, M., Fhaolain, I.N., Jennings, P.V. and O'Cuinn, G. (1990a) Purification and characterisation of a post-proline dipeptidyl aminopeptidase from *Streptococcus cremoris* AM2. *Journal of Dairy Research* 57, 89–99.
Booth, M., Jennings, P.V., Fhaolain, I.N. and O'Cuinn, G. (1990b) Prolidase activity of *Lactococcus lactis* subsp. *cremoris* AM2: partial purification and characterisation. *Journal of Dairy Research* 57, 245–254.
Bordallo, C.J., Schwencke, J. and Rendueles, R. (1984) Localisation of the thermosensitive X-prolyl dipeptidyl aminopeptidase in the vacuolar membrane of *Saccharomyces cerevisiae*. *FEBS Letters* 173, 199–203.
Brock, F.M., Forsberg, C.W. and Buchanan-Smith, J.G. (1982) Proteolytic activity of rumen microorganisms and effects of proteinase inhibition. *Applied and Environmental Microbiology* 44, 561–569.
Broderick, G.A. and Balthrop, J.E. (1979) Chemical inhibition of amino acid deamination by ruminal microbes *in vitro*. *Journal of Animal Science* 49, 1101–1111.
Broderick, G.A. and Wallace, R.J. (1988) Effects of dietary nitrogen source on concentrations of ammonia, free amino acids and fluorescamine-reactive peptides in the sheep rumen. *Journal of Animal Science* 66, 2233–2238.
Broderick, G.A., Wallace, R.J. and McKain, N. (1988) Uptake of small neutral peptides by mixed rumen microorganisms *in vitro*. *Journal of Science Food and Agriculture* 42, 109–118.
Broudiscou, L., Van Nevel, C.J. and Demeyer, D.I. (1990a) Incorporation of soya oil hydrolysate in the diet of defaunated or refaunated sheep: effect on rumen fermentation in vitro. *Archives of Animal Nutrition* 40, 329–337.
Broudiscou, L., Van Nevel, C.J. and Demeyer, D.I. (1990b) Effect of soya oil hydrolysate on rumen

digestion in defaunated and refaunated sheep. *Animal Feed Science and Technology* 30, 51–67.
Callaway, T.R. and Russell, J.B. (1999) Selection of a highly monensin-resistant *Prevotella bryantii* subpopulation with altered outer membrane characteristics. *Applied and Environmental Microbiology* 65, 4753–4759.
Callaway, T.R., Carneiro De Melo, A.M. and Russell, J.B. (1997) The effect of nisin and monensin on ruminal fermentations in vitro. *Current Microbiology* 35, 90–96.
Castillo, A.R., Kebreab, E., Beever, D.E., Barbi, J.H., Sutton, J.D., Kirby, H.C. and France, J. (2001a) The effect of energy supplementation on nitrogen utilization in lactating dairy cows fed grass silage diets. *Journal of Animal Science* 79, 240–246.
Castillo, A.R., Kebreab, E., Beever, D.E., Barbi, J.H., Sutton, J.D., Kirby, H.C. and France, J. (2001b) The effect of protein supplementation on nitrogen utilization in lactating dairy cows fed grass silage diets. *Journal of Animal Science* 79, 247–253.
Chalupa, W. (1972) Metabolic aspects of non-protein nitrogen utilization in ruminant animals. *Federation Proceedings* 31, 1152–1164.
Chalupa, W. (1976) Degradation of amino acids by the mixed rumen microbial population. *Journal of Animal Science* 43, 828–834.
Chalupa, W. (1980) Chemical control of rumen microbial metabolism. In: Ruckebusch, Y. and Thivend, P. (eds) *Digestive Physiology and Metabolism in Ruminants.* MTP Press, Lancaster, UK, pp. 325–347.
Chalupa, W., Clark, J., Opliger, P. and Lavker, R. (1970) Ammonia metabolism in rumen bacteria and mucosa from sheep fed soy protein or urea. *Journal of Nutrition* 100, 161–169.
Chalupa, W., Patterson, J.A., Parish, R.C. and Chow, A.W. (1983) Effects of diaryliodonium chemicals on nitrogen metabolism in growing steers. *Journal of Animal Science* 57, 195–200.
Chan, S.A.T., Toursarkissian, K., Sweeney, J.P. and Jones, T.H.D. (1985) Dipeptidyl aminopeptidases in *Dictyostelium discoideum. Biochemical and Biophysical Research Communications* 127, 962–968.
Cheeke, P.R. (1998) *Natural Toxicants in Feeds, Forages and Poisonous Plants.* Interstate Publishers, Danville, IL, 283 pp.
Chen, G. and Russell, J.B. (1988) Fermentation of peptides and amino acids by a monensin-sensitive ruminal *Peptostreptococcus. Applied and Environmental Microbiology* 54, 2742–2749.
Chen, G. and Russell, J.B. (1989) More monensin-sensitive, ammonia-producing bacteria from the rumen. *Applied and Environmental Microbiology* 55, 1052–1057.
Chen, G.J. and Russell, J.B. (1990) Transport and deamination of amino acids by a Gram-positive, monensin-sensitive ruminal bacterium. *Applied and Environmental Microbiology* 56, 2186–2192.
Chen, G. and Russell, J.B. (1991) Effect of monensin and a protonophore on protein degradation, peptide accumulation and deamination by mixed ruminal microorganisms *in vitro. Journal of Animal Science* 69, 2196–2203.
Chen, G., Sniffen, C.J. and Russell, J.B. (1987a) Concentration and estimated flow of peptides from the rumen of dairy cattle: effects of protein quantity, protein solubility, and feeding frequency. *Journal of Dairy Science* 70, 983–992.
Chen, G., Strobel, H.J., Russell, J.B. and Sniffen, C.J. (1987b) Effect of hydrophobicity on utilization of peptides by ruminal bacteria in vitro. *Applied and Environmental Microbiology* 53, 2021–2025.
Chen, Z., Potempa, J., Polanowski, A., Wikstrom, M. and Travis, J. (1992) Purification and characterization of a 50 kDa cysteine-proteinase (gingipain) from *Porphyromonas gingivalis. Journal of Biological Chemistry* 267, 18896–18901.
Cheng, K.-J. and Wallace, R.J. (1979) The mechanism of passage of endogenous urea through the rumen wall and the role of ureolytic epithelial bacteria in the urea flux. *British Journal of Nutrition* 42, 553–557.
Chikunya, S., Newbold, C.J., Rode, L., Chen, X.B. and Wallace, R.J. (1996) Influence of dietary rumen-degradable protein on bacterial growth in the rumen of sheep receiving different energy sources. *Animal Feed Science and Technology* 63, 333–340.
Coleman, G.S. (1962) The preparation and survival of almost bacteria free suspensions of *Entodinium caudatum. Journal of General Microbiology* 28, 271–281.
Coleman, G.S. (1964) The metabolism of *Escherichia coli* and other bacteria by *Entodinium caudatum. Journal of General Microbiology* 37, 209–223.
Coleman, G.S. (1967) The metabolism of free amino acids by washed suspensions of the rumen ciliate *Entodinium caudatum. Journal of General Microbiology* 47, 433–447.
Coleman, G.S. (1983) Hydrolysis of fraction 1 leaf protein and casein by rumen entodiniomorphid protozoa. *Journal of Applied Bacteriology* 55, 111–118.
Coleman, G.S. and Hall, F.J. (1972) Fine structural studies on digestion of bacterial species in the rumen ciliate, *Entodinium caudatum. Tissue and Cell* 4, 37–48.
Cotta, M.A. and Hespell, R.B. (1986) Protein and amino acid metabolism of rumen bacteria. In: Milligan, L.P., Grovum, W.L. and Dobson, A. (eds) *Control of Digestion and Metabolism in Ruminants.* Prentice-Hall, Englewood Cliffs, New Jersey, pp. 122–136.
Cotta, M.A. and Russell, J.B. (1982) Effect of peptides and amino acids on efficiency of rumen bacterial protein synthesis in continuous culture. *Journal of Dairy Science* 65, 226–234.
Cruz Soto, R., Muhammed, S.A., Newbold, C.J., Stewart, C.S. and Wallace, R.J. (1994) Influence of peptides, amino acids and urea on microbial activity in the rumen of sheep receiving grass hay and

on the growth of rumen bacteria *in vitro*. *Animal Feed Science and Technology* 49,151–161.

Debroas, D. and Blanchart, G. (1993) Interactions between proteolytic and cellulolytic rumen bacteria during hydrolysis of plant cell wall protein. *Reproduction, Nutrition, Development* 33, 283–288.

Debroas, D., Depardon, N. and Blanchart, G. (1998) Study of enzyme activities and physicochemical parameters during hydrolysis of soy peptides by *Prevotella ruminicola*. *Journal of the Science of Food and Agriculture* 78, 453–460.

Denholm, A.M. and Ling, J.R. (1989) *In vitro* metabolism of 2,2′-Diaminopimelic acid from Gram-positive and Gram-negative bacterial cells by ruminal protozoa and bacteria. *Applied and Environmental Microbiology* 55, 212–218.

Dennis, S.M., Nagaraja, T.G. and Dayton, A.D. (1986) Effect of lasalocid, monensin and thiopeptin on rumen protozoa. *Research in Veterinary Science* 41, 251–256.

Depardon, N., Debroas, D. and Blanchart, G. (1995) Breakdown of peptides from a soya bean hydrolysate by rumen bacteria. Simultaneous study of enzyme activities and of two physicochemical parameters: molecular weight and hydrophobicity *Journal of the Science of Food and Agriculture* 68, 25–31.

Depardon, N., Debroas, D. and Blanchart, G. (1996) Breakdown of peptides from a casein hydrolysate by rumen bacteria. Simultaneous study of enzyme activities and physicochemical parameters. *Reproduction, Nutrition, Development* 36, 457–466.

Devant, M., Ferret, A., Gasa, J., Calsamiglia, S. and Casals, R. (2000) Effects of protein concentration and degradability on performance, ruminal fermentation, and nitrogen metabolism in rapidly growing heifers fed high-concentrate diets from 100 to 230 kg body weight. *Journal of Animal Science* 78, 1667–1676.

Devant, M., Ferret, A., Calsamiglia, S., Casals, R. and Gasa, J. (2001) Effect of nitrogen source in high-concentrate, low-protein beef cattle diets on microbial fermentation studied in vivo and in vitro. *Journal of Animal Science* 79, 1944–1953.

Diaz, A., Avendan M. and Escobar, A. (1993) Evaluation of *Sapindus saponaria* as a defaunating agent and its effects on different ruminal digestion parameters. *Livestock Research for Rural Development* 5, 1–10.

Dijkerman, R., Ledeboer, J., Verkley, A.J., Op den Camp, H.J.M., Van der Drift, C. and Vogels, G.D. (1997) The anaerobic fungus *Piromyces* sp. strain E2: nitrogen requirement and enzymes involved in primary nitrogen metabolism. *Archives of Microbiology* 166, 399–404.

Duggan, P.S., Chambers, P.A., Heritage, J. and Forbes, J.M. (2003) Fate of genetically modified maize DNA in the oral cavity and rumen of sheep. *British Journal of Nutrition* 89, 159–166.

Duncan, P.A., White, B.A. and Mackie, R.I. (1992) Purification and properties of NADP-dependent glutamate dehydrogenase from *Ruminococcus flavefaciens* FD-1. *Applied and Environmental Microbiology* 58, 4032–4037.

Erfle, J.D., Sauer, F.D. and Mahadevan, S. (1977) Effect of ammonia concentration on activity of enzymes of ammonia assimilation and on synthesis of amino acids by mixed rumen bacteria in continuous culture. *Journal of Dairy Science* 60, 1064–1072.

Eschenlauer, S.C.P., McEwan, N.R., Onodera, R., Wallace, R.J., Calza, R.E. and Newbold, C.J. (2000) Properties of a lysozyme from the rumen ciliate protozoon *Entodinium caudatum*. *Reproduction, Nutrition, Development* 40, 196–197.

Eschenlauer, S.C.P., McKain, N., Walker, N.D., McEwan, N.R., Newbold, C.J. and Wallace, R.J. (2002) Ammonia production by ruminal microorganisms and enumeration, isolation, and characterization of bacteria capable of growth on peptides and amino acids from the sheep rumen. *Applied and Environmental Microbiology* 68, 4925–4931.

Eugène, M., Archimède, H. and Sauvant, D. (2003) Quantitative meta-analysis on the effects of defaunation of the rumen on growth, intake and digestion in ruminants. *Livestock Production Science* 85, 81–97.

Falconer, M.L. and Wallace, R.J. (1998) Variation in proteinase activities in the rumen. *Journal of Applied Microbiology* 84, 377–382.

Fenn, P. and Leng, R.A. (1988) The effects of bentonite on wool growth of faunated and fauna-free sheep. In: Nolan, J.V., Leng, R.A. and Demeyer, D.I. (eds) *The Roles of Protozoa and Fungi in Ruminant Digestion*. Penambul Books, Armidale, Australia, pp. 1–3.

Ferguson, K.A. (1975) The protection of dietary proteins and amino acids against microbial fermentation in the rumen. In: McDonald, I.W. and Warner, A.C.I. (eds) *Digestion and Metabolism in the Ruminant*. The University of New England Publishing Unit, Armidale, Australia, pp. 448–464.

Flint, H.J. and Scott, K.P. (2000) Genetics of rumen microorganisms: gene transfer, genetic analysis and strain manipulation. In: Cronjé, P.B. (ed.) *Ruminant Physiology, Digestion, Metabolism, Growth and Reproduction*. CAB International, Wallingford, UK, pp. 389–408.

Flint, H.J. and Thomson, A.M. (1990) Deoxyribonuclease activity in rumen bacteria. *Letters in Applied Microbiology* 11, 18–21.

Floret, F., Chaudhary, L.C., Ellis, W.C., El Hassan, S., McKain, N., Newbold, C.J. and Wallace, R.J. (1999) Influence of 1-[(E)-2-(2-methyl-4-nitrophenyl)diaz-1-enyl]pyrrolidine-2-carboxylic acid and diphenyliodonium chloride on ruminal protein metabolism and ruminal microorganisms. *Applied and Environmental Microbiology* 65, 3258–3260.

Ford, A.L. and Milligan, L.P. (1970) Tracer studies of urea recycling in sheep. *Canadian Journal of Animal Science* 50, 129–135.

Forsberg, C.W., Lovelock, L.K.A., Krumholz, L. and Buchanan-Smith, J.G. (1984) Protease activities of rumen protozoa. *Applied and Environmental Microbiology* 47, 101–110.

Forster, R.J., Teather, R.M., Gong, J. and Deng, S.J. (1996) 16S rDNA analysis of *Butyrivibrio fibrisolvens*: phylogenetic position and relation to butyrate-producing anaerobic bacteria from the rumen of white-tailed deer. *Letters in Applied Microbiology* 23, 218–222.

Froetschel, M.A., Groom, W.J., Gaskins, H.R., Leonard, E.S. and Whitacre, M.D. (1983) Effects of avoparcin on ruminal propionate production and amino acid degradation in sheep fed high and low fibre diets. *Journal of Nutrition* 113, 1355–1362.

Gibson, S.A.W. and MacFarlane, G.T. (1988) Studies of the proteolytic activity of *Bacteroides fragilis*. *Journal of General Microbiology* 134, 19–27.

Gilbert, H.J., Hazlewood, G.P., Laurie, J.I., Orpin, C.G. and Xue, G.P. (1992) Homologous catalytic domains in a rumen fungal xylanase: evidence for gene duplication and prokaryotic origin. *Molecular Microbiology* 6, 2065–2072.

Gnanasampanthan, G. (1993) Immune responses of sheep to rumen ciliates and the survival and activity of antibodies in the rumen fluid. PhD thesis, University of Adelaide, Adelaide, Australia.

Goodman, H.J. and Woods, D.R. (1993) Cloning and nucleotide sequence of the *Butyrivibrio fibrisolvens* gene encoding a type III glutamine synthetase. *Journal of General Microbiology* 139, 1487–1493.

Griswold, K.E., White, B.A. and Mackie, R.I. (1999a) Proteolytic activities of the starch-fermenting ruminal bacterium, *Streptococcus bovis*. *Current Microbiology* 39, 180–186.

Griswold, K.E., White, B.A. and Mackie, R.I. (1999b) Diversity of extracellular proteolytic activities among *Prevotella* species from the rumen. *Current Microbiology* 39, 187–194.

Guenoune, D., Landau, S., Amir, R., Badani, H., Devash, L., Wolf, S. and Galili, S. (2002) Resistance of soybean vegetative storage proteins (S-VSPs) to proteolysis by rumen microorganisms. *Journal of Agriculture and Food Chemistry* 50, 2256–2260.

Haborne, J.B. (1989) Biosynthesis and function of antinutritional factors in plants. *Aspects of Applied Biology* 19, 21–28.

Hancock, K.R., Ealing, P.M. and White, D.W. (1994) Identification of sulphur-rich proteins which resist rumen degradation and are hydrolysed rapidly by intestinal proteases. *British Journal of Nutrition* 72, 855–863.

Hausinger, R.P. (1986) Purification of a nickel-containing urease from the rumen anaerobe *Selenomonas ruminantium*. *Journal of Biological Chemistry* 261, 7866–7870.

Hazlewood, G.P. and Edwards, R. (1981) Proteolytic activities of a rumen bacterium, *Bacteroides ruminicola* R8/4. *Journal of General Microbiology* 125, 11–15.

Hazlewood, G.P., Jones, G.A. and Mangan, J.L. (1981) Hydrolysis of leaf fraction 1 protein by the proteolytic rumen bacterium *Bacteroides ruminicola* R8/4. *Journal of General Microbiology* 123, 223–232.

Hazlewood, G.P., Orpin, C.G., Greenwood, Y. and Black, M.E. (1983) Isolation of proteolytic rumen bacteria by use of selective medium containing leaf fraction 1 (ribulose bis phosphate carboxylase). *Applied and Environmental Microbiology* 45, 1780–1784.

Hespell, R.B. (1984) Influence of ammonia assimilation pathways and survival strategy on ruminal microbial growth. In: Gilchrist, F.M.C. and Mackie, R.I. (eds) *Herbivore Nutrition in the Subtropics and Tropics*. Science Press, South Africa, pp. 346–358.

Hino, T. and Russell, J.B. (1985) Effect of reducing-equivalent disposal and NADH/NAD on deamination of amino acids by intact rumen microorganisms and their cell extracts. *Applied and Environmental Microbiology* 50, 1368–1374.

Hino, T. and Russell, J.B. (1987) Relative contributions of ruminal bacteria and protozoa to the degradation of protein *in vitro*. *Journal of Animal Science* 64, 261–270.

Hobson, P.N., McDougall, E.I. and Summers, R. (1968) The nitrogen sources of *Bacteroides amylophilus*. *Journal of General Microbiology* 50, i.

Hobson, P.N., Summers, R., Postgate, J.R. and Ware, D.A. (1973) Nitrogen fixation in the rumen of a living sheep. *Journal of General Microbiology* 77, 225–226.

Holtenius, P. (1957) Nitrite poisoning in sheep, with special reference to the detoxification of nitrite in the rumen. *Acta Agriculturae Scandinavia* 7, 113–163.

Horacek, G.L., Fina, L.R., Tillinghast, H.S. and Gettings, R.L. (1977) Agglutinating immunoglobulins to encapsulated *Streptococcus bovis* in bovine serum and saliva and a possible relation to feedlot bloat. *Canadian Journal of Microbiology* 23, 100–106.

Hostettmann, K. and Marston, A. (1995) *Saponins*. University Press, Cambridge, 548 pp.

Houpt, R.T. (1970) Transfer of urea and ammonia to the rumen. In: Phillipson, A.T. (ed.) *Physiology of Digestion and Metabolism in the Ruminant*. Oriel Press, Newcastle upon Tyne, UK, pp. 119–131.

Hristov, A.N., McAllister, T.A., Van Herk, F.H., Cheng, K.-J., Newbold, C.J. and Cheeke, P.R. (1999) Effect of *Yucca schidigera* on ruminal fermentation and nutrient digestion in heifers. *Journal of Animal Science* 77, 2554–2563.

Hullah, W.A. and Blackburn, T.H. (1971) Uptake and incorporation of amino acids and peptides by *Bacteroides amylophilus. Applied Microbiology* 21, 187–191.

Hume, I.D. (1970) Synthesis of microbial protein in the rumen. III. The effect of dietary protein. *Australian Journal of Agricultural Research* 21, 305–314.

Hungate, R.E. and Stack, R.J. (1982) Phenylpropanoic acid:growth factor for *Ruminococcus albus. Applied and Environmental Microbiology* 44, 79–83.

Ikwuegbu, O.A. and Sutton, J.D. (1982) The effect of varying the amount of linseed oil supplementation on rumen metabolism in sheep. *British Journal of Nutrition* 48, 365–375.

Isozaki, T., Gillin, A.G., Swanson, C.E. and Sands, J.M. (1994) Protein restriction sequentially induces new urea transport processes in rat initial IMCD. *American Journal of Physiology* 266, F756–F761.

Ivan, M., Dayrell, M.D.S., Mahadevan, S. and Hidiroglou, M. (1992) Effects of bentonite on wool growth and nitrogen metabolism in fauna-free and faunated sheep. *Journal of Animal Science* 70, 3194–3202.

Ivan, M., Neill, L. and Entz, T. (2000a) Ruminal fermentation and duodenal flow following progressive inoculations of fauna-free wethers with major individual species of ciliate protozoa or total fauna. *Journal of Animal Science* 78, 750–759.

Ivan, M., Neill, L., Forster, R., Alimon, R., Rode, L.M. and Entz, T. (2000b) Effects of *Isotricha, Dasytricha, Entodinium,* and total fauna on ruminal fermentation and duodenal flow in wethers fed different diets. *Journal of Dairy Science* 83, 776–787.

Ivan, M., Koenig, K.M., Teferedegne, B., Newbold, C.J., Entz, T., Rode, L.M. and Ibrahim, M. (2003) Effects of dietary *Enterolobium cyclocarpum* foliage on the population dynamics of rumen ciliate protozoa in sheep. *Small Ruminant Research* 52, 81–91.

Jarvis, S.C. (1994) The pollution potential and flows of nitrogen to waters and atmosphere from grassland under grazing. In: Dewi, A.P., Axford, R.F.E., Fayez, M.M., Marai, M. and Omed, H.M. (eds) *Pollution in Livestock Production Systems.* CAB International, Wallingford, UK, pp. 227–239.

John, A., Isaacson, H.R. and Bryant, M.P. (1974) Isolation and characteristics of a ureolytic strain of *Selenomonas ruminantium. Journal of Dairy Science* 57, 1003–1014.

Johnson, D.E. (1974) Adaptational responses in nitrogen and energy balance of lambs fed a methane inhibitor. *Journal of Animal Science* 38, 154–157.

Jones, G.A. (1972) Dissimilatory metabolism of nitrate by the rumen microbiota. *Canadian Journal of Microbiology* 18, 1783–1787.

Jones, G.A., McAllister, T.A., Muir, A.D. and Cheng, K.-J. (1994) Effects of sainfoin (*Onobrychis viciifolia* Scop.) condensed tannins on growth and proteolysis by 4 strains of ruminal bacteria. *Applied and Environmental Microbiology* 60, 1374–1378.

Jones, K. and Thomas, J.G. (1974) Nitrogen fixation by rumen contents of sheep. *Journal of General Microbiology* 85, 97–101.

Jouany, J.P. (1996) Effect of rumen protozoa on nitrogen utilization by ruminants. *Journal of Nutrition* 126, 1335S–1346S.

Jouany, J.P. and Thivend, P. (1986) *In vitro* effects of avoparcin on protein degradability and rumen fermentation. *Animal Feed Science and Technology* 15, 215–229.

Jouany, J.P., Ivan, M., Papon, Y. and Lassalas, B. (1992) Effects of *Isotricha, Eudiplodinium, Epidinium + Entodinium* and a mixed population of rumen protozoa on the *in vitro* degradation of fishmeal, soybean meal and casein. *Canadian Journal of Animal Science* 72, 871–880.

Kaufmann, W. and Lüpping, W. (1982) Protected proteins and protected amino acids for ruminants. In: Miller, E.L., Pike, I.H. and Van Es, A.J.H. (eds) *Protein Contribution of Feedstuffs for Ruminants.* Butterworths, London, pp. 36–75.

Kebreab, E., France, J., Mills, J.A.N., Allison, R. and Dijkstra, J. (2002) A dynamic model of N metabolism in the lactating dairy cow and an assessment of impact of N excretion on the environment. *Journal of Animal Science* 80, 248–259.

Kennedy, P.M. and Milligan, L.P. (1980) The degradation and utilization of endogenous urea in the gastrointestinal tract of ruminants: a review. *Canadian Journal of Animal Science* 60, 205–221.

Kim, H.-D., Walker, N.D., McKain, N. and Wallace, R.J. (2001) Purification and properties of a type-II-like dipeptidyl peptidase from the ruminal peptidolytic bacterium, *Prevotella albensis* M384. *Anaerobe* 7, 307–313.

Kirk, J.M., Woodward, C.L., Ellis, W.C. and Ricke, S.C. (2000) Glutamine synthetase and protease enzyme activities and growth response of ruminal bacterium *Prevotella ruminicola* strain B_14 to nitrogen source and concentration. *Journal of Environmental Science and Health B* 35, 103–120.

Koenig, K.M., Newbold, C.J., McIntosh, F.M. and Rode, L.M. (2000) Effects of protozoa on bacterial nitrogen recycling in the rumen. *Journal of Animal Science* 78, 2431–2445.

Komatani, K., Onodera, R., Karrer, K.M., White, B.A. and Morgavi, D.P. (2000) Further trial of molecular cloning of a chitinase gene from a rumen protozoon, *Entodinium caudatum. Proceedings of Japanese Society for Rumen Metabolism and Physiology* 11, 73.

Kopecny, J. and Wallace, R.J. (1982) Cellular location and some properties of proteolytic enzymes of rumen bacteria. *Applied and Environmental Microbiology* 43, 1026–1033.

Kopecny, J., Logar, R.M. and Kobayashi, Y. (2001) Phenotypic and genetic data supporting reclassification of *Butyrivibrio fibrisolvens* isolates. *Folia Microbiology (Praha)* 46, 45–48.

Kowalcyk, J., Ørskov, E.R., Robinson, J.J. and Stewart, C.S. (1977) Effect of fat supplementation on voluntary food intake and rumen metabolism in sheep. *British Journal of Nutrition* 37, 251–257.

Krause, D.O. and Russell, J.B. (1996) An rRNA approach for assessing the role of obligate amino acid fermenting bacteria in ruminal amino acid deamination. *Applied and Environmental Microbiology* 62, 815–821.

Laukova, A. and Koniarova, I. (1995) Survey of urease activity in ruminal bacteria isolated from domestic and wild ruminants. *Microbios* 84, 7–11.

Lazdunski, A.M. (1989) Peptidases and proteases of *Escherichia coli* and *Salmonella typhimurium*. *FEMS Microbiology Reviews* 63, 265–276.

Lee, C.M. and Snyder, S.H. (1982) Dipeptidyl-aminopeptidase III of rat brain. Selective affinity for enkephalin and angiotensin. *Journal of Biological Chemistry* 257, 12043–12050.

Lee, S.S., Ha, J.K. and Cheng, K.-J. (2000) Relative contributions of bacteria, protozoa, and fungi to *in vitro* degradation of orchard grass cell walls and their interactions. *Applied and Environmental Microbiology* 66, 3807–3813.

Leibholz, J. (1969) Effect of diet on the concentration of free amino acids, ammonia and urea in the rumen and blood plasma of the sheep. *Journal of Animal Science* 29, 628–633.

Lenartova, V., Holovska, K., Javorsky, P., Rybosova, E. and Havassy, I. (1987) Glutamate dehydrogenase and glutamine synthetase activity of the bacteria of the sheep's rumen after different nitrogen intake. *Physiologia Bohemoslov* 36, 471–476.

Leng, L., Szanyiova, M. and Boda, K. (1985) The renal response of sheep to a low dietary nitrogen intake. *Physiologia Bohemoslov* 34, 147–154.

Leng, R.A. and Nolan, J.V. (1984) Nitrogen metabolism in the rumen. *Journal of Dairy Science* 67, 1072–1089.

Leng, R.A., Bird, S.H., Klieve, A., Choo, B.S., Ball, F.M., Asefa, G., Brumby, P., Mudgal, V.D., Chaudhry, U.B., Haryono, S.U. and Hendratno, N. (1992) The potential for tree forage supplements to manipulate rumen protozoa to enhance protein to energy ratios in ruminants fed on poor quality forages. In: Speedy, A. and Pugliese, P.L. (eds) *Legume Trees and Other Fodder Trees as Protein Sources for Livestock.* FAO, Rome, pp. 177–191.

Lesk, E.M. and Blackburn, T.H. (1971) Purification of *Bacteroides amylophilus* protease. *Journal of Bacteriology* 106, 394–402.

Lewis, D. (1951) The metabolism of nitrate and nitrite in the sheep. *Biochemical Journal* 48, 175–180.

Ling, J.R. (1990) Digestion of bacterial cell walls in the rumen. In: Hoshino, S., Onodera, R., Minato, H. and Itabashi, H. (eds) *The Rumen Ecosystem: The Microbial Metabolism and its Regulation.* Japan Scientific Societies Press, Springer Verlag, Tokyo, pp. 83–90.

Lloyd, R.J. and Pritchard, G.G. (1991) Characterization of X-prolyl dipeptidyl peptidase from *Lactococcus lactis* subsp. *lactis*. *Journal of General Microbiology* 137, 49–55.

Lockwood, B.C., Coombs, G.H. and Williams, A.G. (1988) Proteinase activity in rumen ciliate protozoa. *Journal of General Microbiology* 134, 2605–2614.

Lockyer, D.R. and Whitehead, D.C. (1990) Volatilization of ammonia from cattle urine applied to grassland. *Soil Biology and Biochemistry* 22, 1137–1142.

Lowe, S.E., Theodorou, M.K., Trinci, A.P.J. and Hespell, R.B. (1985) Growth of anaerobic rumen fungi on defined and semi-defined media lacking rumen fluid. *Journal of General Microbiology* 131, 2225–2229.

Ludden, P.A., Harmon, D.L., Larson, B.T. and Axe, D.E. (2000a) Influence of the novel urease inhibitor N-(n-butyl) thiophosphoric triamide on ruminant nitrogen metabolism: I. *In vitro* urea kinetics and substrate digestion. *Journal of Animal Science* 78, 181–187.

Ludden, P.A., Harmon, D.L., Huntington, G.B., Larson, B.T. and Axe, D.E. (2000b) Influence of the novel urease inhibitor N-(n-butyl) thiophosphoric triamide on ruminant nitrogen metabolism: II. Ruminal nitrogen metabolism, diet digestibility, and nitrogen balance in lambs. *Journal of Animal Science* 78, 188–198.

MacRae, J.C., Bruce, L.A. and Yu, F. (1999) The effect of flavomycin on gastrointestinal leucine metabolism and liveweight gain in lambs. *South African Journal of Animal Science* 29, 243–244.

Madeira, H.M.F., Peng, L. and Morrison, M. (1997) Biochemical and mutational analysis of a gingipain-like peptidase activity from *Prevotella ruminicola* B_14 and its role in ammonia production by ruminal bacteria. *Applied and Environmental Microbiology* 63, 670–675.

Maeng, W.J. and Baldwin, R.L. (1976) Factors influencing rumen microbial growth rates and yields: effect of amino acid additions to a purified diet with nitrogen from urea. *Journal of Dairy Science* 59, 648–655.

Maeng, W.J., Van Nevel, C.J., Baldwin, R.L. and Morris, J.G. (1976) Rumen microbial growth rates and yields: effects of amino acids and protein. *Journal of Dairy Science* 59, 68–79.

Makkar, H.P.S. and Becker, K. (1997) Degradation of quillaja saponins by mixed cultures of rumen microbes. *Letters in Applied Microbiology* 25, 243–245.

Makkar, H.P.S., Sen, S., Blümmel, M. and Becker, K. (1998) Effects of fractions containing saponins *from Yucca schidigera, Quillaja saponaria,* and *Acacia auriculiformis* on rumen fermentation. *Journal of Agricultural and Food Chemistry* 46, 4324–4328.

Mangan, J.L. and West, J. (1977) Ruminal digestion of chloroplasts and the protection of protein by glutaraldehyde treatment. *Journal of Agricultural Science, Cambridge* 89, 3–15.

Marini, J.C., Simpson, K.W., Gerold, A. and Van Amburgh, M.E. (2003) The effect of immunization with jackbean urease on antibody response and nitrogen recycling in mature sheep. *Livestock Production Science* 81, 283–292.

McAllan, A.B. and Smith, R.H. (1973a) Degradation of nucleic acid derivatives by rumen bacteria in vitro. *British Journal of Nutrition* 29, 467–474.

McAllan, A.B. and Smith, R.H. (1973b) Degradation of nucleic acids in the rumen. *British Journal of Nutrition* 29, 331–345.

McDonald, J.K. (1998) Dipeptidyl peptidase II. In: Barrett, A.J., Rawlings, N.D. and Woessner, J.F. (eds) *Handbook of Proteolytic Enzymes.* Academic Press, London, pp. 408–410.

McDonald, J.K. and Barrett, A.J. (1986) *Mammalian Proteases: A Glossary and Bibliography. Vol. 2. Exopeptidases.* Academic Press, London, 357 pp.

McDonald, J.K., Zeitman, B.B., Reilly, T.J. and Ellis, S. (1969a) New observations on the substrate specificity of cathepsin C (dipeptidyl aminopeptidase I) including the degradation of b-corticotropin and other peptide hormones. *Journal of Biological Chemistry* 244, 2693–2709.

McDonald, J.K., Callahan, P.X., Zeitman, B.B. and Ellis, S. (1969b) Inactivation and degradation of glucagon by dipeptidyl aminopeptidase I (cathepsin C) of rat liver. Including a comparative study of secretin degradation. *Journal of Biological Chemistry* 244, 6199–6208.

McEwan, N.R., McIntosh, F.M., Thomas, N.A., Nelson, N., Jouany, J.P., Nsabimana, E., Macheboeuf, D., Mitsumori, M., Nagamine, T., Ushida, K., Hackstein, J.H.P. and Newbold, C.J. (2003) Domain shuffling and evolutionary rates in the polyubiquitin genes of rumen ciliates. *Conference on Gastrointestinal Function, Chicago, 2003,* p. 26. Available at: www.ncaur.usda.gov/fbt/GIF/Tentative%20Agenda

McGuire, M.J., Lipsky, P.E. and Thiele, D.L. (1992) Purification and characterization of dipeptidyl-peptidase I from human spleen. *Archives in Biochemistry and Biophysiology* 295, 280–288.

McInerney, M.J., Mackie, R.I. and Bryant, M.P. (1981) Syntrophic association of a butyrate-degrading bacterium and *Methanosarcina* enriched from bovine rumen fluid. *Applied and Environmental Microbiology* 41, 826–828.

McIntosh, F.M., Stanley, K., Stewart, C.S. and Newbold, C.J. (2002) The role of protozoa in the survival of *E. coli*, verocytotoxin encoding bacteriophage and their VTEC lysogens in the rumen. *Reproduction, Nutrition, Development* 42, 34.

McIntosh, F.M., Williams, P., Losa, R., Wallace, R.J., Beever, D.E. and Newbold, C.J. (2003) Effects of essential oils on ruminal microorganisms and their protein metabolism. *Applied and Environmental Microbiology* 69, 5011–5014.

McKain, N., Wallace, R.J. and Watt, N.D. (1992) Selective isolation of bacteria with dipeptidyl aminopeptidase type I activity from the sheep rumen. *FEMS Microbiology Letters* 95, 169–174.

McSweeney, C.S., Palmer, B., Bunch, R. and Krause, D.O. (1999) Isolation and characterization of proteolytic ruminal bacteria from sheep and goats fed the tannin-containing shrub legume *Calliandra calothyrsus. Applied and Environmental Microbiology* 65, 3075–3083.

McSweeney, C.S., Palmer, B., McNeill, D.M. and Krause, D.O. (2001) Microbial interactions with tannins: nutritional consequences for ruminants. *Animal Feed Science and Technology* 16, 83–93.

Meijer, W., Marugg, J.D. and Hugenholtz, J. (1996) Regulation of proteolytic enzyme activity in *Lactococcus lactis. Applied Environmental Microbiology* 62, 156–161.

Merry, R.J., Smith, R.H. and McAllan, A.B. (1982) Glycosyl ureides in ruminant nutrition. 2. In vitro studies on the metabolism of glycosyl ureides and their free component molecules in rumen contents. *British Journal of Nutrition* 48, 287–318.

Merry, R.J., McAllan, A.B. and Smith, R.H. (1990) *In vitro* continuous culture studies on the effect of nitrogen source on microbial growth and fibre digestion. *Animal Feed Science and Technology* 31, 55–64.

Meyer-Barton, E.C., Klein, J.R., Imam, M. and Plapp, R. (1993) Cloning and sequence analysis of the X-prolyl dipeptidyl aminopeptidase (pepX) from *Lactobacillus delbruekii* subsp *lactis* DSM7290. *Applied Microbiology and Biotechnology* 40, 82–89.

Michel, V., Fonty, G., Millet, L., Bonnemoy, F. and Gouet, P. (1993) In vitro study of the proteolytic activity of rumen anaerobic fungi. *FEMS Microbiology Letters* 110, 5–10.

Milligan, L.P. (1970) Carbon dioxide fixing pathways of glutamic acid synthesis in the rumen. *Canadian Journal of Biochemistry* 48, 463–468.

Min, B.R., Attwood, G.T., Reilly, K., Sun, W., Peters, J.S., Barry, T.N. and McNabb, W.C. (2002) *Lotus corniculatus* condensed tannins decrease *in vivo* populations of proteolytic bacteria and affect nitrogen metabolism in the rumen of sheep. *Canadian Journal of Microbiology* 48, 911–921.

Mineyama, R. and Saito, K. (1991) Purification and characterisation of dipeptidyl peptidase IV from *Streptococcus salivarius* HHT. *Microbiology* 68, 37–52.

Miyagi, T., Javorsky, P., Pristas, P., Karita, S., Sakka, K. and Ohmiya, K. (1998) Partial purification and characterization of RalF40I, a class II restriction endonuclease from *Ruminococcus albus* F-40, which recognizes and cleaves 5′-/GATC-3′. *FEMS Microbiology Letters* 164, 215–218.

Moharrery, A. and Das, T.K. (2001) Correlation between microbial enzyme activities in the rumen fluid of sheep under different treatments. *Reproduction, Nutrition, Development* 41, 513–529.

Moisio, R., Kreula, M. and Virtanen, A.I. (1969) Experiments on nitrogen fixation in cow's rumen. *Suomen Kemistilehti B* 42, 432–433.

Molan, A.L., Attwood, G.T., Min, W.K. and McNabb, W.C. (2001) The effect of condensed tannins from *Lotus pedunculatus* and *Lotus corniculatus* on the growth of proteolytic rumen bacteria in vitro and their possible mode of action. *Canadian Journal of Microbiology* 47, 626–633.

Moon-van der Staay, S.Y., Van der Staay, G.W.M., Javorský, P., Jouany, J.P., Michalowski, T., Nsabimana, E., Macheboeuf, D., Kišidayová, S., McEwan, N.R., Newbold, C.J. and Hackstein, J.H.P. (2002) Diversity of rumen ciliates revealed by 18S ribosomal DNA analysis. *Reproduction, Nutrition, Development* 42, 76.

Morgavi, D.P., Sakurada, M., Tomita, Y. and Onodera, R. (1994) Presence in rumen bacterial and protozoal populations of enzymes capable of degrading fungal cell walls. *Microbiology* 140, 631–636.

Morgavi, D.P., Sakurada, M., Tomita, Y. and Onodera, R. (1996) Electrophoretic forms of chitinolytic and lysozyme activities in ruminal protozoa. *Current Microbiology* 32, 115–118.

Morrison, M. and Mackie, R. I. (1996) Nitrogen metabolism by ruminal microorganisms: current understanding and future perspectives. *Australian Journal of Agricultural Research* 47, 227–246.

Naga, M.A. and El-Shazly, K. (1968) The metabolic characterization of the ciliate protozoan *Eudiplodinium medium* from the rumen of buffalo. *Journal of General Microbiology* 53, 305–315.

Nagaraja, T.G., Newbold, C.J., Van Nevel, C.J. and Demeyer, D.I. (1997) Manipulation of ruminal fermentation. In: Hobson, P.N. and Stewart, C.S. (eds) *The Rumen Microbial Ecosystem.* Chapman & Hall, London, pp. 523–632.

Nagasawa, T., Uchida, T. and Onodera, R. (1992) Exopeptidase activity of mixed rumen ciliate protozoa. *Animal Science and Technology* 63, 481–487.

Nagy, J.G. and Tengerdy, R.P. (1968) Antibacterial action of essential oils of *Artemisia* as an ecological factor. II. Antibacterial action of the volatile oils of *Artemisia tridentata* (big sagebrush) on bacteria from the rumen of mule deer. *Applied Microbiology* 16, 441–444.

Nardi, M., Chopin, M.C., Chopin, A., Cals, M.M. and Gripon, J.-C. (1991) Cloning and DNA sequence analysis of an X-prolyl dipeptidyl aminopeptidase gene from *Lactococcus lactis* subsp. *lactis* NCDO 763. *Applied and Environmental Microbiology* 57, 45–50.

Navas-Camacho, .A, Laredo, M.A., Cuesta, A., Anzola, H. and Leon, J. C. (1993) Effect of supplementation with a tree legume forage on rumen function. *Livestock Research for Rural Development* 5, 58–71.

Newbold, C.J. and Hillman, K. (1990) The effect of ciliate protozoa on the turnover of bacterial and fungal protein in the rumen of sheep. *Letters in Applied Microbiology* 11, 100–102.

Newbold, C.J. and Jouany, J.P. (1997) The contribution of individual genera in a mixed protozoal population to the breakdown of bacteria in the rumen. *Reproduction, Nutrition, Development* 88, 46.

Newbold, C.J., McKain, N. and Wallace, R.J. (1989) The role of protozoa in ruminal peptide metabolism. In: Lloyd, D., Coombs G.H. and Paget, T.A.P. (eds) *Biochemistry and Molecular Biology of Anaerobic Protozoa.* Harwood Academic Publishers, London, pp. 42–56.

Newbold, C.J., Wallace, R.J. and McKain, N. (1990) Effects of the ionophore tetronasin on nitrogen metabolism by rumen microorganisms in vitro. *Journal of Animal Science* 68, 1103–1109.

Newbold, C.J., Wallace, R.J. and Watt, N.D. (1992) Properties of ionophore-resistant *Bacteroides ruminicola* enriched by cultivation in the presence of tetronasin. *Journal of Applied Bacteriology* 72, 65–70.

Newbold, C.J., el Hassan, S.M., Wang, J., Ortega, M.E. and Wallace, R.J. (1997) Influence of foliage from African multipurpose trees on activity of rumen protozoa and bacteria. *British Journal of Nutrition* 78, 237–249.

Newbold, C.J., Martin, H.C., McIntosh, F.M., Onodera, R., Wallace, R.J. and Calza, R.E. (1999) Breakdown of bacterial cell walls by the rumen ciliate protozoan *Entodinium caudatum. South African Journal of Animal Science* 29, 114–115.

Newbold, C.J., Teferedegne, B., Kim, H.-S., Zuur, G. and Lobley, G.E. (2000a) Effects of protozoa on nitrogen metabolism in the rumen of sheep. *Reproduction, Nutrition, Development* 40, 199 (Abstract).

Newbold, C.J., McEwan, N.R., Eschenlauer, S.C.P., Nelson, N., Travis, A.J., Duval, S.M., Chareyon, E.L.N. and Wallace, R.J. (2000b) Characterisation of a glutamate dehydrogenase from the rumen ciliate *Entodinium caudatum. Proceedings of Japanese Society for Rumen Metabolism and Physiology* 11, 74.

Newbold, C.J., Stewart, C.S. and Wallace, R.J. (2001) Developments in rumen fermentation – the scientist's view. In: Garnsworthy, P.C. and Wiseman, J. (eds) *Recent Advances in Animal Nutrition 2001.*

Nottingham University Press, Nottingham, UK, pp. 251–279.

Nikolic, J.A., Pavlicevic, A., Zeremski, D. and Negovanovic, D. (1980) Adaptation to diets containing significant amounts of non-protein nitrogen. In: Ruckebusch, Y. and Thivend, P. (eds) *Digestive Physiology and Metabolism in Ruminants*. MTP Press Ltd, Lancaster, UK, pp. 603–620.

Nolan, J.V. and Stachiw, S. (1979) Fermentation and nitrogen dynamics in merino sheep given a low quality roughage diet. *British Journal of Nutrition* 42, 63–80.

Nugent, J.H.A. and Mangan, J.L. (1981) Characteristics of the rumen proteolysis of fraction I (18S) leaf protein from lucerne (*Medicago sativa* L). *British Journal of Nutrition* 46, 39–58.

Nugent, J.H.A., Jones, W.T., Jordan, D.J. and Managan, J.L. (1983) Rates of proteolysis in the rumen of the soluble proteins casein, fraction 1 (18S) leaf protein, bovine serum albumin and bovine submaxillary mucoprotein. *British Journal of Nutrition* 50, 357–368.

Ogasawara, W., Ochiai, K., Ando, K., Yano, K., Yamasaki, M., Okada, H. and Morikawa, Y. (1996a) A Novel dipeptidyl aminopeptidase from *Pseudomonas* spp. strain WO24. *Journal of Bacteriology* 178, 1283–1288.

Ogasawara, W., Kobayashi, G., Okada, H. and Morikawa, Y. (1996b) Two types of novel dipeptidyl aminopeptidases from *Pseudomonas* sp. strain WO24. *Journal of Bacteriology* 178, 6288–6295.

Ogasawara, W., Inanobe, N., Ochiaia, K., Ando, K., Okada, H. and Morikawa, Y. (1997) Purification and characterisation of dipeptidyl aminopeptidase from *Aureobacterium* sp. WO26. *Bioscience, Biotechnology, and Biochemistry* 61, 146–151.

Oh, H.K., Sakai, T., Jones, M.B. and Longhurst, W.M. (1967) Effect of various essential oils isolated from Douglas fir needles upon sheep and deer rumen microbial activity. *Applied Microbiology* 15, 777–784.

Oh, H.K., Jones, M.B. and Longhurst, W.M. (1968) Comparison of rumen microbial inhibition resulting from various essential oils isolated from relatively unpalatable plant species. *Applied Microbiology* 16, 39–44.

Okuuchi, K., Tomita, Y. and Onodera, R. (1993) Synthesis of tryptophan from indoleacetic acid by mixed rumen bacteria and a mixture of rumen bacteria and protozoa. *Animal Science Technology (Japan)* 64, 1182–1186.

Onodera, R. (1990) Amino acid and protein metabolism by ciliate protozoa. In: Hoshino. S., Onodera, R, Minato, H. and Itabashi, H. (eds) *The Rumen Ecosystem: The Microbial Metabolism and its Regulation*. Japanese Scientific Societies Press, Springer Verlag, Tokyo, pp. 33–42.

Onodera, R. and Goto, Y. (1990) The metabolism of branched chain amino acids by starved rumen protozoa. *Japanese Journal of Zootechnical Science* 60, 843–849.

Onodera, R. and Kandatsu, M. (1970) Amino acid and protein metabolism of rumen ciliate protozoa. 4. Metabolism of casein. *Japanese Journal of Animal Science* 41, 307–313.

Onodera, R. and Kandatsu, M. (1973) Synthesis of lysine from a,e-diaminopimelic acid by mixed ciliated rumen protozoa. *Nature New Biology* 244, 31–32.

Onodera, R. and Yakiyama, M. (1990) Coagulation and degradation of casein by entodiniomorphid and holotrich rumen protozoa. *Japanese Journal of Animal Science* 61, 533–539.

Onodera, R., Nakagawa, Y., and Kandatsu, M. (1977) Ureolytic activity of the washed cell suspension of rumen ciliate protozoa. *Agricultural and Biological Chemistry* 41, 2177–2182.

Onodera, R., Yamaguchi, Y. and Morimoto, S. (1983) Metabolism of arginine, citrulline, ornithine and proline by rumen protozoa. *Agricultural and Biological Chemistry* 47, 821–828.

Onodera, R., Takashima, H. and Ling, J. R. (1991) In vitro production of lysine from 2,2′-diaminopimelic acid by rumen protozoa. *Journal of Protozology* 38, 421–425.

Orpin, C.G. and Greenwood, Y. (1986) Nutritional and germination requirements of the rumen chytridomycete *Neocallimastix frontalis*. *Transactions of the British Mycological Society* 86, 103–109.

Orpin, C.G. and Joblin, K.N. (1997) The rumen anaerobic fungi. In: Hobson, P.N. and Stewart, C.S. (eds) *The Rumen Microbial Ecosystem*. Chapman & Hall, London, pp. 140–195.

Orskov, E.R., Mills, C.F. and Robinson, J.J. (1980) The use of whole blood for the protection of organic materials from degradation in the rumen. *Proceedings of the Nutrition Society* 39, 60A.

Ouwerkerk, D., Klieve, A.V. and Forster, R.J. (2002) Enumeration of *Megasphaera elsdenii* in rumen contents by real-time Taq nuclease assay. *Journal of Applied Microbiology* 92, 753–758.

Pal, D.T., Kamra, D.N., Pathak, N.N. and Bisht, G.S. (1998) Influence of rumen ciliate protozoa and protein level on rumen fermentation, enzyme activities and blood metabolites in crossbred cattle calves. *Journal of Applied Animal Research* 13, 153–159.

Paster, B.J., Russell, J.B., Yang, C.M.J., Chow, J.M., Woese, C.R. and Tanner, R. (1993) Phylogeny of the ammonia-producing ruminal bacteria *Peptostreptococcus anaerobius, Clostridium sticklandii* and *Clostridium aminophilum* sp. nov. *International Journal of Systematic Bacteriology* 43, 107–110.

Patte, J.-C. (1996) Biosynthesis of threonine and lysine. In: Neidhardt, F.C. (ed.) *Escherichia coli and Salmon-*

ella. Cellular and Molecular Biology. ASM Press, Washington, DC, pp. 528–541.

Prins, R.A., Van Rheenen, D.L. and Van't Klooster, A.T. (1983) Characterisation of microbial proteolytic enzymes in the rumen. *Antonie van Leeuwenhoek* 49, 585–595.

Pristas, P., Vanat, I. and Javorsky, P. (1995) Isolation and characterization of a new restriction endonuclease, Sru30DI, from *Selenomonas ruminantium. Gene* 158, 139–140.

Pristas, P., Molnarova, V., and Javorsky, P. (2001) Restriction and modification systems of ruminal bacteria. *Folia Microbiology (Praha)* 46, 71–72.

Ramsak, A., Peterka, M., Tajimab, K., Martin, J.C., Wood, J., Johnstone, M.E., Aminov, R.I., Flint, H.J. and Avgustin, G. (2000) Unravelling the genetic diversity of ruminal bacteria belonging to the CFB phylum. *FEMS Microbiology and Ecology* 33, 69–79.

Reilly, K., Carruthers, V.R. and Attwood, G.T. (2002) Design and use of 16S ribosomal DNA-directed primers in competitive PCRs to enumerate proteolytic bacteria in the rumen. *Microbial Ecology* 43, 259–270.

Reis, P.J., Tunks, D.A. and Munro, S.G. (1990) Effects of the infusion of amino acids into the abomasum of sheep, with emphasis on the relative value of methionine, cysteine and homocysteine for wool growth. *Journal of Agricultural Science, Cambridge* 114, 59–68.

Ritzhaupt, A., Breves, G., Schröder, B., Winckler, C.G. and Shirazi-Beechey, S.P. (1997) Urea transport in gastrointestinal tract of ruminants: effect of dietary nitrogen. *Biochemical Society Transactions* 25, S490.

Ritzhaupt, A., Wood, I.S., Jackson, A.A., Moran, B.J. and Shirazi-Beechey, S.P. (1998) Isolation of a RT-PCR fragment from human colon and sheep rumen RNA with nucleotide sequence similarity to human and rat urea transporter isoforms. *Biochemical Society Transactions* 26, S122.

Roffler, R.E. and Satter, L.D. (1975) Relationship between ruminal ammonia and non protein nitrogen utilization by ruminants. I. Development of a model for predicting nonprotein nitrogen utilisation by cattle. *Journal of Dairy Science* 58, 1880–1888.

Rooke, J.A. and Armstrong, D.G. (1989) The importance of the form of nitrogen on microbial protein synthesis in the rumen of cattle receiving grass silage and continuous intrarumen infusion of sucrose. *British Journal of Nutrition* 61, 113–121.

Russell, J.B. and Martin, S.A. (1984) Effects of various methane inhibitors on the fermentation of amino acids by mixed rumen microorganisms in vitro. *Journal of Animal Science* 59, 1329–1338.

Russell, J.B. and Robinson, P.H. (1984) Composition and characteristics of strains of *Streptococcus bovis. Journal of Dairy Science* 67, 1525–1531.

Russell, J.B. and Strobel, H.J. (1987) Concentration of ammonia across cell membranes of mixed ruminal bacteria. *Journal of Dairy Science* 70, 970–976.

Russell, J.B. and Wilson, D.B. (1988) Potential opportunities and problems for genetically altered rumen microorganisms. *Journal of Nutrition* 118, 271–279.

Russell, J.B., Bottje, W.G. and Cotta, M.A. (1981) Degradation of protein by mixed cultures of rumen bacteria: identification of *Streptococcus bovis* as an actively proteolytic rumen bacterium. *Journal of Animal Science* 53, 242–252.

Russell, J.B., Strobel, H.J. and Chen, G. (1988) Enrichment and isolation of a ruminal bacterium with a very high specific activity of ammonia production. *Applied and Environmental Microbiology* 54, 872–877.

Russell, J.B., Onodera, R. and Hino, T. (1991) Ruminal protein fermentation: new perspectives on previous contradictions. In: Tsuda, T., Sasaki, Y. and Kawashima, R. (eds) *Physiological Aspects of Digestion and Metabolism in Ruminants.* Academic Press, San Diego, California, pp. 681–697.

Russell, J.B., O'Connor, J.D., Fox, D.G., Van Soest, P.J. and Sniffen, C.J. (1992) A net carbohydrate and protein system for evaluating cattle diets. I. Ruminal fermentation. *Journal of Animal Science* 70, 3551–3561.

Russi, J.P., Wallace, R.J. and Newbold, C.J. (2002) Influence of the pattern of peptide supply on microbial activity in the rumen simulating fermenter (RUSITEC). *British Journal of Nutrition* 88, 73–80.

Rychlik, J.L. and Russell, J.B. (2000) Mathematical estimations of hyper-ammonia producing ruminal bacteria and evidence for bacterial antagonism that decreases ruminal ammonia production. *FEMS Microbiology and Ecology* 32, 121–128.

Rychlik, J.L. and Russell, J.B. (2002) Bacteriocin-like activity of *Butyrivibrio fibrisolvens* JL5 and its effect on other ruminal bacteria and ammonia production. *Applied and Environmental Microbiology* 68, 1040–1046.

Rychlik, J.L., LaVera, R. and Russell, J.B. (2002) Amino acid deamination by ruminal *Megasphaera elsdenii* strains. *Current Microbiology* 45, 340–345.

Sahu, D.S., Prasad, K., Agrawal, I.S. and Verma, M.L. (2000) The effect of chemical defaunation on rumen bacterial, protozoal population and enzyme activities in the cattle and buffalo rumen. *Indian Journal of Animal Science* 70, 1074–1076.

Sakurada, M., Morgavi, D.P., Tomita, Y. and Onodera, R. (1994) Ureolytic activity of anaerobic rumen fungi, *Piromyces* sp. OTS3 and *Neocallimastix* sp. OTS4. *Animal Science and Technology* 65, 950–955.

Sales, M., Lucas, F. and Blanchart, G. (2000) Effects of ammonia and amino acids on the growth and proteolytic activity of three species of rumen bacteria: *Prevotella albensis, Butyrivibrio fibrisolvens,* and *Streptococcus bovis. Current Microbiology* 40, 380–386.

Sales-Duval, M., Lucas, F. and Blanchart, G. (2002) Effects of exogenous ammonia or free amino acids on proteolytic activity and protein breakdown products in *Streptococcus bovis*, *Prevotella albensis*, and *Butyrivibrio fibrisolvens*. *Current Microbiology* 44, 435–443.

Salter, D.N., Daneshvar, K. and Smith, R.H. (1979) The origin of nitrogen incorporated into compounds in the rumen bacteria of steers given protein- and urea-containing diets. *British Journal of Nutrition* 41, 197–209.

Sauer, F.D., Erfle, J.D. and Mahadevan, S. (1975) Amino acid biosynthesis in mixed rumen cultures. *Biochemical Journal* 150, 357–372.

Scheifinger, C., Russell, N. and Chalupa, W. (1976) Degradation of amino acids by pure cultures of rumen bacteria. *Journal of Animal Science* 43, 821–827.

Schwab, C.G. (1995) Protected proteins and amino acids for ruminants. In: Wallace, R.J. and Chesson, A. (eds) *Biotechnology in Animal Feeds and Animal Feeding*. VCH, Weinheim, Germany, pp. 115–141.

Schwartz, H.M. (1967) The rumen metabolism of non-protein nitrogen. In: Briggs, M.H. (ed.) *Urea as a Protein Supplement*. Pergamon, London, pp. 95–109.

Shepherd, J.L., Newbold, C.J., Hillman, K. and Fenlon, D.R. (2000) The effect of protozoa on the survival and flow of listeria from the rumen of sheep. *Reproduction, Nutrition, Development* 40, 216 (Abstract).

Shu, Q., Gill, H.S., Hennessy, D.W., Leng, R.A., Bird, S.H. and Rowe, J.B. (1999) Immunisation against lactic acidosis in cattle. *Research in Veterinary Science* 67, 65–71.

Sinclair, L.A., Garnsworthy, P.C., Newbold, J.R. and Buttery, P.J. (1995) Effects of synchronizing the rate of dietary energy and nitrogen release in diets with a similar carbohydrate composition on rumen fermentation and microbial protein synthesis in sheep. *Journal of Agricultural Science, Cambridge* 124, 463–472.

Skrivanova, V. and Marounek, M. (1990) The influence of *Lactobacillus acidophilus* probiotics on efficiency and parameters of rumen fluid in calves. *Zivocisna Vyroba* 35, 87–94.

Smith, C.J. and Bryant, M.P. (1979) Introduction to metabolic activities of intestinal bacteria. *American Journal of Clinical Nutrition* 32, 149–157.

Smith, C.J., Hespell, R.B. and Bryant, M.P. (1981) Regulation of urease and ammonia assimilatory enzymes in *Selenomonas ruminantium*. *Applied and Environmental Microbiology* 42, 89–96.

Smith, K.A. and Frost, J.P. (2000) Nitrogen excretion by farm livestock with respect to land spreading requirements and controlling nitrogen losses to ground and surface waters. Part 1. Cattle and sheep. *Bioresource Technology* 71, 173–181.

Smith, R.H. and McAllan, A.B. (1970) Nucleic acid metabolism in the ruminant. 2. Formation of microbial nucleic acids in the rumen in relation to the digestion of food nitrogen, and the fate of dietary nucleic acids. *British Journal of Nutrition* 24, 545–556.

Somerville, H.J. (1968) Enzymatic studies on the biosynthesis of amino acids from lactate by *Peptostreptococcus elsdenii*. *Biochemical Journal* 108, 107–119.

Somerville, H.J. and Peel, J.L. (1967) Tracer studies on the biosynthesis of amino acids from lactate by *Peptostreptococcus elsdenii*. *Biochemical Journal* 105, 299–310.

Spears, J.W. and Hatfield, E.E. (1978) Nickel for ruminants. I. Influence of dietary nickel on ruminal urease activity. *Journal of Animal Science* 47, 1345–1350.

Spears, J.W., Smith, C.J. and Hatfield, E.E. (1977) Rumen bacterial urease requirement for nickel. *Journal of Dairy Science* 60, 1073–1076.

Stack, R.J. and Cotta, M.A. (1986) Effect of phenylpropanoic acid on growth of and cellulose utilization by cellulolytic rumen bacteria. *Applied and Environmental Microbiology* 52, 209–210.

Stahl, D.A., Flesher, B., Mansfield, H.R. and Montgomery, L. (1988) Use of phylogenetically based hybridization probes for studies of ruminal microbial ecology. *Applied and Environmental Microbiology* 54, 1079–1084.

Stewart, C.S. and Richardson, A.J. (1989) Enhanced resistance of anaerobic rumen fungi to the ionophores monensin and lasalocid in the presence of methanogenic bacteria. *Journal of Applied Bacteriology* 66, 85–93.

Stewart, C.S., Grossley, M.V. and Garrow, S.H. (1983) The effect of avoparcin on laboratory cultures of rumen bacteria. *European Journal of Applied Microbiology and Biotechnology* 17, 292–297.

Stewart, C.S., Flint, H.J. and Bryant, M.P. (1997) The rumen bacteria. In: Hobson, P.N. and Stewart, C.S. (eds) *The Rumen Microbial Ecosystem*. Chapman & Hall, London, pp. 10–72.

Strydom, E., Mackie, R.I. and Woods, D.R. (1986) Detection and characterization of extracellular proteases in *Butyrivibrio fibrisolvens* H17C. *Applied Microbiology and Biotechnology* 24, 214–217.

Tachi, H., Ito, H. and Ichishima, E. (1992) An X-prolyl dipeptidyl aminopeptidase from *Aspergillus oryzae*. *Phytochemistry* 31, 3707–3709.

Tajima, K., Aminov, R.I., Nagamine, T., Ogata, K., Nakamura, M., Matsui, H. and Benno, Y. (1999) Rumen bacterial diversity as determined by sequence analysis of 16S rDNA libraries. *FEMS Microbiology Ecology* 29, 159–169.

Tajima, K., Arai, S., Ogata, K., Nagamine, T., Matsui, H., Nakamura, M., Aminov, R.I. and Benno, Y. (2000) Rumen bacterial community transition during adaptation to a high grain diet. *Anaerobe* 6, 273–284.

Tajima, K., Aminov, R.I., Nagamine, T., Matsui, H., Nakamura, M. and Benno, Y. (2001) Diet-dependent shifts in the bacterial population of the rumen revealed with real-time PCR. *Applied and Environmental Microbiology* 67, 2766–2774.

Tamminga, S. and Verstegen, M.W.A. (1996) Implications of nutrition of animals on environmental pollution. In: Garnsworthy, P.C., Haresign, W. and Cole, D.J.A. (eds) *Recent Advances in Animal Nutrition.* Nottingham University Press, Nottingham, UK, pp. 121–134.

Tan, Z.L., Nagaraja, T.G. and Chengappa, M.M. (1994) Biochemical and biological characterization of ruminal *Fusobacterium necrophorum. FEMS Microbiology Letters* 120, 81–86.

Tan, Z.L., Nagaraja, T.G. and Chengappa, M.M. (1996) *Fusobacterium necrophorum* infections: virulence factors, pathogenic mechanism and control measures. *Veterinary Research Communications* 20, 113–140.

Teferedegne, B. (2000) New perspectives on the use of tropical plants to improve ruminant nutrition. *Proceedings of the Nutrition Society* 59, 209–214.

Teferedegne, B., McIntosh, F., Osuji, P.O., Odenyo, A.A., Wallace, R.J. and Newbold, C.J. (1999) Influence of foliage from different accessions of the sub-tropical leguminous tree, *Sesbania sesban*, on ruminal protozoa in Ethiopian and Scottish sheep. *Animal Feed Science and Technology* 78, 11–20.

Thalib, A., Widiawati, Y., Hamid, H., Suherman, D. and Sabrani, M. (1995) The effect of saponins from *Sapindus rarak* fruit on rumen microbes and host animal growth. *Annales de Zootechnie* 44, 161.

Theodorou, M.K., Davies, D.R. and Orpin, C.G. (1994) Nutrition and survival of anaerobic fungi. In: Mountfort, D.O. and Orpin, C.G. (eds) *Anaerobic Fungi.* Marcel Dekker, New York, pp. 107–128.

Theodorou, M.K., Merry, R.J. and Thomas, H. (1996) Is proteolysis in the rumen of grazing animals mediated by plant enzymes? *British Journal of Nutrition* 75, 507–510.

Tomlinson, A.P., Powers, H.H., Van Horn, H.H., Nordstedt, R.A. and Wilcox, C.J. (1996) Dietary protein effects on nitrogen excretion and manure characteristics of lactating cows. *Transactions of the American Society of Agricultural Engineers* 39, 1441–1448.

Tsubota, H. and Hoshino, S. (1969) Transaminase activity in sheep rumen content. *Journal of Dairy Science* 52, 2024–2028.

Umbarger, H.E. (1969) Regulation of amino acid metabolism. *Annual Review of Biochemistry* 38, 323–370.

Ushida, K. and Jouany, J.P. (1985) Effect of protozoa on rumen protein degradation in sheep. *Reproduction, Nutrition, Developement* 25, 1075–1081.

Ushida, K. and Jouany, J.P. (1986) Influence des protozoaires sur la degradation des proteines measurée in vitro et in sacco. *Reproduction, Nutrition, Development* 26, 293–294.

Van Gylswyk, N.O. (1990) Enumeration and presumptive identification of some functional groups of bacteria in the rumen of dairy cows fed grass silage-based diets. *FEMS Microbiology and Ecology* 73, 243–254.

Van Horn, H.H., Newton, G.L. and Kunkle, W.E. (1996) Ruminant nutrition from an environmental perspective: factors affecting whole-farm nutrient balance. *Journal of Animal Science* 74, 3082–3102.

Van Nevel, C.J. and Demeyer, D.I. (1977) Effect of monensin on rumen metabolism in vitro. *Applied and Environmental Microbiology* 34, 251–257.

Van Nevel, C.J. and Demeyer, D.I. (1990) Effects of antibiotics, a deaminase inhibitor and sarsaponin on nitrogen metabolism of rumen contents in vitro. *Animal Feed Science and Technology* 31, 323–348.

Van Nevel, C.J. and Demeyer, D.I. (1992) Influence of antibiotics and a deaminase inhibitor on volatile fatty acids and methane production from detergent washed hay and soluble starch by rumen microbes *in vitro. Animal Feed Science and Technology* 37, 21–31.

Virtanen, A.I. (1966) Milk production of cows on protein-free feed. *Science* 153, 1603–1614.

Walker, N.D. (2001) Characterisation of peptide transport and hydrolysis by the ruminal bacterium *Prevotella albensis.* PhD thesis, University of Aberdeen.

Walker, N.D., McEwan, N.R. and Wallace, R.J. (2003) Cloning and functional expression of dipeptidyl peptidase IV from the ruminal bacterium *Prevotella albensis* M384^{T}. *Microbiology* 149, 2227–2234.

Wallace, R.J. (1979) Effect of ammonia concentration on the composition, hydrolytic activity and nitrogen metabolism of the microbial flora of the rumen. *Journal of Applied Bacteriology* 47, 443–455.

Wallace, R.J. (1983) Hydrolysis of ^{14}C-labelled proteins by rumen micro-organisms and by proteolytic enzymes prepared from rumen bacteria. *British Journal of Nutrition* 50, 345–355.

Wallace, R.J. (1985) Synergism between different species of proteolytic rumen bacteria. *Current Microbiology* 12, 59–64.

Wallace, R.J. (1994) Amino acid and protein synthesis, turnover, and breakdown by ruminal microorganisms. In: Asplund, J.M. (ed.) *Principles of Protein Nutrition of Ruminants.* CRC Press, Boca Raton, Florida, pp. 71–111.

Wallace, R.J. (1996) Rumen microbial metabolism of peptides and amino acids. *Journal of Nutrition* 126, 1326S–1334S.

Wallace, R.J. and Brammall, M.L. (1985) The role of different species of bacteria in the hydrolysis of protein in the rumen. *Journal of General Microbiology* 131, 821–832.

Wallace, R.J. and Falconer, M.L. (1992) *In vitro* studies of conditions required to protect protein from ruminal degradation by heating in the presence of sugars. *Animal Feed Science and Technology* 37, 129–141.

Wallace, R.J. and Joblin, K.N. (1985) Proteolytic activity of a rumen anaerobic fungus. *FEMS Microbiology Letters* 29, 119–125.

Wallace, R.J. and McKain, N. (1989) Analysis of peptide metabolism by ruminal microorganisms. *Applied and Environmental Microbiology* 55, 2372–2376.

Wallace, R.J. and McKain, N. (1990) A comparison of methods for determining the concentration of extracellular peptides in rumen fluid of sheep. *Journal of Agriculture and Science* 114, 101–105.

Wallace, R.J. and McKain, N. (1991) A survey of peptidase activity in rumen bacteria. *Journal of General Microbiology* 137, 2259–2264.

Wallace, R.J. and McKain, N. (1996) Influence of 1,10-phenanthroline and its analogues, other chelators, and transition metal ions on dipeptidase activity of the rumen bacterium, *Prevotella ruminicola*. *Journal of Applied Bacteriology* 81, 42–47.

Wallace, R.J. and McPherson, C.A. (1987) Factors affecting the rate of breakdown of bacterial protein in rumen fluid. *British Journal of Nutrition* 58, 313–323.

Wallace, R.J. and Newbold, C.J. (1991) Effects of bentonite on fermentation in the rumen simulation technique (Rusitec) and on rumen ciliate protozoa. *Journal of Agriculture and Science, Cambridge,* 116, 163–168.

Wallace, R.J., Cheng, K.-J., Dinsdale, D. and Ørskov, E.R. (1979) An independent microbial flora of the epithelium and its role in the ecomicrobiology of the rumen. *Nature* 279, 424–426.

Wallace, R.J., Broderick, G.A. and Brammall, M.L. (1987) Microbial protein and peptide metabolism in rumen fluid from faunated and ciliate-free sheep. *British Journal of Nutrition* 58, 87–93.

Wallace, R.J., McKain, N. and Newbold, C.J. (1990a) Metabolism of small peptides in rumen fluid. Accumulation of intermediates during hydrolysis of alanine oligomers, and comparison of peptidolytic activities of bacteria and protozoa. *Journal of the Science of Food and Agriculture* 50, 191–199.

Wallace, R.J., Newbold, C.J. and McKain, N. (1990b) Influence of ionophores and energy inhibitors on peptide metabolism by rumen bacteria. *Journal of Agricultural Science, Cambridge* 115, 285–290.

Wallace, R.J., Newbold, C.J. and McKain, N. (1990c) Patterns of peptide metabolism by rumen microorganisms. In: Hoshino, S., Onodera, R., Minato, H. and Itabashi, H. (eds) *The Rumen Ecosystem. The Microbial Metabolism and its Regulation.* Japanese Scientific Societies Press, Springer Verlag, Tokyo, pp. 43–50.

Wallace, R.J., McKain, N. and Broderick, G.A. (1993a) Breakdown of different peptides by *Prevotalla (Bacteroides) ruminicola* and mixed microorganisms from the sheep rumen. *Current Microbiology* 26, 333–336.

Wallace, R.J., Frumholtz, P.P. and Walker, N.D. (1993b) Breakdown of N-terminally modified peptides and an isopeptide by rumen microorganisms. *Applied and Environmental Microbiology* 59, 3147–3149.

Wallace, R.J., Arthaud, L. and Newbold, C.J. (1994) Influence of *Yucca shidigera* extract on ruminal ammonia concentrations and ruminal microorganisms. *Applied and Environmental Microbiology* 60, 1762–1767.

Wallace, R.J., Kopecny, J., Broderick, G.A., Walker, N.D., Sichao, L., Newbold, C.J. and McKain, N. (1995) Cleavage of di- and tripeptides by *Prevotella ruminicola. Anaerobe* 1, 335–343.

Wallace, R.J., Newbold, C.J. and McKain, N. (1996) Inhibition by 1,10-phenanthroline of the breakdown of peptides by rumen bacteria and protozoa. *Journal of Applied Bacteriology* 80, 425–430.

Wallace, R.J., McKain, N., Broderick, G.A., Rode, L.M., Walker, N.D., Newbold, C.J. and Kopecny, J. (1997a) Peptidases of the rumen bacterium, *Prevotella ruminicola. Anaerobe* 3, 35–42.

Wallace, R.J., Onodera, R. and Cotta, M.A. (1997b) Metabolism of nitrogen-containing compounds. In: Hobson, P.N. and Stewart, C.S. (eds) *The Rumen Microbial Ecosystem.* Chapman & Hall, London, pp. 283–328.

Wallace, R.J., Min, W.K., Witt, M.W., Grant, G., MacRae, J.C., Maeng, W.J., Newbold, C.J. and Rode L.M. (1998) Uptake of acetylated peptides from the small intestine in sheep and their nutritive value in rats. *British Journal of Nutrition* 80, 101–108.

Wallace, R.J., Wallace, S.J. and McKain, N. (2000a) Proteolytic activity of ruminal digesta during the feeding cycle in sheep receiving grass hay/concentrate or maize silage/concentrate diets. *Letters in Applied Microbiology* 30, 317–319.

Wallace, R.J., Wallace, S.J.A. and McKain, N. (2000b) An assessment of the contribution of plant proteinases to proteolytic digestion in the rumen of sheep. *British Society for Animal Science, Winter Meeting,* p. 5. Available at: www.bsas.org.UK/meetings/annlproc/Pdf2000/005.pdf

Wallace, R.J., Newbold, C.J., Bequette, B.J., MacRae, J.C. and Lobley, G.E. (2001) Increasing the flow of protein from ruminal fermentation. *Asian-Australasian Journal of Animal Science* 14, 885–893.

Wallace, R.J., McEwan, N.R., McIntosh, F.M., Teferedegne, B. and Newbold, C.J. (2002) Natural products as manipulators of rumen fermentation. *Asian-Australasian Journal of Animal Science* 15, 1458–1468.

Wallace, R.J., McKain, N., McEwan, N.R., Miyagawa, E., Chaudhary, L.C., King, T.P., Walker, N.D., Apajalahti, J.H. and Newbold, C.J. (2003) *Eubacterium pyruvativorans*, a novel non-saccharolytic anaerobe from the rumen which ferments pyruvate and amino acids, forms caproate and utilises acetate and propionate. *International Journal of Systematic and Evolutionary Microbiology* 53, 965–970.

Wang, H., McKain, N., Walker, N.D. and Wallace, R.J. (2003) Influence of dipeptidyl peptidase inhibitors on growth, peptidase activity and ammonia production by ruminal microorganisms. *Current Microbiology* 49, 115–122.

Wang, Y., McAllister, T.A., Newbold, C.J., Rode, L.M., Cheeke, P.R. and Cheng, K.-J. (1998) Effects of *Yucca schidigera* extract on fermentation and degradation of steroidal saponins in the rumen simulation technique (RUSITEC). *Animal Feed Science and Technology* 74, 143–153.

Ward, D.M., Weller, R. and Bateson, M.M. (1990) 16S rRNA reveal numerous uncultured microorganisms in a natural community. *Nature* 345, 63–65.

Wen, Z. and Morrison, M. (1996) The NAD(P)H-dependent glutamate dehydrogenase activities of *Prevotella ruminicola* $B_1$4 can be attributed to one enzyme (GdhA), and gdhA expression is regulated in response to the nitrogen source available for growth. *Applied and Environmental Microbiology* 62, 3826–3833.

Wen, Z. and Morrison, M. (1997) Glutamate dehydrogenase activity profiles for type strains of ruminal *Prevotella* spp. *Applied and Environmental Microbiology* 63, 3314–3317.

Wetherall, J.A., Armstrong, D.G., Finlayson, H.J. and Rooke, J.A. (1995) Reduction of proteolysis during ensilage of perennial ryegrass by protease inhibitors. *Journal of the Science of Food and Agriculture* 68, 497–505.

Whetstone, H.D., Davis, C.L. and Bryant, M.P. (1981) Effect of monensin on breakdown of protein by ruminal microorganisms in vitro. *Journal of Animal Science* 53, 803–809.

Willems, A., Amat-Marco, M. and Collins, M.D. (1996) Phylogenetic analysis of *Butyrivibrio* strains reveals three distinct groups of species within the *Clostridium* subphylum of the Gram-positive bacteria. *International Journal of Systematic Bacteriology* 46, 195–199.

Williams, A.G. and Coleman, G.S. (1992) *The Rumen Protozoa.* Springer-Verlag, New York, 441 pp.

Williams, A.P. and Cockburn, J.E. (1991) Effect of slowly and rapidly degraded protein sources on the concentration of amino acids and peptides in the rumen of steers. *Journal of the Science of Food and Agriculture* 56, 303–314.

Williams, M., Gagnon, M.C., Doucet, A. and Beauregard, M. (2002) Design of high essential amino acid proteins: two design strategies for improving protease resistance of the nutritious MB-1 protein. *Journal of Biotechnology* 94, 245–254.

Witt, M.W., Newbold, C.J. and Wallace, R.J. (1998) Influence of dietary acetylated peptides on fermentation and peptidase activities in the sheep rumen. *Journal of Applied Microbiology* 84, 847–851.

Wood, J., Scott, K.P., Avgustin, G., Newbold, C.J. and Flint, H.J. (1998) Estimation of the relative abundance of different *Bacteroides* and *Prevotella* ribotypes in gut samples by restriction enzyme profiling of PCR-amplified 16S rRNA gene sequences. *Applied and Environmental Microbiology* 64, 3683–3689.

Yang, C.M.J. and Russell, J.B. (1992) Resistance of proline-containing peptides to ruminal degradation *in vitro*. *Applied and Environmental Microbiology* 58, 3954–3958.

Yanke, L.J., Dong, Y., McAllister, T.A., Bae, H.D. and Cheng, K.J. (1993) Comparison of amylolytic and proteolytic activities of ruminal fungi grown on cereal grains. *Canadian Journal of Microbiology* 39, 817–820.

Yoon, I.K. and Stern, M.D. (2003) Influence of various levels of *Lactobacillus acidophilus* supplementation on fermentation by rumen microorganisms in continuous culture. *Proceedings of the 21st Biennial Conference on Gastrointestinal Function, Chicago,* p. 35. Available at: www.ncaur.usda.gov/fbt/ GIF/ Tentative%20Agenda

Zhu, W.Y., Kingston-Smith, A.H., Troncoso, D., Merry, R.J., Davies, D.R., Pichard, G., Thomas, H. and Theodorou, M.K. (1999) Evidence of a role for plant proteases in the degradation of herbage proteins in the rumen of grazing cattle. *Journal of Dairy Science* 82, 2651–2658.

4 Factors Affecting the Efficiency of Nitrogen Utilization in the Rumen

A.N. Hristov[1] and J.-P. Jouany[2]
[1]*Department of Animal and Veterinary Science, University of Idaho, Moscow, Idaho, USA*
[2]*INRA, URH-DIMA, Research Centre de Clermont-Theix, France*

4.1 Introduction

The purpose of animal nutrition is to supply the right amount of nutrients that the animal needs to meet a pre-set production target. Any shortfall in nutrient provision will reduce animal productivity, while over-supply of nutrients will add wastefully to farmers' costs. Also, unutilized nutrients are eventually catabolized and excreted into the environment in faeces and urine. Close matching of

©CAB International 2005. *Nitrogen and Phosphorus Nutrition of Cattle* (eds E. Pfeffer and A.N. Hristov)

nitrogen (N) supply to requirements is an important objective for nutritionists because proteins are the most expensive ingredients in animal diets, and because the amounts of N excreted in animal manure have increased markedly during the last decades, causing unacceptable air and water pollution. Curbing this N pollution first requires an improved knowledge of the animal N needs, specifically amino acid needs. Nutritionists have made great progress in this area in recent years, but controlling pollution due to animal manure remains a serious challenge.

The ruminant animal is unique in its ability to convert feed N into microbial protein. Ruminant nutritionists must feed both the rumen microorganisms and their host, and the requirements of each are different. The composition of the amino acid mixture entering the duodenum of ruminants is different from the amino acid composition of the diet because most of the former is made up of microbial proteins flowing from the rumen and of dietary proteins that have been partly degraded in the rumen. Ammonia (NH_3) is the main product of protein catabolism in the rumen and is also the principal substrate for microbial protein synthesis (MPS). Peptides can also accumulate when ruminants are fed diets rich in proteins with high degradability (Chen *et al.*, 1987; Broderick and Wallace, 1988; Wallace and McKain, 1990; Williams and Cockburn, 1991; Wallace *et al.*, 1993). Ammonia in excess is absorbed through the rumen wall, mostly metabolized into urea in the liver, and excreted in urine. Some of the blood plasma urea-N (PUN) can be recycled to the rumen through saliva and rumen mucosa if NH_3 is needed for microbial growth. Therefore, a better control of ruminal N, particularly NH_3, metabolism is an obvious way to achieve an improvement in the efficiency of N utilization by ruminants and to limit the excretion of nitrogenous compounds that result in environmental pollution around animal production areas.

4.2 Dietary Crude Protein and Ruminally Degradable Protein Effects on Ammonia Concentration, Nitrogen Utilization and Ammonia Losses from Manure

Ammonia is a major source of N for the ruminal bacteria, particularly cellulolytics (Russell *et al.*, 1992). Ruminal bacteria derive between 38% (Table 4.1) and 70–80% (Mathison and Milligan, 1971; Oldham *et al.*, 1980; Leng and Nolan, 1984; Hristov and Broderick, 1996; Koenig *et al.*, 2000) of their N from NH_3-N and adequate NH_3 level is the first recommendation for optimization of MPS in the rumen (see Chapter 2). As a proportion of N intake, the irreversible loss of N from the NH_3-N pool can be as low as 23% (Table 4.1) or as high as 88% (Oldham *et al.*, 1980). The relatively low efficiency of utilization of dietary N for milk protein synthesis in the dairy cow (19–20%; Tamminga, 1992; MacRae *et al.*, 1995) is due in large part to the wasteful process of intraruminal N cycling. Tamminga (1992) estimated that up to 50% of the dietary N is lost to the dairy cow through urinary N excretion. Of this 50%, approximately 30% is lost due to inefficient N metabolism in the rumen. Therefore, the efficiency of NH_3 utilization in the rumen is a central factor determining the economic cost and environmental impact of ruminant production. As ruminal NH_3 levels correlate positively ($r = 0.57$) with the concentration of milk urea N (MUN) in dairy cows (Broderick and Clayton, 1997; Chapter 9) and increased NH_3 levels in the intestine have resulted in increased non-protein N (NPN) content of milk (Moorby and Theobald, 1999), improving the utilization of NH_3-N in the rumen has the potential of reducing MUN content and consequently enhancing the processing quality of milk (Bachmann and Jans, 1995; Martin *et al.*, 1997; Chapter 2).

Ruminal NH_3 concentration is a function of both rate of ruminal degradability and concentration of ruminally degradable dietary protein (RDP) over microbial needs and dietary energy available to the ruminal microorganisms. Certain bioactive compounds and fatty acids (FA) can also impact specific groups of rumen microorganisms and consequently alter RDP/NH_3 utilization in the rumen. The effect of dietary crude protein (CP), excess RDP and ruminally undegradable protein (RUP) supplementation on ruminal fermentation and production, particularly in dairy cows, has been covered extensively in the literature (Armentano *et al.*, 1993; Christensen *et al.*, 1993; Santos *et al.*, 1998; Kebreab *et al.*, 2001; Reynal and Broderick, 2003). Increasing CP content of the diet may result in greater milk production (Armentano *et al.*, 1993; Tomlinson *et al.*, 1994; Powers *et al.*, 1995; Wu and Satter, 2000),

Table 4.1. Effects of carbohydrate source on ruminal fermentation and milk N efficiency in lactating dairy cows (data from Hristov *et al.*, 2005).

	Carbohydrate source[1]				
	GLU	STA	NDF	MIX	SE
DMI (kg/day)	21.8	21.4	22.6	22.5	0.48
Rumen fermentation					
pH	6.00^{c}	6.19^{b}	6.41^{a}	6.05^{c}	0.049
NH_3 (mM)	8.5^{c}	9.6^{c}	16.4^{a}	12.4^{b}	1.05
NH_3-N[2] (g)	5.6^{bAB}	7.2^{abB}	12.3^{aA}	11.6^{aA}	1.24
Total VFA (mmol/l)	122.3^{bAB}	139.1^{aAB}	135.9^{aB}	145.2^{aA}	3.08
Acetate	74.0^{b}	95.4^{a}	94.5^{a}	97.0^{a}	2.45
Propionate	22.1	23.0	23.2	24.0	1.28
Butyrate	22.2^{a}	15.0^{c}	12.4^{d}	18.9^{b}	0.81
iso-Butyrate	0.63^{d}	1.29^{bc}	1.57^{a}	1.22^{c}	0.081
iso-Valerate	0.61^{bAB}	1.73^{aA}	1.68^{aAB}	13.7^{aB}	0.126
MN[3] flow (g/day)	197^{a}	185^{a}	153^{b}	188^{a}	9.7
MNNH$_3$-N[4] flow (g/day)	75^{b}	111^{a}	74^{b}	89^{ab}	10.4
MN efficiency[5] (g/kg)	15^{aAB}	15^{aAB}	12^{bB}	14^{abA}	0.8
^{15}N data					
Rumen NH_3 irreversible loss (g N/day)	230^{b}	343^{a}	320^{a}	294^{ab}	26.7
% of N intake	36^{bB}	55^{aAB}	49^{abA}	45^{abAB}	4.0
Flux (g N/day)	350^{b}	487^{a}	533^{a}	525^{a}	39.1
% of N intake	55^{b}	77^{a}	80^{a}	80^{a}	5.9
NH_3-N recycled (g/day)	119	144	213	231	40.4
Proportion of irreversible loss trapped into MN	33^{a}	33^{a}	23^{b}	30^{a}	2.4
Bacterial N from NH_3-N (%)	38.3^{bB}	60.1^{aAB}	48.3^{bA}	47.3^{bA}	3.51
Milk protein N from bact. N (%)	44.8	46.3	46.3	43.3	1.91
Milk protein N from NH_3-N (%)	17.1^{bA}	27.7^{aA}	22.4^{abB}	20.8^{bAB}	1.82
N partitioning (g N/day)					
N intake	634	630	662	651	17.4
Urine N	274^{bB}	286^{bAB}	339^{aAB}	317^{abA}	18.5
% of N intake	44	46	51	49	3.8
Faecal N	160^{a}	155^{a}	88^{b}	136^{a}	12.2
% of N intake	25^{aAB}	24^{abA}	13^{cAB}	21^{bB}	1.6
Total N (urinary and faecal)	435	441	427	453	18.6
% of N intake	69	71	65	70	3.9
Milk N	97	89	90	83	9.7
% of N intake	15^{aA}	14^{abB}	14^{bAB}	13^{bAB}	1.2
Faecal C:N	16.9^{bc}	16.4^{c}	25.2^{a}	17.5^{b}	0.36
PUN[6] (mg/100 ml)	21.2	21.8	27.0	23.8	1.99
MUN[6] (mg/100 ml)	16.9^{c}	17.4^{c}	22.4^{a}	19.7^{b}	1.19

[1]GLU (maize dextrose), STA (maize starch), NDF (white oat fibre), MIX (25% each of apple pectin, GLU, STA and NDF). All carbohydrates given intraruminally at 20% of DMI. SE, standard error.

[2]Ammonia-N pool size estimated from rumen evacuation data and ammonia concentration in ruminal fluid.

[3]MN, microbial N flow to the duodenum.

[4]MNNH$_3$-N, microbial N derived from NH_3-N flow to the duodenum.

[5]Efficiency of microbial N synthesis, g MN/kg OM truly digested in the rumen.

[6]PUN and MUN, plasma and milk urea-N, respectively.

[a,b,c,d]Means not having same superscripts differ at $P < 0.05$.

[A,B]Means not having same superscripts differ at $P < 0.1$.

but also leads to increased concentration of ruminal NH_3 and PUN and consequently greater urinary N losses (Armentano *et al.*, 1993; Christensen *et al.*, 1993; Metcalf *et al.*, 1996; Castillo *et al.*, 2001a). High-yielding dairy cows did not benefit from increased CP concentration in the diets as RUP increased from 4.5% to 8.4% (corresponding to 17.3% and 20.9% CP, respectively) in the study by Komaragiri and Erdman (1997). Although CP concentration of milk increased (only a trend at $P < 0.1$ was observed for milk protein yield) with the high CP diet, efficiency of utilization of dietary N for milk protein synthesis (MilkNE = milk protein N yield/N intake) remained unchanged (30% and 29%, low and high CP diets, respectively). Wu and Satter (2000) demonstrated that diets with CP concentration of 16.0% to 17.5% could be fed to high-yielding dairy cows (~11,000 kg/308 days) without affecting milk yield. MilkNE progressively decreased from 28.8% to 25.9%, 24.1% and 24.7% with 15.4–16.0%, 17.4–16.0%, 17.4–17.9% and 19.3–17.9% CP diets (lactation weeks 1 to 16 and 17 to 44, respectively). Although, as exemplified by Tamminga (1992), ruminal N loss is the greatest single contributor to urinary N losses, metabolic losses, indigestible microbial N (MN), losses in maintenance and inefficient conversion of absorbed amino acids into milk protein comprise up to 72% of the urinary N losses in the dairy cow. Nitrogen excreted as urinary urea (representing from 60% to 94% of the total urinary N in cattle; Bristow *et al.*, 1992) is quickly converted to NH_3 and lost through evaporation or, when applied to cropland, converted to nitrate in the aerobic topsoil; substantial amounts of nitrate will leach into groundwater supplies.

Increased dietary CP concentration usually leads to increased urinary N losses. Külling *et al.* (2001) reported increased urinary N excretion from 0.84 to 3.02 g/kg excreta with increasing CP concentration of the diet from 12.5% to 17.5% (dry matter (DM), basis). Similarly, Castillo *et al.* (2001a) observed proportional increase in urinary N losses in dairy cows with elevating level and degradability of dietary CP. Increasing CP concentration of the diet from 15.8% to 18.3% resulted in an increase in urinary N losses from 58% to 63% (as percentage of total losses), PUN (15 vs. 19 mg/100 ml) and MUN (13 vs. 16 mg/100 ml) concentrations with no effect on milk or milk protein yield (Hristov *et al.*, 2004a).

Dietary RDP can be used for MPS provided energy is not limiting; in most feeding systems, MPS is assumed to be energy-dependent (NKJ Protein Group, 1985; Tamminga *et al.*, 1994; GfE, 2001; NRC, 2001). If not utilized for MPS, RDP will most likely be degraded to NH_3 and detoxified in the liver (Lobley *et al.*, 1995), although a small proportion may by-pass the rumen and contribute to the duodenal amino acid and peptide flow (Choi *et al.*, 2002). Direct comparison of RDP (solvent soybean meal, SSBM) vs. proteins with greater resistance to ruminal degradation (fishmeal and expeller soybean meal, ESBM) showed no effect of protein degradability on PUN or MilkNE (Calsamiglia *et al.*, 1992). Increasing RDP concentration of the diet from 9.5% to 11.7% (13.9% and 16.0% CP, respectively) resulted in increased NH_3 concentration in the rumen (by 72%), did not affect milk or milk protein yields, but decreased MilkNE from 30% to 25%, respectively (Armentano *et al.*, 1993). Christensen *et al.* (1993) fed high-yielding cows (37–40 kg/day milk yield) diets varying in CP (from 16.4% to 19.6%) and RUP concentrations. Milk and milk protein yields were not affected by treatment, but MilkNE dropped from 36 (16.4% CP diets) to 30–33% (19.2–19.6% CP diets). At the same time, PUN increased with increasing CP level and protein degradability of the diets. In a similar design (cows produced from 34 to 39 kg/day of milk), Cunningham *et al.* (1996) reported linear increase in ruminal NH_3 concentration with increasing dietary CP level (from 14.4, basal diet to 16.4–18.4%, supplemented diets) and degradability. Microbial N flow and efficiency of MPS in the rumen were unaffected by treatment. Milk yield was also similar between treatments except that increasing RUP concentration increased milk protein yield (Trial 1 only). MilkNE, however, was decreased from 35 (basal diet, 14.4% CP) to 30–33% (diets with 16.4% to 18.4% CP) in Trial 1; only the diet with 16.4% CP and high RUP concentration produced MilkNE of 35%. In Trial 2, MilkNE apparently decreased from 30–31% (16.4% CP diets) to 27–28% (18.4% CP diets) with no effect of RUP level. Milk or milk protein yields of very high-yielding cows (45–48 kg/day) were sustained with 14.7–14.9% CP, but milk protein yield was increased (trend at $P < 0.1$) as undegradability (and amino acid profile) of dietary CP increased (Bach *et al.*, 2000). MilkNE was greater (40–44% vs. 35–37%) and MUN concentration was lower with the low CP diets. Amino acid profile/RUP concentration appeared to increase MilkNE; 40% vs. 37%, for the high-quality amino acid profile/high RUP concentration diets and low-quality amino acid profile/low RUP diets, respectively. Reynal and Broderick (2003) supplemented a basal diet (15.8% CP) with protein feeds having various degradabilities in the rumen and observed increased ruminal NH_3 concentrations with RDP (on average by 35%) or RUP (average of 19%) supplements. Duodenal amino acid flow and milk and milk protein yields were increased by protein supplementation, but MilkNE decreased from 27.2 in the unsupplemented control to 23.2–24.6% in the protein-supplemented diets. Broderick (2003) reported greater milk yield with 16.7% CP vs. 15.1% CP diets, but MilkNE linearly decreased from 30.3% to 27.0%, respectively. The diet with the highest CP concentration, 18.4%, had the lowest MilkNE, 23.9%. Urinary N losses and MUN concentration also linearly increased with increasing CP content of the diet. Excess of dietary RDP above requirements (NRC, 2001) did not increase MPS or the efficiency of MPS in the rumen, but reduced the efficiency of utilization of ruminal NH_3-N for milk protein synthesis in dairy cows (Fig. 4.1; Hristov *et al.*, 2004a). Similarly, Holthausen (2002) found lesser recovery of intravenously dosed ^{15}N-urea into milk protein of dairy cows receiving a urea-supplemented diet (2.3% urea, DM basis) compared to the unsupplemented control diet. Milk urea N concentration and urinary N excretion were elevated with the urea diet compared to the control.

Ammonia emissions from human activities are an environmental issue of growing public concern (Cowling and Galloway, 2001) and in developed countries farm animals are the greatest contributor to these emissions (US data; Kerchner *et al.*, 2000). Feeding, particularly dietary CP level, environment and type of manure handling system all contribute to NH_3 losses from cattle operations. Monteny and Erisman (1998) suggested that NH_3 emissions from cubicle dairy houses can be reduced by up to 50% through flushing of floors

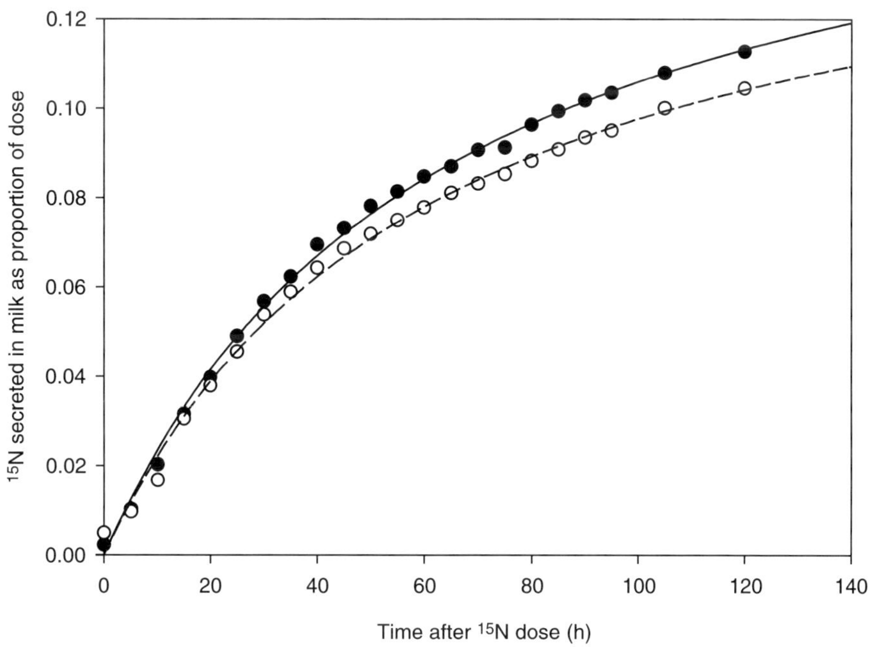

Fig. 4.1. Effect of CP and RDP concentration of the diet on the cumulative secretion of NH_3-^{15}N in milk protein as proportion of ^{15}N dosed intraruminally (data from Hristov *et al.*, 2004a). Closed circles and solid line, low protein diet (15.8% CP), measured and predicted, respectively; open circles and dashed line, high CP diet (18.3% CP), measured and predicted, respectively.

with water or diluted formaldehyde, optimizing feeding strategies and slurry acidification. Külling *et al.* (2001) investigated the relationships between CP level in the diet, type of manure system and NH_3, nitrous oxide and methane emissions from dairy operations. The authors demonstrated that at 17.5% CP in the diet, N losses from manure after 7 weeks of storage were from 21% (slurry) to 108% (urine-rich slurry) greater as compared to the N losses from manure from cows fed 12.5% CP. Across manure systems, rates of NH_3 emissions were 163 and 42 $\mu g/m^2/s$ for the two CP levels, respectively. At the same time, N_2O emissions were on average 205.7 and 35.4 $ng/m^2/s$, respectively. McGinn *et al.* (2002) found a positive relationship between dietary CP intake and NH_3-N content of beef cattle manure. Although *in vitro* data from this particular study did not indicate that NH_3-N content of manure caused significant difference in NH_3 emissions, the lowest emission rate for surface manure was associated with the lowest dietary CP level. Thus, the possibility exists that reducing CP level of the diet in combination with balancing carbohydrate (CHO)/protein degradability in the rumen and accounting for the effects of manure pH and ambient temperatures can significantly reduce NH_3 losses from dairy and beef cattle operations.

Some feeding systems for ruminants regulate dietary N utilization through ruminal availability of feed proteins (the Dutch DVE/OEB and the Scandinavian AAT-PBV, for example). In the DVE/OEB system (Tamminga *et al.*, 1994), reducing the degradable protein balance (OEB) in the rumen to 0.4 kg/day resulted in a steady decrease in N losses (Berentsen and Giesen, 1996). Utilizing a combination of prediction equations (urine volume) and actual analyses (urine composition), De Boer *et al.* (2002) demonstrated the importance of OEB in reducing N losses by dairy cows. Increasing OEB from 0 (maximal utilization of RDP) to 1000 g/cow per day resulted in a linear increase in urinary N excretion. The approach was further developed as NH_3 emissions from manure were predicted based on urinary N/urea excretion (affected by feed characteristics), manure pH and environmental inputs such as temperature, floor area and air velocity (Monteny *et al.*, 2002). Compared with feeding excess RDP, the problem of how energy availability affects the extent and efficiency of RDP/NH_3 utilization in the rumen is more complex.

4.3 Dietary Carbohydrate Availability

Ammonia concentration in the rumen can vary greatly depending on diet, time of feeding and feeding frequency, animal and other unknown factors. This variation can result in decreased efficiency of microbial NH_3 capture and eventually, in N wastage. The extent to which NH_3 is utilized in the rumen depends primarily on the rate of release and the balance of CHO and N availability. Carbohydrate availability determines the rate of microbial growth in the rumen (Isaacson *et al.*, 1975; Strobel and Russell, 1986; Hoover and Stokes, 1991) and efficiency of utilization of ruminal NH_3 (Russell *et al.*, 1983; Newbold and Rust, 1992; Hristov *et al.*, 1997; Heldt *et al.*, 1999). If energy is limiting, ruminal microorganisms degrade feed proteins to NH_3 (see Chapters 2 and 3) and NH_3 uptake is suppressed (Nocek and Russell, 1988; Hristov *et al.*, 1997). Carbohydrate supplementation and source, starch degradability and synchronization of ruminal energy and N release may be key factors in improving the efficiency of ruminal NH_3 and overall dietary N utilization in ruminants.

4.3.1 Carbohydrate supplementation and source

Earlier studies investigating the effects of CHO supplementation and type on ruminal fermentation were focused on the possibility of using NPN (urea) as a substitute for feed protein in cattle diets. Mills *et al.* (1944) reported that ruminal NH_3- and urea-N concentrations were decreased and protein concentration in ruminal contents (linked to microbial protein in this case) was increased, when molasses was added to a timothy hay-based diet. The efficiency of dietary NPN utilization was further improved when starch replaced molasses as a CHO supplement. Belasco (1956) concluded that urea utilization *in vitro* was dependent on the amount and type of CHO present in the incubation media. Compared to cellulose, starch promoted a slightly greater utilization of urea; xylan and pectin were inferior compared to starch. More recently, the target of CHO supplementation was improving dietary CP and NH_3 utilization through manipulation of ruminal microbial fermentation.

Carbohydrate supplementation has consistently reduced ruminal NH_3 concentration and MPS in the rumen was often enhanced. Supplementation of an all-grass silage diet with barley, starch, various sugars, or sugars and buffer resulted in reduced NH_3 concentration in the rumen of sheep and goats (Chamberlain *et al.*, 1985). The authors related the NH_3 effect of barley/starch vs. that of sugars to the effect of CHO supplementation on ruminal protozoa, i.e. no effect, or reduction in protozoal counts would lead to a further decrease in NH_3 concentration; increased protozoal counts with starch supplementation may offset the effect of starch on NH_3 by increasing intraruminal recycling of bacterial protein. Rooke *et al.* (1987) showed that an intraruminal infusion of glucose resulted in a 45% decrease in ruminal NH_3-N concentration. The efficiency of ruminal fermentation was significantly improved as MPS was increased by 29% as a result of the CHO infusion. Huhtanen (1987) reported a linear increase in MPS in cattle with increasing levels of intraruminal CHO (sucrose or xylose) infusion from 0 to 450 and 900 g/day. Ruminal NH_3 concentration was reduced in a quadratic manner and sucrose had a greater effect than xylose despite the fact that volatile fatty acids (VFA) concentration was similar between the two CHO sources. Intraruminal treatment of dairy cows with 1 kg/day sucrose decreased, irrespectively of the application method (pulse dose or continuous infusion), NH_3 concentration and increased MPS compared with the unsupplemented, grass silage/barley grain diet (Khalili and Huhtanen, 1991). Chamberlain *et al.* (1993) found a significant reduction in ruminal NH_3 concentration in sheep fed an all-grass silage diet with various sugars and starch (4.7% of dietary DM) compared to the unsupplemented control. Microbial N production (as estimated from urinary purine derivatives) was increased by the CHO addition. Surprisingly, Feng *et al.* (1993) reported increased ruminal NH_3 levels and decreased MPS in dairy cows fed a 39% non-structural CHO (NSC) diet compared to a 29% NSC diet (diets had similar estimated concentrations of RDP). Estimated MilkNE was not affected by treatment (26.7% and 26.6%, high and low NSC diets, respectively). In this trial, ingredient composition rather than NSC concentration was most likely the factor determining the ruminal responses. Increasing the level of CHO intake from 0.15% to 0.30% of liveweight (LW) decreased ruminal NH_3 concentration in beef steers, but type of CHO (pure starch, glucose or fibre) had no effect on NH_3 concentration in ruminal fluid (Heldt *et al.*, 1999). The greatest concentration of VFA (and lactate) in the rumen was associated with the starch treatment and the lowest with the glucose. Maltodextrin supplementation of a grass-silage/grain concentrate diet reduced ruminal NH_3 concentrations in dairy cows (Kim *et al.*, 1999a). Microbial N flow (based on urinary purine derivatives excretion) followed the ruminal NH_3 pattern. MilkNE was not affected. In a similar design, with infusion of sucrose instead of maltodextrin, however, ruminal NH_3 concentrations or MPS in non-lactating dairy cows were not affected by the sugar addition (Kim *et al.*, 1999b). Oh *et al.* (1999) found reduced NH_3 concentrations with starch and further reduction with sucrose supplementation (both CHO fed at 250 g/day), but no effect of CHO source on MPS in sheep. In a second trial, Oh *et al.* (1999) observed a reduction in NH_3 concentration associated with a higher MPS when greater (500 g/day) barley starch supplementation was given. Sucrose supplementation of up to 5% of dietary DM reduced NH_3 concentration *in vitro*, but no corresponding effect was observed *in vivo* (McCormick *et al.*, 2001). MilkNE of dairy cows did not appear to be affected by treatment (23.6%, 24.3%, 25.6% and 24.1%; SSBM, SSBM with 5% sucrose, ESBM and ESBM with 5% sucrose, respectively). Replacing barley grain with sugarbeet molasses (0% to 60% of dietary DM) linearly reduced NH_3 concentrations in the rumen of bulls fed a wheat straw/barley grain/sunflower meal basal diet (Araba *et al.*, 2002).

Broderick *et al.* (2002) fed starch sources with different degradabilities in the rumen (high-moisture ear maize, HMEC vs. cracked shell maize, i.e. more rapidly vs. more slowly degradable starch) at 38% of diet DM and a 50:50 combination of HMEC and citrus pulp to early lactation dairy cows. At a similar CP concentration of the diet (19%, 18.6% and 18.9%), MilkNE was lower with the HMEC/citrus pulp diet (21.5%) than with the HMEC (26.0%) and the cracked shell maize (25.3%) diets. Blood plasma urea N and MUN were not affected by the CHO source. In a corresponding trial, ruminal NH_3 concentration was lowest for the HMEC diet (12.8 mmol/l) followed by the HMEC/citrus pulp and the cracked shell maize diets (15.2 and 18.5 mmol/l, respectively).

Sannes *et al.* (2002) compared starch vs. sugar in diets for lactating dairy cows and reported a significant reduction in ruminal NH_3 concentration and urinary urea-N excretion with the sugar addition (supplemented at 3.2% of dietary DM, partially replacing ground maize). Strong relationships between ruminal NH_3 concentration, urinary N excretion and MilkNE were observed. A reduction in MPS associated with the sucrose treatment, as measured through urinary purine derivatives excretion, could not be attributed to any of the observed effects in this study. MilkNE was unaffected by sucrose addition (23.5% vs. 23.2%, control and sucrose-supplemented diets, respectively). Lee *et al.* (2003) infused increasing amounts of water-soluble CHO (WSC) in the rumen simulation technique RUSITEC (3.75, 4.69, 5.63 and 6.56 g/day per vessel) and observed a quadratic decrease in NH_3 concentration and a quadratic increase in MPS. *In vivo*, the same group (Lee *et al.*, 2002) reported a 47% decrease in NH_3 concentration in the rumen of steers fed high- (243 g/kg DM) versus low-WSC (161 g/kg DM) grass diets. Microbial N flow to the duodenum was increased with the high-WSC diet (by 27%). Richardson *et al.* (2003) reported reduced MPS (based on urinary purine derivatives excretion) in lambs receiving an unmolassed sugarbeet pulp concentrate compared to a barley-based concentrate. No main effects of energy source on ruminal NH_3 concentration were reported. Compared to intake of non-structural carbohydrates and fermentable starch, greater intake of ruminally fermentable neutral detergent fibre (NDF) decreased NH_3 concentration in the rumen of lactating dairy cows and increased the efficiency of NH_3-N utilization for milk protein synthesis, but had no effect on urinary or faecal N losses (Hristov and Ropp, 2003).

Apart from reduction in NH_3 concentration, CHO supplementation or type has a variable effect on urinary N excretion, milk composition and MilkNE. Starch, sucrose and xylose decreased urinary N losses in sheep compared to beet pulp, but expressed as a proportion of N intake, the effect of CHO was minimal (Huhtanen and Robertson, 1988). Martin *et al.* (2000) reported stimulation of ruminal fermentation (decreased pH and increased VFA) *in vitro* with inclusion of sugar and malate, but NH_3 concentration was not measured. Solomon *et al.* (2000) found slightly reduced milk protein yield and MilkNE (29.3% versus 31.4%, respectively) in dairy cows receiving dry citrus pulp compared with maize grain (fed at 24% of dietary DM). Carbohydrate source had no effect on MUN. Leiva *et al.* (2000) fed dairy cows diets contrasting in NDF and starch (from citrus pulp or hominy, respectively) and reported no effects on ruminal fermentation (NH_3 concentration was not measured) and N partitioning, but reported increased milk yield with the high-starch diet. MilkNE appeared to be slightly increased ($P=0.132$) and MUN decreased by the starch diet (21.8% vs. 24.1%, citrus pulp and hominy diets, respectively). Castillo *et al.* (2001b) investigated the effect of high-NDF (wheat middlings), low- or high-degradability starch (maize versus barley) and sugar/soluble fibre (molassed sugarbeet pulp) supplements on dietary N utilization in lactating dairy cows and found no effect of sugar/soluble fibre vs. starch supplementation on urinary N excretion. MilkNE was not affected by treatment, although the maize-based supplement had numerically greater efficiency (30%) than the other treatments (27.8%, 26.9% and 26.1%, high-NDF, high starch degradability and sugar/soluble fibre supplements, respectively). Substituting 2.7% ground maize with sucrose did not result in any significant changes in milk and milk components yields, MilkNE (30.8% vs. 31.2%, maize and sucrose, respectively), or MUN in early lactation cows (Ordway *et al.*, 2002). When administered through the drinking water, 1% or 2% glucose solutions reduced ruminal NH_3, branched-chain VFA and PUN concentrations in fresh dairy cows (Osborne *et al.*, 2002a), but 5% glucose solution had no effect on body weight gain or feed intake in dairy calves (Osborne *et al.*, 2002b).

Hristov *et al.* (2005) investigated the effect of various CHO sources on utilization of ruminal NH_3-N in dairy cows. Late-lactation cows (217 ± 35.2 days in milk) were fed an all lucerne hay basal diet. Carbohydrates [maize dextrose, GLU; maize starch, STA; fibre, NDF (white oat fibre); and a CHO mix (25% of each: apple pectin, GLU, STA and NDF, MIX)] were introduced into the rumen during feeding (twice daily) at 20% of dietary DM intake. Nitrogen-15 was pulse-dosed into the rumen and ^{15}N-enrichments of ruminal NH_3, bacterial and milk protein-N were followed for 30 and 120 h, respectively. Fermentation and N utilization data from this trial are presented in Table 4.1. Microbial N flow to the duodenum and the efficiency of MPS in the rumen were increased by GLU, MIX (significant effect on MN only) and STA compared to NDF. Concentration of NH_3 in ruminal fluid was the lowest with GLU (which also had the smallest NH_3-N pool size) and STA and was also reduced by MIX compared to NDF. The least amount of N irreversibly lost from the ruminal NH_3-N pool and flux was associated with GLU. The proportion of NH_3-N irreversibly lost that was incorporated into MN was the least for NDF. Based on the respective areas under the ^{15}N-enrichment curves, more bacterial N was formed from NH_3-N with STA compared with the other CHO, which is reflected in a greater proportion of milk protein-N derived from NH_3-N for STA compared to GLU and MIX. Provision of ruminally available CHO (GLU, STA) shifted N losses from the urine to faeces; less N was lost with the urine for GLU and STA and MilkNE was slightly greater for GLU compared with the other CHO sources. Milk urea N concentration was reduced by GLU and STA compared to NDF, and faecal C:N ratio was dramatically greater for the latter treatment. It appears that the two CHO with positive effects on ruminal NH_3-N utilization had different modes of action. Overall, GLU resulted in more rapid fermentation, more efficient utilization of dietary amino acids and reduced NH_3 production in the rumen, which decreased PUN concentration and urinary N losses. Addition of starch to the basal diet enhanced production, microbial utilization and transfer of ruminal NH_3-N into milk protein-N, which also resulted in reduced PUN concentration, urinary N losses and lowered MUN concentration compared to the NDF treatment.

Readily available energy can also be supplied to the ruminal microorganisms through the forage component of the diet. For over 30 years it has been known that concentration of fermentable carbohydrates in forage crops varies diurnally (Bowden *et al.*, 1968; Holt and Hilst, 1969; Lechtenberg *et al.*, 1971). More recently, Fisher *et al.* (1999) reported greater concentration of total nonstructural carbohydrates (TNC) and mono- and disaccharides in tall fescue hay harvested at sundown compared to sunup. *In vitro* digestibility of DM was also greater with the pm than with the am hay. Similar compositional changes were observed with legumes (Owens *et al.*, 2002). Palatability and DMI increased with increasing TNC concentration of grasses (Tava *et al.*, 1995; Fisher *et al.*, 1999; Mayland *et al.*, 2000; Downing, 2003) and lucerne (Fisher *et al.*, 2002), although no effect on DMI was reported by Miller *et al.* (2001). Provision of a greater amount of fermentable substrate in the rumen with high-TNC forages may enhance the utilization of ruminal NH_3 (Lee *et al.*, 2002) and reduce urinary N excretion in ruminants. Indeed, high-WSC (16.5% of DM) perennial ryegrass reduced urinary N output and increased MilkNE in lactating dairy cows compared to low-WSC (12.6%) forage (Miller *et al.*, 2001). Data reported by Downing (2003), however, indicated no effect of WSC concentration of green chop on MUN content of milk.

Reduction in ruminal NH_3 concentration is the most likely effect of provision of readily fermentable carbohydrates (sugars or starch) in the rumen (Table 4.2). Although more variable, MPS may also increase with CHO supplementation. Fewer studies reported urinary N losses, but in four out of six trials urinary N excretion was reduced due to CHO supplementation. Effect on MilkNE and MUN concentration in lactating dairy cows is less consistent. Reducing urinary N loss is a key in improving the efficiency of conversion of dietary N into milk protein-N, but other cow- and diet-related factors play important roles in this process. In a meta-analysis involving 846 observations (diets) from 256 feeding trials, Hristov *et al.* (2004b) reported that MilkNE averaged 24.7 ± 0.14%, varying considerably between diets (minimum and maximum of 13.7% and 39.8%, respectively). This variability highlights the potential for improving the efficiency of utilization of dietary N in dairy cows and in cattle in general. Results from the Hristov *et al.* (2004b) analysis indicated that diets producing high MilkNE contained more maize and cereal silages and concentrate and less lucerne forage. For example, only 29% of the diets with MilkNE greater than 30% (77 diets; average efficiency of 33%) contained lucerne silage, compared to 50% for all diets in the data set. Maize silage was fed with 74% of the high efficiency diets compared to 57% of all diets; average concentration of maize silage in dietary DM was similar between diets: 35% and 32%, respectively. The high MilkNE diets more often contained maize grain (77%, compared to 61% for all diets) and barley grain (29% vs. 15%, respectively). Average CP concentration of the high MilkNE diets was 15.8% while the average for all diets was 17.8%. Compared with the average milk yield per cow from all diets, cows produced more

Table 4.2. Effect of carbohydrate supplementation and source on ruminal ammonia concentration, microbial protein synthesis, urinary N losses and milk N efficiency (selected *in vivo* trials only).

Carbohydrate	Ammonia	MPS[a]	Urine N	Species	MilkNE[a]	MUN[a]	Reference
Sugar/starch	↓	(↑)[b]	N/R[c]	Cattle	N/A[c]	N/A	Mills *et al.* (1944)
Starch/sugar	↓	N/R	N/R	Sheep/goats	N/A	N/A	Chamberlain *et al.* (1985)
Sugar	↓	↑	N/R	Cattle	N/A	N/A	Huhtanen (1987)
Sugar	↓	↑	N/R	Cattle	N/A	N/A	Rooke *et al.* (1987)
Starch/sugar	N/R	N/R	↓	Sheep	N/A	N/A	Huhtanen and Robertson (1988)
Sugar	↓	↑	N/R	Cattle	N/A	N/A	Khalili and Huhtanen (1991)
Sugar/starch	↓	↑	N/R	Sheep	N/A	N/A	Chamberlain *et al.* (1993)
Maltodextrin	↓	↑	N/R	Dairy cows	=	N/R	Kim *et al.* (1999a)
Sugar	=	=	N/R	Dairy cows	N/R	N/R	Kim *et al.* (1999b)
Sugar/starch	↓	=	N/R	Sheep	N/A	N/A	Oh *et al.* (1999)
Starch (Exp. 3)	↓	↑	N/R	Sheep	N/A	N/A	Oh *et al.* (1999)
Starch vs. soluble fibre	N/R	N/R	=	Dairy cows	(↑)[†]	↓	Leiva *et al.* (2000)
Sugar	=	N/R	N/R	Dairy cows	=	=	McCormick *et al.* (2001)
Sugar/starch	N/R	N/R	=	Dairy cows	=	N/R	Castillo *et al.* (2001b)
Starch[d] vs. soluble fibre	↓	N/R	N/R	Dairy cows	↑	=	Broderick *et al.* (2002)
Sugar	↓	↓	↓	Dairy cows	=	(↓)[‡]	Sannes *et al.* (2002)
Sugar	↓	↑	N/R	Cattle	N/A	N/A	Lee *et al.* (2002)
Molasses vs. starch	↓	N/R	N/R	Cattle	N/A	N/A	Araba *et al.* (2002)
Sugar vs. starch	N/R	N/R	N/R	Dairy cows	=	=	Ordway *et al.* (2002)
Sugar[e]	↓	N/R	N/R	Dairy cows	N/R	N/R	Osborne *et al.* (2002a)
Sugar	↓	↑	↓	Dairy cows	↑	=	Hristov *et al.* (2005)
Starch	↓	(↑)[§]	↓		=	↓	Hristov *et al.* (2005)
Starch vs. soluble fibre	=	↑	N/R	Lambs	N/A	N/A	Richardson *et al.* (2003)
Starch	N/R	↑	↓	Dairy cows	↑	↓	Broderick (2003)

[a]MPS, microbial protein synthesis; MilkNE, milk protein-N efficiency (milk protein-N/N intake) (in most cases MilkNE data were not statistically evaluated by the original authors); MUN, milk urea-N.
[b]Protein concentration in ruminal contents was used as a measure of MPS.
[c]N/R, not reported; N/A, not applicable.
[d]High-moisture ear maize vs. high-moisture ear maize plus dried citrus pulp (50:50).
[e]Glucose supplied with the drinking water.
[†]$P = 0.132$.
[‡]$P = 0.057$.
[§]$P = 0.110$.
↓, decrease; ↑, increase (only effects significant at $P < 0.05$ are reported); =, no change ($P > 0.05$).

milk in trials where MilkNE was high: 35.2 vs. 31.9 kg/day (milk protein concentration was not different between the groups, 3.10% and 3.11%, respectively). Extremely high N utilization efficiencies were associated with comparatively low CP intake and very high milk yields (data by Bach *et al.*, 2000, for example). Thus, high MilkNE was more often found with diets in which maize silage rather than lucerne silage was the main forage ingredient, CP concentration was low and cows had greater milk yields. Satter *et al.* (2002) calculated that with diets in which low DM lucerne silage is the only forage, NRC (2001) predicted 20.8% dietary CP was needed to meet the requirements of a high-producing, non-pregnant dairy cow; whereas with diets in which the forage was lucerne hay and maize silage (1:1), the requirements of the same cow would be met at 15.8%

CP. Wilkerson *et al.* (1997) reported proportionally greater urinary N losses and lower MilkNE with low-yielding cows (<20 kg/day) than with high-yielding cows (>20 kg/day milk): 37.9% vs. 34.6% and 22.0% vs. 29.7%, respectively. It has to be noted, however, that the high-yielding cows excreted a total of 143 g/day per 1000 kg LW more N than the low-yielding cows. As a proportion of intake, the total N excreted in urine and faeces was not dramatically different between the two groups: 72.6% vs. 68.9%, low- and high-producing cows, respectively. In general, it can be concluded that MilkNE/N losses are closely related to N intake. As demonstrated by Hof *et al.* (1994), below a certain level of protein in the diet, there is little that can be done to reduce N losses from dairy cows. The potential remains with diets having N content above the threshold required to maintain desired milk protein yield (16 g DVEc/MJ NELc, as expressed by Hof *et al.*, 1994).

4.3.2 Starch supplementation of the diets and rate of starch degradation in the rumen

Diets providing higher levels of ruminally fermentable energy should, in theory, enhance NH_3 capture and promote greater MPS in the rumen (Ørskov *et al.*, 1971). Kennelly *et al.* (1999) fed diets with concentrate:forage ratios of 50:50 and 75:50 and observed a significant reduction in ruminal NH_3 concentration with the high-concentrate diet. Milk protein yield increased with the increase in concentrate feeding as did MilkNE (26.9% vs. 30.7% and 26.1% vs. 34.1%, 50:50 vs. 75:50 concentrate:forage with or without buffer, respectively). Increasing the concentrate (HMEC and SSBM) proportion in the diet of lactating dairy cows from 20% to 65% (replacing lucerne silage) linearly increased digestibilities of dietary DM, organic matter (OM) and CP and milk and milk protein yields (Valadares *et al.*, 2000). Milk urea N concentration decreased in a quadratic manner and MilkNE had a similar (estimated) quadratic response with the 65% concentrate diet having the greatest MilkNE (23.0%, 21.0%, 22.3% and 25.0%, for 20%, 35%, 50% and 65% concentrate, respectively). Increasing dietary concentrate from 35% to 50% increased MPS by 39%, NH_3 by 18% and had no effect on ruminal VFA concentration (Khorasani *et al.*, 2001b). Supplementation of an all-grass silage diet with barley grain (at 29% of DM) increased N intake but did not significantly affect NH_3 concentration in the rumen (3.4 vs. 4.4 mmol/l, silage and barley-supplemented diets, respectively; Ahvenjarvi *et al.*, 2002). Total non-ammonia nitrogen (NAN) flow and MPS were increased and overall N balance was improved with the barley supplementation. Estimated MilkNE was unaffected by the additional starch (28.5% vs. 28.4%, all-silage and barley-supplemented diets, respectively). Broderick (2003) reported a linear increase in milk yield and MilkNE (24.9%, 26.9% and 29.5%) with increasing non-fibre CHO (NFC) concentration of the diet (from 37.1% to 41.3%, and 45.8%). Urinary N losses and MUN concentration linearly decreased and excretion of urinary purine derivatives linearly increased with increasing dietary NFC.

Type of energy concentrate and degree of processing affect rate and extent of starch degradation in the rumen and can have a considerable effect on fermentation and NH_3 utilization. Studies examining the effects of grain source and processing on the performance of growing cattle and dairy cows have been comprehensively reviewed (Huntington, 1997; Owens *et al.*, 1997; Theurer *et al.*, 1999; Firkins *et al.*, 2001). The following is a summary of the effects of grain sources with varying rate and extent of ruminal starch degradability and processing on ruminal NH_3 concentration, MPS and overall N utilization by the ruminant. Zinn (1993) reported increased MPS with barley vs. maize (fed at 74% of diet DM) in feedlot cattle. Steam-rolling of barley or degree of processing (coarse vs. thin), however, did not increase MPS. With maize, tempering of the grain did not affect MPS, but steam-flaking increased MPS compared to tempered or dry-rolled maize (Zinn *et al.*, 1998). Across different levels of supplemental RDP, starch degradability (maize vs. barley) did not affect NH_3 levels in the rumen of beef cattle but increased the duodenal flow and efficiency of MPS (Martin-Orue *et al.*, 2000). In a finishing diet setting (86% temper-rolled barley), processing barley from coarse (458 kg/l; 13.5%/h rate of ruminal DM degradation) to medium (422 kg/l; 11.6%/h rate of ruminal DM degradation) increased MPS by 49%; no further improvement was achieved with greater degrees of processing (Beauchemin *et al.*, 2001).

Ruminal NH_3 concentration was not affected by treatment.

In dairy cows, barley, fed at approximately 30% of dietary DM, produced lower NH_3 concentrations in the rumen, but did not seem to affect MilkNE compared to maize (31.5% vs. 30.2%, respectively; Casper *et al.*, 1990). A later publication by Casper *et al.* (1999) reported no effect of matching ruminal degradabilities of NSC and CP on rumen fermentation or production parameters, but the barley-based diets reduced ruminal NH_3 concentration compared to the maize-based diets (grains were fed at approximately 42% and 39% of dietary DM, barley and maize, respectively). Microbial N outflow from the rumen and VFA concentrations were not affected by grain type. Calculated MilkNE was similar between treatments (20.9%, 21.5%, 22.6% and 20.0%, maize/SSBM, maize/ESBM, barley/SSBM and barley/ESBM, respectively); except that the barley/SSBM diet had a slightly higher MilkNE due to lower N intake (10% lower than the maize/SSBM diet). Roasting of barley to reduce ruminal starch degradability did not improve MilkNE in a study by Robinson and McNiven (1994). Gradual substitution of ground-shelled maize with steam-rolled barley (fed at 39% to 49% of dietary DM) reduced linearly ruminal NH_3 concentrations and had no effect on MPS in lactating dairy cows (Overton *et al.*, 1995). Although milk protein yield was linearly reduced, as a result of decreased DM/CP intake, MilkNE increased in a quadratic manner with increasing proportion of barley in the diet (22.7%, 25.0%, 25.1%, 25.8% and 24.6%, for 0%, 25%, 50%, 75% and 100% barley, respectively).

Tempering of dry-rolled barley increased milk protein yield and MilkNE in dairy cows; no effect on production was observed if whole barley was tempered (Christen *et al.*, 1996). Inclusion of barley vs. maize in the diet of lactating dairy cows (50% of DM) had no effect on ruminal NH_3 concentration or MilkNE (23.8% vs. 23.3%, barley and maize, respectively), but increased MPS (Yang *et al.*, 1997). In an attempt to synchronize rumen availability of energy and protein, Shabi *et al.* (1998) concluded that the rate of OM degradability in the rumen (cracked vs. expanded maize) was the factor mostly responsible for ruminal N utilization; NH_3 concentration was reduced with increasing ruminal OM/starch degradability. Microbial N flow to the duodenum and MilkNE (23.3% on average) were similar among treatments. Processing of maize grain (ground vs. rolled) did not affect ruminal NH_3 concentration, MPS or MilkNE in lactating dairy cows (Knowlton *et al.*, 1998). Feeding dry vs. high-moisture maize, however, resulted in increased ruminal NH_3 concentration. No effects on MPS or MilkNE were observed.

Summarizing six studies with dry-rolled vs. steam-flaked maize (and sorghum), Theurer *et al.* (1999) indicated a trend for increased MPS with the latter treatment. With both grains, the effect was due to increased starch degradation in the rumen with steam-flaking vs. dry-rolling. Processing of barley grain increased its rate of ruminal DM degradability from 2.7% to 9.2%/h (coarse and flat, respectively); the effect on ruminal NH_3 concentration in lactating dairy cows, however, was minimal (3.29 vs. 3.15 mmol/l, respectively; Yang *et al.*, 2001). Cows produced more milk protein on the diets with greater barley degradability, but N intake data were not published and MilkNE could not be calculated in this study. In a trial with similar objectives, Yang *et al.* (2000) found no effect of the degree of processing on MPS in lactating cows fed barley (43% of dietary DM) processed from 463 (coarse) to 317 (flat) kg/l volume weight. Ruminal NH_3 concentration was linearly decreased with increasing degree of processing. Milk and milk protein yields were increased and MilkNE appeared to be increased with increasing degree of barley processing (25.0%, 25.3%, 28.1% and 28.2%, coarse, medium, medium-flat and flat, respectively). In a study by Castillo *et al.* (2001b), supplementing the diet with barley vs. a maize-based concentrate increased urinary N excretion; MilkNE was not affected by treatment. The authors speculated that the level of urinary N excretion is related to the rate of degradation of barley protein, which resulted in a greater proportion of N being excreted via urinary urea than with the maize-based concentrate. Khorasani *et al.* (2001a), substituting barley (fed at 37% of dietary DM) with maize, observed no effect on ruminal NH_3 concentrations or MilkNE (27.1%, 27.4% and 27.2%, barley, maize/barley (50:50) and maize, respectively). Ruminal (or abomasal) infusion of 1.5 kg starch/day in cows receiving a basal diet with 29% NSC content did not affect PUN or ruminal NH_3 concentrations and had no effect on MilkNE (29.1%, 28.2% and 27.6%, control, ruminal and abomasal starch infusion, respectively;

Arieli *et al.*, 2001). Callison *et al.* (2001) processed maize to different particle sizes (fine-, medium- and coarse-ground) or steam-rolled it. Fed to lactating dairy cows at 36.6% of DM, coarse-ground maize increased (quadratic effect) ruminal NH_3 concentration compared to fine- or medium-ground maize, reflecting a similar pattern in true ruminal OM digestibility. Steam-rolling reduced NH_3 concentration (in a linear manner) compared to grinding. No effect of processing on MPS was reported. Estimated MilkNE was slightly lower with the coarse-ground vs. fine-ground maize diets (25.1%, 24.7%, 23.8%, 22.0% and 25.6%, fine-, medium- and coarse-ground, 50:50 steam-rolled and coarse-ground and steam-rolled maize, respectively). Processing to increase ruminal degradability of starch (HMEC vs. cracked shelled maize) reduced ruminal NH_3 concentration, but had no effect on milk protein yield, MilkNE (26.0 vs. 25.3, high-moisture and cracked shelled maize, respectively) and PUN or MUN (Broderick *et al.*, 2002). Increasing ruminal degradability of dietary NSC through replacing ground with steam-flaked maize reduced ruminal NH_3 concentration and increased MPS in lactating dairy cows (Harvatine *et al.*, 2002). Concentration of MUN was lowered and milk protein yield and MilkNE were not affected by the processing method.

Heat treatment (expansion) increased fractional rate of degradation of both barley and maize starch in dairy cows (determined through ruminal evacuation) and caused a reduction in ruminal NH_3 concentration (Tothi *et al.*, 2003). Compared to barley, cows fed the maize-based diet had a slightly lower concentration of NH_3 in the rumen. *In vitro*, barley with high amylopectin content (waxy barley) appeared to enhance NH_3 incorporation by ruminal bacteria (Hristov *et al.*, 2002a) but *in vivo* results showed no advantage of replacing maize with barley in the diet of lactating dairy cows (Foley *et al.*, 2004). In this latter study, cows were fed 40% (DM basis) steam-rolled maize, normal steam-rolled barley or waxy, high-amylopectin steam-rolled barley (at 75:25 barley:maize ratio). The barley diets increased NH_3 (5.5, 8.2 and 7.4 mmol/l, maize, normal and waxy barleys, respectively) and had no effect on pH or VFA concentration in the rumen. Microbial protein flow to the duodenum was not different among treatments and urinary N loss (as proportion of N intake) was greater with the waxy barley compared to the maize or normal barley diets. Proportions of bacterial N derived from ruminal NH_3-N (56–63%), milk protein-N derived from bacterial N (46–50%) and milk protein-N derived from ruminal NH_3-N (27–29%) were not affected by grain type. Milk urea N concentration was lower for the maize compared to the normal and waxy barley diets (10.4, 13.8 and 13.7 mg/100 ml, respectively). Cows on the normal barley diet consumed more N, had milk protein yields similar to the maize diet, and as a result, tended ($P < 0.1$) to have lower MilkNE (17% vs. 20%, respectively). Rate of starch digestion in the rumen was greater for the normal barley compared to the maize and the waxy barley.

Increasing the proportion of concentrate (up to the 65–70% range) in the diet of lactating dairy cows would most likely reduce NH_3 concentration in the rumen and improve overall N utilization and MilkNE. Unlike CHO supplementation, the effect of starch degradability, as affected by type of grain and degree of processing, on ruminal NH_3 concentration is less distinct (Table 4.3). In the case of barley vs. maize, the level of NH_3 in the rumen is a function of the greater concentration and degradability of barley protein (Herrera-Saldana *et al.*, 1990), which will produce increased ruminal NH_3 concentrations and greater availability of energy from barley vs. maize to utilize this NH_3 for cell protein synthesis by the ruminal microorganisms. Processing either decreased or did not affect ruminal NH_3 concentration. In some studies, increasing starch availability resulted in increased MPS, but in others MPS was not affected. Most data with lactating dairy cows indicate no relationship between ruminal effects of dietary starch and MilkNE.

4.3.3 Synchronization of ruminal energy and nitrogen release

The rumen synchrony concept (Johnson, 1976) implies that MPS (and presumably NH_3 utilization) in the rumen will be maximized if availabilities of energy and protein are synchronized. Synchrony can be achieved by changing the composition of the dietary CHO and N fractions, by altering the relative times of feeding of the dietary ingredients, or by a combination of both approaches (for reviews on this topic see Sauvant and van Milgen, 1995 and Dewhurst *et al.*, 2000

Table 4.3. Effect of ruminal starch availability on ruminal ammonia concentration, microbial protein synthesis, urinary N losses and milk N efficiency (selected *in vivo* trials only).

Starch source	Ammonia	MPS[a]	Urine N	Species	MilkNE[a]	MUN[a]	Reference
Roasting of barley	N/R[b]	N/R	N/R	Dairy cows	=	N/R	Robinson and McNiven (1994)
SRB[c] vs. GSC	↓	=	N/R[c]	Dairy cows	(↑)*	N/R	Overton *et al.* (1995)
Tempering of barley	N/R	N/R	N/R	Dairy cows	(↑)[†]	N/R	Christen *et al.* (1996)
Barley vs. maize	=	↑	N/R	Dairy cows	=	N/R	Yang *et al.* (1997)
Expanded vs. cracked maize	↓	=	N/R	Dairy cows	=	N/R	Shabi *et al.* (1998)
Ground vs. rolled maize	=	=	N/R	Dairy cows	=	N/R	Knowlton *et al.* (1998)
or dry vs. HMC[c]	↑	=	N/R	Dairy cows	=	N/R	Knowlton *et al.* (1998)
SFC vs. TRC[c]	N/R	↑	N/R	Cattle	N/A	N/A	Zinn *et al.* (1998)
Barley vs. maize	↓	=	N/R	Dairy cows	=	N/R	Casper *et al.* (1999)
Barley vs. maize	=	↑	N/R	Cattle	N/A[b]	N/A	Martin-Orue *et al.* (2000)
Processing of barley	↓	=	N/R	Dairy cows	(↑)[†]		Yang *et al.* (2000)
Barley vs. maize	=	N/R	N/R	Dairy cows	=	N/R	Khorasani *et al.* (2001a)
Barley vs. maize	N/R	N/R	↑	Dairy cows	=	N/R	Castillo *et al.* (2001b)
Maize particle size	↓	=	N/R	Dairy cows	(↑)[†]	N/R	Callison *et al.* (2001)
SRC vs. CGC[c]	↓	=	N/R	Dairy cows	(↑)[†]	N/R	Callison *et al.* (2001)
Processing of barley	=	N/R	N/R	Dairy cows	N/R	N/R	Yang *et al.* (2001)
Processing of barley	=	(↑)[‡]	N/R	Cattle	N/A	N/A	Beauchemin *et al.* (2001)
HMEC vs. CSC[c]	↓	N/R	N/R	Dairy cows	=	=	Broderick *et al.* (2002)
SFC[c] vs. ground maize	↓	↑	N/R	Dairy cows	=	↓	Harvatine *et al.* (2002)
Barley[d] vs. maize	↑	=[d]	↑	Dairy cows	(↓)[§]	↑	Foley *et al.* (2004)

[a]MPS, microbial protein synthesis; MilkNE, milk protein-N efficiency (milk protein-N/N intake), in most cases MPNE data were not statistically evaluated by the original authors; MUN, milk urea-N.
[b]N/R, not reported; N/A, not applicable.
[c]SRB, steam-rolled barley; GSC, ground-shelled maize; HMC, high-moisture maize; SFC, steam-flaked maize; TRC, tempered rolled maize; HMEC, high-moisture ear maize; CSC, cracked shelled maize.
[d]High-amylopectin (waxy) or normal barley.
*Calculated. Seemingly, a quadratic response; data not statistically evaluated.
[†]Calculated. Data not statistically evaluated.
[§]Trend at $P < 0.1$.
[‡]MPS was increased from coarse to medium processing but no further improvement was reported with the greater degrees processing.
↓, decrease; ↑, increase (only effects significant at $P < 0.05$ are reported); =, no change ($P > 0.05$).

and Chapter 3). As indicated by Thomas (1973, referring to Blackburn, 1965), maximum utilization of NH_3 derived from NPN can be expected when CHO fermentation occurs simultaneously with NH_3 release. Reports by Sinclair *et al.* (1993, 1995) demonstrated the advantage of synchronous (to the release of dietary energy) release of N in the rumen on reducing peak NH_3 concentrations and enhancing the efficiency of MPS. Henning *et al.* (1993) provided the rumen of sheep with energy (soluble CHO) and N (urea and sodium caseinate) at patterns representing fast or slow synchronous and asynchronous release. Gradual (vs. pulse) release of N was more effective in preventing rapid increase in ruminal NH_3 concentrations than the pattern of energy supplementation. Microbial N outflow and the efficiency of MPS in the rumen were not affected by the synchronous release of energy and N, but efficiency of MPS was improved with the continuous release of energy. The study failed to confirm earlier *in vitro* results from the same group suggesting energy supply pattern can significantly affect growth rate and efficiency of growth of ruminal bacteria (Henning *et al.*, 1991). Kim *et al.* (1999a) supplemented a grass-silage/grain concentrate diet with maltodextrin (2 kg/day) either infused continuously into the rumen or given synchronously during feeding or asynchronously, 6 h after feeding. Continuous intraruminal infusion of maltodextrin reduced ruminal NH_3 concentration compared to the unsupplemented control and synchronous or asynchronous delivery. Providing the supplement at the time of feeding also reduced ruminal NH_3 levels compared to the asynchronous treatment. Plasma urea N concentrations were lower in all maltodextrin treatments, but synchrony had no effect on this variable. It appeared that MPS was increased by the synchronous delivery of the supplemental CHO. Synchronous release of maltodextrin had no effect on MilkNE (calculated at 21.8%, 21.6%, 22.7% and 23.8%, basal diet, continuous, synchronous and asynchronous supplementation, respectively).

Witt *et al.* (2000) failed to demonstrate a response in milk production or composition and rumen fermentation in lactating ewes fed diets providing three 'synchrony index' (N to OM release in the rumen) levels and to confirm their previous findings with growing lambs (Witt *et al.*, 1999a,b). In the Witt *et al.* (1999a) study, however, PUN levels were decreased in animals fed the synchronous as compared to the asynchronous diets. Rate of energy release or the synchrony index had no effect on urinary N excretion or MPS (estimated based on excretion of purine derivatives). Sinclair *et al.* (2000) reported a dramatic increase in plasma NH_3 concentrations in cattle fed starch (barley)- or soluble fibre (unmolassed sugarbeet pulp)-based supplements when dietary N was more rapidly released in the rumen (asynchronous release) vs. a more slow release (synchronous). Carbohydrate source did not affect plasma NH_3 levels. In a similar design, Richardson *et al.* (2003) found no synchrony or energy source main effects on ruminal NH_3 concentration. Synchronization did not affect MPS (based on urinary purine derivatives excretion). The unmolassed sugarbeet pulp supplement significantly reduced MPS compared to the barley-based concentrate. Synchronous vs. asynchronous provision of ruminally available energy at the time of feeding reduced NH_3 concentration 2 h after feeding. The overall effect was only numerically decreased NH_3 concentration for the synchronous feeding regime (20 mg N/l). Estimated MilkNE was not affected by treatment (25.1% and 24.9%, synchronous and asynchronous, respectively). Sheep fed an orchardgrass diet were treated intraruminally, twice daily with 120 g wheat starch at the beginning of each meal or 3 h after the meal (Remond *et al.*, 2002). Overall, arterial urea and NH_3 concentrations, ruminal NH_3 concentration and ruminal net urea and NH_3-N fluxes were not different between treatments. The authors concluded that time of starch supplementation had no effect on N utilization in the rumen. Independent of the time of supply, starch fermentation resulted in an increase in net urea transfer across the ruminal epithelium. When energy was available, NH_3 concentration was low and urea transfer was high, providing the ruminal microorganisms with an additional N source.

Feeding two diets with negative (−30) or positive (+30) ruminal N balance (OEB) in patterns designed to provide synchronous (diets fed at the same time) vs. 12-h or 24-h asynchronous (feeding either the negative or the positive OEB diet or two consecutive feedings of the same diet, respectively) release of energy and N in the rumen did not affect ruminal NH_3 concentrations over a 48-h period in double-muscled Belgian Blue bulls (Valkeners *et al.*, 2004). Although the asynchronous feeding resulted in greater postprandial NH_3 and PUN concentrations within 48 h after feeding, overall PUN concentration, urinary and faecal N losses and N retention were unaffected by treatment. Cows grazing high-protein pasture (19.8% CP) are more likely to benefit from synchronous delivery of dietary N and energy (Nielsen *et al.*, 2003). Supplementation of grass pasture with low- (11% CP) vs. high-protein (17% CP) concentrate supplement reduced MUN concentration (3.0 vs. 4.2 mmol/l) and manure N excretion, but also decreased milk and milk protein yields (25.2 vs. 27.0 kg/day and 838 vs. 900 g/day, respectively).

Results suggest rapid (asynchronous) release of dietary N as the main factor in increasing ruminal NH_3 and PUN concentrations after feeding, irrespective of the synchrony of energy release. Sauvant and van Milgen (1995) also indicated that higher N degradation rates result in increased ruminal ammonia levels. These authors suggested that reduction in ruminal ammonia concentrations could be achieved either by increasing the rate of CHO degradation or decreasing the rate of N degradation. Indeed, data summarized here (Henning *et al.*, 1993; Kim *et al.*, 1999a; Sinclair *et al.*, 2000, for example) support this concept. Synchronization rarely affected MPS and its efficiency in the rumen. It is likely that ruminal microorganisms have the ability to overcome periods of nutrient shortages and undergo compensatory growth when substrate is available (Sauvant and van Milgen, 1995). Urea-N recycling to the rumen is an important physiological function of the ruminant animal, providing a N source for the ruminal microorganisms in times when N intake/NH_3 production is low; in cattle, from 25% to 53% of the recycled urea was degraded in the rumen (Bunting *et al.*, 1989; Huntington, 1989). Liver ureagenesis is highly and positively correlated to N intake and the body urea pool acts as a source or a sink for N, thus normalizing short-term variations in RDP supply from the diet (Huntington and Archibeque, 1999). It is also possible, as noted by Dewhurst *et al.* (2000), that effects attributed to synchrony may simply be effects specific to the individual CHO and protein fractions of the diet. Increasing the proportion of one fraction in the diet inevitably decreases the proportion of the other fraction(s), thus confounding attempts to determine the relative importance of the different CHO or N fractions (Armentano and Pereira, 1997). Data on overall N balance and partitioning and efficiency of dietary N use for milk or meat production in relation to the synchrony of N and energy release in the rumen are scarce.

4.4 Effect of Dietary Lipids and Individual Fatty Acids on Nitrogen Metabolism in the Rumen

Supplementation of ruminant, particularly dairy cow, diets with sources of free or protected long-chain FA is aimed at: (i) increasing the energy density of the diet with minimal disruption of ruminal fermentation and digestion; and (ii) manipulation of the FA composition of milk and meat (Jenkins, 1993). Dietary lipids undergo dramatic transformations in the rumen and can have profound effects on ruminal protozoa, overall microbial activities, fermentation and digestion of nutrients (Jenkins, 1993; Doreau and Ferlay, 1994) and intake (Faverdin, 1999). Despite the relatively lower deaminative activities of ruminal ciliates, as compared to ruminal bacteria (Hino and Russell, 1987), reduced protozoal populations are usually associated with decreased ruminal NH_3 concentration (Williams and Coleman, 1992), mostly resulting from a reduction in protozoal fermentation of bacterial proteins (Broderick *et al.*, 1991) and from inhibited deamination of dietary amino acids (Jouany *et al.*, 1988). As the specific inhibitory effect of certain FA on ruminal protozoa is well documented (Williams and Coleman, 1992; Doreau and Ferlay, 1995; Jouany, 1996), lipids can influence N metabolism in the rumen and the overall efficiency of utilization of dietary N. Reviews on the effects of dietary lipids on N metabolism and ruminal CHO fermentation were provided by Doreau and Ferlay (1995) and Nagaraja *et al.* (1997), respectively. The implications of feeding unsaturated FA on MPS were also discussed by Firkins (1996).

Typically, ruminant diets contain up to 5–7% lipids on DM basis, a significant portion of which is in the form of triacylglycerols, which are hydrolysed during lipolysis to generate free FA and glycerol. Then, the FA are metabolized by the ruminal microorganisms; unsaturated FA are saturated and isomerized and/or incorporated into microbial matter (Hawke and Robertson, 1964). Because of the anaerobic conditions in the rumen, FA provide little energy to the ruminal microorganisms (Nagaraja *et al.*, 1997). Therefore, inclusion of fats in the diet will generally reduce the energy available for microbial growth (Firkins, 1996), and in a number of studies dietary fats have been shown to inhibit ruminal microbial activities and digestion (Jenkins, 1993). Tallow (C18:1, C16:0 and C18:0) or soy oil (C18:2 and C18:1) did not affect MPS in steers, but (due to decreased ruminal degradability of dietary OM) efficiency of MPS in the rumen was increased compared to the Ca-soap treatment (Jenkins and Palmquist, 1984). Addition of lecithin (C18:2, C16:0 and C18:1) or maize oil (C18:2, C18:1 and C16:0) to the diet of sheep reduced ruminal NH_3 concentration, had no effect on MPS, and increased the true efficiency of MPS (Jenkins and Fotouhi, 1990). Feeding of two levels of animal-vegetable fat (predominantly C18:1, C18:2 and C16:0) to lactating dairy cows had no effect on ruminal NH_3 and VFA concentrations, or MPS and MPS efficiency (Ohajuruka *et al.*, 1991). Rapeseed oil (C18:1, C18:2 and C16:0) effectively decreased ruminal NH_3 and butyrate concentrations and increased the efficiency of MPS in the rumen (but not MPS), although protozoal numbers were not decreased (Tesfa, 1993). In dairy cows, however, two levels of rapeseed oil (78 and

167 g/kg DM) or tallow (167 g/kg DM) had only a marginal effect on ruminal NH_3 concentration (a significant reduction was found only in one of the four sampling times) and no effect on VFA concentration, ruminal or total tract nutrient digestion, or MPS and the efficiency of MPS (Doreau *et al.*, 1991). In a later study, Doreau *et al.* (1993) reported that supplementation of a dairy diet with 0.99 kg/day of rapeseed oil reduced total tract OM digestion, but had no effect on ruminal fermentation (NH_3 or VFA concentration and digestibility of OM and fibre fractions), MPS and the efficiency of MPS. The cows receiving the fat-supplemented diets (Ca-salts of rapeseed oil was one of the treatments) had dramatically higher calculated MilkNE than the control (29.5, 27.5 and 15.6 for Ca-salts, rapeseed oil and control, respectively); the effect, however, could hardly be attributed to oil treatment since the control diet had a greater N content.

Pantoja *et al.* (1994) investigated the effects of fat supplementation of dairy cow diets with saturated or unsaturated FA sources (saturated tallow, predominantly C16:0 and C18:0; tallow, C18:1, C18:0 and C16:0; and animal-vegetable fat, C18:1, C18:2 and C16:0) in combination with different fibre sources. Ammonia concentration in the rumen was not reported and VFA concentration was reduced by fat supplementation with no effect of degree of FA saturation. Microbial protein synthesis and the efficiency of MPS were not affected by fat addition but increased with FA saturation. Calculated MilkNE was 22.8% for the control and slightly increased (25.4%) with tallow and two of the animal-vegetable fat diets (24.2% and 24.1%, low forage and soyhulls diets, respectively). A combination of tallow (low ratio of unsaturated to saturated FA) and yellow grease (high ratio of unsaturated to saturated FA) in the diet of dairy cows resulted in different degrees of saturation of the total diet FA (Avila *et al.*, 2000). Microbial N flow was increased with increasing degree of FA unsaturation, but efficiency of MPS and NH_3 and VFA concentrations were not affected by fat supplementation or FA saturation ratio. Oldick and Firkins (2000) reported a linear decrease ($P = 0.10$) in ruminal protozoa in heifers with increasing degree of unsaturation of dietary fats. Ammonia concentration in the rumen and MPS were unaffected by fat supplementation or FA composition. The efficiency of MPS was increased with the fat-supplemented diets compared to the control. Scollan *et al.* (2001) fed whole linseed and fish oils (213 and 54 g/kg, respectively) to steers and found no effect on ruminal NH_3 concentration; total VFA concentration was increased compared to a protected FA source. At two levels of supplementation (2% and 4%) of tallow or choice white grease (C18:1 and C16:0) to the diet of lactating dairy cows, Onetti *et al.* (2001) reported reduced ruminal NH_3 concentration with the lower level of fat, decreased protozoal counts and a typical reduction in acetate:propionate ratios. Fat supplementation reduced DMI, milk yield and milk protein yield (no source of fat effect was found with the greater fat supplementation). Milk N efficiency appeared to be slightly increased with 2% fat but remained the same as the control with 4% fat. In sheep, linoleic acid-rich sunflower oil reduced ruminal protozoal counts and NH_3 concentration (Ivan *et al.*, 2001). Oil had no effect on daily gain or efficiency of feed conversion. An example of the complex effects of fat on production and N efficiency of dairy cows is the study by Jenkins (1999). Oleamide, a source of oleic acid partially resistant to ruminal biohydrogenation, added at increasing dietary levels of 0% to 5%, linearly reduced DMI, CP intake and milk and milk protein yields. This, however, had a positive effect on MilkNE; compared to the control (29.1%), MilkNE was increased (calculated from the original data) to 31.0%, 31.4%, 32.2%, 31.2% and 32.8% for 1%, 2%, 3%, 4% and 5% oleamide, respectively. Thus, if accompanied by a reduction in N intake, a reduction in milk/milk protein yields may not affect, or may even increase MilkNE in dairy cows. At lower levels of production, however, the balance between N intake and N secreted with milk protein may be shifted and MilkNE reduced by fat supplementation (Kalscheur *et al.*, 1997); MilkNE was 21.8%, 20.2%, 21.6% and 19.9% with control, high oleic acid, food grade sunflower oil and vegetable shortening, respectively. In this experiment, fat addition or fat source had no effect on MPS. *In vitro*, tallow, maize oil (C18:2 and C18:1) and yellow grease (C18:1, C16:0 and C18:0/C18:2) did not affect ruminal fermentation, but the potassium soaps of their FA decreased gas production and VFA and NH_3 concentrations (Getachew *et al.*, 2001).

Medium-chain saturated FA are known to strongly inhibit ruminal protozoa (Newbold and Chamberlain, 1988; Matsumoto *et al.*, 1991).

Matsumoto *et al.* (1991) reported inhibition of ruminal protozoal counts by C8:0, C10:0, C12:0 and C14:0 FA in goats; protozoa were completely eradicated after 2 days of feeding C10:0 and C12:0 and after 3 days of feeding C14:0. Ha *et al.* (2001) found inhibition of cellulose digestion, fibrolytic activities of a pure culture of a rumen fungus (*Neocallimatix frontalis*), and VFA concentration with C8:0 and C10:0 FA, *in vitro*. Studies have indicated a strong inhibitory effect of coconut oil (C12:0 is a major component of coconut oil) on rumen ciliates (Sutton *et al.*, 1983; Dohme *et al.*, 1999, 2000; Machmüller and Kreuzer, 1999; Sutter *et al.*, 2000) and a significant decrease in ruminal NH_3 concentration in the RUSITEC (Dohme *et al.*, 1999). In a series of *in vitro* incubations, Hristov *et al.* (2004c) investigated the effects of a range of FA (from C6 to C18) on ruminal fermentation, protozoal counts and microbial activities. Capric (C10:0) and lauric (C12:0) acids (applied at 0.063%, 0.125% and 0.25% and 0.25%, 0.5% and 1% final media concentration, respectively) had dramatic effects on ruminal fermentation. At all three application levels, both FA completely eradicated ruminal protozoa, decreased bacterial incorporation of N from ^{15}N-casein, and significantly shifted concentrations of fermentation end products compared to the control. The two FA decreased NH_3, butyrate and branched-chain VFA and increased total free amino acids and soluble protein concentrations. Bacterial proteolytic activities were also inhibited. The lower FA levels did not affect ruminal polysaccharide-degrading activities but, at concentrations of 0.25% and 1% (capric and lauric acids, respectively), xylanase and amylase activities were decreased. Data from this study suggested that C10 and C12 saturated FA blocked proteolysis and deamination of amino acids. Capric acid is found in small amounts in coconut and palm-kernel oils (approximately 8% and 7% of the total FA, respectively), but lauric acid is a major constituent of these oils. Another medium-chain saturated FA, caprylic acid (C8), was also effective in eliminating rumen protozoa at the highest application level (0.25%). At that concentration, caprylic acid showed properties similar to the C10 and C12 acids in decreasing NH_3 and increasing free amino acids concentrations. Myristic acid (C14:0) did not affect NH_3 concentrations, but at 0.5% reduced protozoal counts and incorporation of ^{15}N similarly to C8:0 and C12:0.

Eradication of ruminal protozoa usually results in increased microbial protein flow to the duodenum, although responses are variable and often not statistically significant (Williams and Coleman, 1992; Jouany, 1996; Koenig *et al.*, 2000). In comparing linseed (high in unsaturated C18 FA) and coconut (high in C12–C14 saturated FA) oils, Sutton *et al.* (1983) observed stronger antiprotozoal properties with the coconut oil. With both types of oils the authors found changes in VFA proportions (typically associated with reduced protozoal numbers) and increased MPS and efficiency of MPS in sheep. Ruminal NH_3 concentrations were not reported. The experiments of Newbold and Chamberlain (1988) indicated stronger antiprotozoal properties of C18 unsaturated acids (supplied as linseed oil) than C12:0–C14:0 acids (from coconut oil). Broudiscou *et al.* (1994) attempted to separate the effect of protozoa presence (defaunation) from the effect of linseed oil (C18:3 and C18:2) on ruminal fermentation. Sheep, defaunated by freezing the ruminal contents and repeatedly washing the empty rumen, or refaunated sheep, were given 65 g/day of linseed oil. Although both defaunation and oil reduced NH_3 concentration in the rumen, protozoa had by far the stronger effect: 55–57% vs. 26–22% reduction, defaunation and oil, respectively. Both treatments reduced fibre degradation in the rumen and increased MPS and the efficiency of MPS. Although the observed increase in efficiency of MPS due to oil supplementation with the defaunated sheep could be explained by the reduced ruminal OM digestion, no satisfactory explanation was provided for the increased MPS with this treatment combination. In the Hristov *et al.* (2004c) study, a reduction in protozoal numbers with all unsaturated C18 FA was observed; particularly strong effects were produced by linolenic and linoleic acids (results were confirmed for linoleic acid *in vivo* with beef cattle; Hristov *et al.*, 2002b). These FA did not have a negative effect on bacterial proteolytic activity, although they significantly reduced the incorporation of ^{15}N into protozoal protein and, consequently, the proportion of protozoal protein originating from bacterial N. As compared to the medium-chain saturated FA, the effects of the C18 unsaturated acids on ruminal fermentation were less pronounced; the increase in free amino acid concentration was of a lower magnitude and there was no effect on total VFA production, acetate:propionate ratio was decreased, and (similar to the C8:0–C14:0 FA) butyrate and valerate concentrations were reduced. All C18 unsaturated FA enhanced the polysaccharide-degrading (particularly amylase) activities of the incubation media and C18:3 and C18:2 reduced NH_3 concentrations compared to the blank. Following up on these *in vitro* results, Hristov *et al.* (2004d) treated dairy cows intraruminally, twice a day, with 240 g/day sodium laurate (representing approximately 1% of daily DMI or 0.3% of the ruminal contents). Treatment had no effect on ruminal pH and total VFA concentration and composition, NH_3 concentration, NH_3-N pool size and irreversible loss of NH_3-N. Compared to the control (water), protozoal counts, carboxymethylcellulase and xylanase activities of ruminal fluid and flow of MPS were reduced (by 90%, 40%, 36% and 12%, respectively) by sodium laurate. Despite the apparently depressed microbial activities in the rumen, ruminal and total tract digestion of nutrients were not affected. Milk yield, fat-corrected milk yield, milk fat and protein concentrations and yields and MUN concentration were not affected by treatment. The transfer of ruminal ^{15}N-NH_3 into bacterial or milk protein was also not affected by the sodium laurate. In a follow-up study (Faciola *et al.*, 2004), sodium laurate or lauric acid (given intraruminally once a day at 160 g/day) dramatically reduced ruminal protozoal counts (by 91% and 94%, respectively) and ruminal concentrations of ammonia and free amino acids (by 30% and 60% and by 62% and 37%, respectively) in dairy cows. The inconsistency observed between *in vitro* and *in vivo* ruminal effects of lipids and specific FA may be due to the fact that dietary FA are preferentially adsorbed on to the feed particles (Harfoot *et al.*, 1974), and consequently their effect may be diminished *in vivo* compared to *in vitro*.

The possibility of using medium-chain saturated and long-chain unsaturated FA (and lipids in general) to manipulate NH_3 utilization in the rumen has to be approached cautiously so that fibre degradation and microbial protein synthesis in the rumen are not impaired (Doreau *et al.*, 1993; Dohme *et al.*, 2000); most data with defaunated/reduced fauna animals indicate decreased OM and fibre degradability in the rumen (see Williams and Coleman, 1992). Sutton *et al.* (1983) observed a dramatic decrease in ruminal degradation of dietary NDF with coconut oil compared to the basal diet (12% vs. 50%, respectively). Oldick and Firkins (2000) found decreased ruminal degradation of NDF with fat addition, independent of the degree of saturation, although only a numerical decrease in protozoal counts was reported. Similarly, Faichney *et al.* (2002) observed a dramatic decrease in ruminal neutral-detergent insoluble OM degradability with increasing dietary level of free FA. Machmüller *et al.* (2000) found only a slight, numerical decrease in total tract apparent digestion of dietary NDF with coconut oil (treatment that resulted in a 72% numerical reduction in protozoal counts) compared to the control. Of the other oils tested, mostly rich in C18 unsaturated FA, only sunflower oil reduced total tract fibre digestion in sheep, but protozoal counts were unaffected. Earlier work by Machmüller and Kreuzer (1999) reported a reduction in apparent OM digestibility by 2.5% coconut oil and concomitant reduction in ruminal protozoa (by 88%), with no significant effect on NDF digestibility.

In their review on the effects of dietary oils on N metabolism in the rumen, Doreau and Ferlay (1995) concluded that, generally, NH_3 concentration in the rumen is reduced and microbial (as well as non-microbial) N flow to the small intestine is not influenced by fat addition. Evidently, the effect on ruminal NH_3 concentration is variable; out of the 42 studies listed, NH_3 concentration was reduced in 15, was not changed in 22 and was increased in five. In most of the studies reviewed, protozoal counts in the rumen were decreased by fat addition (statistical evaluation of the differences was not provided); in 11 out of 45 observations, protozoal counts were either increased or remained unchanged compared to the controls. The authors indicated that the effect of dietary lipids on the efficiency of MPS in the rumen depends on the nature of the FA and can be linked to depression of the protozoal population. Medium-chain saturated FA, particularly lauric acid, have a powerful effect on ruminal protozoa and fermentation and can potentially be used to regulate NH_3 utilization and methane production in the rumen. The effects of these FA, however, need to be thoroughly tested and verified *in vivo* and an optimal dose determined with high-yielding dairy cows or intensively growing cattle.

There is also a considerable interest in regulating methane production in the rumen using dietary lipids. A survey of 37 diets reported a decrease in methane production from dairy cows by increasing the degree of unsaturation of the dietary FA (Giger-Reverdin *et al.*, 2003). Methane production in the rumen of sheep was reduced with fish oil supplementation (Fievez *et al.*, 2003). Fish oils contain considerable amounts of FA longer than 18 carbons (Opstvedt, 1984), shown to inhibit microbial activities (Galbraith *et al.*, 1971) and MPS in the rumen (Hoover *et al.*, 1989; Fievez *et al.*, 2001). Palm kernel oil, coconut oil and high lauric acid-rapeseed oil (all rich in C12:0) reduced methane release in the RUSITEC (Dohme *et al.*, 1999, 2000). Rapeseed, sunflower seed, linseed and particularly coconut oil decreased methane production in lambs to various extents (Machmüller *et al.*, 2000). Long-chain unsaturated FA (rapeseed oil, C18:1; sunflower oil, C18:2; and linseed oil, C18:3) were also effective in reducing methane production *in vitro* (Jalc and Cerešnáková, 2001).

Essential oils (EO, defined as 'highly volatile substance isolated by a physical process from an odoriferous plant of a single botanical species'; *Encyclopædia Britannica*, 2003) have been investigated as manipulators of ruminal fermentation (Oh *et al.*, 1967, 1968) and may present a natural alternative to feed antibiotics (McEwan *et al.*, 2002a,b; Wallace *et al.*, 2002; Chapter 3). These compounds have been shown to possess strong antibacterial activities against human pathogens, which predictably attracted the attention of ruminant nutritionists (Losa and Brufau, 2001; Wallace *et al.*, 2002). A product containing EO from marjoram (*Majorana hortensis*) did not affect total tract nutrient digestibility in sheep, but appeared to increase protozoal protein in ruminal fluid (Kozelov *et al.*, 2001). In dairy cows, EO inhibited deamination of amino acids and reduced ruminal ammonia concentration (Wallace *et al.*, 2002). Further investigations revealed no effect of EO on proteolysis or ruminal peptidase activities, but a depression of microbial colonization of fibrous substrates and amylolytic activities was suggested (Wallace *et al.*, 2002). In sheep, Newbold *et al.* (2004) reported a numerical, 20% reduction in the rate of degradation of soybean N and a significant reduction in the degradation rate of soybean DM *in sacco* with EO (CRINA® RUMINANTS, CRINA S.A., Gland, Switzerland) compared to the control, but the rates of degradation of rapeseed meal, or ryegrass hay were unaffected. Proteolytic or peptidolytic activities and MPS in the rumen (as measured by urinary excretion of purine derivatives) were not affected by EO, but bacterial deaminative activities were reduced by 25% *in vitro*. Essential oils did not affect ammonia and VFA concentrations, or protozoal numbers. From this study it was concluded that EO might inhibit certain key ammonia-hyperproducing species (McIntosh *et al.*, 2003), or inhibit certain deaminative reactions in all species. In heifers fed a high-concentrate diet, the CRINA EO product slightly reduced *in situ* CP degradation of lupin seeds, green peas and sunflower meal (Molero *et al.*, 2004). With the low-concentrate diet, EO produced a significant reduction in soybean and sunflower meals CP degradation after a prolonged adaptation period, suggesting that adaptation to treatment/diet might be an important factor in manifesting the effect of EO. Hristov *et al.* (2004e) investigated the ruminal effects of 40 EO and reported increased VFA concentration (particularly acetate) and inhibited deamination, comparable to that of monensin, by a number of oils, but no effect on overall ammonia concentration *in vitro*.

4.5 Effect of Other Bioactive Agents

The effects of ionophore antibiotics on ruminal fermentation have been well documented (Owens, 1980; Nagaraja *et al.*, 1997; Chapter 3) and include increased propionate production and decreased methanogenesis, protein degradation, deamination of amino acids and NH_3 concentration, as well as decreased lactic acid production and, therefore, reduced risk of acidosis (Nagaraja *et al.*, 1997). Ionophore antibiotics are extensively used in the USA in growing cattle, and their production and health benefits to lactating dairy cows have been recently discussed (McGuffey *et al.*, 2001; Ipharraguerre and Clark, 2003). A monensin-mediated decrease in ruminal NH_3 concentration was reported with (Yang and Russell, 1993) or without (Haïmoud *et al.*, 1996) an associated increase in MPS in the rumen, although lack of effect is not uncommon (Ruiz *et al.*, 2001). Similarly, Kobayashi *et al.* (1990, 1992) and McAllister *et al.* (1994) reported suppressed ruminal NH_3 concentration and protozoal numbers by salinomycin, but these effects were not observed in an *in vivo* trial with cattle (Hristov *et al.*, 2000). A reduction in MPS was reported for lasalocid and cationomycin (Gomez *et al.*, 1991). Another antibiotic, abierixin, did not influence ruminal NH_3 concentration or MPS in sheep (Gomez *et al.*, 1990). *In vitro*, monensin and salinomycin applied at 2.5 to 10 and 1.25 to 5 ppm final concentration in the medium linearly reduced NH_3, increased free amino acid concentration, had no effect on protozoal counts and reduced bacterial predation by protozoa (Hristov *et al.*, 2003a). With the EU drive to completely phase out feed antibiotics as growth promoters, including monensin sodium and salinomycin sodium (EU Agriculture Council Press Release IP/03/1058, 22 July 2003), however, and legislative initiatives in the US Senate (Preservation of Antibiotics for Medical Treatment Act of 2003, 21 July 2003), the future of non-therapeutic use of antibiotics in animal production is uncertain.

A number of naturally occurring plant compounds possess inhibitory activities against ruminal microorganisms and can be used to manipulate ruminal fermentation/NH_3 utilization in cattle (Wallace *et al.*, 2002; Chapter 3). Plants rich in secondary metabolites such as *Acacia aneura*, *Brachychiton populneum*, *Chamaecytisus palmensis*, *Flindersia maculosa*, *Leucaena leucocephala*, *Sesbania sesban* and *Vernonia amyedalina* have all reduced, to a different extent, protozoal activity *in vitro* (Newbold *et al.*, 1997). The plant with the strongest antiprotozoal properties, *S. sesban*, however, had no effect on NH_3 concentration, and, as indicated by the authors, ruminal microflora could adapt to detoxify the antiprotozoal agent (identified to be associated with the saponin-containing fraction of the plant); protozoal populations recovered within 14 days following treatment. Hristov *et al.* (2003a) reported a significant reduction in ruminal NH_3 concentrations by a yucca (*Yucca schidigera*) product *in vitro*; applied at 0.2% and 0.4%, yucca reduced NH_3 concentration by 15% and 27%, respectively. Similar effects were observed with saponin treatments *in vitro* and *in vivo* (Kil *et al.*, 1994; Wallace *et al.*, 1994; Hussain and Cheeke, 1995; Makkar *et al.*, 1998; Hristov *et al.*, 1999). Wilson *et al.* (1998), however, found no effect on ruminal NH_3, PUN or MUN concentrations in lactating dairy cows fed 9 g/day *Y. schidigera* extract. The effect of yucca on NH_3 is twofold: glycofractions have been known to bind NH_3 (Headon *et al.*, 1991) and reductions in protozoal counts and activities usually result in decreased deamination and recycling of bacterial proteins within the rumen (Williams and Coleman, 1992). As indicated by Wallace *et al.* (1994), the NH_3-binding potential of yucca is negligible when NH_3 concentrations are as high as in the rumen and, therefore, the observed ruminal effects result primarily from inhibition of protozoa. Yucca powder did not affect protozoal counts in the Hristov *et al.* (2003a) study, but reduced them by 42% *in vivo* (Hristov *et al.*, 1999). Another saponin-containing plant, *Quillaja saponaria*, had no effect on NH_3, but reduced protozoal counts (by 54% relative to the control; Hristov *et al.*, 2003a), suggesting that *Q. saponaria* extract might be effective as a defaunating agent for cattle fed high grain diets. *In vivo*, Baah *et al.* (2002) reported that ruminal protozoal counts were reduced by 61% in cattle fed *Quillaja* extract at 60 g/day per head, compared to the control. Work by Teferedegne (2000), however, indicated that the effect of saponin-containing plants on protozoa in the rumen may be transient rather than permanent, and inactivation by saliva may occur *in vivo*. Clearly, the effect of these compounds has to be verified in long-term *in vivo* trials.

Another group of bioactive compounds of interest in manipulating ruminal fermentation are polyphenolic substances collectively termed tannins. Tannins include the chemically distinct hydrolysable tannins (HT) and condensed tannins (CT). The effects of CT on ruminal fermentation and microbial activities have been well documented (Barry, 1989; Bae *et al.*, 1993; Chesson and Forsberg, 1997). The topic has also been excellently covered in a review by Makkar (2003). In summary, tannins: (i) reduce DMI through their astringent properties and perhaps lowered rate of digestion; (ii) enhance the efficiency of MPS in the rumen (at low dietary concentrations); (iii) increase molar proportion of propionate; (iv) inhibit protozoal counts; (v) reduce protein degradability through their protein-binding properties; and (vi) if absorbed, HT (and CT) can damage organs, such as liver, kidneys and spleen (Makkar, 2003). As indicated by Makkar (2003), responses can vary among type (CT vs. HT), form (free vs. bound) and sources (different trees) of tannins and presence of other antinutrient factors in the diet. Hristov *et al.* (2003a) observed dramatic effects on ruminal fermentation and microbial activities exerted by tannic acid (also known as – gallotannin, which is a HT) *in vitro*. All levels of inclusion of tannic acid (0.1–0.4%) drastically reduced polysaccharide-degrading activities and concentration of NH_3, as well as total and individual VFA and increased concentrations of soluble proteins and reducing sugars. Tannic acid is toxic, and high concentrations can cause ruminal stasis (Cheeke and Shull, 1985; Zhu and Flippich, 1995), although there is evidence that ruminal bacteria can degrade it to yield gallic acid, pyrogallol and resorcinol as fermentation end products (Murdiati *et al.*, 1992; Skene and Brooker, 1995; Singh *et al.*, 2001). In the study by Hristov *et al.* (2003a), tannic acid was one of the few additives that reduced protozoal counts. It also decreased protozoal incorporation of ^{15}N, but not that by bacteria. Makkar (2003) also suggested more efficient synthesis of microbial protein (^{15}N was used as a bacterial marker) *in vitro* when tannins were present in the incubation media. Reductions in protozoal counts in association with consumption of CT by ruminants, e.g. *Lotus corniculatus* grazed by sheep (Wang *et al.*, 1996) and *Quebracho* powder fed to cattle at 0.6% of dietary DM (Baah *et al.*, 2002), have been attributed to the astringent nature of the CT. The decrease in polysaccharide-degrading activities in the Hristov *et al.* (2003a) study may be indicative of inhibition of bacterial growth, but incorporation of N by bacteria was not impaired by tannic acid. Reduced enzymatic activity may have been due to the propensity of tannins to bind specifically to soluble proteins (Cheeke and Shull, 1985; Santos *et al.*, 2000). As with saponins, there is a need to investigate the effects of tannins on ruminal and total tract N metabolism and efficiency of dietary N utilization/N losses in long-term studies with high-yielding lactating cows or growing cattle.

4.6 Effect of Protozoa on Rumen Microbial Protein Synthesis and Flow of Amino Acids of Microbial Origin in the Small Intestine

Research to improve N utilization by ruminants has generally taken two approaches. First, to optimize MPS, and second, to limit the degradation of microbial and dietary proteins in the rumen and to supply the large amount of amino acids to the ruminants and reduce N outputs. Protozoa make up 40% to 50% of total microbial biomass in the rumen, and their ability to assimilate and convert both dietary and microbial proteins plays a significant role in the N economy of the ruminant. Control of the rumen protozoa population may thus offer a way to improve N retention in ruminants (see also Chapter 3).

4.6.1 Predatory action of protozoa on bacteria

Rumen protozoa have no significant ability to synthesize amino acids *de novo*. Only the holotrichs, which account for less than 10% of the total rumen ciliate population, are able to incorporate some ^{14}C-labelled monosaccharides into their cellular proteins (Williams and Harfoot, 1976; Williams, 1979). Rumen protozoa must, therefore, use pre-formed amino acids originating from engulfed bacteria and engulfed protozoa and from plant proteins or chloroplasts to synthesize their own proteins. The entodiniomorphid protozoa are efficient in taking up insoluble and particulate matter in suspension in ruminal fluid. The optimum size of engulfed particles is about 1 μm, indicating that protozoa take up bacteria selectively. Coleman (1975) reported that, *in vitro*, a single protozoon can take up 10^2–10^4 bacteria hourly. Assuming a bacteria concentration of 10^9/ml in the rumen content, the author estimated that predation could almost renew the entire bacterial-biomass in a few hours in a rumen harbouring a high concentration of protozoa (10^5–10^6 cells/ml). Although protozoa grown *in vitro* ingest more bacteria than those grown *in vivo* (Coleman and Sandford, 1979), such an extent of bacterial predation by protozoa is enough to explain the increase in bacterial biomass observed after the elimination of protozoa from the rumen and changes in the population densities of culturable bacteria (Eadie and Hobson, 1962; Kurihara *et al.*, 1968, 1978; Eadie and Gill, 1971; Eugène *et al.*, 2004a). Entodiniomorphid ciliates can take up other small particles such as starch grains, oil droplets, polystyrene latex beads, black palladium (Coleman and Hall, 1969) and bentonite (Forster and Leng, 1989; Wallace and Newbold, 1991). A competitive and selective ingestion process can regulate bacterial uptake. For example, starch grains are quickly engulfed and can completely fill protozoa, thus limiting bacteria engulfment, which is a slow, continuous process.

Little is known about the engulfment and digestion of bacteria by holotrich ciliates. Although they display some preference for taking up soluble compounds, viable and dead bacteria have been observed within the holotrich cells in digestive vacuoles (Gutierrez and Hungate, 1957; Gutierrez, 1958; Stern *et al.*, 1977) and in the endoplasm (Williams and Coleman, 1992). It is considered that bacteria are essential for the growth of holotrichs that cannot be cultivated without viable bacteria (Gutierrez and Hungate, 1957; Clarke and Hungate, 1966). Living intracellular bacteria may be involved in the metabolism of holotrichs or entodiniomorphs (White, 1969).

After engulfment, bacteria are stored and digested in large vesicles each containing five to seven bacteria (Williams and Coleman, 1992). Up to 31% of the protozoal volume can be occupied by engulfed bacteria (Coleman, 1967). Experiments carried out with rumen protozoa cultures, and bacteria originating from other habitats, showed that the rate of bacterial engulfment by *Entodinium* spp. is pH-dependant, optimum in the pH range 6–7 and unimportant below pH 5.5 and above pH 8.0, protozoa being metabolically inactive at such extreme pH values (Coleman, 1972). Some of the engulfed bacteria can survive for 1 h and can use maltose and glucose released *in situ* from the digestion of starch to synthesize a glucose-based polysaccharide capsule that protects them from the protozoal enzymes. The number of engulfed bacteria and their digestion rate increase in starved protozoa. Gram-positive bacteria are more extensively killed and digested by protozoa than Gram-negative bacteria. Digestion of the former is complete in 3–6 h. The cell contents are digested first, leaving the cell walls apparently intact, which are then slowly broken down into small fragments (Coleman and Hall, 1972).

Mixed protozoal preparations of ciliates harbour proteolytic enzymes that are involved in the degradation of bacterial proteins (Abou Akkada and Howard, 1962; Shinchi and Kandatsu, 1981; Brock *et al.*, 1982; Forsberg *et al.*, 1984). The presence of proteolytic enzymes has been described in entodiniomorphs (Coleman, 1983) and holotrichs (Lockwood *et al.*, 1988, 1989). The specific proteinase activity is higher in *Dasytricha ruminantium* than in *Isotricha* spp.

In vitro digestion of ^{14}C-labelled bacteria by *Entodinium caudatum* showed that only 50% of engulfed bacterial proteins are incorporated into the protozoal proteins. The other half of the bacterial proteins is released into the rumen medium (Harmeyer, 1971; Coleman, 1972; Coleman and Laurie, 1977; Owen and Coleman, 1977; Coleman and Sandford, 1979) in the form of short peptides and amino acids that can then be utilized or degraded to NH_3 by bacteria. The presence of such NAN in the rumen juice can stimulate the growth of certain rumen microbes in faunated animals (Maeng and Baldwin, 1976; Russell *et al.*, 1983; Argyle and Baldwin, 1989; Rooke and Armstrong, 1989; Cecava *et al.*, 1991; Fujimaki *et al.*, 1992; Kaur *et al.*, 1992; Russi *et al.*, 2002) although a positive effect of peptide or amino acid supplementation on microbial synthesis has not always been observed (Cruz Soto *et al.*, 1994; Wallace *et al.*, 1998; Fu *et al.*, 2001). Ranilla *et al.* (2000a,b) showed that bacteria associated to solid particles, which are mainly made up of cellulolytic organisms, assimilate more pre-formed amino acids or short peptides than liquid-associated bacteria. According to Chikunya *et al.* (1996), only diets with high energy content supporting a high bacterial growth rate respond to peptide supplementation. An extra supply of N in peptide form, however, can increase microbial yield of fibre-digesting bacteria (Ranilla *et al.*, 2001), especially when ruminants are fed high-energy diets.

4.6.2 Effect of protozoa on qualitative composition of bacterial population and consequences on the biological value of bacterial proteins

Few studies have been carried out on the evolution of bacterial species in relation to the protozoal population. The comparative ability of protozoa to engulf either solid-adherent or liquid-associated bacteria is still being debated. The preference of ciliates for starch granules and sugars over fibre should result in a selective uptake of amylolytic and free bacteria, while cellulolytic bacteria attached to large fibre particles should be saved. Nutritional competition between bacteria and protozoa for starch use can also explain the decrease in the amylolytic bacteria population when protozoa are present. Kurihara *et al.* (1978) found that the number of cellulolytic bacteria decreased, while that of amylolytic bacteria increased with defaunation. The higher growth rate of the latter over the former may be responsible for the positive effect of defaunation on MPS.

Also, protozoa host certain types of bacteria. Imai and Ogimoto (1978) observed that bacterial cells are trapped between the cilia of holotrichs rather than attaching to the cell wall. These bacteria were identified as cocci in chains belonging to *Streptococcus bovis* and *Ruminococcus albus* species. Methanogenic bacteria were also found closely associated with the external ciliate pellicle (Vogels *et al.*, 1980; Stumm *et al.*, 1982; Krumholz *et al.*, 1983). Protozoa have a significant oxygen scavenging effect in the rumen (Williams, 1986; Jouany *et al.*, 1999), and therefore protect methanogens, which are strictly anaerobic organisms. Physically associated with protozoal cells, methanogens use the hydrogen released by protozoa to convert carbon dioxide (CO_2) into methane (CH_4). The number of methanogenic bacteria associated with protozoa is large when the hydrogen partial pressure is low in the rumen (Stumm *et al.*, 1982), probably because these bacteria are efficient hydrogen users. Such interrelationships between protozoa and bacteria highlight the effects of defaunation on the rumen bacterial population. Furthermore, Preston and Leng (1986) noted that the number of anaerobic fungi was increased by defaunation. The real impact of this alteration of the balance of bacterial population on the biological value of bacterial proteins is difficult to assess. The results of Cockburn and Williams (1984) and Czerkawski (1976) suggest that diet has little effect on the amino acid composition of mixed protozoa and mixed bacteria isolated from rumen content. Even large and small bacteria harvested from the rumen contents of sheep fed different diets had the same amino acid composition (Czerkawski, 1976). This probably indicates that qualitative changes in the bacterial population have no significant impact on the bacterial amino acid profile at the intestinal level.

4.6.3 Quantitative aspects of the ruminal turnover of microbial proteins

The turnover of microbial protein in the rumen can result from autolysis of bacteria, protozoa and fungi and engulfment and digestion of bacteria by protozoa. According to Wallace and McPherson (1987), predation by small entodiniomorphid ciliates is by far the main cause of bacterial protein turnover in the rumen, accounting for 88% of total bacterial protein turnover *in vitro*. The rate of bacterial breakdown in the presence of protozoa ranged from 5.3% to 28.6%/h, depending on the tested bacteria, while autolysis of bacteria due to nutrient starvation was lower than 3%/h for all the bacterial species. The effects of other lytic factors such as bacteriophages and mycoplasmas, and endogenous proteolysis were minor.

In 1972, Nolan and Leng, using isotope dilution techniques with [^{15}N]ammonium sulphate, [^{15}N]-urea and [^{14}C]urea, indicated that in sheep fed chaffed lucerne hay in a steady feeding condition, 42% of dietary N entering the rumen was degraded to NH_3. They showed that 30% of the N entering the ruminal NH_3 pool was recycled through the 'ruminal NH_3 $\rightarrow$ microbial protein $\rightarrow$ amino acids $\rightarrow$ NH_3' pathways to the NH_3 pool and was then mostly excreted in urine. Recycling of microbial N was measured through the NH_3 pool only, which thus underestimates the total recycling. A two-compartment model consisting of the NH_3 pool and the amino acids or peptides derived from the hydrolysis of proteins and incorporated into bacterial and protozoal protein without entering the NH_3 pool, has been suggested by Firkins *et al.* (1992). The authors calculated that 76% to 90% of total microbial synthesis was recycled in the rumen compared to 30% to 50% reported by Nolan and Leng (1972) and Nolan and Stachiw (1979) when only the NH_3 pool was considered. A three-compartment model where the NAN pool was divided into fast- and slow-turnover microbial NAN pools was proposed by Oldick *et al.* (2000), but did not improve the fit of the data obtained from the two-compartment model.

4.6.3.1 Effect of protozoa on total microbial synthesis and turnover of bacterial proteins in the rumen

Microbial synthesis rate is set by the amount of fermentable carbohydrates, the ATP yield during fermentation and the efficiency of ATP use for synthesis of microbial matter. Protozoa have a positive effect on the former since they improve the digestion of organic matter in the rumen (see Jouany *et al.*, 1988). As a consequence, total ATP production is probably stimulated by the presence of protozoa. Owing to their slow growth rate and their long retention time in the rumen, protozoa have high maintenance energy needs and their growth efficiency is low compared to that of bacteria. These conflicting factors affecting ATP yield and ATP utilization for protozoal growth explain why protozoa have no significant effect on total microbial synthesis including the recycled fraction of microbial proteins (Demeyer and Van Nevel, 1979; Koenig *et al.*, 2000).

Teather *et al.* (1984) investigated protozoal and bacterial protein levels in lactating dairy cows and found a highly significant negative correlation ($P < 0.001$) between the total population of bacteria and the protozoal biomass. Thus any factor influencing the concentration of protozoa in the rumen will have a large impact on the bacterial biomass and, perhaps, on the amount of recycled microbial protein.

Availability of starch and sugars has a major effect on the number of protozoa in the rumen (Jouany, 1989). Dijkstra (1994) confirmed that the proportion of protozoal N in the total microbial N is related to the amount of dietary starch and soluble sugars at low or moderate DMI. In a situation where a large number of protozoa are established in the rumen and the dietary supply of starch is not extreme (30% maximum of the dietary DMI), ingestion and storage of starch granules in the protozoal endoplasm just after animal feeding allows the rumen pH to stay above 6.0 all day and enables a stable microbial population to be maintained (Ushida *et al.*, 1991). However, because protozoa do not control their rate of uptake of soluble sugar (Williams, 1979) or starch (Williams and Coleman, 1988), they burst when a large amount of concentrate is given to the ruminants. Thus with an excessive supply of dietary starch to the rumen, protozoal cells lyse and so tend to increase the turnover of microbial N in the rumen. The proportion of concentrate in the diet and the level of feed intake are, therefore, the two main factors controlling the population of protozoa. As shown by Eadie *et al.* (1970), providing cattle with *ad libitum* access to a pelleted barley diet can eliminate all the protozoa from the rumen. Lysis of protozoa may be due to intracellular accumulation of acidic products (Prins and van Hoven, 1977) and the severe rumen pH drop associated with an increase in lactic and propionic acid concentrations, because protozoa are sensitive to these environmental conditions (Hino, 1981; Kobayashi and Itabashi, 1986). Starch-digesting bacteria and lactobacilli in such low pH conditions will grow quickly at the expense of cellulolytic bacteria, thus increasing the risk of acidosis and the associated instability of the bacterial population. Furthermore, amylolytic bacteria tend to have a higher proteolytic activity than cellulolytic ones (Siddons and Paradine, 1981). This means that more dietary proteins will be degraded in the rumen, which will again increase the NH_3 production and the subsequent turnover of N.

Also, a high level of intake by ruminants accelerates the turnover rate of rumen content (Owens and Goetsch, 1986) and, owing to the long division time of protozoa (10 to 38 h), it decreases the number of protozoa by washing them out of the rumen. Such harmful effects on protozoa with high-starch diets fed *ad libitum* will have a detrimental action on pH, on the stability and biodiversity of bacterial population and finally on the main digestive rumen functions including MPS. Some lytic factors related to protozoal death, such as toxic dietary compounds or amounts of swallowed oxygen can be involved with high intake of concentrate diets (Coleman, 1985; Leng, 1989). The estimates reported by Firkins *et al.* (1992) confirm that rumen microbial protein recycling was highest (90% vs. 76%) when DMI was the lowest (4.7 vs. 10 kg/day). Likewise, the simulations carried out by Dijkstra *et al.* (1998) showed a decrease in ruminal N turnover when ruminants were fed different diets [all grass; all maize silage; hay (33%) and commercial concentrate (67%); hay (33%), commercial concentrate (33%) and barley (33%)] at increasing levels of DMI. The authors concluded that simulated recycling of microbial N occurs significantly for diets containing high amounts of starch, low amounts of RDP or both, such conditions being also favourable to the growth of protozoa. In sheep fed a diet of dried grass, Newbold *et al.* (2000) computed from the addition of a pulse dose of [^{15}N]ammonium chloride in the rumen that defaunation decreased the intraruminal bacterial N recycling by 87% (0.8 vs. 6.6 g N/day). Demeyer and Van Nevel (1979) calculated the total growth of microbial cells *in vitro* from the rate of incorporation of ^{32}P for up to 4 h and the net rate of N incorporation to derive the net growth during the same incubation period. They showed that elimination of protozoa increased the net synthesis of microbial proteins by 30%, the latter being considered as the amount of microbial proteins flowing to the duodenum, whereas the total microbial growth including the recycled proteins remained unchanged. This resulted in the net outflow of microbial proteins and was then validated in comparative studies carried out *in vivo* with defaunated and refaunated animals (see Jouany, 1996). According to Koenig *et al.* (2000), defaunation has more effect on increasing the size of the bacterial pool in the rumen and raising the outflow of bacterial N than on decreasing the amount of bacterial N recycling within the rumen. For whatever reason, defaunation has a clearly beneficial effect on the supply of well-balanced proteins of bacterial origin to the duodenum of ruminants (Table 4.4).

Defaunation both stimulates ruminal outflow of microbial proteins and decreases OM digestion in the rumen (Jouany *et al.*, 1988). Both effects explain the significant increase in the efficiency of MPS expressed as the amount of microbial protein flowing to the duodenum per unit of fermented or digested OM (Table 4.4). The energy used for protein synthesis is lost after the microbial proteins have been degraded in the rumen. The utilization of degraded microbial matter for a new synthesis entails energy expenditure in the rumen (Hespell and Bryant, 1979). Dijkstra *et al.* (1998) calculated that 4.7 and 12.1 g of OM are necessary to synthesize 1 g of microbial protein flowing to the duodenum at 35% and 75% recycling rates, respectively.

In a quantitative meta-analysis carried out from 19 *in vivo* trials collected from comparative studies on sheep, Eugène *et al.* (2004b) gave a clear indication of the positive effect of defaunation on MPS efficiency (39.9 vs. 28.1 g microbial N/kg OM digested in the rumen) and on the duodenal flow of microbial protein (17.6 vs. 15.7 g microbial N/day) (Fig. 4.3). Therefore, defaunation will have a higher relative effect when the flow of microbial protein is low in faunated animals, i.e. in poor diets with low protein content and supplemented with available energy such as sugar or starch. Also, the response of animal growth to defaunation is maximal with the same type of diet (Bird and Leng, 1984). However, high-yielding animals are usually fed well-balanced diets and may not benefit as much from complete defaunation, but may benefit from reduced fauna.

4.6.3.2 Recycling of protozoal proteins in the rumen

^{14}C-labelled protozoa obtained after incubation with [^{14}C-methyl]choline can be returned to the rumen to estimate the size of the protozoal pool in the rumen, their apparent turnover rate (Leng, 1982; Punia *et al.*, 1992), and the amount of protozoal N reaching the duodenum. Leng *et al.* (1981), who developed this technique, assessed the pool of protozoa as 24–46 g of N in the rumen of zebu bulls fed on sugarcane and weighing 350 kg. Using a similar technique, Faichney *et al.* (1997) evaluated the rumen protozoal N pool as 6.9 and 14.9 g in an adult sheep fed hay and a mixed diet, respectively. About 65% of protozoa were associated with the liquid phase, and 35% were located in the matrix of solid particles. The long half-time of large protozoa compared with the half-time of liquid or solid rumen digesta indicates that protozoa are selectively retained in the rumen (Leng and Nolan, 1984). The authors estimated that nearly two-thirds of the total protozoal biomass recycles in the rumen. They computed that 23 g protozoal proteins were synthesized per day, which means that 14 g were degraded in the rumen and 9 g travelled to the duodenum of young bulls or steers fed sugarcane- or molasses-based diets. In steers and heifers fed a mixture of molasses and urea added to two levels of oaten chaff, Ffoulkes and Leng (1988) confirmed that 74% of protozoal N was irreversibly lost from the rumen. Thus, only 26% of protozoa apparently entered the small intestine. Applying the ^{15}N dilution technique to sheep, Koenig *et al.* (2000), using a mixture of molasses and urea added to two levels of oaten chaff, found that total microbial N recycling was 38% higher than that of bacterial N, indicating that nearly 75% of protozoal N was recycled in the rumen, assuming protozoal N accounted for 50% of all microbial N.

Internal markers have also been used to evaluate the contribution of protozoa to the total microbial biomass at the rumen level and in the digesta flowing out of the rumen, and to estimate the outflow rate of the rumen ciliates. Aminoethylphosphonic acid has been considered as a specific marker of protozoal phospholipids (Horiguchi and Kandatsu, 1959; Abou Akkada *et al.*, 1968; Cockburn and Williams, 1984) as was phosphatidyl choline (Coleman *et al.*, 1980; Leng, 1982). The amino acid profile method was used to identify the origin of amino acids in the duodenum and quantify the protozoal proteins (Evans *et al.*, 1975; Cottle and Nolan, 1982). Counting protozoa in digesta samples taken simultaneously from the rumen and the omasal canal (Weller and

Table 4.4. Effect of defaunation on the duodenal flow of nitrogenous compounds in sheep (g/day) and microbial protein yield (g N/kg OMDR)[a].

Diet	Non-ammonia N		Microbial or bacterial N		Microbial protein yield		Reference
	D[a]	F[a]	D	F	D	F	
Red clover	31.7	29.4	19.3(a)[b]	18.0	40.5	35.6	Lindsay and Hogan (1972)
Lucerne	21.3	18.3	14.0(a)[b]	12.0	35.1	28.1	Lindsay and Hogan (1972)
Straw (75) + beet pulp (15) + groundnut meal (9) + urea (1)	28.2	22.8	18.5(a)[b]	13.7	64.1	33.0*	Collombier (1981)
NaOH-treated straw (75) + beet pulp (15) + groundnut meal (9) + urea (1)	25.4	25.3	16.3(a)[b]	16.0	38.0	38.1	Jouany and Thivend (1983)
Maize silage (48) + shelled maize (47) + minerals (4) + urea (1)	17.4	15.6	ND	ND	ND	ND	Veira *et al.* (1983)
Lucerne hay (70) + barley grain (30)	32.8	23.4	17.7(a)[b]	12.1	60.6	26.9*	Ushida *et al.* (1984)
			18.1(b)[b]	15.8	63.0	34.2*	
NaOH-treated straw (80) + hay (10) + concentrate (10)	ND	ND	ND	ND	49.6[c]	34.9[c]	Kayouli *et al.* (1986)
Maize stover (46) + molasses (10) + maize grain (30)	ND	ND	16.1(a)[b]	7.8*	42.7	27.4*	Meyer *et al.* (1986)
+ minerals (3) + urea (1)	ND	ND		16.9(c)	12.2		
NH_3-treated straw (89) + fishmeal (6) + NH_4 sulphate (3)	16.1	12.8	7.8(a)[b]	4.9	24.4	15.2*	Ushida *et al.* (1990)
+ minerals (2)			8.4(b)	6.0			
NH_3-treated straw (72) + maize grain (18) + fishmeal (6)	19.3	13.3	10.2(a)[b]	5.9	35.7	16.1*	Ushida *et al.* (1990)
+ NH_4 sulphate (3) + minerals (1)			9.3(b)	6.6			
Grass hay (69) + concentrate (31)	ND	ND	5.77(b)[b]	5.02	ND	ND	Han *et al.* (1999)
Maize silage (50) + haylage (42) + soya meal (7) + mineral (1)	29.1	22.8	24.3(a)[b]	13.6[d]	ND	ND	Ivan *et al.* (2000)
	28.2	21.2	25.6(a)[b]	13.7[d]	ND	ND	
Barley (58) + lucerne haylage (23) + beet pulp (15) + supplement (3)	16.3	19.4	17.3	10.8*	37.8[f]	20.0[f,]*	Koenig *et al.* (2000)
Dried grass	–	–	13.3	8.9	–	–	Newbold *et al.* (2000)

[a]OMDR, organic matter apparently digested in the rumen; D, defaunated; F, faunated.
[b]Microbial N determined with (a) Diamino Pimelic Acid (DAPA); (b) purine bases; (c) ^{35}S.
[c]Individual data.
[d]Values were expressed as g/kg OM intake.
[e]Non-microbial N.
[f]g N/kg OM truly digested in the rumen.
ND, not determined; *effect of defaunation was significant ($P < 0.05$).

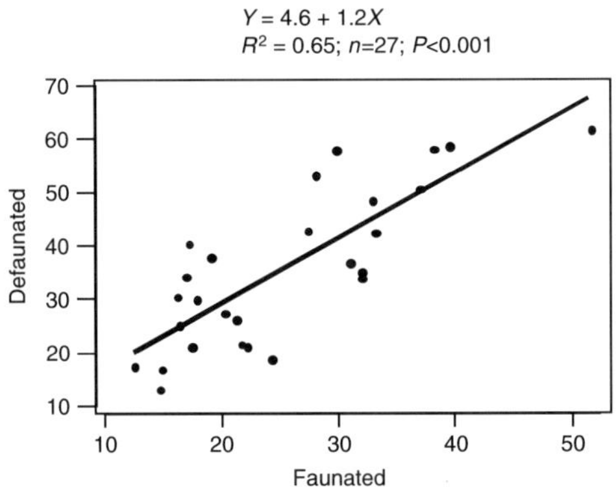

Fig. 4.2. Effect of defaunation on rumen microbial yield (g microbial N/kg organic matter apparently digested in rumen) (Eugène, 2002).

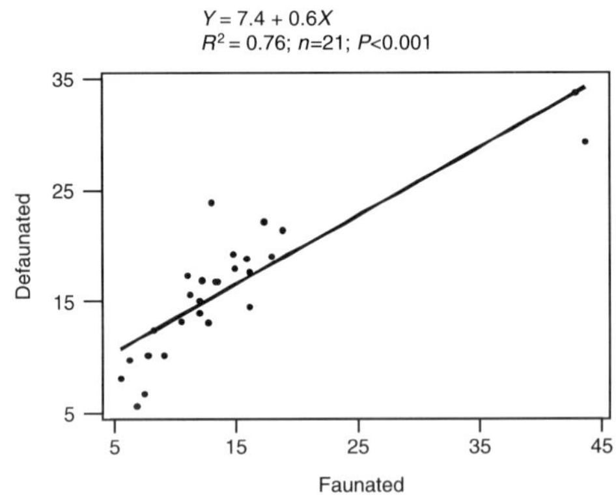

Fig. 4.3. Effect of defaunation on duodenal flow of microbial N (g/day) in sheep (Eugène, 2002).

Pilgrim, 1974; Collombier *et al.*, 1984) in animals fitted with rumen and omasal cannulae was also used to quantify the ruminal sequestration of protozoal cells. It is now accepted that only 20–40% of protozoa leave the rumen (Weller and Pilgrim, 1974; Jouany, 1978; Harrison *et al.*, 1979; Collombier *et al.*, 1984) and that protozoa remaining in the rumen die and lyse (Table 4.5). The decrease in bacterial protein flow in the duodenum due to the presence of protozoa in faunated animals is therefore not quantitatively compensated for by the duodenal supply of protozoal proteins in the same animals. Thus, the synthesis of protozoal nitrogenous compounds and their breakdown in the rumen contribute significantly to N turnover within the rumen.

4.6.4 Effect of protozoa on dietary nitrogen degradation

Little information is available on the utilization of dietary proteins by protozoa. There is some evidence that entodiniomorphid protozoa take up and digest insoluble proteins, but do not efficiently utilize the soluble fraction of proteins. Testing the effect of different proteins on the growth of entodiniomorphid protozoa, Michalowski (1989) and Muszynski *et al.* (1985) showed that these ciliates do not metabolize soluble proteins and do not grow unless insoluble proteins are supplied. *In vitro* studies with mixed A-type protozoa (Ushida and Jouany, 1985) or mixed B-type protozoa (Jouany *et al.*, 1992) indicated that the degradation

Table 4.5. Passage of protozoa from the rumen.

Diet	Animals	Protozoa in duodenum/ protozoa in rumen fluid (per ml)	Protozoal N/microbial N in duodenum	Reference
Wheaten-hay (75) + lucerne hay (25)	Sheep	0.20	ND[a]	Weller and Pilgrim (1974)
Wheaten-hay (68) + lucerne (23) + wheat grain (9)	–	–	ND	–
Crushed lucerne hay (100)	–	–	ND	–
Ground and pelleted lucerne hay (100)	–	–	ND	–
Five diets differing in carbohydrate sources (cellulose, starch, inulin, sucrose, lactose)	Goat	0.10–0.55[b]	0.16–0.58	Jouany (1978)
Six diets differing in N and energy sources	Calves	ND	0.50	Smith *et al.* (1978)
Semi-purified diet: starch (57) + cellulose (25) + supplement (18)	Sheep	ND	0.38–0.48	Harrison *et al.* (1979)
Dehydrated lucerne (62) + NaOH-treated straw (25) + whey (13)	Sheep	0.20	0.20	Collombier (1981)
Maize silage + hay + concentrate	Cows	ND	0.14	Allam *et al.* (1982)
Maize silage (54) + maize grain (44) + urea (1) + supplement (1)	Steer	ND	0.54	Steinhour *et al.* (1982)
Maize silage (55) + maize (39) + soybean meal (4) + supplement (2)	–	ND	0.37	Steinhour *et al.* (1982)
NaOH-treated straw (75) + beet pulp (15) + groundnut cake (9) + urea (1)	Sheep	0.20	–	Jouany and Thivend (1983)
Straw (57) + tapioca (40) + urea (3)	Steers	ND	0.11	Cockburn and Williams (1984)
Straw (56) + tapioca (24) + groundnut meal (20)	–	ND	0.15	–
Straw (53) + tapioca (36) + casein (11)	–	ND	0.11	–
Straw (53) + tapioca (36) + formaldehyde-treated casein (11)	–	ND	0.08	–
Maize stover (46) + ground maize (30) + molasses (10) + fishmeal (7) + supplement (7)	Sheep	ND	0.36	Meyer *et al.* (1986)
Lucerne hay (65) + barley (30) + wheat straw (3) + supplement (2)	Sheep	ND	0.21	Ushida *et al.* (1986)
NaOH-treated straw (67) + beet pulp (15)+ cakes (14) + urea (1) + supplement (3)	–	ND	0.04	–
Orchard grass (100)	Sheep	ND	0.04–0.07	Faichney *et al.* (1997)
Orchard grass (60) + concentrate (40)	–	ND	0.10–0.14	–
Hay (69) + concentrate (31)	–	ND	0.17–0.23	Han *et al.* (1999)
	Mean	0.40	0.23 ± 0.17	–

[a]ND, not determined.

[b]Individual values: 0.30 for *Polyplastron*; 0.35 for *Entodinium*; 0.16 for *Dasytricha*; 0.70 for *Isotricha*; 0.80 for *Epidinium*.

of four different protein sources (fishmeal, soybean meal, lupin and groundnut meal) up to the NH_3 stage was 20% greater in ruminal contents obtained from faunated than from defaunated sheep. Entodiniomorphid ciliates had no significant effect on the degradation of soluble casein (Onodera and Kandatsu, 1970; Jouany *et al.*, 1992). Studies carried out with simplified rumen fauna such as the entodiniomorphs *Eudiplodinium medium* alone, or *Epidinium* spp. and *Entodinium* spp. in mixture, gave the same rate of protein degradation as defaunated rumen content or even less (Jouany *et al.*, 1992). Like entodiniomorphs, holotrichs show multiple forms of protease (Lockwood *et al.*, 1988, 1989). Holotrichs are able to take up and degrade soluble casein (Onodera and Kandatsu, 1970). Onodera and Yakiyama (1990) indicate that entodiniomorphid and holotrich protozoa coagulate soluble casein before degrading it. This may indicate that insolubility is a prerequisite for proteins to be metabolized by rumen protozoa. *In situ* studies carried out with the nylon bag technique showed that degradability of the insoluble fraction of soybean meal and the rate of degradation were both increased by 11% when a mixed A-type protozoa was inoculated into a defaunated rumen (Ushida and Jouany, 1985).

Dietary protein degradation in the rumen, chiefly the insoluble fraction, is, therefore, increased by the presence of protozoa. It is tempting to ascribe this effect to a direct action of proteolytic activity of protozoa, but changes in bacterial populations and increase in ruminal digesta retention times observed after refaunation of defaunated rumen may also be involved.

4.6.5 Effect of protozoa on the intestinal supply of amino acids

Elimination of rumen protozoa has a positive effect on both main sources of intestinal NAN, microbial proteins and rumen undigested dietary proteins. This effect is greater when animals are fed diets more favourable to protozoal growth (Ushida *et al.*, 1991). Veira *et al.* (1984), Ivan *et al.* (1991) and Hsu *et al.* (1991) confirmed that defaunation stimulates the amino acid flow to the duodenum, of both essential and non-essential amino acids.

Some differences exist in amino acid composition between rumen protozoa and rumen bacteria. Protozoal proteins are richer in lysine, glutamic acid and aspartic acid, while bacterial proteins have higher alanine and glycine contents (Bergen *et al.*, 1968; Williams and Dinusson, 1973; Czerkawski, 1976; Jouany, 1978; Cockburn and Williams, 1984). Such differences are used to discriminate between the two types of microbial proteins when the amino acid profile method is applied. Weller and Pilgrim (1974) calculated that the amount of protozoal N leaving the rumen represented less than 2% of dietary N intake. The authors concluded that the contribution of protozoa is too small to significantly affect the composition of the protein mixture entering the intestine. However, given that protozoa represent nearly 23% of all microbial proteins entering the duodenum (Table 4.5) and that the digestibility of protozoal protein is greater than that of bacterial protein (McNaught *et al.*, 1954), this contribution cannot be totally ignored. Thus the highest observed increase for alanine and the lowest for lysine in the intestinal flow of amino acids in defaunated ruminants are very likely due to the disappearance of protozoal proteins for the benefit of bacterial proteins (Hsu *et al.*, 1991; Ivan *et al.*, 1991). For the same reason, defaunation significantly stimulated ($P < 0.05$) the intestinal flow of the total non-essential amino acids, but had no effect on the flow of essential amino acids (Hsu *et al.*, 1991). An additional supply of methionine as a result of defaunation was also noted by Veira *et al.* (1983).

4.7 Effect of Protozoa on Nitrogen Losses in Faeces and Urine

In ruminants, faecal N is largely made up of undigested dietary N, which is mainly linked to the protein and cell wall structure of plants, and of microbial and endogenous N, which are both closely related to the amount of NDF intake. Thus, faecal N losses are less variable than urinary N losses. Most *in vivo* studies have shown that defaunation lowers cell-wall carbohydrate digestion in the rumen. A quantitative meta-analysis carried out on 15 experiments indicated that on average, the extent of ruminal NDF disappearance in ruminants was 11% less in defaunated ruminants, whereas the decrease was only 5.7% when digestibility was estimated over the whole digestive tract (Eugène *et al.*, 2004b). This result shows that a shift occurs in digestion of cellulosic material from the rumen to the large intestine of defaunated animals. ATP produced during plant cell wall degradation and fermentation in the hindgut induces growth of the bacterial population and consequently increases the excretion of microbial proteins in the animal faeces as indicated in Table 4.6. Faecal N losses due to defaunation will therefore be maximal in animals fed high-forage diets or diets rich in low-digestible starch inducing significant starch digestion in the large intestine. Increasing the level of intake of such diets will increase the rumen bypass of potentially digestible material and extend their digestion in the hindgut with, as a consequence, greater excretion of faecal N. The same effect of defaunation on faecal N outputs is found in grazing animals.

On the contrary, urinary excretion is easier to manage through the protozoal population. It tends to decrease with defaunation (Table 4.6). This result is explained by a lower PUN concentration and a weaker NH_3 production in the digestive tract (Table 4.6), resulting from less dietary and microbial protein degradation in the rumen and, to a lesser degree, from an increase in NH_3 capture for MPS at both the rumen and large intestine sites in defaunated animals. Urea and NH_3 are small molecules that diffuse easily across membranes between blood and tissues towards lower physiological concentrations (Houpt, 1970). Blood plasma urea concentration is under physiological control, and its constancy is a function of homeostatic mechanisms. The blood urea pool originates mainly from the digestive NH_3 pool and from the metabolism by the animal tissues of absorbed amino acids, while output is under the control of kidney filtration through urea excretion in urine. Owing to the decrease in NH_3 production, some deficiency in ruminal NH_3-N for bacterial growth can occur in the defaunated rumen when NH_3 concentrations are near or below 3.6 mmol/l (Luther *et al.*, 1966; Kurihara *et al.*, 1978; Itabashi *et al.*, 1982, 1984; Ushida *et al.*, 1986; Punia *et al.*, 1987; Chapter 2). In these circumstances, some urea can return to the rumen and to the large intestine across the digestive mucosa and through saliva for the rumen, and this regulation of blood urea concentration competes with urea-N output in urine. Also, a greater intestinal supply of well-balanced amino acids arising from defaunation may reduce N catabolism in the animal tissues and so improve N retention in animals and decrease the amount of N excreted in urine. It can be concluded that defaunation is a sensitive way to reduce N excretion in urine, chiefly when animals are fed diets rich in RDP exceeding the capacity of NH_3 uptake of microorganisms and favouring the growth of protozoa, as shown in the meta-analysis carried out by Eugène (2004b). The positive impact of defaunation will also be significant on animal production in all the feeding situations where the amino acid needs of ruminants are not fully covered, i.e. when animals are fed diets poor in rumen undegradable protein content and rich in energy.

The impact of defaunation on N excretion by high-yielding animals fed diets rich in energy and with a well-balanced energy:N ratio is more difficult to assess. The level of intake is important in determining the effects of defaunation with regard to N outputs because it controls the protozoal population, the retention time of fluids and solids in rumen digesta and the associated bacterial biomass, as well as the degradation of dietary proteins. As a consequence, the passage of dietary proteins and microbial N to the small intestine of dairy cows is linearly related to the amount of OM intake (Clark *et al.*, 1992). Also, increasing feed intake supplies more energy, more N and additional nutrients for microbial growth. The higher rate of growth coupled with faster travel of microbes to the intestine may reduce the turnover of N within the rumen because of decreased cell lysis and cell uptake by protozoa. Increasing the outflow rate of rumen digesta beyond the generation time of protozoa will be detrimental to the protozoa population in the rumen, mainly for the large entodiniomorphs, and this explains why the amount of NAN flow to the intestine still rose even after the total microbial synthesis had peaked. Also, the passage of rumen-undegraded carbohydrates to the intestine increases with the level of intake, and so more potentially digestible OM enters the duodenum. For instance, up to 50% of ingested starch can escape from the rumen (Poncet *et al.*, 1995) and about 20% can be digested in the hindgut, because the amylolytic activity in the small intestine is limited. As a result, microbial digestion of starch and cell wall carbohydrates in the large intestine can lead to

Table 4.6. Effect of defaunation on ruminal ammonia concentration and N excretion in faeces and urine.

		Ruminal NH_3-N (mg N/l)		Faecal N (g N/day)		Urinary N (g N/day)		Total excreted N (g N/day)	
Reference	Species	F[a]	D[a]	F	D	F	D	F	D
Lindsay and Hogan (1972)	Sheep	271	220	5.4	5.9	ND[b]	ND	ND	ND
		208	193	7.2	8.2	ND	ND	ND	ND
Ikwuegbu and Sutton (1982)	Sheep	167	71	3.0	3.6	7.2	6.1	10.2	9.7
Rowe *et al.* (1985)	Sheep	104	104	3.6	5.3	ND	ND	ND	ND
		ND	ND	8.3	9.2	10.7	9.3	19.0	18.5
		ND	ND	10.7	11.8	12.8	11.5	23.5	23.3
Kreuzer and Kirchgessner (1986, 1989)	Sheep	ND	ND	5.5	5.0	11.2	10.2	16.7	15.2
		ND	ND	5.5	6.2	10.1	8.1	15.6	14.3
		ND	ND	6.0	6.3	8.5	8.0	14.5	14.3
		ND	ND	7.7	7.7	10.8	11.4	18.5	19.1
		ND	ND	7.4	7.6	10.5	10.7	17.9	18.3
Ushida *et al.* (1986)	Sheep	191	98*	8.0	8.6*	ND	ND	ND	ND
		140	54*	7.4	8.0*	ND	ND	ND	ND
Punia *et al.* (1987)	Cattle	190	171*	26.8	29.6	44.2	39.4	71.0	69.0
Hsu *et al.* (1991)	Sheep	235	154	6.6	7.3	ND	ND	ND	ND
Jouany and Ushida (1999)	Sheep	2.5	0.82[c]*	6.0	7.8*	11.9	10.6*	179	18.4
Koenig *et al.* (2000)	Sheep	322	172*	6.1	8.6*	10.3	10.3	16.4	18.9
Newbold *et al.* (2000)	Sheep	2.0	0.63[d]	–	–	–	–	–	–

[a]F, faunated animals; D, defaunated animals.
[b]ND, not determined.
[c]Total ammonia N pool (g N) determined by rumen emptying method, 5 h after animals have been fed.
[d]Ammonia pool (g N/day) estimated from a pulse dose of [^{15}N]ammonium chloride.
*Significant effect of defaunation ($P < 0.05$).

large amounts of microbial N in faeces when animals are fed diets with high energy content and properly supplied with N. Excretion of urinary N by high-yielding animals will be reduced by defaunation for the same reasons as discussed above. Only the origins of inputs into the blood urea pool are slightly changed in these animals, because their diets are usually rich in protected dietary proteins and their body proteins turnover more quickly. Thus, more urea originates from the body metabolism while less comes from the ruminal metabolism of proteins in such animals.

Of course, the effect of defaunation on N excretion will be greater with diets favouring the growth of protozoa, i.e. diets rich in starch or sugars and available insoluble proteins (Jouany *et al.*, 1988; Jouany, 1989). However, protozoal cells can degenerate and burst in diets containing large amounts of easily degradable carbohydrates. In extreme dietary conditions such as whole pelleted barley rations fed *ad libitum*, ruminants can even lose all their rumen protozoa (Eadie *et al.*, 1970).

A dynamic model of N metabolism in lactating dairy cows was used to predict N excretion in urine, faeces and milk (Kebreab *et al.*, 2002). The authors showed that reducing CP concentration in the diet to about 16% could lessen rumen NH_3 production by 20% and decrease urinary N excretion to the same extent. Faecal N output is less sensitive to N intake than urinary N output. In the same way, an increase in dietary energy concentration could stimulate NH_3 uptake by bacteria and potentially reduce both rumen NH_3 concentration and urinary N excretion by up to 25% per cow. However, as discussed above, the model indicates that the lower excretion in urine is compensated for by a higher faecal N excretion when the dietary energy level is increased. Hence the effect of increasing energy intake on total N excretion is lower than that observed on the urinary excretion alone. A similar relationship between faecal and urinary N has been noted after rumen defaunation.

4.8 Reservations in Respect of Defaunation

No reliable technique is currently available to defaunate the rumen. The most commonly used agents, such as surfactants (Abou Akkada *et al.*, 1968; Orpin, 1977; Bird and Leng, 1978, 1984; Bird *et al.*, 1979), are not specific against protozoa and can have short-term detrimental effects on bacteria (Orpin, 1977; Eadie and Shand, 1981) and the host animal. Thus, there is a need for new antiprotozoal agents that are safe for animals, consumers and the environment. Probably, a significant reduction of protozoa may prove to be equally satisfactory, with regard to N excretion, to complete defaunation. This can be achieved by lipid supplementation with sources rich in long-chain polyunsaturated FA such as linseed or sunflower oils, or medium-chain saturated FA from coconut or palm kernel oils (Newbold and Chamberlain, 1988; Broudiscou *et al.*, 1990; Matsumoto *et al.*, 1991). The defaunating efficiency of individual FA is difficult to assess. Hristov *et al.* (2004c) ranked the toxic effect of pure FA salts against protozoa and consequences on some rumen functions in an *in vitro* study; C10:0, C12:0 and C18:3 FA induced a complete defaunation, while C18:2, C14:0 and C18:1 eliminated 88%, 75% and 45% (medium levels) of protozoa, respectively. However, Abel *et al.* (2002) observed no effect of caprylic acid (C8:0) or capric acid (C10:0) on protozoa population in the RUSITEC. Regarding the effect of level of FA unsaturation, no clear conclusion can be given. Broudiscou *et al.* (1990, 1994) indicated that linseed oil and soybean oil rich in C18:3 and C18:2, respectively, decreased the number of protozoa *in vivo* by 66% and 60%, whereas Fievez *et al.* (2003) noted no significant effect of addition of eicosapentaenoic acid (C20:5) or docosahexaenoic acid (C22:6) on rumen fauna. The nycthemeral mean concentration of FA in rumen content is likely the most significant factor explaining the effect of lipid supplementation on rumen microbes. So, the rhythm of addition could be more important than the amount of lipid supplementation. More studies need to be undertaken to understand the real impact of lipids on protozoa and the subsequent effect on rumen N metabolism. Some plant extracts such as saponins (Teferedegne *et al.*, 1999; Muetzel *et al.*, 2000; Ningrat *et al.*, 2002) or essential oils (Newbold *et al.*, 2004) can be also considered as promising natural agents for such purposes.

In addition to their digestive and nutritional contribution, protozoa have positive effects on the health and general welfare of the host animal. Therefore, elimination of protozoa can have side effects. For example, it makes the ruminants more sensitive to lactic acidosis especially in high-yielding animals (Whitelaw *et al.*, 1972; Newbold *et al.*, 1986; Ushida *et al.*, 1991), to copper toxicity (Ivan, 1989), to legume bloat (Clarke, 1965; Clarke and Reid, 1974), to plant toxins (Shiroma and Akashi, 1976; Yoshida *et al.*, 1982; Tangendjaja *et al.*, 1983), to mycotoxins (Yiannikouris and Jouany, 2002) and to pathogens such as *Escherichia coli* (McIntosh *et al.*, 2000). These aspects have to be considered when rumen defaunation is envisaged.

4.9 Conclusions

A significant proportion of nitrogen losses in ruminants has its origin in the rumen. Overfeeding of ruminally degradable protein, uncontrolled proteolysis, peptide degradation and deamination of amino acids all contribute to the inefficiency of nitrogen utilization by the ruminal microbiota. Energy availability is, however, the major constraint determining utilization of dietary nitrogen and ammonia for synthesis of microbial protein in the rumen. Provision of energy, through carbohydrate supplementation or increased carbohydrate availability, reduces ammonia concentration and has often increased microbial protein synthesis in the rumen, but effects on overall nitrogen losses, particularly with urine, and efficiency of conversion of feed nitrogen into milk protein are yet to be determined. Long-chain unsaturated and medium-chain saturated fatty acids and bioactive substances such as saponins, tannins and essential oils are perhaps the instruments of choice in manipulating ruminal fermentation to achieve more complete utilization of ammonia and to reduce the environmental nitrogen load from cattle operations. Although all these compounds operate chiefly at the rumen level, they differ in their mode of action. Tannins can chemically bind to dietary proteins, and proteolysis is then slowed down. Some direct effect of tannins on rumen microbial community, more especially on protozoa, has been also demonstrated. On the contrary, the effect of the other additives is due to their antimicrobial activity. Some of them such as fatty acids or saponins have antiprotozoal properties, while others such as essential oils or some antibiotics currently used as growth promoters are chiefly active against bacteria. Considering the position of consumers and the national regulation authorities in developed countries, it is likely that more attention will be given in the future to 'natural additives' than chemicals or antibiotics to control the ruminal degradation of dietary nitrogenous compounds.

References

Abel, H.J., Immig, I. and Harman, E. (2002) Effect of adding caprylic and capric acid to grass on fermentation characteristics during ensiling and in the artificial rumen system RUSITEC. *Animal Feed Science and Technology* 99, 65–72.

Abou Akkada, A.R. and Howard, B.H. (1962) The biochemistry of rumen protozoa. 5. The nitrogen metabolism of *Entodinium. Biochemical Journal* 82, 313–320.

Abou Akkada, A.R., Bartley, E.E., Berube, R., Fina, L.R., Meyer, R.M., Hendricks, D. and Julius, F. (1968) Simple method to remove completely ciliate protozoa of adult ruminants. *Applied Microbiology* 16, 1475–1477.

Ahvenjarvi, S., Vanhatalo, A. and Huhtanen, P. (2002) Supplementing barley or rapeseed meal to dairy cows fed grass-red clover silage. I. Rumen degradability and microbial flow. *Journal of Animal Science* 80, 2176–2187.

Allam, S.M., Rohr, K., Brandt, M. and Lebzien, P. (1982) Beiträge zur Quantifizierung der N-Umsetzungen in den Vormägen von Milchkühen. 3. 2,6-Diaminopimelinsäure im Vergleich zu ^{15}N als Bezugssubstanz für die Bestimmung des Mikrobenstickstoffs im Duodenalchymus von Milchkühen. *Zeitschrift für Tiererphysiologie, Tierernährung und Futtermittelkunde* 48, 58–64.

Araba, A., Byers, F.M. and Guessous, F. (2002) Patterns of rumen fermentation in bulls fed barley/molasses diets. *Animal Feed Science and Technology* 97, 53–64.

Argyle, J.L. and Baldwin, R.L. (1989) Effects of amino acids and peptides on rumen microbial growth yields. *Journal of Dairy Science* 72, 2017–2027.

Arieli, A., Abramson, S., Mabjeesh, S.J., Zamwel, S. and Bruckental, I. (2001) Effect of site and source of energy supplementation on milk yield in dairy cows. *Journal of Dairy Science* 84, 462–470.

Armentano, L. and Pereira, M. (1997) Measuring the effectiveness of fiber by animal response trials. *Journal of Dairy Science* 80, 1416–1425.

Armentano, L.E., Bertics, S.J. and Riesterer, J. (1993) Lack of response to addition of degradable protein to a low protein diet fed to midlactation dairy cows. *Journal of Dairy Science* 76, 3755–3762.

Avila, C.D., DePeters, E.J., Perez-Monti, H., Taylor, S.J. and Zinn, R.A. (2000) Influences of saturation ratio of supplemental dietary fat on digestion and milk yield in dairy cows. *Journal of Dairy Science* 83, 1505–1519.

Baah, J., Hristov, A.N., McAllister, T.A., Ivan, M., Koenig, K.M. and Rode, L.M. (2002) Evaluation of quillaja extract, quebracho tannin and safflower oil as selective defaunating agents in cattle. *Journal of Animal Science* 80 (Suppl. 1), 404.

Bach, A., Huntington, G.B., Calsamiglia, S. and Stern, M.D. (2000) Nitrogen metabolism of early lactation cows fed diets with two different levels of protein and different amino acid profiles. *Journal of Dairy Science* 83, 2585–2595.

Bachmann, H.P. and Jans, F. (1995) Fütterung kann Käsequalität beeinflussen. *Schweizerische Milchzeitung* 121, 3.

Bae, H.D., McAllister, T.A., Yanke, J., Cheng, K.-J. and Muir, A.D. (1993) Effects of condensed tannins on endoglucanase activity and filter paper digestion by *Fibrobacter succinogenes* S85. *Applied and Environmental Microbiology* 59, 2132–2138.

Barry, T.N. (1989) Condensed tannins: their role in the ruminant protein and carbohydrate digestion and possible effects upon the rumen ecosystem. In: Nolan, J.V., Leng, R.A. and Demeyer, D.I. (eds) *The Roles of Protozoa and Fungi in Ruminant Digestion*. Penambul Books, Armidale, New South Wales, pp. 153–169.

Beauchemin, K.A., Yang, W.Z. and Rode, L.M. (2001) Effects of barley grain processing on the site and extent of digestion of beef feedlot finishing diets. *Journal of Animal Science* 79, 1925–1936.

Belasco, I.J. (1956) The role of carbohydrates in urea utilization, cellulose digestion and fatty acid formation. *Journal of Animal Science* 15, 496–508.

Berentsen, P.B.M. and Giesen, G.W.J. (1996) Economic aspects of feeding dairy cows to contain environmental pollution. In: Phillips, C.J.C. (ed.) *Progress in Dairy Science*. CAB International, Wallingford, UK, pp. 355–375.

Bergen, W.G., Purser, D.B. and Cline, J.H. (1968) Determination of limiting amino acids of rumen isolated microbial protein fed to rat. *Journal of Dairy Science* 51, 1698–1700.

Bird, S.H. and Leng, R.A. (1978) The effects of defaunation of the rumen on the growth of cattle on low-protein high-energy diets. *British Journal of Nutrition* 40, 163–167.

Bird, S.H and Leng, R.A. (1984) Further studies on the effects of the presence or absence of protozoa in the rumen on live-weight gain and wool growth of sheep. *British Journal of Nutrition* 52, 607–611.

Bird, S.H., Hill, M.K. and Leng, R.A. (1979) The effects of defaunation of the rumen on the growth of lambs on low-protein-high-energy diets. *British Journal of Nutrition* 42, 81–87.

Blackburn, T.H. (1965) Nitrogen metabolism in the rumen. In: Dougherty, R.W. (ed.) *Physiology of Digestion in the Ruminant*. Butterworths, Washington, DC, pp. 322–334.

Bowden, D.M., Taylor, D.K. and Davis, W.E.P. (1968) Water-soluble carbohydrates in orchardgrass and mixed forages. *Canadian Journal of Plant Science* 48, 9–15.

Bristow, A.W., Whitehead, D.C. and Cockburn, J.E. (1992) Nitrogenous constituents in the urine of cattle, sheep and goats. *Journal of the Science of Food and Agriculture* 59, 387–394.

Brock, F.M., Forsberg, C.W. and Buchanan-Smith, J.G. (1982) Proteolytic activity of rumen microorganisms and effects of proteinase inhibitors. *Applied Environmental Microbiology* 44, 561–569.

Broderick, G.A. (2003) Effects of varying dietary protein and energy levels on the production of lactating dairy cows. *Journal of Dairy Science* 86, 1370–1381.

Broderick, G.A. and Clayton, M.K. (1997) A statistical evaluation of animal and nutritional factors influencing concentration of milk urea nitrogen. *Journal of Dairy Science* 80, 2964–2971.

Broderick, G.A. and Wallace, R.J. (1988) Effect of dietary nitrogen source on concentrations of ammonia, free amino acids and fluorescamine-reactive peptides in the sheep rumen. *Journal of Animal Science* 66, 2233–2238.

Broderick, G.A., Wallace, R.J. and Ørskov, E.R. (1991) Control of rate and extent of protein degradation. In: Tsuda, T., Sasaki, Y. and Kawashima, R. (eds) *Physiological Aspects of Digestion and Metabolism in Ruminants*. Academic Press, San Diego, California, pp. 541–592.

Broderick, G.A., Mertens, D.R. and Simons, R. (2002) Efficacy of carbohydrate sources for milk production by cows fed diets based on alfalfa silage. *Journal of Dairy Science* 85, 1767–1776.

Broudiscou, L., Van Nevel, C.J. and Demeyer, D.I. (1990) Incorporation of soya oil hydrolysate in the diet of defaunated or refaunated sheep: effect on rumen fermentation *in vitro*. *Archives of Animal Nutrition* 40, 329–337.

Broudiscou, L., Pochet, S. and Poncet, C. (1994) Effect of linseed supplementation on feed degradation and microbial synthesis in the rumen of ciliate-free and refaunated sheep. *Animal Feed Science and Technology* 49, 189–202.

Bunting, L.D., Boling, J.A. and MacKown, C.T. (1989) Effect of dietary protein level on nitrogen metabolism in the growing bovine. I. Nitrogen recycling and intestinal protein supply in calves. *Journal of Animal Science* 67, 810–819.

Callison, S.L., Firkins, J.L., Eastridge, M.L. and Hull, B.L. (2001) Site of nutrient digestion by dairy cows fed corn of different particle sizes or steam-rolled. *Journal of Dairy Science* 84, 1458–1467.

Calsamiglia, S., Hongerholt, D.D., Crooker, B.A., Stern, M.D., Hartnel, G.F. and Hintz, R.L. (1992) Effect of fishmeal and expeller-processed soybean meal fed to dairy cows receiving bovine somatotropin (Sometribove). *Journal of Dairy Science* 75, 2454–2462.

Casper, D.P., Schingoethe, D.J. and Eisenbeisz, W.A. (1990) Response of early lactation dairy cows fed diets varying in source of nonstructural carbohydrate and crude protein. *Journal of Dairy Science* 73, 1039–1050.

Casper, D.P., Maiga, H.A., Brouk, M.J. and Schingoethe, D.J. (1999) Synchronization of carbohydrate and protein sources on fermentation and passage rates in dairy cows. *Journal of Dairy Science* 82, 1779–1790.

Castillo, A.R., Kebreab, E., Beever, D.E., Barbi, J.H., Sutton, J.D., Kirby, H.C. and France, J. (2001a) The effect of protein supplementation on nitrogen utilization in lactating dairy cows fed grass silage diets. *Journal of Animal Science* 79, 247–253.

Castillo, A.R., Kebreab, E., Beever, D.E., Barbi, J.H., Sutton, J.D., Kirby, H.C. and France, J. (2001b) The effect of energy supplementation on nitrogen utilization in lactating dairy cows fed grass silage diets. *Journal of Animal Science* 79, 240–246.

Cecava, M.J., Merchen, N.R., Berger, L.L., Mackie, R.I. and Fahey, G.C. (1991) Effects of dietary energy level and protein source on nutrient digestion and ruminal nitrogen metabolism in steers. *Journal of Animal Science* 69, 2230–2243.

Chamberlain, D.G., Thomas, P.C., Wilson, W., Newbold, C.J. and MacDonald, J.C. (1985) The effect of carbohydrate supplements on ruminal concentrations of ammonia in animals given diets of grass silage. *Journal of Agricultural Science, Cambridge* 104, 331–340.

Chamberlain, D.G., Robertson, S. and Choung, J.-J. (1993) Sugars versus starch as supplements to grass silage: effects on ruminal fermentation and the supply of microbial protein to the small intestine, estimated from the urinary excretion of purine derivatives, in sheep. *Journal of the Science of Food and Agriculture* 63, 189–194.

Cheeke, P.R. and Shull, L.R. (1985) *Natural Toxicants in Feed and Poisonous Plants*, 2nd edn. AVI Publishing Company, Westport, Connecticut, pp. 329–381.

Chen, G., Sniffen, C.J. and Russell, J.B. (1987) Concentration and estimated flow of peptides from the rumen of dairy cattle: effects of protein quantity, protein solubility and feeding frequency. *Journal of Dairy Science* 70, 983–992.

Chesson, A. and Forsberg, C.W. (1997) Metabolism of nitrogen-containing compounds. In: Hobson, P.N. and Stewart, C.S. (eds) *The Rumen Microbial Ecosystem*. Blackie Academic and Professional, London, pp. 329–381.

Chikunya, S., Newbold, C.J., Rode, L., Chen, X.B. and Wallace, R.J. (1996) Influence of dietary rumen-degradable protein on bacterial growth in the rumen of sheep receiving different energy sources. *Animal Feed Science and Technology* 63, 333–340.

Choi, C.W., Vanhatalo, A., Ahvenjärvi, S. and Huhtanen, P. (2002) Effects of several protein supplements on flow of soluble non-ammonia nitrogen from the forestomach and milk production in dairy cows. *Animal Feed Science and Technology* 102, 15–33.

Christen, S.D., Hill, T.M. and Williams, M.S. (1996) Effects of tempered barley on milk yield, intake, and digestion kinetics of lactating Holstein cows. *Journal of Dairy Science* 79, 1394–1399.

Christensen, R.A., Lynch, G.L. and Clark, J.H. (1993) Influence of amount and degradability of protein on production of milk and milk components by lactating Holstein cows. *Journal of Dairy Science* 76, 3490–3496.

Clark, J.H., Klusmeyer, T.H. and Cameron, M.R. (1992) Microbial protein synthesis and flows of nitrogen fractions to the duodenum of dairy cows. *Journal of Dairy Science* 75, 2304–2323.

Clarke, R.T.J. (1965) Role of rumen ciliates in bloat in cattle. *Nature, London* 205, 95–96.

Clarke, R.T.J. and Hungate, R.E. (1966) Culture of the rumen holotrich ciliate *Dasytricha ruminantium* Schuberg. *Applied Microbiology* 14, 340–345.

Clarke, R.T.J. and Reid, C.S.W. (1974) Foamy bloat of cattle. A review. *Journal of Dairy Science* 57, 753–785.

Cockburn, J.E. and Williams, A.P. (1984) The simultaneous estimation of the amounts of protozoal, bacterial, and dietary nitrogen entering the duodenum of steers. *British Journal of Nutrition* 51, 111–132.

Coleman, G.S. (1967) The metabolism of free amino acids by washed suspensions of the rumen ciliate *Entodinium caudatum*. *Journal of General Microbiology* 47, 433–447.

Coleman, G.S. (1972) The metabolism of starch, glucose, amino acids, purines, pyrimidines and bacteria by the rumen ciliate *Entodinium simplex*. *Journal of General Microbiology* 71, 117–131.

Coleman, G.S. (1975) The inter-relationships between rumen ciliate protozoa and bacteria. In: McDonald, I.W. and Warner, A.C. (eds) *Digestion and Metabolism in the Ruminant*. University of New England Publishing Unit, Armidale, Australia, pp. 149–164.

Coleman, G.S. (1983) Hydrolysis of fraction 1 leaf protein and casein by rumen entodiniomorphid protozoa. *Journal of Applied Bacteriology* 55, 111–118.

Coleman, G.S. (1985) Possible causes of the high death rate of ciliate protozoa in the rumen. *Journal of Agricultural Science, Cambridge* 105, 39–43.

Coleman, G.S. and Hall, F.J. (1969) Electron microscopy of the rumen ciliate *Entodinium caudatum* with special reference to the engulfment of bacteria and other particulate matter. *Tissue and Cell* 1, 607–618.

Coleman, G.S. and Hall, F.J. (1972) Fine structural studies on the digestion of bacterial species in the rumen ciliate *Entodinium caudatum*. *Tissue and Cell* 4, 37–48.

Coleman, G.S. and Laurie, J.I. (1977) The metabolism of starch, glucose, amino acids, purines, pyrimidines and bacteria by the rumen ciliate *Polyplastron multivesiculatum*. *Journal of General Microbiology* 98, 29–37.

Coleman, G.S. and Sandford, D.C. (1979) The engulfment and digestion of mixed rumen bacteria and individual bacteria species by single and mixed species of rumen ciliate protozoa grown *in vivo*. *Journal of Agricultural Science, Cambridge* 92, 729–742.

Coleman, G.S., Dawson, B.M.C. and Grime, D.W. (1980) The rate of passage of ciliate protozoa from the ovine rumen. *Proceedings of the Nutrition Society* 39, 6A.

Collombier, J. (1981) Contribution à l'étude du rôle des protozoaires ciliés du rumen dans l'apport d'azote microbien entrant dans le duodénum du ruminant. PhD thesis, University of Clermont II, France 85 pp.

Collombier, J., Golière, C.A., Sénaud, J., Grain, J. and Thivend, P. (1984) Etude du rôle des protozoaires ciliés du rumen dans l'apport d'azote microbien entrant dans le duodénum du ruminant par l'estimation directe de la quantité de ciliés sortant du rumen. *Protistologica* 20, 431–436.

Cottle, D.J. and Nolan, J.V. (1982) The use of amino acid profiles to measure microbial protein outflow from the rumen. *Proceedings of the Australian Society of Animal Production* 14, 588–591.

Cowling, E.B. and Galloway, J.N. (2001) Challenges and opportunities facing animal agriculture: optimizing nutrient management in the atmosphere and biosphere of the Earth. Symposium paper presented at the Joint Annual Meeting of the ADSA, ASAS, PSA, and AMSA, Indianapolis, Indiana, July 24–28, 2001. Available at: http://www.fass.org/fass01/pdfs/Cowling.pdf

Cruz Soto, R., Muhammed, S.A., Newbold, C.J., Stewart, C.S. and Wallace, R.J. (1994) Influence of peptides, amino acids and urea on microbial activity in the rumen of sheep receiving grass hay and on the growth of rumen bacteria *in vitro*. *Animal Feed Science and Technology* 49, 151–161.

Cunningham, K.D., Cecava, M.J., Johnson, T.R. and Ludden, P.A. (1996) Influence of source and amount of dietary protein on milk yield by cows in early lactation. *Journal of Dairy Science* 79, 620–630.

Czerkawski, J.W. (1976) Chemical composition of microbial matter in the rumen. *Journal of the Science of Food and Agriculture* 27, 621–632.

De Boer, I.J.M., Smits, M.C.J., Mollenhorst, H., Van Duinkerken, G. and Monteny, G.J. (2002) Prediction of ammonia emission from dairy barns using feed characteristics. I. Relation between feed characteristic and urinary urea concentration. *Journal of Dairy Science* 85, 3382–3388.

Demeyer, D.I. and Van Nevel, C.J. (1979) Effect of defaunation on the metabolism of rumen microorganisms. *British Journal of Nutrition* 42, 515–524.

Dewhurst, R.J., Davies, D.R. and Merry, R.J. (2000) Microbial protein supply from the rumen. *Animal Feed Science and Technology* 85, 1–21.

Dijkstra, J. (1994) Simulation of the dynamics of protozoa in the rumen. *British Journal of Nutrition* 72, 679–699.

Dijkstra, J., France, J. and Tamminga, S. (1998) Quantification of the recycling of microbial nitrogen in the rumen using a mechanistic model of rumen fermentation processes. *Journal of Agricultural Science, Cambridge* 130, 81–94.

Dohme, F., Machmüller, A., Estermann, B.L., Pfister, P., Wasserfallen, A. and Kreuzer, M. (1999) The role of the rumen ciliate protozoa for methane suppression caused by coconut oil. *Letters in Applied Microbiology* 29, 1–6.

Dohme, F., Machmüller, A., Wasserfallen, A. and Kreuzer, M. (2000) Comparative efficiency of various fats rich in medium-chain fatty acids to suppress ruminal methanogenesis as measured with RUSITEC. *Canadian Journal of Animal Science* 80, 473–482.

Doreau, M. and Ferlay, A. (1994) Digestion and utilization of fatty acids by ruminants. *Animal Feed Science and Technology* 45, 379–396.

Doreau, M. and Ferlay, A. (1995) Effect of dietary lipids on nitrogen metabolism in the rumen: a review. *Livestock Production Science* 43, 97–110.

Doreau, M., Legay, F. and Bauchart, D. (1991) Effect of source and level of supplemental fat on total and ruminal organic matter and nitrogen digestion in dairy cows. *Journal of Dairy Science* 74, 2233–2242.

Doreau, M., Ferlay, A. and Elmeddah, Y. (1993) Organic matter and nitrogen digestion by dairy cows fed calcium salts of rapeseed oil fatty acids or rapeseed oil. *Journal of Animal Science* 71, 499–504.

Downing, T. (2003) Non-structural carbohydrates in cool season grasses. In: *Proceedings of the 38th Pacific Northwest Animal Nutrition Conference*. Portland, Oregon, pp. 169–177.

Eadie, J.M. and Gill, J.C. (1971) The effect of the absence of rumen ciliate protozoa on growing lambs fed on a roughage–concentrate diet. *British Journal of Nutrition* 26, 155–167.

Eadie, J.M. and Hobson, P.N. (1962) Effect of the presence or absence of rumen ciliate protozoa on the total rumen bacterial count in lambs. *Nature, London* 193, 503–505.

Eadie, J.M. and Shand, W.J. (1981) The effect of Synperonic NP9 upon ciliate-free and faunated sheep. *Proceedings of the Nutrition Society* 40, 113A.

Eadie, J.M., Hyldegaard-Jensen, J., Mann, S.O., Reid, R.S. and Whitelaw, F.G. (1970) Observations on the microbiology and biochemistry of the rumen in cattle given different quantities of a pelleted barley ration. *British Journal of Nutrition* 24, 157–177.

Encyclopædia Britannica (2003) "Essential Oil", Encyclopædia Britannica Premium Service, 02 October 2003.

Eugène, M. (2002) Effets de la défaunation de ruminants sur les performances de production, en fonction de la ration ingérée. Etude des variations de la protéosynthèse et de la cellulolyse microbienne ruminale. PhD thesis, Institut National Agronomique Paris-Grignon, France, 122 pp.

Eugène, M., Archimède, H., Michalet-Doreau, B., Fonty, G. and Sauvant, D. (2004a) Effects of defaunation on microbial activities in the rumen of rams consuming mixed diet (fresh *Digitaria decumbens* grass and concentrate). *Animal Research* 53, 187–200.

Eugène, M., Archimède, H. and Sauvant, D. (2004b) Quantitative meta-analysis on the effects of defaunation of the rumen on growth, intake and digestion in ruminants. *Livestock Production Science* 85, 81–97.

Evans, R.A., Axford, A.F.E. and Offer, N.W. (1975) A method for estimating the quantities of microbial and dietary proteins flowing in the duodenal digesta of ruminants. *Proceedings of the Australian Society of Animal Production* 34, 65A–66A.

Faciola, A.P., Broderick, G.A., Hristov, A.N. and Leão, M.I. (2004) Effect of two forms of lauric acid on ruminal protozoa and fermentation pattern in dairy cows. *Journal of Dairy Science* 87 (Suppl. 1), 53.

Faichney, G.J., Poncet, C., Lassalas, B., Jouany, J.P., Millet, L., Doré, J. and Brownlee, A.G. (1997) Effect of concentrates in a hay diet on the contribution of anaerobic fungi, protozoa, and bacteria to nitrogen in rumen and duodenal digesta in sheep. *Animal Feed Science and Technology* 64, 193–213.

Faichney, G.J., Gordon, G.L.R., Welch, R.J. and Rintoul, A.J. (2002) Effect of dietary free lipid anaerobic fungi and digestion in the rumen of sheep. *Australian Journal of Agricultural Research* 53, 519–527.

Faverdin, P. (1999) The effect of nutrients on feed intake in ruminants. *Proceedings of the Nutrition Society* 58, 523–531.

Feng, P., Hoover, W.H., Miller, T.K. and Blauwiekel, R. (1993) Interactions of fiber and nonstructural carbohydrates on lactation and ruminal function. *Journal of Dairy Science* 76, 1324–1333.

Ffoulkes, D. and Leng R.A. (1988) Dynamics of protozoa in the rumen of cattle. *British Journal of Nutrition* 59, 429–436.

Fievez, V., Danneels, M. and Demeyer, D.I. (2001) Effect of rumen fish oil injection on *in vivo* digestion and fermentation characteristics: differences in animal response. *Proceedings of the Nutrition Society* 60, 75A.

Fievez, V., Dohme, F., Danneels, M., Raes, K. and Demeyer, D.I. (2003) Fish oils as potent rumen methane inhibitors and associated effects on rumen fermentation in vitro and in vivo. *Animal Feed Science and Technology* 104, 41–58.

Firkins, J.L. (1996) Maximizing microbial protein synthesis in the rumen. *Journal of Nutrition* 126, 1347S–1354S.

Firkins, J.L., Weiss, W.P. and Piwonka, E.J. (1992) Quantification of intraruminal recycling of microbial nitrogen using nitrogen-15. *Journal of Animal Science* 70, 3223–3233.

Firkins, J.L., Eastridge, M.L., St-Pierre, N.R. and Noftsger, S.M. (2001) Effects of grain variability and processing on starch utilization by lactating dairy cattle. *Journal of Animal Science* 79, E218–E238.

Fisher, D.S., Mayland, H.F. and Burns, J.C. (1999) Variation in ruminants' preference for tall fescue hays cut either at sundown or at sunup. *Journal of Animal Science* 77, 762–768.

Fisher, D.S., Mayland, H.F. and Burns, J.C. (2002) Variation on ruminant preference for alfalfa hays cut at sunup and sundown. *Crop Science* 42, 231–237.

Foley, A.E., Hristov, A.H., Ropp, J.K., Melgar, A., Etter, R.P., Hunt, C.W. and Huber, K. (2004) Effect of barley and its amylopectin content on ruminal fermentation and nitrogen utilization in lactating dairy cows, *Journal of Dairy Science* 87 (Suppl. 1), 467.

Forsberg, C.W., Lovelock, L.K.A., Krumholz, L. and Buchanan-Smith, J.G. (1984) Protease activities of rumen protozoa. *Applied Environmental Microbiology* 47, 101–110.

Forster, R.J. and Leng, R.A. (1989) The effect on sheep of varying rumen protozoal population and bentonite supplementation in the drinking water. In: Nolan, J.V., Leng R.A. and Demeyer D.I. (eds) *The Roles of Protozoa and Fungi in Ruminant Digestion.* University of New England Publishing Unit, Armidale, Australia, pp. 331–332.

Fu, C.J., Felton, E.E.D., Lehmkuhler, J.W. and Kerley, M.S. (2001) Ruminal peptide concentration to optimize microbial growth and efficiency. *Journal of Animal Science* 79, 1305–1312.

Fujimaki, T., Kobayashi, Y., Wakita, M. and Hoshimo, S. (1992) Amino acid supplements: a least combination that increases microbial yields of washed cell suspension from goat rumen. *Journal of Animal Physiology and Animal Nutrition* 67, 41–50.

Galbraith, H., Miller, T.B., Paton, A.M. and Thompson, J.K. (1971) Antibacterial activity of long chain fatty acids and the reversal with calcium, magnesium, ergocalciferol and cholesterol. *Journal of Applied Bacteriology* 34, 803–813.

Getachew, G., DePeters, E.J., Robinson, P.H. and Taylor, S.J. (2001) In vitro rumen fermentation and gas production: influence of yellow grease, tallow, corn oil and their potassium soaps. *Animal Feed Science and Technology* 93, 1–15.

GfE [Ausschuss für Bedarfsnormen der Gesellschaft für Ernährungsphysiologie] (2001) *Empfehlungen zur Energie- und Nährstoffversorgung der Milchkühe und Aufzuchtrinder.* Energie- und Nährstoffbedarf landwirtschaftlicher Nutztiere, No. 8. DLG-Verlag, Frankfurt am Main, Germany, 135 pp.

Giger-Reverdin, S., Morand-Fehr, P. and Tran, G. (2003) Literature survey of the influence of dietary fat composition on methane production in dairy cattle. *Livestock Production Science* 82, 73–79.

Gomez, L., Jouany, J.P., Hillaire, M.C. and Bogaert, C. (1990) Effect of a new ionophore antibiotic, abierixin, on the digestion of a roughage-based diet measured in sheep; comparison with results obtained in RUSITEC. *Animal Feed Science and Technology* 31, 29–42.

Gomez, L., Bogaert, C., Jouany, J.P. and Lassalas, B. (1991) The influence of lasalocid and cationomycin on nitrogen digestion in sheep: comparison of methods for estimating microbial nitrogen. *Canadian Journal of Animal Science* 71, 389–399.

Gutierrez, J. (1958) Observations on bacterial feeding by the rumen ciliate *Isotricha prostoma*. *Journal of Protozoology* 5, 122–126.

Gutierrez, J. and Hungate, R.E. (1957) Interrelationships between certain bacteria and the rumen ciliate *Dasytricha ruminantium*. *Science* 126, 511.

Ha, J.K., Lee, S.S., Gao, Z., Kim, C.-H., Kim, S.W., Ko, J.Y. and Cheng, K.-J. (2001) The effect of saturated fatty acids on cellulose digestion by the rumen anaerobic fungus, *Neocallimatix frontalis* C5-1. *Asian-Australasian Journal of Animal Science* 14, 941–946.

Haïmoud, D.A., Bayourthe, C., Moncoulon, R. and Vernay, M. (1996) Avoparcin and monensin effects on digestive function in cows fed a high forage diet. *Journal of the Science of Food and Agriculture* 70, 181–189.

Han, C.Y., Lu, D.X., Hu, M. and Tan, Z.L. (1999) Influence of controlling protozoa on the degradation and utilization of dietary fibre and protein in the rumen and nitrogenous flow entering the duodenum of sheep. *Asian-Australasian Journal of Animal Science* 12, 1241–1245.

Harfoot, C.G., Crouchman, M.L., Noble, R.C. and Moore, J.H. (1974) Competition between food particles and rumen bacteria in the uptake of long-chain fatty acids and triglycerides. *Journal of Applied Bacteriology* 37, 633–641.

Harmeyer J. (1971) Der Aminosäurenstoffwechsel isolierter Pansenprotozoenarten (*Isotricha prostoma* und *I. Intestinalis*). Exkretion von Aminossäuren. *Zeitschrift für Tierphysiologie, Tierernährung und Futtermittelkunde* 28, 75–85.

Harrison, D.G., Beever, D.E. and Osbourn, D.F. (1979) The contribution of protozoa to the protein entering the duodenum of sheep. *British Journal of Nutrition* 41, 521–527.

Harvatine, D.I., Firkins, J.L. and Eastridge, M.L. (2002) Whole linted cottonseed as a forage substitute fed with ground or steam-flaked corn: digestibility and performance. *Journal of Dairy Science* 85, 1976–1987.

Hawke, J.C. and Robertson, J.A. (1964) Studies on rumen metabolism. II. In vivo hydrolysis and hydrogenation of lipid. *Journal of the Science of Food and Agriculture* 15, 283–289.

Headon, D.R., Buggle, K., Nelson, A. and Killeen, G. (1991) Glycofractions of the Yucca plant and their role in ammonia control. In: Lyons, T.R. (ed.) *Biotechnology in the Feed Industry, Proceedings of the 7th Alltech Symposium.* Alltech Technical Publications, Nicholasville, Kentucky, pp. 95–108.

Heldt, J.S., Cochran, R.C., Mathis, C.P., Woods, B.C., Olson, K.C., Titgemeyer, E.C., Nagaraja, T.G., Vanzant, E.S. and Johnson, D.E. (1999) Effects of level and source of carbohydrate and level of degradable intake protein on intake and digestion of low-quality tallgrass-prairie hay by beef steers. *Journal of Animal Science* 77, 2846–2854.

Henning, P.H., Steyn, D.G. and Meissner, H.H. (1991) The effect of energy and nitrogen supply pattern on rumen bacterial growth in vitro. *Animal Production* 53, 165–175.

Henning, P.H., Steyn, D.G. and Meissner, H.H. (1993) Effect of synchronization of energy and nitrogen supply on ruminal characteristics and microbial growth. *Journal of Animal Science* 71, 2516–2528.

Herrera-Saldana, R.E., Huber, T.J. and Poore, M.H. (1990) Dry matter, crude protein, and starch degradability of five cereal grains. *Journal of Dairy Science* 73, 2386–2393.

Hespell, R.B. and Bryant, M.P. (1979) Efficiency of rumen microbial growth: influence of some theoretical and experimental factors on Y_{ATP}. *Journal of Animal Science* 49, 1640–1659.

Hino, T. (1981) Action of monensin on rumen protozoa. *Japanese Journal of Zootechnical Science* 52, 171–179.

Hino, T. and Russell, J.B. (1987) Relative contributions of ruminal bacteria and protozoa to the degradation of protein in vitro. *Journal of Animal Science* 64, 261–270.

Hof, G., Tamminga, S. and Lenaers, P.J. (1994) Efficiency of protein utilization in dairy cows. *Livestock Production Science* 38, 169–178.

Holt, D.A. and Hilst, A.R. (1969) Daily variation in carbohydrate content of selected forage crops. *Agronomy Journal* 61, 239–242.

Holthausen, A. (2002) Einfluss der Rohproteinaufnahme von Rindern auf deren N-Bilanzen. Dissertation, Agricultural Faculty, University of Bonn, Germany, 100 pp.

Hoover, W.H. and Stokes, S.R. (1991) Balancing carbohydrates and proteins for optimum rumen microbial yield. *Journal of Dairy Science* 74, 3630–3644.

Hoover, W.H., Miller, T.K., Stokes, S.R. and Thayne, W.V. (1989) Effects of fishmeals on rumen bacterial fermentation in continuous culture. *Journal of Dairy Science* 72, 2991–2998.

Horiguchi, M. and Kandatsu, M. (1959) Isolation of 2-aminoethane phosphonic acid from rumen protozoa. *Nature, London* 184, 901–902.

Houpt, T.R. (1970) Transfer of urea and ammonia to the rumen. In: Phillipson, A.T. (ed.) *Physiology of Digestion and Metabolism in the Ruminant.* Oriel Press, Newcastle upon Tyne, UK, pp. 119–131.

Hristov, A.N. and Broderick, G.A. (1996) Synthesis of microbial protein in ruminally cannulated cows fed alfalfa silage, alfalfa hay or corn silage and fitted with only ruminal cannulae. *Journal of Dairy Science* 79, 1627–1637.

Hristov, A.N. and Ropp, J.K. (2003) Effect of dietary carbohydrate composition and availability on utilization of ruminal ammonia nitrogen for milk protein synthesis in dairy cows. *Journal of Dairy Science* 86, 2416–2427.

Hristov, A.N., McAllister, T.A. and Cheng, K.J. (1997) Effect of carbohydrate level and ammonia availability on utilization of α-amino nitrogen by mixed ruminal microorganisms in vitro. *Proceedings, Western Section, American Society of Animal Science* 48, 186–189.

Hristov, A.N., McAllister, T.A., Van Herk, F.H., Cheng, K.-J., Newbold, C.J. and Cheeke, P.R. (1999) Effect of *Yucca schidigera* on ruminal fermentation and nutrient digestion in heifers. *Journal of Animal Science* 77, 2554–2563.

Hristov, A.N., McAllister, T.A., Olson, M.E., Yanke, L.J. and Shelford, J.A. (2000) Effect of Tween 80 and salinomycin on ruminal fermentation and nutrient digestion in steers fed a diet containing 70% barley. *Canadian Journal of Animal Science* 80, 363–372.

Hristov, A.N., Ropp, J.K. and Hunt, C.W. (2002a) Effect of barley and its amylopectin content on ruminal fermentation and bacterial utilization of ammonia-N in vitro. *Animal Feed Science and Technology* 99, 25–36.

Hristov, A.N., Kennington, L.R., McGuire, M.A., Hunt, C.W. and Ropp, J.K. (2002b) Effect of linoleic or oleic acid-rich oils on ruminal fermentation, nutrient digestibility and performance of finishing cattle. *Journal of Dairy Science* 85 (Suppl. 1), 271.

Hristov, A.N., Ivan, M., Neill, L. and McAllister, T.A. (2003) Evaluation of several potential bioactive agents for reducing protozoal activity in vitro. *Animal Feed Science and Technology* 105, 163–184.

Hristov, A.N., Etter, R.P., Ropp, J.K. and Grandeen, K.L. (2004a) Effect of dietary crude protein level and degradability on ruminal fermentation and nitrogen utilization in lactating dairy cows. *Journal of Animal Science* 82, 3219–3229.

Hristov, A.N., Price, W.J. and Shafii, B. (2004b) A meta-analysis examining the relationship among dietary factors, dry matter intake, and milk yield and milk protein yield in dairy cows. *Journal of Dairy Science* 87, 2184–2196.

Hristov, A.N., Ivan, M. and McAllister, T.A. (2004c) In vitro effects of individual fatty acids on protozoal numbers and on fermentation products in ruminal fluid from cattle fed a high concentrate, barley-based diet. *Journal of Animal Science* 82, 2693–2704.

Hristov, A.N., Grandeen, K.L., Ropp, J.K. and McGuire, M. (2004d) Effect of sodium laurate on ruminal fermentation and utilization of ruminal ammonia nitrogen for milk protein synthesis in dairy cows. *Journal of Dairy Science* 87, 1820–1831.

Hristov, A.N., Ropp, J.K. and Melgar, A. (2004e) Effect of essential oils on ruminal fermentation in vitro. *Journal of Dairy Science* 87 (Suppl. 1), 334.

Hristov, A.N., Ropp, J.K., Grandeen, K.L., Abedi, S., Etter, R.P., Melgar, A. and Foley, A.E. (2005) Effect of carbohydrate source on ruminal fermentation and nitrogen utilization in lactating dairy cows. *Journal of Animal Science* 83 (in press).

Hsu, J.T., Fahey, G.C., Berger, L.L., Mackie, R.I. and Merchen, N.R. (1991) Manipulation of nitrogen digestion by sheep using defaunation and various nitrogen supplementation regimens. *Journal of Animal Science* 69, 1290–1299.

Huhtanen, P. (1987) The effects of intraruminal infusions of sucrose and xylose on nitrogen and fibre digestion in the rumen intestines of cattle receiving diets of grass silage and barley. *Journal of Agricultural Science in Finland* 59, 405–424.

Huhtanen, P. and Robertson, S. (1988) The effect of dietary inclusion of starch, sucrose and xylose on the utilization of dietary energy in sheep. *Animal Feed Science and Technology* 21, 11–21.

Huntington, G.B. (1989) Hepatic urea synthesis and site and rate of urea removal from blood of beef steers fed alfalfa hay or a high concentrate diet. *Canadian Journal of Animal Science* 69, 215–223.

Huntington, G.B. (1997) Starch utilization by ruminants: from basics to the bunk. *Journal of Animal Science* 75, 852–867.

Huntington, G.B. and Archibeque, S.L. (1999) Practical aspects of urea and ammonia metabolism in ruminants. *Proceedings of the American Society of Animal Science*, 1–11.

Hussain, I. and Cheeke, P.R. (1995) Effect of dietary *Yucca schidigera* extract on rumen and blood profiles of steers fed concentrate- or roughage-based diets. *Animal Feed Science and Technology* 51, 231–242.

Ikwuegbu, O.A. and Sutton, J.D. (1982) The effect of varying the amount of linseed oil supplementation on metabolism in sheep. *British Journal of Nutrition* 48, 365–375.

Imai, S. and Ogimoto, K. (1978) Scanning electron and fluorescent microscopic studies on the attachment of spherical bacteria to ciliate protozoa in the ovine rumen. *Japanese Journal of Veterinary Science* 40, 9–19.

Ipharraguerre, I.R. and Clark, J.H. (2003) Usefulness of ionophores for lactating dairy cows: a review. *Animal Feed Science and Technology* 106, 39–57.

Isaacson, H.R., Hinds, F.C., Bryant, M.P. and Owens, F.N. (1975) Efficiency of energy utilization by mixed rumen bacteria in continuous culture. *Journal of Dairy Science* 58, 1645–1659.

Itabashi, H., Kobayashi, T., Morii, R. and Okamoto, S. (1982) Effects of ciliate protozoa on the concentrations of ruminal and duodenal volatile fatty acids and plasma glucose and insulin after feeding. *Bulletin of the National Institute of Animal Industries (Irabaki)* 39, 21–32.

Itabashi, H., Kobayashi, T. and Matsumoto, M. (1984) The effects of rumen ciliate protozoa on energy metabolism and some constituents in rumen fluid and blood plasma of goats. *Japanese Journal of Zootechnical Science* 55, 248–256.

Ivan, M. (1989) The involvement of rumen protozoa in copper metabolism and animal health. In: Nolan, J.V., Leng, R.A. and Demeyer, D.I. (eds) *The Roles of Protozoa and Fungi in Ruminant Digestion.* University of New England Publishing Unit, Armidale, Australia, pp. 199–210.

Ivan, M., Hidiroglou, M. and Petit, H.V. (1991) Duodenal flow of nitrogen following protozoal inoculation of fauna-free sheep fed a diet supplemented with casein or soybean meal. *Canadian Journal of Animal Science* 71, 791–801.

Ivan, M., Neill, L. and Entz, T. (2000) Ruminal fermentation and duodenal flow following progressive inoculations of fauna-free wethers with major individual species of ciliate protozoa or total fauna. *Journal of Animal Science* 78, 750–759.

Ivan, M., Mir, P.S., Koenig, K.M., Rode, L.M., Neill, L., Entz, T. and Mir, Z. (2001) Effects of dietary sunflower seed oil on rumen protozoa population and tissue concentration of conjugated linoleic acid in sheep. *Small Ruminant Research* 41, 215–227.

Jalc, D. and Cerešňáková, Z. (2001) Effect of plant oils and aspartate on rumen fermentation *in vitro*. *Journal of Animal Physiology and Animal Nutrition* 85, 378–384.

Jenkins, T.C. (1993) Lipid metabolism in the rumen. *Journal of Dairy Science* 76, 3851–3863.

Jenkins, T.C. (1999) Lactation performance and fatty acid composition of milk from Holstein cows fed 0 to 5% oleamide. *Journal of Dairy Science* 82, 1525–1531.

Jenkins, T.C. and Fotouhi, N. (1990) Effects of lecithin and corn oil on site of digestion, ruminal fermentation and microbial protein synthesis in sheep. *Journal of Animal Science* 68, 460–466.

Jenkins, T.C. and Palmquist, D.L. (1984) Effect of fatty acids or calcium soaps on rumen and total nutrient digestibility of dairy cows. *Journal of Dairy Science* 67, 978–986.

Johnson, R.R. (1976) Influence of carbohydrate solubility on non-protein nitrogen utilization in the ruminant. *Journal of Animal Science* 43, 184–191.

Jouany, J.P. (1978) Contribution à l'étude des protozoaires ciliés du rumen: leur dynamique, leur rôle dans la digestion et leur intérêt pour le ruminant. PhD thesis, University of Clermont II, France 195 + 152 pp. (2 volumes).

Jouany, J.P. (1989) Effect of diet on populations of rumen protozoa in relation to fibre digestion. In: Nolan, J.V., Leng, R.A. and Demeyer, D.I. (eds) *The Roles of Protozoa and Fungi in Ruminant Digestion.* University of New England Publishing Unit, Armidale, Australia, pp. 59–74.

Jouany, J.P. (1996) Effect of rumen protozoa on nitrogen utilization by ruminants. *Journal of Nutrition* 126, 1335S–1346S.

Jouany, J.P. and Thivend, P. (1983) Influence of protozoa on nitrogen digestion in the ruminant. In: *Fourth Symposium on Protein Metabolism and Nutrition, Clermont-Ferrand, France,* Vol. 2. Versailles, France, Les Colloques de l'INRA. INRA Editions, France, pp. 287–290.

Jouany, J.P. and Ushida, K. (1999) The role of protozoa in feed digestion–review. *Asian-Australasian Journal of Animal Science* 12, 113–128.

Jouany, J.P., Demeyer, D.I. and Grain, J. (1988) Effect of defaunating the rumen. *Animal Feed Science and Technology* 21, 229–265.

Jouany, J.P., Ivan, M., Papon, Y. and Lassalas, B. (1992) Effects of *Isotricha, Eudiplodinium, Epidinium + Entodinium* and a mixed population of rumen protozoa on the *in vitro* degradation of fishmeal, soybean meal and casein. *Canadian Journal of Animal Science* 72, 871–880.

Jouany, J.P., Mathieu, F., Senaud, J., Bohatier, J., Bertin, G. and Mercier, M. (1999) Influence of protozoa and fungal additives on ruminal pH and redox potential. *South African Journal of Animal Science* 29, 64–66.

Kalscheur, K.F., Teter, B.B., Piperova, L.S. and Erdman, R.A. (1997) Effect of fat source on duodenal flow of *trans*-C18:1 fatty acids and milk fat production in dairy cows. *Journal of Dairy Science* 80, 2115–2126.

Kaur, G., Singh, S. and Sareen, V.K. (1992) Effect of dietary amino acids on *in vitro* rumen bacterial protein synthesis in buffaloes. *Acta Veterineria Hungarica* 40, 33–38.

Kayouli, C., Van Nevel, C.J., Dendooven, R. and Demeyer, D.I. (1986) Effect of defaunation and refaunation of the rumen on rumen fermentation and N-flow in the duodenum of sheep. *Archiv für Tierernährung* 36, 827–838.

Kebreab, E., France, J., Beever, D.E. and Castillo, A.R. (2001) Nitrogen pollution by dairy cows and its mitigation by dietary manipulation. *Nutrient Cycling in Agroecosystems* 60, 275–285.

Kebreab, E., France, J., Mills, J.A.N., Allison, R. and Dijkstra, J. (2002) A dynamic model of N metabolism in the lactating dairy cow and an assessment of impact of N excretion on the environment. *Journal of Animal Science* 80, 248–259.

Kennelly, J.J., Robinson, B. and Khorasani, G.R. (1999) Influence of carbohydrate source and buffer on rumen fermentation characteristics, milk yield, and milk composition in early-lactation of Holstein cows. *Journal of Dairy Science* 82, 2486–2496.

Kerchner, M., Halloran, R., Thomas, J. and Hicks, B. (2000) Airsheds and watersheds. III. A shared resources workshop. The significance of ammonia to coastal and estuarine areas. Nov. 15–16, 2000. The BayCenter at Ruddertowne Dewey Beach, Delaware. [Available at www.chesapeakebay.net]

Khalili, H. and Huhtanen, P. (1991) Sucrose supplement in cattle given grass silage-based diet. 1. Digestion of organic matter and nitrogen. *Animal Feed Science and Technology* 33, 247–261.

Khorasani, G.R., Okine, E.K. and Kennelly, J.J. (2001a) Effects of substituting barley grain with corn on ruminal fermentation characteristics, milk yield, and milk composition of Holstein cows. *Journal of Dairy Science* 84, 2760–2769.

Khorasani, G.R., Okine, E.K. and Kennelly, J.J. (2001b) Effects of forage source and amount of concentrate on rumen and intestinal digestion of nutrients in late-lactation cows. *Journal of Dairy Science* 84, 1156–1165.

Kil, J.Y., Cho, N.K., Kim, B.S., Lee, S.R. and Maeng, W.J. (1994) Effects of yucca extract addition on the in vitro fermentation characteristics of feed and feces, and on the milk yields in lactating cows. *Korean Journal of Animal Science* 36, 698–709.

Kim, K.H., Choung, J.J. and Chamberlain, D.G. (1999a) Effects of varying the degree of synchrony of energy and nitrogen release in the rumen on the synthesis of microbial protein in lactating dairy cows consuming a diet of grass silage and a cereal-based concentrate. *Journal of the Science of Food and Agriculture* 79, 1441–1447.

Kim, K.H., Oh, Y.-G., Choung, J.-J. and Chamberlain, D.G. (1999b) Effects of varying degrees of synchrony of energy and nitrogen release in the rumen on the synthesis of microbial protein in cattle consuming grass silage. *Journal of the Science of Food and Agriculture* 79, 833–838.

Knowlton, K.F., Glenn, B.P. and Erdman, R.A. (1998) Performance, ruminal fermentation, and site of starch digestion in early lactation cows fed corn grain harvested and processed differently. *Journal of Dairy Science* 81, 1972–1984.

Kobayashi, T. and Itabashi, H. (1986) Effect of intraruminal VFA infusion on the protozoal populations of the rumen. *National Institute of Animal Industry* 44, 47–54.

Kobayashi, Y., Wakita, M., Sakauchi, R. and Hoshino, S. (1990) Effects of ionophores on rumen microbes and host animal nutrition. In: Hoshino, S., Onodera, R., Minato, H. and Itabashi, H. (eds) *The Rumen Ecosystem: The Microbial Metabolism and its Regulation.* Japan Scientific Society Press, Tokyo, pp. 179–186.

Kobayashi, Y., Wakita, M. and Hoshino, S. (1992) Effects of the ionophore salinomycin on nitrogen and long-chain fatty acid profiles of digesta in the rumen and the duodenum of sheep. *Animal Feed Science and Technology* 36, 67–76.

Koenig, K.M., Newbold, C.J., McIntosh, F.M. and Rode, L.M. (2000) Effects of protozoa on bacterial nitrogen recycling in the rumen. *Journal of Animal Science* 78, 2431–2445.

Komaragiri, M.V.S and Erdman, R.A. (1997) Factors affecting body tissue mobilization in early lactation dairy cows. 1. Effect of dietary protein on mobilization of body fat and protein. *Journal of Dairy Science* 80, 929–937.

Kozelov, L., Iliev, F., Profirov, J., Nikolov, I.S., Ganev, G., Modeva, T. and Krasteva, M. (2001) The effect of supplementing sheep with Ropadiar on digestibility and fermentation in the rumen. *Zhivotnov"dni Nauki* 38, 152–154.

Kreuzer, M. and Kirchgessner, M. (1986) Die Bedeutung der Protozoen in Pansen von Hammeln für Verdaulichkeit, N-Bilanz und N-Fraktionen in Kot und Harn bei unterschiedlichen Gehalten und Arten von Stärke in der Ration. *Journal of Animal Physiology and Animal Nutrition* 56, 217–231.

Kreuzer, M. and Kirchgessner, M. (1989) Effect of rumen protozoa on metabolism and retention in ruminants, with special reference to diet characteristics. In: Nolan, J.V., Leng, R.A. and Demeyer, D.I. (eds) *The roles of Protozoa and Fungi in Ruminant Digestion.* University of New England Publishing Unit, Armidale, Australia, pp. 189–197.

Krumholz, L.R., Forsberg, C.W. and Veira, D.M. (1983) Association of methanogenic bacteria with rumen protozoa. *Canadian Journal of Microbiology* 29, 676–680.

Külling, D.R., Menzi, H., Kröber, T.F., Neftel, A., Sutter, F., Lischer, P. and Kreuzer, M. (2001) Emissions of ammonia, nitrous oxide and methane from different types of dairy manure during storage as affected by dietary protein content. *Journal of Agricultural Science, Cambridge* 137, 235–250.

Kurihara, Y., Eadie, J.M., Hobson, P.N. and Mann, S.O. (1968) Relationship between bacteria and ciliate protozoa in the sheep rumen. *Journal of General Microbiology* 51, 267–288.

Kurihara, Y., Takechi, T. and Shibata, F. (1978) Relationship between bacteria and ciliate protozoa in the rumen of sheep fed on a purified diet. *Journal of Agricultural Science, Cambridge* 90, 373–381.

Lechtenberg, V.L., Holt, D.A. and Youngberg, H.W. (1971) Diurnal variation in non-structural carbohydrates, *in vitro* digestibility, and leaf to stem ratio of alfalfa. *Agronomy Journal* 63, 719–724.

Lee, M.R.F., Harris, L.J., Moorby, J.M., Humphreys, M.O., Theodorou, M.K., MacRae, J.C. and Scollan, N.D. (2002) Rumen metabolism and nitrogen flow to the small intestine in steers offered *Lolium perenne* containing different levels of water-soluble carbohydrate. *Animal Science* 74, 587–596.

Lee, M.R.F., Merry, R.J., Davies, D.R., Moorby, J.M., Humphreys, M.O., Theodorou, M.K., MacRae, J.C. and Scollan, N.D. (2003) Effect of increasing availability of water-soluble carbohydrates on in vitro rumen fermentation. *Animal Feed Science and Technology* 104, 59–70.

Leiva, E., Hall, M.B. and Van Horn, H.H. (2000) Performance of dairy cattle fed citrus pulp or corn products as sources of neutral detergent-soluble carbohydrates. *Journal of Dairy Science* 83, 2866–2875.

Leng, R.A. (1982) Dynamics of protozoa in the rumen of sheep. *British Journal of Nutrition* 48, 399–415.

Leng, R.A. (1989) The dynamics of protozoa in the rumen. In: Nolan, J.V., Leng, R.A. and Demeyer, D.I. (eds) *The Roles of Protozoa and Fungi in Ruminant Digestion.* University of New England Publishing Unit, Armidale, Australia, pp. 51–58.

Leng, R.A. and Nolan, J.V. (1984) Nitrogen metabolism in the rumen. *Journal of Dairy Science* 67, 1072–1089.

Leng, R.A., Gill, M., Kempton, T.J., Rowe, J.B., Nolan, J.V., Stachiw, S.J. and Preston, T.R. (1981) Kinetics of large ciliate protozoa in the rumen of cattle given sugar cane diets. *British Journal of Nutrition* 46, 371–384.

Lindsay, J.R. and Hogan, J. (1972) Digestion of two legumes and rumen bacterial growth in defaunated sheep. *Australian Journal of Agricultural Research* 23, 321–330.

Lobley, G.E., Connell, A., Lomax, M.A., Brown, D.S., Milne, E., Calder, A.G. and Farningham. D.A.H. (1995) Hepatic detoxification of ammonia in the ovine liver: possible consequences for amino acid catabolism. *British Journal of Nutrition* 73, 667–685.

Lockwood, B., Coombs, G.H. and Williams, A.G. (1988) Proteinase activity in rumen ciliate protozoa. *Journal of General Microbiology* 134, 2605–2614.

Lockwood, B., North, M.J., Williams, A.G. and Coombs, G.H. (1989) Proteinases of anaerobic protozoa: trichomonads and rumen ciliates. In: Lloyd, D., Coombs, G.H. and Paget, T.A. (eds) *Biochemistry and Molecular Biology of Anaerobic Protozoa.* Harwood Academic Press, Chur, Switzerland, pp. 66–77.

Losa, R. and Brufau, J. (2001) The use of essential oils in animal nutrition. *Cahiers Options Mediterraneennes* 54, 39–44.

Luther, R., Trenkle, A. and Burroughs, W. (1966) Influence of rumen protozoa on volatile acid production and ration digestibility in lambs. *Journal of Animal Science* 25, 1116–1122.

Machmüller, A. and Kreuzer, M. (1999) Methane suppression by coconut oil and associated effects on nutrient and energy balance in sheep. *Canadian Journal of Animal Science* 79, 65–72.

Machmüller, A., Ossowski, D.A. and Kreuzer, M. (2000) Comparative evaluation of the effects of coconut oil, oilseeds and crystalline fat on methane release, digestion and energy balance in lambs. *Animal Feed Science and Technology* 85, 41–60.

MacRae, J.C., Lobley, G.E., Bequette, B.J. and Backwell, F.R.C. (1995) Protein metabolism in specific organs. In: Nunes, A.F., Portugal, A.V., Costa, J.P. and Robeiro, J.R. (eds) *Proceedings of the 7th EAAP Symposium on Protein Metabolism and Nutrition.* EAAP Publication No. 81 Estacao Zootechnica Nacional, J. Vale de Santarem, Portugal, pp. 297–308.

Maeng, W.J. and Baldwin, R.L. (1976) Factors influencing rumen microbial growth rates and yields: effects of amino acid additions to a purified diet with nitrogen from urea. *Journal of Dairy Science* 59, 648–655.

Makkar, H.P.S. (2003) Effects and fate of tannins in ruminant animals, adaptation to tannins, and strategies to overcome detrimental effects of feeding tannin-rich feeds. *Small Ruminant Research* 49, 241–256.

Makkar, H.P.S., Sen, S., Blümmel, M. and Becker, K. (1998) Effects of fractions containing saponins from *Yucca schidigera*, *Quillaja saponaria*, and *Accacia auriculoformis* on rumen fermentation. *Journal of Agricultural and Food Chemistry* 46, 4324–4328.

Martin, B., Coulon, J.B., Chamba, J.F. and Bugaud, C. (1997) Effect of milk urea content on characteristics of matured Reblochon cheese. *Lait* 77, 505–514.

Martin, S.A., Sullivan, H.M. and Evans, J.D. (2000) Effect of sugars and malate on ruminal microorganisms. *Journal of Dairy Science* 83, 2574–2579.

Martin-Orue, S.M., Balcells, J., Vincente, F. and Castrillo, C. (2000) Influence of dietary rumen-degradable protein supply on rumen characteristics and carbohydrate fermentation in beef cattle offered high-grain diets. *Animal Feed Science and Technology* 88, 59–77.

Mathison, G.W. and Milligan L.P. (1971) Nitrogen metabolism in sheep. *British Journal of Nutrition* 25, 351–366.

Matsumoto, M., Kobayashi, T., Takenaka, A. and Itabashi, H. (1991) Defaunation effects of medium-chain fatty acids and their derivatives on goat rumen protozoa. *The Journal of General and Applied Microbiology* 37, 439–445.

Mayland, H.F., Shewmaker, G.E., Harrison, P.A. and Chatterton, N.J. (2000) Nonstructural carbohydrates in tall fescue cultivars: relationship to animal preference. *Agronomy Journal* 92, 1203–1206.

McAllister, T.A., Moustafa, S.M.S., Cheng, K.-J., Newbold, C.J., McKain, N. and Wallace, R.J. (1994) Effect of salinomycin on fermentation and nitrogen metabolism in the artificial rumen. *Canadian Journal of Animal Science* 74, 575–578.

McCormick, M.E., Redfearn, D.D., Ward, J.D. and Blouin, D.C. (2001) Effect of protein source and soluble carbohydrate addition on rumen fermentation and lactation performance of Holstein cows. *Journal of Dairy Science* 84, 1686–1697.

McEwan N.R., Graham R.C., Wallace R.J., Losa R., Williams P. and Newbold C.J. (2002a) Effect of essential oils on ammonia production by rumen microbes. *Reproduction, Nutrition, Development* 42 (Suppl. 1), S65.

McEwan N.R., Graham R.C., Wallace R.J., Losa R., Williams P. and Newbold C.J. (2002b) Effect of essential oils on protein digestion in the rumen. *Reproduction, Nutrition, Development* 42 (Suppl. 1), S65–S66.

McGinn, S.M., Koenig, K.M. and Coates, T. (2002) Effect of diet on odorant emissions from cattle manure. *Canadian Journal of Animal Science* 82, 435–444.

McGuffey, R.K., Richardson, L.F. and Wilkinson, J.I.D. (2001) Ionophores for dairy cattle: current status and future outlook. *Journal of Dairy Science* 84 (E. Suppl.), E194–E203.

McIntosh, F.M., Stanley, K.N., Stewart, C.S. and Newbold, C.J. (2000) The role of protozoa in the survival of *E. coli*, Verocytotoxin encoding bacteriophage and their VTEC lysogens in the rumen. *Reproduction, Nutrition, Development* 42 (Suppl. 1), S34.

McIntosh, F.M., Williams, P., Losa, R., Wallace, R.J., Beever, D.E. and Newbold, C.J. (2003) Effects of essential oils on ruminal metabolism and their protein metabolism. *Applied and Environmental Microbiology* 69, 5011–5014.

McNaught, M.L., Owen, E.C., Henry, K.M. and Kon, S.K. (1954) The utilization of non protein nitrogen in the bovine rumen. 8. The nutritive value of the proteins of preparations of dried rumen bacteria, rumen protozoa and brewers yeast for rats. *Biochemical Journal* 56, 151–156.

Metcalf, J.A., Wray-Cahen, D., Chettle, E.E., Sutton, J.D., Beever, D.E., Crompton, L.A., MacRae, J.C., Bequette, B.J. and Backwell, F.R.C. (1996) The effect of dietary crude protein as protected soybean meal on mammary metabolism in the lactating dairy cow. *Journal of Dairy Science* 79, 603–611.

Meyer, J.H.F., Van der Walt, S.I. and Schwartz, H.M. (1986) The influence of diet and protozoal numbers on the breakdown and synthesis of protein in the rumen of sheep. *Journal of Animal Science* 62, 509–520.

Michalowski, T. (1989) Importance of protein solubility and nature of dietary nitrogen for the growth of rumen ciliates *in vitro*. In: Nolan, J.V., Leng, R.A. and Demeyer, D.I. (eds) *The Roles of Protozoa and Fungi in Ruminant Digestion*. University of New England Publishing Unit, Armidale, Australia, pp. 223–232.

Miller, L.A., Moorby, J.M., Davies, D.R., Humphreys, M.O., Scollan, N.D., MacRae, J.C. and Theodorou, M.K. (2001) Increased concentration of water-soluble carbohydrate in perennial ryegrass (*Lolium perenne* L.): milk production from late-lactation dairy cows. *Grass and Forage Science* 56, 383–394.

Mills, R.C., Lardinois, C.C., Rupel, I.W. and Hart, E.B. (1944) Utilization of urea and growth of heifer calves with corn molasses or cane molasses as the only readily available carbohydrate in the ration. *Journal of Dairy Science* 27, 571–578.

Molero, R., Ibars, M., Calsamiglia, S., Ferret, A. and Losa, R. (2004) Effects of a specific blend of essential oil compounds on dry matter and crude protein degradability in heifers fed diets with different forage to concentrate ratios. *Animal Feed Science and Technology* 114, 91–104.

Monteny, G.J. and Erisman, J.W. (1998) Ammonia emission from dairy cow buildings: a review of measurement techniques, influencing factors and possibilities for reduction. *Netherlands Journal of Agricultural Science* 46, 225–247.

Monteny, G.J., Smits, M.C.J., Van Duinkerken, G., Mollenhorst, H and De Boer, I.J.M. (2002) Prediction of ammonia emission from dairy barns using feed characteristics part II: Relation between urinary urea concentration and ammonia emission. *Journal of Dairy Science* 85, 3389–3394.

Moorby, J.M. and Theobald, V.J. (1999) Short communication: the effect of duodenal ammonia infusions on milk production and nitrogen balance of the dairy cows. *Journal of Dairy Science* 82, 2440–2442.

Muetzel, S., Hoffmann, E. and Becker, K. (2000) In vitro supplementation with *Sesbania* leaves enhances the growth of ruminal bacteria and depresses eukaryotic populations. *Reproduction, Nutrition, Development* 40, 221.

Murdiati, T.B., McSweeney, C.S. and Lowry, J.B. (1992) Metabolism in sheep of gallic acid, tannic acid and hydrolysable tannin from *Terminalia oblongata*. *Australian Journal of Agricultural Research* 43, 1307–1319.

Muszynski, P., Michalowski, T. and Szczepkowski, P. (1985) The influence of protein on the number of rumen ciliates *Entodinium caudatum* and *Diplodinium affine* in vitro. *Acta Protostologica* 24, 307–317.

Nagaraja, T.G., Newbold, C.J., Van Nevel, C.J. and Demeyer, D.I. (1997) Manipulation of ruminal fermentation. In: Hobson, P.N. and Steward, C.S. (eds) *The Rumen Microbial Ecosystem*. Blackie Academic and Professional, London, pp. 523–623.

Newbold, C.J. and Chamberlain, D.G. (1988) Lipids as rumen-defaunating agents. *Proceedings of the Nutrition Society* 47, 154A.

Newbold, J.R. and Rust, S.R. (1992) Effect of asynchronous nitrogen and energy supply on growth of ruminal bacteria in batch culture. *Journal of Animal Science* 70, 538–546.

Newbold, C.J., Chamberlain, D.G. and Williams, A.G. (1986) The effects of defaunation on the metabolism of lactic acid in the rumen. *Journal of the Science of Food and Agriculture* 37, 1083–1090.

Newbold, C.J., El Hassan, S.M., Wang, J., Ortega, M.E. and Wallace, R.J. (1997) Influence of foliage from African multipurpose trees on activity of rumen protozoa and bacteria. *British Journal of Nutrition* 78, 237–249.

Newbold, C.J., Teferedegne, B., Kim, H.S., Zuur, G. and Lobley, G.E. (2000) Effects of protozoa on nitrogen metabolism in the rumen of sheep. *Reproduction, Nutrition, Development* 40, 199.

Newbold, C.J., McIntosh, F.M., Williams, P., Losa, R. and Wallace, R.J. (2004) Effects of a specific blend of essential oil compounds on rumen fermentation. *Animal Feed Science and Technology* 114, 105–112.

Nielsen, N.M., Kristensen, T., Nørgaard, P. and Hansen, H. (2003) The effect of low protein supplementation to dairy cows grazing clover grass during half of the day. *Livestock Production Science* 81, 293–306.

Ningrat, R.W.S., Garnsworthy, P.C. and Newbold, C.J. (2002) Saponin fractions in *Sapindus rarak*: effects on rumen microbes. *Reproduction, Nutrition, Development* 42, S82–S83.

NKJ Protein Group (1985) Introduction of the Nordic Protein Evaluation System for ruminants into practice and future research requirements. Proposals by the NKJ protein group. *Acta Agriculturae Scandinavica* 25 (Suppl.), 216–220.

Nocek, J.E. and Russell, J.B. (1988) Protein and energy as an integrated system. Relationship of ruminal protein and carbohydrate availability to microbial synthesis and milk production. *Journal of Dairy Science* 71, 2070–2107.

Nolan, J.V. and Leng, R.A. (1972) Dynamic aspects of ammonia and urea metabolism in sheep. *British Journal of Nutrition* 27, 177–194.

Nolan, J.V. and Stachiw, S. (1979) Fermentation and nitrogen dynamics in Merino sheep given a low-quality-roughage diet. *British Journal of Nutrition* 42, 63–80.

NRC [National Research Council] (2001). *Nutrient Requirements of Dairy Cattle,* 7th rev. edn. National Academy Press, Washington, DC, 381 pp.

Oh, H.K., Sakai, T., Jones, M.B. and Longhurst, W.M. (1967) Effect of various essential oils isolated from Douglas fir needles upon sheep and deer rumen microbial activity. *Applied Microbiology* 15, 777–784.

Oh, H.K., Jones, M.B. and Longhurst, W.M. (1968) Comparison of rumen microbial inhibition resulting from various essential oils isolated from relatively unpalatable plant species. *Applied Microbiology* 16, 39–44.

Oh, Y.-G., Kim, K.H., Kim, J.-H., Choung, J.-J. and Chamberlain, D.G. (1999) The effect of the form of nitrogen in the diet on ruminal fermentation and the yield of microbial protein in sheep consuming diets of grass silage supplemented with starch or sucrose. *Animal Feed Science and Technology* 78, 227–237.

Ohajuruka, O.A., Wu, Z. and Palmquist, D.L. (1991) Ruminal metabolism, fiber, and protein digestion by lactating cows fed calcium soap or animal-vegetable fat. *Journal of Dairy Science* 74, 2601–2609.

Oldham, J.D., Bruckental, I. and Nissenbaum, A. (1980) Observations on rumen ammonia metabolism in lactating dairy cows. *Journal of Agricultural Science Cambridge* 95, 235–238.

Oldick, B.S. and Firkins, J.L. (2000) Effects of degree of fat saturation on fiber digestion and microbial protein synthesis when diets are fed twelve times daily. *Journal of Animal Science* 78, 2412–2420.

Oldick, B.S., Firkins J.L. and Kohn R.A. (2000) Compartmental modelling with nitrogen-15 to determine effects of degree of fat saturation on intraruminal N recycling. *Journal of Animal Science* 78, 2421–2430.

Onetti, S.G., Shaver, R.D., McGuire, M.A. and Grummer, R.R. (2001) Effect of type and level of dietary fat on rumen fermentation and performance of dairy cows fed corn silage-based diets. *Journal of Dairy Science* 84, 2751–2759.

Onodera, R. and Kandatsu, M. (1970) Amino acid and protein metabolism of rumen ciliate protozoa. 4. Metabolism of casein. *Japanese Journal of Zootechnical Science* 41, 307–313.

Onodera, R. and Yakiyama, M. (1990) Coagulation and degradation of casein by entodiniomorphid and holotrich rumen protozoa. *Japanese Journal of Zootechnical Science* 61, 533–539.

Opstvedt, J. (1984) Fish fats. In: Wiseman, J. (ed.) *Fats in Animal Nutrition*. Proceedings of the 37th Nottingham Easter School. Butterworths, Boston, Massachusetts, pp. 53–82.

Ordway, R.S., Ishler, V.A. and Varga, G.A. (2002) Effects of sucrose supplementation on dry matter intake, milk yield, and blood metabolites of periparturient Holstein dairy cows. *Journal of Dairy Science* 85, 879–888.

Orpin, C.G. (1977) Studies on the defaunation of the ovine rumen using dioctyl sodium sulphosuccinate. *Journal of Applied Bacteriology* 43, 309–318.

Ørskov, E.R., Fraser, C. and McDonald, I. (1971) Digestion of concentrates in sheep. 3. Effects of rumen fermentation of barley and maize on protein digestion. *British Journal of Nutrition* 26, 477–486.

Osborne, V.R., Leslie, K.E. and McBride, B.W. (2002a) Effect of supplementing glucose in drinking water on the energy and nitrogen status of the transition dairy cows. *Canadian Journal of Animal Science* 82, 427–433.

Osborne, V.R., McBride, B.W. and Hacker, R.R. (2002b) The effect of additional lighting and glucose supplemented drinking water on the performance of dairy calves. *Journal of Dairy Science* 85 (Suppl. 1), 229.

Overton, T.R., Cameron, M.R., Elliott, J.P., Clark, J.H. and Nelson, D.R. (1995) Ruminal fermentation and passage of nutrients to the duodenum of lactating cows fed mixtures of corn and barley. *Journal of Dairy Science* 78, 1981–1998.

Owen, R.W. and Coleman, G.S. (1977) The uptake and utilisation of bacteria, amino acids and carbohydrates by the rumen ciliate *Entodinium longinucleatum* in relation to the source of amino acids for protein synthesis. *Journal of Applied Bacteriology* 43, 67–74.

Owens, F.N. (1980) Ionophore effect on utilization and metabolism of nutrients – ruminants. In: *Proceedings of the Georgia Nutritional Conference.* University of Georgia, Athens, Georgia, p. 17.

Owens, F.N. and Goetsch, A.L. (1986) Digesta passage and microbial protein synthesis. In: Milligan, L.P., Grovum, W.L. and Dobson, A. (eds) *Control of Digestion and Metabolism in Ruminants.* Prentice-Hall, Englewood Cliffs, New Jersey, pp. 196–223.

Owens, F.N., Secrist, D.S., Hill, W.J. and Gill, D.R. (1997) The effect of grain source and grain processing on performance of feedlot cattle: a review. *Journal of Animal Science* 75, 868–879.

Owens, V.N., Albrecht, K.A. and Muck, R.E. (2002) Protein degradation and fermentation characteristics of unwilted red clover and alfalfa silage harvested at various times during the day. *Grass and Forage Science* 57, 329–341.

Pantoja, J., Firkins, J.L., Eastridge, M.L. and Hull, B.L. (1994) Effects of fat saturation and source of fiber on site of nutrient digestion and milk production by lactating dairy cows. *Journal of Dairy Science* 77, 2341–2356.

Poncet, C., Michalet-Doreau, B., McAllister, T. and Rémond, D. (1995) Dietary compounds escaping rumen digestion. In: Journet, M., Grenet, E., Farce, M.-H., Theriez, M. and Demarquilly, C. (eds) *Recent Developments in the Nutrition of Herbivores.* INRA Editions, Versailles, France, pp. 167–204.

Powers, W.J., Van Horn, H.H., Harris, B. and Wilcox, C.J. (1995) Effects of variable sources of distillers dried grains plus solubles on milk yield and composition. *Journal of Dairy Science* 78, 388–396.

Preston, T.R. and Leng, R.A. (1986) *Matching Livestock Production Systems to Available Resources.* International Livestock Centre for Africa, Addis Ababa, Ethiopia, pp. 40–42.

Prins, R.A. and van Hoven, W. (1977) Carbohydrate fermentation by the rumen ciliate *Isotricha prostoma. Protistologica* 13, 549–556.

Punia, B.S., Leibholz, J. and Faichney, G.J. (1987) The role of rumen protozoa in the utilization of *paspalum* (*Paspalum dilatatum*) hay by cattle. *British Journal of Nutrition* 57, 395–406.

Punia, B.S., Leibholz, J. and Faichney, G.J. (1992) Rate of production of protozoa in the rumen and the flow of protozoal nitrogen to the duodenum in sheep and cattle given a pelleted diet of lucerne hay and barley. *Journal of Agricultural Science, Cambridge* 118, 229–236.

Ranilla, M.J., Carro, M.D., López, S., Valdés, C., Newbold, C.J. and Wallace, R.J. (2000a) ^{15}N-ammonia incorporation by ruminal bacteria growing on different N sources. *Reproduction, Nutrition, Development* 40, 199 (Abstract).

Ranilla, M.J., Carro, M.D., López, S., Valdés, C., Newbold, C.J. and Wallace, R.J. (2000b) Effect of N source on in vitro microbial growth. *Reproduction, Nutrition, Development* 40, 199 (Abstract).

Ranilla, M.J., Carro, M.D., López, S., Newbold, C.J. and Wallace, R.J. (2001) Influence of nitrogen source on the fermentation of fibre from barley straw and sugar beet pulp by ruminal micro-organisms *in vitro. British Journal of Nutrition* 86, 717–724.

Remond, D., Nozière, P. and Poncet, C. (2002) Effect of time of starch supply to the rumen on the dynamics of urea and ammonia net flux across the rumen wall of sheep. *Animal Research* 51, 3–13.

Reynal, S.M. and Broderick, G.A. (2003) Effects of feeding dairy cows protein supplements of varying ruminal degradability. *Journal of Dairy Science* 86, 835–843.

Richardson, J.M., Wilkison, R.G. and Sinclair, L.A. (2003) Synchrony of nutrient supply to the rumen and dietary energy source and their effects on the growth and metabolism of lambs. *Journal of Animal Science* 81, 1332–1347.

Robinson, P.H. and McNiven, M.A. (1994) Influence of flame roasting and feeding frequency of barley on performance of dairy cows. *Journal of Dairy Science* 77, 3631–3643.

Rooke, J.A. and Armstrong, D.G. (1989) The importance of the form of nitrogen on microbial protein synthesis in the rumen of cattle receiving grass silage and continuous intrarumen infusions of sucrose. *British Journal of Nutrition* 61, 113–121.

Rooke, J.A., Lee, N.H. and Armstrong, D.G. (1987) The effects of intraruminal infusion of urea, casein, glucose syrup and a mixture of casein and glucose syrup on nitrogen digestion in the rumen of cattle receiving grass-silage diets. *British Journal of Nutrition* 57, 89–98.

Rowe, J.B., Davies, A. and Broome, A.W.J. (1985) Quantitative effects of defaunation on rumen fermentation and digestion in sheep. *British Journal of Nutrition* 54, 105–119.

Ruiz, R., Albrecht, G.L., Tedeschi, L.O., Jarvis, G., Russell, J.B. and Fox, D.G. (2001) Effect of monensin on the performance and nitrogen utilization of lactating dairy cows consuming fresh forage. *Journal of Dairy Science* 84, 1717–1727.

Russell, J.B., Sniffen, C.J. and Van Soest, P.J. (1983) Effect of carbohydrate limitation on degradation and utilization of casein by mixed rumen bacteria. *Journal of Dairy Science* 66, 763–775.

Russell, J.B., O'Connor, J.D., Fox, D.G., Van Soest, P.J. and Sniffen, C.J. (1992) A net carbohydrate and protein system for evaluating cattle diets. I. Ruminal fermentation. *Journal of Animal Science* 70, 3551–3561.

Russi, J.P., Wallace, R.J. and Newbold, C.J. (2002) Influence of the pattern of peptide supply on microbial activity in the rumen simulating fermenter (RUSITEC). *British Journal of Nutrition* 88, 73–80.

Sannes, R.A., Messman, M.A. and Vagnoni, D.B. (2002) Form of rumen-degradable carbohydrate and nitrogen on microbial protein synthesis and protein efficiency of dairy cows. *Journal of Dairy Science* 85, 900–908.

Santos, F.A.P., Santos, J.E.P., Theurer, C.B. and Huber, J.T. (1998) Effects of rumen-undegradable protein on dairy cow performance: a 12-year literature review. *Journal of Dairy Science* 81, 3182–3213.

Santos, G.T., Oliveira, R.L., Petit, H.V., Cecato, U., Zeoula, L.M., Rogolon, L.P., Damasceno, J.C., Branco, A.F. and Bett, V. (2000) Effect of tannic acid on composition and ruminal degradability of bermudagrass and alfalfa silages. *Journal of Dairy Science* 83, 2016–2020.

Satter, L.D., Klopfenstein, T.J. and Erickson, G.E. (2002) The role of nutrition in reducing nutrient output from ruminants. *Journal of Animal Science* 80 (E. Suppl. 2), E143–E156.

Sauvant, D. and van Milgen, J. (1995) Dynamic aspects of carbohydrate and protein breakdown and the associated microbial matter synthesis. In: Engelhardt, W.V., Leonhard-Marek, S., Breves, G. and Giesecke, D. (eds) *Ruminant Physiology: Digestion, Metabolism, Growth and Reproduction.* Ferdinand Enke Verlag, Stuttgart, Germany, pp. 71–91.

Scollan, N.D., Dhanoa, M.S., Choi, N.J., Maeng, W.J., Enser, M. and Wood, J.D. (2001) Biohydrogenation and digestion of long chain fatty acids in steers fed on different sources of lipid. *Journal of Agricultural Science, Cambridge* 136, 345–355.

Shabi, Z., Arieli, A., Bruckental, I., Aharoni, Y., Zamwel, S., Bor, A. and Tagari, H. (1998) Effect of the synchronization of the degradation of dietary crude protein and organic matter and feeding frequency on ruminal fermentation and flow of digesta in the abomasum of dairy cows. *Journal of Dairy Science* 81, 1991–2000.

Shinchi, S. and Kandatsu, M. (1981) On some properties of extracellular and intracellular proteolytic activity of rumen ciliate protozoa. *Japanese Journal of Zootechnical Science* 52, 861–868.

Shiroma, S. and Akashi, A. (1976) Degradation of mimosine in *Leucaena leucocephala* de Wit by goat rumen microorganisms. *Japanese Journal of Zootechnical Science* 47, 739–747.

Siddons, R.C. and Paradine, J. (1981) Effect of diet on protein degradation activity in the sheep rumen. *Journal of the Science of Food and Agriculture* 32, 973–981.

Sinclair, K.D., Sinclair, L.A. and Robinson, J.J. (2000) Nitrogen metabolism and fertility in cattle. I. Adaptive changes in intake and metabolism to diets differing in their rate of energy and nitrogen release in the rumen. *Journal of Animal Science* 78, 2659–2669.

Sinclair, L.A., Garnsworthy, P.C., Newbold, J.R. and Buttery, P.J. (1993) Effect of synchronizing the rate of dietary energy and nitrogen release on rumen fermentation and microbial protein synthesis in sheep. *Journal of Agricultural Science, Cambridge* 120, 251–263.

Sinclair, L.A., Garnsworthy, P.C., Newbold, J.R. and Buttery, P.J. (1995) Effects of synchronizing the rate of dietary energy and nitrogen release in diets with a similar carbohydrate composition on rumen fermentation and microbial protein synthesis in sheep. *Journal of Agricultural Science, Cambridge* 124, 463–472.

Singh, B., Bhat, T.K. and Sharma, O.P. (2001) Biodegradation of tannic acid in an in vitro ruminal system. *Livestock Production Science* 68, 259–262.

Skene, I.K. and Brooker, J.D. (1995) Characterization of tannin acylhydrolase activity in the ruminal bacterium *Selenomonas ruminantium. Anaerobe* 1, 321–327.

Smith, R.H., McAllan, A.B., Hewitt, D. and Lewis, P.E. (1978) Estimation of amount of microbial and dietary nitrogen compounds entering the duodenum of cattle. *Journal of Agricultural Science, Cambridge* 90, 557–568.

Solomon, R., Chase, L.E., Ben-Ghedalia, D. and Bauman, D.E. (2000) The effect of nonstructural carbohydrate and addition of full fat extruded soybeans on the concentration of conjugated linoleic acid in the milk fat of dairy cows. *Journal of Dairy Science* 83, 1322–1329.

Steinhour, W.D., Stokes, M.R., Clark, J.H., Rogers, J.A. and Davis, C.L. (1982) Estimation of the proportion of non-ammonia-nitrogen reaching the lower gut of the ruminant derived from bacterial and protozoal nitrogen. *British Journal of Nutrition* 48, 417–431.

Stern, M.D., Hoover, W.H. and Leonard, J.B. (1977) Ultrastructure of rumen holotrichs by electron microscopy. *Journal of Dairy Science* 60, 911–918.

Strobel, H.J. and Russell, J.B. (1986) Effect of pH and energy spilling on bacterial protein synthesis by carbohydrate-limited cultures of mixed rumen bacteria. *Journal of Dairy Science* 69, 2941–2947.

Stumm, C.K., Gijzen, H.J. and Vogels G.D. (1982) Association of methanogenic bacteria with ovine rumen ciliates. *British Journal of Nutrition* 47, 95–99.

Sutter, F., Casutt, M.M., Ossowski, D.A., Scheeder, M.R.L. and Kreuzer, M. (2000) Comparative evaluation of rumen-protected fat, coconut oil and various oilseeds supplemented to fattening bulls. 1. Effects on growth, carcass and meat quality. *Archives of Animal Nutrition* 53, 1–23.

Sutton, J.D., Knight, R., McAllan, A.B. and Smith, R.H. (1983) Digestion and synthesis in the rumen of sheep given diets supplemented with free and protected oils. *British Journal of Nutrition* 49, 419–432.

Tamminga, S. (1992) Nutrition management of dairy cows as a contribution to pollution control. *Journal of Dairy Science* 75, 345–357.

Tamminga, S., Van Straalen, W.M., Subnel, A.P.J., Meijer, R.G.M., Steg, A., Wever, C.J.G. and Blok, M.C. (1994) The Dutch protein evaluation system: the DVE/OEB system. *Livestock Production Science* 40, 139–155.

Tangendjaja, B., Hogan, J.P. and Wills, R.B.H. (1983) Degradation of mimosine by rumen contents: effects of feed composition and *Leucaena* substrates. *Australian Journal of Agricultural Research* 34, 289–293.

Tava, A., Berardo, N., Cunico, C., Romani, M. and Odoardi, M. (1995) Cultivar differences and seasonal changes of primary metabolites and flavor constituents in tall fescue in relation to palatability. *Journal of Agricultural and Food Chemistry* 43, 98–101.

Teather, R.M., Mahadevan, S., Erfle, J.D. and Sauer, F.D. (1984) Negative correlation between protozoal and bacterial levels in rumen samples and its relation to the determination of dietary effects on the rumen microbial population. *Applied Environmental Microbiology* 47, 566–570.

Teferedegne, B. (2000) New perspectives on the use of tropical plants to improve ruminant nutrition. *Proceedings of the Nutrition Society* 59, 209–214.

Teferedegne, B., McIntosh, F., Osuji, P.O., Odenyo, A., Wallace, R.J. and Newbold, C.J. (1999) Influence of foliage from different accessions of the sub-tropical leguminous tree, *Sesbania sesban*, on ruminal protozoa in Ethiopian and Scottish sheep. *Animal Feed Science and Technology* 78, 11–20.

Tesfa, A.T. (1993) Effects of rape-seed oil supplementation on digestion, microbial protein synthesis and duodenal microbial amino acid composition in ruminants. *Animal Feed Science and Technology* 41, 313–328.

Theurer, C.B., Huber, J.T., Delgado-Elorduy, A. and Wanderley, R. (1999) Invited review: summary of steam-flaking corn or sorghum grain for lactating dairy cows. *Journal of Dairy Science* 82, 1950–1959.

Thomas, P.C. (1973) Microbial protein synthesis. *Proceedings of the Nutrition Society* 32, 85–91.

Tomlinson, A.P., Van Horn, H.H., Wilcox, C.J. and Harris, B. (1994) Effects of undegradable protein and supplemental fat on milk yield and composition and physiological responses of cows. *Journal of Dairy Science* 77, 145–156.

Tothi, R., Lund, P., Weisbjerg, M.R. and Hvelplund, T. (2003) Effect of expander processing on fractional rate of maize and barley starch degradation in the rumen of dairy cows estimated using rumen evacuation and in situ techniques. *Animal Feed Science and Technology* 104, 71–94.

Ushida, K. and Jouany, J.P. (1985) Effect of protozoa on rumen protein degradation in sheep. *Reproduction, Nutrition, Development* 25, 1075–1081.

Ushida, K., Jouany, J.P., Lassalas, B. and Thivend, P. (1984) Protozoal contribution to nitrogen digestion in sheep. *Canadian Journal of Animal Science* 64 (Suppl.), 20–21.

Ushida, K., Jouany, J.P. and Thivend, P. (1986) Role of protozoa in nitrogen digestion in sheep given two isonitrogenous diets. *British Journal of Nutrition* 56, 407–419.

Ushida, K., Kayouli, C., De Smet, S. and Jouany, J.P. (1990) Effect of defaunation on protein and fibre digestion in sheep fed on ammonia-treated straw-based diets with or without maize. *British Journal of Nutrition* 64, 765–775.

Ushida, K., Jouany, J.P. and Demeyer, D.I. (1991) Effects of presence or absence of rumen protozoa on the efficiency of utilization of concentrate and fibrous feeds. In: Tsuda, T., Sasaki, Y. and Kawashima, R. (eds) *Physiological Aspects of Digestion and Metabolism in Ruminants.* Academic Press, San Diego, California, pp. 625–654.

Valadares, F., Broderick, G.A., Valadares, R.F.D. and Clayton, M.K. (2000) Effect of replacing alfalfa silage with high moisture corn on nutrient utilization and milk production. *Journal of Dairy Science* 83, 106–114.

Valkeners, D., Thewis, A., Piron, F. and Beckers, Y. (2004) Effect of imbalance between energy and nitrogen supplies on microbial protein synthesis and nitrogen metabolism in growing double-muscled Belgian Blue bulls. *Journal of Animal Science* 82, 1818–1825.

Veira, D.M., Ivan, M. and Jui, P.Y. (1983) Rumen ciliate protozoa: effects on digestion in the stomach of sheep. *Journal of Dairy Science* 66, 1015–1022.

Veira, D.M., Ivan, M. and Jui, P.Y. (1984) The effect of ciliate protozoa on the flow of amino acids from the stomach of sheep. *Canadian Journal of Animal Science* 64 (Suppl.), 22–23.

Vogels, G.D., Hoppe, W.F. and Stumm, C.K. (1980) Association of methanogenic bacteria with rumen ciliates. *Applied Environmental Microbiology* 40, 608–612.

Wallace, R.J. and McKain, N. (1990) A comparison of methods for determining the concentration for extracellular peptides in rumen fluid of sheep. *Journal of Agricultural Science, Cambridge* 114, 101–105.

Wallace, R.J. and McPherson, C.A. (1987) Factors affecting the rate of breakdown of bacterial protein in rumen fluid. *British Journal of Nutrition* 58, 313–323.

Wallace, R.J. and Newbold, C.J. (1991) Effects of bentonite on fermentation in the rumen simulation technique (RUSITEC) and on rumen ciliate protozoa. *Journal of Agricultural Science, Cambridge* 116, 163–168.

Wallace, R.J., Newbold, C.J., Watt, N.D., Buchan, V. and Brown, D.S. (1993) Amino acid composition of degradation-resistant peptides in extracellular rumen fluid of sheep. *Journal of Agricultural Science, Cambridge* 120, 129–133.

Wallace, R.J., Arthaud, L. and Newbold, C.J. (1994) Influence of *Yucca shidigera* extract on ruminal ammonia concentrations and ruminal microorganisms. *Applied and Environmental Microbiology* 60, 1762–1767.

Wallace, R.J., Atasoglu, C. and Newbold, C.J. (1998) Role of peptides in rumen microbial metabolism. In: *Proceedings of the 8th World Conference on Animal Production.* Seoul National University Publishing Unit, Korea, pp. 581–586.

Wallace, R.J., McEwan, N.R., McIntosh, F.M., Teferedegne, B. and Newbold, C.J. (2002) Natural products as manipulators of rumen fermentation. *Asian-Australasian Journal of Animal Science* 15, 1458–1468.

Wang, Y., Douglas, G.B., Waghorn, G.C., Barry, T.N., Foote, A.N. and Purchas, R.W. (1996) Effects of condensed tannins upon the performance of lambs grazing *Lotus corniculatus* and lucerne (*Medicago sativa*). *Journal of Agricultural Science, Cambridge* 126, 87–98.

Weller, R.A. and Pilgrim, A.F. (1974) Passage of protozoa and volatile fatty acids from the rumen of the sheep and from a continuous *in vitro* fermentation system. *British Journal of Nutrition* 32, 341–351.

White, R.W. (1969) Viable bacteria inside the rumen ciliate *Entodinium caudatum*. *Journal of General Microbiology* 56, 403–408.

Whitelaw, F.G., Eadie, J.M., Mann, S.O. and Reid, R.S. (1972) Some effects of rumen ciliate protozoa in cattle given restricted amounts of a barley diet. *British Journal of Nutrition* 27, 425–437.

Wilkerson, V.A., Mertens, D.R. and Caster, D.P. (1997) Prediction of excretion of manure and nitrogen by Holstein dairy cows. *Journal of Dairy Science* 80, 3193–3204.

Williams, A.G. (1979) The selectivity of carbohydrate assimilation in the anaerobic rumen ciliate *Dasytricha ruminantium*. *Journal of Applied Bacteriology* 47, 511–520.

Williams, A.G. (1986) Rumen holotrich ciliate protozoa. *Microbiology Review* 50, 25–49.

Williams, A.P. and Cockburn, J.E. (1991) Effect of slowly and rapidly degraded protein sources on the concentrations of amino acids and peptides in the rumen of steers. *Journal of the Science of Food and Agriculture* 56, 303–314.

Williams, A.G. and Coleman G.S. (1988) Rumen ciliate protozoa. In: Hobson, P.N. (ed.) *The Rumen Microbial Ecosystem.* Elsevier Applied Science Publishers, Amsterdam, The Netherlands, pp. 77–128.

Williams, A.G. and Coleman, G.S. (1992) *The Rumen Protozoa*, 1st edn. Springer-Verlag, New York, 441 pp.

Williams, A.G. and Harfoot, C.G. (1976) Factors affecting the uptake and metabolism of soluble carbohydrates by the rumen ciliate *Dasytricha ruminantium* isolated from ovine rumen contents by filtration. *Journal of General Microbiology* 96, 125–136.

Williams, P.P. and Dinusson, W.E. (1973) Amino acid and fatty acid composition of bovine ruminal bacteria and protozoa. *Journal of Animal Science* 36, 151–155.

Wilson, R.C., Overton, T.R. and Clark, J.H. (1998) Effects of *Yucca shidigera* extract and soluble protein on performance of cows and concentrations of urea nitrogen in plasma and milk. *Journal of Dairy Science* 81, 1022–1027.

Witt, M.W., Sinclair, L.A., Wilkinson, R.G. and Buttery, P.J. (1999a) The effects of synchronizing the rate of dietary energy and nitrogen supply to the rumen on the metabolism and growth of ram lambs given food at a restricted level. *Animal Science* 69, 627–636.

Witt, M.W., Sinclair, L.A., Wilkinson, R.G. and Buttery, P.J. (1999b) The effects of synchronizing the rate of dietary energy and nitrogen supply to the rumen on the production metabolism of sheep: food characterization and growth and metabolism of ewe lambs given food *ad libitum*. *Animal Science* 69, 223–235.

Witt, M.W., Sinclair, L.A., Wilkinson, R.G. and Buttery, P.J. (2000) The effects of synchronizing the rate of dietary energy and nitrogen supply to the rumen on milk production and metabolism of ewes offered grass silage based diets. *Animal Science* 71, 187–195.

Wu, Z. and Satter, L.D. (2000) Milk production during the complete lactation of dairy cows fed diets containing different amounts of protein. *Journal of Dairy Science* 83, 1042–1051.

Yang, C.-M.J. and Russell, J.B. (1993) The effect of monensin supplementation on ruminal ammonia accumulation in vivo and the numbers of amino acid-fermenting bacterial. *Journal of Animal Science* 71, 3470–3476.

Yang, W.Z., Beauchemin, K.A., Koenig, K.M. and Rode, L.M. (1997) Comparison of hull-less barley, barley, or corn for lactating cows: effects on extent of digestion and milk production. *Journal of Dairy Science* 80, 2475–2486.

Yang, W.Z., Beauchemin, K.A. and Rode, L.M. (2000) Effects of barley grain processing on extent of digestion and milk production of lactating cows. *Journal of Dairy Science* 83, 554–568.

Yang, W.Z., Beauchemin, K.A. and Rode, L.M. (2001) Effects of grain processing, forage to concentrate ratio, and forage particle size on rumen pH and digestion by dairy cows. *Journal of Dairy Science* 84, 2203–2216.

Yiannikouris, A. and Jouany, J.P. (2002) Mycotoxins in feeds and their fate in animals: a review. *Animal Research* 51, 81–99.

Yoshida, J., Nakamura, Y. and Nakamura, R. (1982) Effects of protozoal fraction and lactate on nitrate metabolism of microorganisms in sheep rumen. *Japanese Journal of Zootechnical Science* 53, 677–685.

Zhu, J. and Flippich, L.J. (1995) Acute intra-abomasal toxicity of tannic acid in sheep. *Veterinary and Human Toxicology* 37, 50–54.

Zinn, R.A. (1993) Influence of processing on the comparative feeding value of barley for feedlot cattle. *Journal of Animal Science* 71, 3–10.

Zinn, R.A., Alvarez, E.G., Montaño, M.F., Plascencia, A. and Ramirez, J.E. (1998) Influence of tempering on the feeding value of rolled corn in finishing diets for feedlot cattle. *Journal of Animal Science* 76, 2239–2246.

5 Whole-animal Nitrogen Balance in Cattle

J.L. Firkins[1] and C. Reynolds[2]
[1]Department of Animal Sciences, Ohio State University, Columbus, Ohio, USA
[2]College of Food, Agriculture and Environmental Sciences, Wooster, Ohio, USA

5.1 Importance of Nitrogen Balance

5.1.1 Environmental issues

Disposal of animal manure has dramatically escalated to be one of the foremost research problems for dairy nutrition and management in the USA (Van Horn and Hall, 1997; Meyer and Mullinax, 1999; Nelson, 1999) and internationally (Kohn *et al.*, 1997; Kuipers *et al.*, 1999; Castillo *et al.*, 2000, 2001b). Major concerns include environmental consequences of nitrogen (N) and phosphorus (P) loading. The greatest environmental effect of N loss in manure is a result of rapid conversion of urea to ammonia (NH_3) and its subsequent volatilization (James *et al.*, 1999; Nelson, 1999), which affects the acidity of precipitation (Van Horn and Hall, 1997), formation of long-lasting aerosols (James *et al.*, 1999) and reduction of the N:P ratio of the residual manure below plant requirements (Van Horn *et al.*, 1996; Van Horn and Hall, 1997). We note that reduction of excreted P (see Chapter 7) influences this ratio favourably. Mechanical costs of reducing NH_3 volatilization are up to US$20/kg of N surplus reduced (Kuipers *et al.*, 1999). Therefore, the most effective solution to reduce this problem entails more efficient N capture in the form of body components, milk

©CAB International 2005. *Nitrogen and Phosphorus Nutrition of Cattle* (eds E. Pfeffer and A.N. Hristov)

and wool. This review will focus on improvement of whole-body N balance (i.e. the retention of dietary N in body tissue or milk) but also specific components of N excretion with respect to dietary N inputs for beef and dairy cattle.

5.1.2 Beef and dairy cattle production systems

5.1.2.1 Beef cattle

Much of the beef production in the USA is concentrated in large feedlots. The majority (80% to 90%) of N fed to feedlot cattle is excreted, with 50% to 75% of that excretion in the form of urinary N (Satter *et al.*, 2002). Therefore, relatively modest improvements in N efficiency can be magnified considerably, particularly if the amount of N lost as urinary urea can be reduced. Reduction of N loss in manure has important ramifications with regard to manure distribution as fertilizer relative to the cropland used to support grain production (Klopfenstein and Erickson, 2002), but this topic is beyond the scope of this chapter.

In recent reviews, Klopfenstein and Erickson (2002) and Satter *et al.* (2002) discussed implementation of metabolizable protein systems to feedlot cattle according to their changing growth phases in order to improve the efficiency of dietary N utilization. 'Metabolizable protein' is defined as the protein reaching the small intestine and digested therein. Clearly, improvements in gain: feed ratio should reduce N excreted per animal because of fewer days on feed. Even if feed efficiency is not increased, however, feeding protein according to requirements should allow less N excretion because cattle protein requirements decrease proportionately with increasing maturity. In both scenarios, 'N balance' (N intake − N lost in faeces and urine; also termed 'tissue N retention') was similar between treatments when calculated as g/day, but decreasing N inputs decreased N lost into the environment by 12% to 21%. However, few studies have continued during the growth phase all the way to finishing. Compensatory growth could potentially make up for any short-term limitations in daily gain (i.e. metabolizable protein slightly below requirements) for cattle retained in the feedlot until finishing (Firkins and Fluharty, 2000), so the potential to reduce N loss might be even greater in such systems.

In order for phase feeding to be adopted more widely, beef producers need to be confident in the supply and requirements of metabolizable protein for cattle under various circumstances (Klopfenstein and Erickson, 2002). Aside from various factors affecting protein requirements (NRC, 2000), the supply is influenced by the ruminal degradability and intestinal digestibility of protein sources. Concomitantly, feed libraries and analyses are improving in both precision and accuracy (Stern *et al.*, 1997). In addition, systems are developing that will improve prediction of requirements of metabolizable amino acids for beef cattle (see Chapter 2). However, Klopfenstein and Erickson (2002) cited studies in which the ruminal degraded protein (RDP) requirements were estimated to be 6.3%, 8.3% and 10.0% of dry matter (DM) for feedlot cattle fed dry rolled, steam-flaked and high-moisture maize, respectively, due to increased ruminal availability of carbohydrate to support microbial protein production. If RDP limited growth of amylolytic microbes, digestion of starch in the small and large intestines could compensate (Firkins *et al.*, 2001) and theoretically could increase efficiency of energy utilization (Harmon and McLeod, 2001). Transfers of N among gut and blood urea pools could have a strong influence on N efficiency in beef feedlot cattle (see Section 5.2.1).

5.1.2.2 Grazing beef cattle

Many beef cow/calf operations in the USA rely heavily on grazing of poor quality grass. Nitrogen recycling between the blood and gut helps compensate for low protein intake, but protein supplementation to increase RDP has improved productivity in some studies (Firkins and Fluharty, 2000). In this regard, studies with lactating beef cows have found that the benefit of RDP supplementation is not reduced when the frequency of RDP feeding is reduced to as little as once every 4 days (Coleman and Wyatt, 1982; Krehbiel *et al.*, 1998). This may reflect recycling of non-protein N (NPN) between the gut and body pools (Krehbiel *et al.*, 1998) or the short-term deposition of amino acids in labile protein pools (Waterlow, 1999). In addition, because of the low energy availability for microbial protein production, some researchers have reported responses to supplementation of rumen undegraded protein (RUP). Moreover, the requirement by the cow for metabolizable protein might be lower than the rumen microbes' requirements for RDP. The low cost/low intensity management of these operations compared with the large area of land usage probably lowers the potential benefit of improved protein usage from an environmental standpoint and will not receive further attention in this review.

5.1.2.3 Dairy heifers

Dairy calves are primarily raised intensively and must be evaluated differently from beef calves. Mammary development can have residual effects on lifetime milk production, promoting the concept of target weight gains (NRC, 2001). Because heifers need to be grown for about 2 years before milk production, needing more replacements to meet the demands for milk production would have a profoundly negative impact on N usage. With economic pressure to increase growth rate, having an adequate supply of metabolizable protein for heifers might be very important in reducing the negative effects of energy for rapid growth (Whitlock *et al.*, 2002). Despite a lack of differences in average daily gain, feed efficiency improved with increasing concentration of crude protein (CP) in the diet, and structural growth measurements tended to improve (Gabler and Heinrichs, 2003). Even if CP requirements are targeted for weight gain, it seems likely that CP concentration in the field will not decrease, particularly with those producers adopting accelerated growth programmes. Increasing CP intake above requirements for growth primarily increased the loss of N in urine (James *et al.*, 1999). Future research is needed to better document the requirements of metabolizable protein and amino acids for dairy heifers so that N input can be decreased reliably without potentially impacting lifetime milk production.

5.1.2.4 Dairy cows

For lactating cows, the secretion of high amounts of protein into milk prioritizes the importance of the amount and profile of amino acids reaching the duodenum. Nitrogen balance can be influenced positively by improved ration balancing to capture more dietary N as milk protein. Considerable research has been done to increase milk protein concentration, particularly when dietary fat is fed (Wu and Huber, 1994). In fact, the NRC (2001) elaborated on the complexities involved with synchronizing ruminal fermentation with microbial protein synthesis for incorporation into a system to meet requirements for metabolizable amino acids (see Section 5.3.1 and Chapter 2). With gaining emphasis on formulating rations to meet requirements for specific amino acids and with improved advances in technology (e.g. reproductive aids and bovine somatotropin) and housing systems, there is a strong potential to multiply a modest gain in efficiency of N utilization per cow to reduce environmental impact. Satter *et al.* (2002) noted that increased milk production will dilute maintenance N costs, so restricting protein supply below requirements to reduce manure N excretion should be avoided; yet still they emphasized that feeding more than about 17.5% CP (18.5% in certain circumstances) on a DM basis to high-yielding cows would only divert more dietary N into urine if the diets have maize silage to dilute the high RDP in legume silages and have protein balanced for degradability. In Europe, excess protein is often fed to lactating dairy cows because it is relatively inexpensive, provides a safety margin against drops in forage CP concentration, and generally increases milk yield through effects on intake of grass silage and other forages. Generally, changes in milk and milk protein yield with differences in dietary CP below 14% to 15% of DM have been attributed to metabolic effects of metabolizable protein supply, whereas, above this threshold, changes in milk yield are typically accompanied by changes in dry matter intake (DMI) attributed to effects on rumen or total tract digestion (Clark and Davis, 1980; Oldham and Smith, 1980; Reynolds, 2000).

Currently, various amino acid supplements are available on the market, but further research is needed to determine when to use them economically, with impending environmental regulations ostensibly increasing their economic feasibility. Balancing to meet metabolizable lysine requirements using conventional protein sources and then supplementing rumen-protected methionine could reduce the total CP fed, potentially reducing N excretion by 13% to 20% compared with current practice (Satter *et al.*, 2002). Sloan (1997) further discussed this ‘ideal protein’ concept, noting that there is only a modest (2% to 5%) gain in efficiency of conversion of dietary CP into milk protein by meeting the requirement for a single limiting amino acid. For example, conversion of rumen-protected methionine into milk methionine might be only about 10% efficient because it is used for other bodily functions. Therefore, any real gain in

N balance relative to N input will primarily be accentuated via decreased N intake. Moreover, as discussed in Chapter 2, lysine and methionine supplies might not actually be limiting or might only be near- or co-limiting (Sloan, 1997; Vanhatalo *et al.*, 1999; Hvelplund *et al.*, 2001).

Nitrogen balance probably changes the most during the period from late gestation into the first few weeks of lactation (the 'transition period'). The splanchnic tissues [portal-drained viscera (PDV; the gastrointestinal tract, pancreas, spleen and associated adipose) plus liver], mammary gland and fetus increase protein synthesis at a time when DMI might be insufficient to meet protein requirements (Bell *et al.*, 2000). Labile protein reserves might be mobilized to balance shortfalls in supply of protein or limiting amino acids but also provide gluconeogenic precursors. In this regard, mRNA and activity for liver pyruvate carboxylase (Greenfield *et al.*, 2000) and alanine use for glucose synthesis by hepatocytes *in vitro* (Drackley *et al.*, 2001) increase immediately after calving, supporting the concept of increased amino acid use for glucose synthesis. However, body protein deficit is usually relatively modest except in the first days of lactation (Grummer, 1995; Table 5.1), and the glucogenic requirements for amino acids may be less of a metabolic priority than hypothesized (Reynolds *et al.*, 2003). When dietary RUP was increased pre- and post-calving, body protein mobilization (assessed by deuterium oxide dilution) accounted for only about 7% of the energy lost or gained (Komaragiri and Erdman, 1997). In the most comprehensive slaughter balance study conducted in dairy cows of which we are aware (Gibb *et al.*, 1992), the amount of body protein lost in the first 8 weeks postpartum was relatively small (5.6 kg), especially when compared to the amount of body fat lost (37.4 kg), in relatively low-yielding cows fed grass silage. Much of this body protein loss occurred in the first 2 weeks postpartum (2.7 kg), which equated to a loss of 31 g N/day. In transition dairy cows catheterized for measurements of splanchnic nutrient flux, the potential gluconeogenic contribution of alanine, as well as lactate and glycerol, was greatest 10 days after calving, but the required contribution of other amino acids was lowest at this time. Indeed, increases in net liver removal of these glucose precursors and volatile fatty acids between 9 days before calving and 10 days after calving could account for all of the increase in the measured release of glucose by the liver (Reynolds *et al.*, 2003).

The NRC (2001) reviewed protein requirements for transition cows, suggesting increases in protein requirements for heifers, but not cows, in late gestation compared with previous requirements. In one study (Putnam and Varga, 1998), increasing dietary CP and RUP concentrations prepartum to multiparous cows did tend ($P = 0.09$) to increase N balance, but even the cows fed diets lower than the NRC (1989) requirements still had positive N balance on days −12 to −5 relative to expected calving, and no response in milk production postpartum was detected. In other studies, feeding supplemental protein before calving increased milk or

Table 5.1. Nitrogen and metabolizable energy (ME) in dairy cows during the first 8 weeks of lactation[a].

	Week								
	1	2	3	4	5	6	7	8	SE
N intake (g/day)[b]	394	440	462	458	447	435	428	426	6.5
Milk N (g/day)[c]	183	183	186	182	172	168	165	164	2.1
Urinary N (g/day)	87	102	90	96	81	76	76	67	3.7
N retained (g/day)	−19	−1	15	4	23	17	10	20	5.0
ME intake (MJ/day)[b]	164	191	199	199	202	198	199	200	2.9
Milk energy (MJ/day)[c]	104	104	106	103	97	95	94	93	1.2
Energy retained (MJ/day)[c]	−64	−36	−36	−33	−24	−22	−21	−21	2.8
Urinary energy (MJ/day)	8.1	9.1	9.3	9.5	8.0	8.3	8.2	8.5	0.28
Urinary energy (% DE)	4.24	4.22	4.13	4.16	3.49	3.64	3.60	3.68	NA

[a]Data from Sutter and Beever (2002). Their data were used to calculate urinary energy as a percentage of digestible energy. (DE), so a SE was not available.
[b]Week 1 < week 2 ($P < 0.05$).
[c]Linear effect of week in lactation ($P < 0.01$).

milk protein yield after calving in heifers (Van Saun *et al.*, 1993; Santos *et al.*, 2001) but decreased DMI or milk yield in multiparous cows (Hartwell *et al.*, 2000; Santos *et al.*, 2001).

To further account for body energy and protein retention during early lactation, Sutter and Beever (2002) performed a series of weekly total collections of faeces and urine, combined with respiration calorimetry to assess the energy status for multiparous cows. Although too variable for statistical significance, N balance was only negative for the first 2 weeks of lactation and primarily only in the first week (Table 5.1), supporting the conclusion that N balance should reach a nadir on about day 7 of lactation (Bell *et al.*, 2000). Body tissue energy balance was negative throughout the study but increased linearly, apparently primarily because of a linear decrease in milk energy secretion (except for week 1). The authors concluded that N mobilization by labile reserves might be more important for relocation within body tissues (e.g. gut and liver) than for milk protein and the changes in body weight might not reflect primarily differences in water repletion of tissues or increases in the weight of the gut plus contents. Increases in the CP content of splanchnic tissues were relatively small (0.85 kg) compared to body protein loss in the first 8 weeks of lactation (Gibb *et al.*, 1992), but a large portion (46%) of this change in splanchnic protein content occurred in the first 2 weeks postpartum. Urinary energy was not affected by week in lactation, and the energy lost in urine accounted for about 4% of digestible energy (Table 5.1). Therefore, with proper balancing of RDP and RUP, some mobilization and repletion of body protein seems to help transition the cow to lactation, but the impact on tissue N balance must be relatively minor.

Based on preceding results, gut metabolism and whole-body urea transfer probably have potentially large impact on efficiency of whole-body N balance, so further attention will be given to these subjects in this review.

5.2 Whole-animal Nitrogen Fluxes and Nitrogen Balance

5.2.1 Nitrogen exchange among tissues

As discussed by Lapierre and Lobley (2001), ruminants have adapted their metabolism to rely on large fluxes of N exchanging between the blood and digestive tract. They calculated that 40% to 80% of the blood urea N (BUN) produced in the liver enters the digestive tract instead of being excreted into the urine. In the rumen, ureolysis, proteolysis and deamination of amino acids is considerable, as would be expected based on the diversity of microbial enzymes responsible (Wallace *et al.*, 1997). Despite the ruminal pH being considerably lower than the pK_a of ammonia/ammonium (Satter *et al.*, 2002), causing a low proportion to be in the unionized form for absorption (Leng and Nolan, 1984), absorption of NH_3-N from the rumen and intestines is extensive (Parker *et al.*, 1995). Consequently, considerable cycling of BUN back to the digestive tract might be needed for positive N balance for many species, including man (Waterlow, 1999), but particularly for ruminants (Lapierre and Lobley, 2001).

In the past decade, considerable research has been done with the double ^{15}N-urea infusion technique, which has been well described by Lobley *et al.* (2000). Using this approach, Lapierre and Lobley (2001) generalized that approximately one-third of BUN actually gets excreted into the urine, with two-thirds (40% to 80%) being cycled back to the digestive tract. Of the NH_3-N produced from urea that gets transferred to the gut, about 10% is excreted as faecal N, 40% is absorbed and converted back to BUN and 50% is incorporated into microbial protein in the rumen, which is subsequently absorbed from the small intestine. The latter flux (50% of the two-thirds) is high, in part, because of multiple entry rather than entry via a single pass. Microbial protein synthesized using N from BUN ranges from 8% to 38% (Lapierre and Lobley, 2001). Because of the eventual loss of urinary N from urea, though, these authors calculated that upper limits for N retention as body tissue or milk would, therefore, be 50% to 60% of dietary N or 70% to 90% of apparently digested N. Based on a regression of literature from cattle with indwelling blood catheters for the measurement of splanchnic flux, they reported a prediction of urea-N synthesis by the liver (g/day) = 0.80 (N intake, g/day) −30 (r^2 = 0.45). In multicatheterized cattle, regression of net liver removal of NH_3-N and release of BUN on digested N gave slopes of 0.68 and 0.90, respectively (Reynolds, 1995), but in most cases these data came from cattle fed protein well in excess of their requirements. In a more recent integration of these (Reynolds, 1995) and more recent observations from the University of Reading (Fig. 5.1; Reynolds, 2002, 2003), the relationship between daily N intake and liver urea release for 304 individual measurements had a slope of 0.65 ($R^2 = 0.64$). In this case, the data set included observations from dairy cows fed varying levels of dietary protein, at various stages of lactation and levels of production, and in some cases receiving abomasal infusions of casein or amino acids. For the same data set, the relationship between N intake and net PDV release of NH_3-N had a slope of 0.42 ($R^2 = 0.84$; Fig. 5.2). Although N intake is a major determinant of PDV absorption of NH_3-N and liver BUN release, other factors are also important. We note that both variables are likely correlated (increasing DMI should be related to N intake and also to overall net flux of all metabolites), so the relatively low R^2 for the prediction of liver urea release (Fig. 5.1) documents the considerable amount of variation remaining to be explained (Lapierre and Lobley, 2001). As emphasized by Reynolds (2002), much of the remaining variation could be attributed to the amount of dietary N absorbed relative to requirements, which ultimately determines the extent of N excretion in urine (Waterlow, 1999).

In contrast to urine N, faecal N excretion is determined by amounts of indigestible N consumed and endogenous N losses, which to a large extent are determined by capture of urea-N as microbial protein in the hindgut. In lactating dairy cows, abomasal starch infusion increased faecal N excretion (Reynolds *et al.*, 2001). Concomitant decreases in faecal pH likely reflect increased starch fermentation in the hindgut, which would explain the increase in faecal N concentration and excretion observed. On the other hand, changing steers from a high-concentrate to a high-lucerne diet, at similar metabolizable energy in-

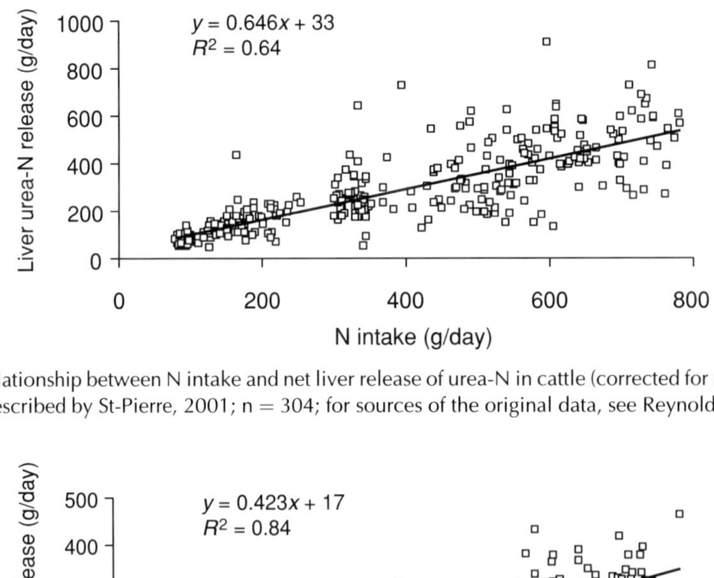

Fig. 5.1. Relationship between N intake and net liver release of urea-N in cattle (corrected for random effects of study as described by St-Pierre, 2001; n = 304; for sources of the original data, see Reynolds, 2003).

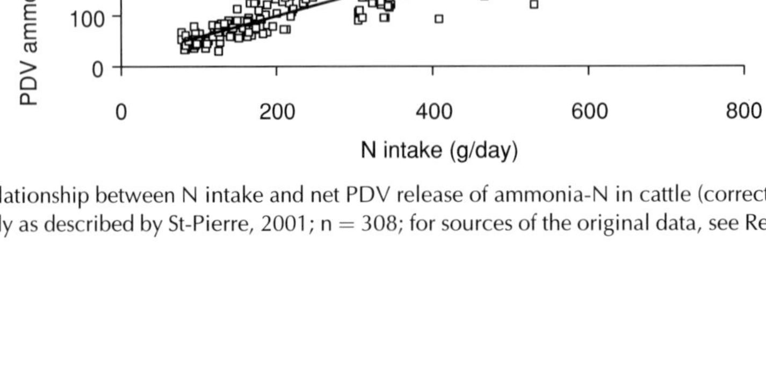

Fig. 5.2. Relationship between N intake and net PDV release of ammonia-N in cattle (corrected for random effects of study as described by St-Pierre, 2001; n = 308; for sources of the original data, see Reynolds, 2003).

take, markedly increased the transfer of BUN to the mesenteric-drained viscera (Reynolds and Huntington, 1988; Huntington, 1989). This increase in BUN transfer to the post-ruminal digestive tract was likely a consequence of increased fermentation of fibre in the hindgut.

Increased absorption of glucose from starch digested in the small intestine may also increase the efficiency of ingested N that is retained as body protein (Obitsu *et al.*, 2000; Reynolds *et al.*, 2001) the latter of which is dependent on insulin (Bergen, 1978). In lactating dairy cows, infusion of starch into the abomasum increased tissue energy balance, and over half of the increase in energy retention was attributable to greater protein deposition (Reynolds *et al.*, 2001). Increasing the amount of starch digested in the rumen or hindgut decreases net absorption of NH_3 by the PDV in dairy cows (Reynolds *et al.*, 1998; Delgado-Elorduy *et al.*, 2002a) and, in some studies, increased urea-N transfer from blood to the rumen (Delgado-Elorduy *et al.*, 2002b). Presumably, these changes in N cycling reflect increases in NH_3 utilization for microbial protein synthesis in the rumen or hindgut.

5.2.2 Energetic cost of urea

If each mole of urea produced in the liver requires four moles of ATP (McBride and Kelly, 1990), then it would be logical that the energetic cost of urea production could exert a strong regulatory constraint against BUN fluxes back and forth from the gut, especially for grazing cattle consuming large amounts of RDP (Kolver and Muller, 1998; Stockdale and Roche, 2002). Supplementation of grain should decrease ruminal NH_3-N concentration, in part because of decreased N intake (Bargo *et al.*, 2003), but grain supplementation is markedly recent in evolutionary terms. Waterlow (1999) commented that, although four moles of ATP are consumed, six moles could be produced per mole of urea synthesis (two moles of NADH produced from oxidative deamination of glutamate and regeneration of aspartate from fumarate). Therefore, urea synthesis might not be as critical as previously thought, especially because none of these amino acids is essential. Ammonia infusion into the duodenum increased urinary N excretion but did not affect N balance or yield of any milk components (Moorby and Theobald, 1999). Similarly, feeding steers urea markedly increased net PDV absorption of NH_3 and liver urea synthesis, without significant effects on net liver oxygen consumption, glucose release or amino acid metabolism (Maltby *et al.*, 1993). In a methodical series of studies in sheep, Lobley and colleagues have explored effects of increased NH_3 absorption on liver metabolism and similarly found no significant deleterious effects on liver metabolism of oxygen, glucose or amino acids (see Lobley *et al.*, 1995, 1996; Milano *et al*, 2000; Milano and Lobley, 2001; Reynolds, 2003). Despite ranging from 67 to 102 g/day of urinary N excretion (data not shown), N excretion in urine is a minor proportion of digestible energy intake (Table 5.1).

The concept of a high 'penalty' for NH_3 absorption and urea recycling needs to be evaluated within this context because it goes against the apparent adaptation toward urea cycling (previous section) and the discovery of urea transporters (Waterlow, 1999; Lapierre and Lobley, 2001) in the mammalian gut and other peripheral tissues. Yet, despite their presence, their role in BUN recycling is not clear (Marini and Van Amburgh, 2003). These latter authors suggested that BUN could be passively transferred through the epithelial cells, so the transporter's role could also be to efflux urea back into the blood before bacterial hydrolysis in the rumen during times of high N availability. Oba and Allen (2003b) reported that ammonium combined to make propionate more potent to depress feed intake, and such a situation of high NH_3-N and propionate would seem to occur only when N intake was excessive. Although it has been proposed that there is a deamination cost involved with high urea fluxes (Parker *et al.*, 1995), the energetic cost of ureagenesis appears now to be more a consequence of the metabolism of amino acids absorbed in excess of requirements rather than a cost of NH_3 absorption and conversion to urea *per se* (Reynolds, 2003) or possibly only in extremely high availability of ruminal NH_3-N (Milano *et al.*, 2000).

5.2.2.1 Importance of the rumen for nitrogen capture

Direct quantifiable relationships between ruminal N metabolism and urinary N excretion are limited, but available data support the concept that the rumen is a major mediator of N retention.

Al-Dehneh *et al.* (1997) reported that the ratio of ^{15}N enrichments in urinary N and BUN was constant by 40 h after the start of infusion of ^{15}N-urea into the jugular vein, implying interrelated N metabolism. Kennedy and Milligan (1980) reported that the transfer of BUN to ruminal NH_3-N was inversely proportional to the ruminal NH_3-N concentration and was increased with increasing grain or degradable carbohydrate inclusion in the diet. Although this could be a result of increased microbial growth and N capture (Delgado-Elorduy *et al.*, 2002b), greater BUN recycling for higher grain diets also could reflect increased energy for body protein retention, which could reduce the catabolism of absorbed amino acids. Whitelaw *et al.* (1991) added a urease inhibitor to the rumen of maintenance-fed sheep. This decreased the irreversible loss rate of BUN by 33% without affecting N intake or urinary N excretion. This limited work is interpreted to suggest that the eventual trapping of BUN for metabolic usage and not as urinary N will depend largely on N capture as microbial N in the rumen as well as the metabolic requirements for metabolizable protein (Lapierre and Lobley, 2001).

5.2.3 Ruminal and intraruminal nitrogen recycling

Wallace *et al.* (1997) cited a model described by Nolan (1975) to conclude that 'ammonia overflow leads to inefficient N retention'. The biological importance of such recycling is extensive and is the subject of Chapters 3 and 4, but modelling efforts will be discussed briefly herein within the context of their role in whole-body N metabolism.

Firkins (1996) reviewed quantitative studies that characterized flux among either chemical [i.e. non-ammonia N (NAN)] or biological (bacterial, protozoal or combined) pools in the rumen. Biological pools are more mechanistic but might be difficult to repeatably fractionate for subsequent determination of specific activity of a tracer. For instance, protozoa-enriched samples are typically based on sedimentation yet probably are significantly contaminated with bacteria (Sharp *et al.*, 1998). Chemical pools are more systematically differentiated but require appropriate independent biological data collection such that the flux rates among those pools have important mechanistic interpretation. Faichney *et al.* (1997) derived a complicated model evaluating protozoa-mediated turnover based on the abundance of protozoal RNA, which was characterized as the difference of signals from eukaryotic minus fungal probes. More recently, Oldick *et al.* (2000) documented extensive recycling of microbial protein in the rumen, but chemical precipitation techniques could not differentiate the recycling of microbial protein from a slowly turning over compartment as opposed to the exchange of NAN from a rapidly turning over compartment. They suggested that rapidly exchanging NAN probably has a smaller impact on efficiency of microbial growth and N capture for metabolizable protein. Direct experimental approaches quantifying intraruminal N recycling typically involved the use of multiple (and often radioactive) tracers, used fractionation procedures that might be difficult to systematically repeat, and/or incorporated by difference calculations that compound variation (which was typically ignored). Therefore, more attention to over parameterization needs to be given using considerations such as those of Oldick *et al.* (2000), so that models can be used in experimental designs with enough statistical power to explain interactions among treatments. Conversely, Dijkstra *et al.* (2002) recently discussed important modelling considerations with regard to mechanistic models. Given the large importance of microbial N capture (previous section), more quantitative work is needed in this area to decrease variability among feeding conditions in order to stimulate the adoption of lower protein diets in the field to decrease N excretion by cattle.

5.2.4 Measurement of nitrogen balance

Although objectives of individual researchers might be to compare treatment differences within a study, it is no longer sufficient to ignore known errors in measurement of N balance; the absolute measurements of multiple studies are being used to either derive or evaluate models with increasing frequency. Martin (1966) and Johnson (1986), among others, have clearly identified and quantified losses of N and sources of experimental error in measuring N balance in ruminants. More recently, Spanghero and Kowalski (1997) described major routes of N loss that accumulate to overestimate the by-difference calculation of N balance. From 35 published trials that they surveyed, about one-fifth did not determine N in faeces on a wet basis, leading to underestimation of N excretion. Methods to capture urinary N were variable or not even reported. In some studies that they cited, equivalents of NH_3 excreted might have exceeded the equivalents of acid added in the urine collection vessels. Several studies did not account for NPN in milk. Despite corrections that they applied to literature data, tissue N balance still had a median of 10.2 g/day, which they estimated to correspond to about 255 g/day of body weight gain. The median was significantly higher than the mean, indicating a skewed distribution of data. To account for differences among studies, they calculated deviations of individual treatment means compared with the mean from each experiment; from these data, they suggested that N balance was overestimated with increasing N availability for metabolism. Faecal N excretion can vary considerably from day to day, and we note that, although the appropriate number of days is likely variable (Schneider and Flatt, 1975), collection periods in the literature (we have noted some as low as 2 days) might be too short. Moreover, N balance data in Table 5.1 document variability among weeks, at least in early lactation. Readers are referred to Castillo *et al.* (2000) for a comprehensive review of dietary factors influencing efficiency of N capture in milk relative to excretion in urine and faeces.

Nitrogen balance can be used to evaluate amino acid requirements for growing cattle (Wessels and Titgemeyer, 1997; Greenwood and Titgemeyer, 2000), although the reader is referred to Chapter 2, this volume, for a more comprehensive review. Moreover, Iburg and Lebzien (2000) noted that amino acid requirements for dairy cattle really should be calculated at zero tissue N balance, which is an assumption that probably should be verified experimentally in more studies. As diets approach and then exceed the requirements for limiting amino acids, then N balance could be fluctuating from negative to positive. In short-term experiments for which milk protein is the response criterion, the degree of response could be mediated in part by tissue protein mobilization. Such reasoning could help to explain the variation in metabolizable lysine and methionine requirements determined by break-point analysis (NRC, 2001).

Manipulation of microbial populations can influence N retention. McGuffey *et al.* (2001) reported that ionophores increased N digestibility by about 3.5 percentage units and that several individual studies documented increasing N retention as a percentage of N intake. Besides increasing the efficiency of beef cattle growth, prepartum feeding of ionophores could increase N retention for dairy cattle (Plaizier *et al.*, 2000), which might positively influence transition to lactation. Defaunation of the rumen had mixed effects on N retention (Jouany, 1996), but the practical importance of elimination or reduction of protozoa in the rumen in actual growing conditions is the subject of Chapter 4.

5.3 Models to Balance Supply and Requirements of Protein

5.3.1 Supply models

Several systems have been developed by leading research institutions in the USA and Europe [see review (Dijkstra *et al.*, 1998)]. Although much improved, the new Dairy NRC (2001) still empirically predicts microbial protein flow from the rumen based on intake of total digestible nutrients (TDN), with the TDN concentration being discounted progressively with increasing DMI and with increasing TDN concentration (excluding high-fat diets). In the NRC (2001) system and many others, requirements for RUP are calculated by difference (after accounting for intestinal digestibility) of the animal's estimated protein requirements minus predicted duodenal flows of microbial and endogenous protein, therefore compounding variation associated with the prediction of microbial protein flow.

5.3.2 Prediction of microbial protein supply

Microbial protein is extremely well balanced with amino acids relative to meat or milk protein (NRC, 2001). RDP normally is much cheaper than RUP (St-Pierre and Glamocic, 2000), even if incomplete conversion of RDP to microbial protein (NRC, 2001) is accounted for. Although TDN includes fat and protein that provide relatively little energy to support microbial protein synthesis, this mechanistic problem (Kebreab *et al.*, 2002) probably is of relatively minor importance for empirical prediction by the NRC (2001).

Two separate equations were justified for the prediction of microbial protein flow to the duodenum based on net energy for lactation (NEL) intake for cattle fed diets with or without fat (Oldick *et al.*, 1999), yet visual inspection of the fitted lines documents that the use of separate equations makes a relatively modest impact at intakes that would be seen in production situations. Fat should decrease protozoal numbers and increase efficiency of microbial protein synthesis (Doreau and Ferlay, 1995; Firkins, 1996). Also, RDP intake was relatively static in most experiments from which the empirical relationship was determined and for which it would be used.

Although the NRC (2001) system ignores the sites of carbohydrate digestion, again this important mechanistic problem might have a statistically minor impact on prediction of microbial protein flow because microbial efficiency probably decreases with increasing ruminal availability of carbohydrate. Satter *et al.* (2002) logically concluded that 'finely ground high moisture shelled maize, through its ability to support microbial growth and protein synthesis, may be the cheapest "protein source" we have'. However, this generalization was not substantiated by experimental data (Firkins *et al.*, 2001; Oba and Allen, 2003a). In fact, when other factors were equalized, cows fed high-moisture maize, despite higher ruminal starch degradability, actually had numerically lower microbial protein flow to the duodenum than those fed maize grain processed in other ways and having lower ruminal starch digestibility (Firkins *et al.*, 2001). In a recent study (Harvatine *et al.*, 2002), replacing ground maize with steam-flaked maize increased microbial N flow to the duodenum by 15%; despite the 36% greater true ruminal starch digestibility, ruminal pH did not decrease, apparently because DMI decreased such that intake of truly digestible organic matter only increased by 7% with steam-flaking. However, in the same study, progressive replacement of forage with whole linted cottonseed linearly increased DMI and microbial N flow; however, ruminal pH and efficiency of microbial protein synthesis were depressed linearly. Clearly, the amount of ruminally available carbohydrate is impacted as much, or more, by changes in total DMI as by the fermentability of the carbohydrate in the diet fed. Increased carbohydrate degradation (g/day) can decrease microbial efficiency by factors directly related to low pH (Russell and Wilson, 1996) or because of increased energy spilling (metabolic wasting of high energy phosphate bonds), particularly if RDP becomes limiting (Wells and Russell, 1996). These results (Harvatine *et al.*, 2002) demonstrate that an empirical prediction using a constant efficiency clearly leads to inaccuracies that contribute to variation. However, they also document the importance of DMI prediction or determination as well as the need to predict carbohydrate fermentation and ruminal pH and its effects on microbial efficiency. Prediction of ruminal pH is very difficult (Allen, 1997) and is interpreted to be a major roadblock for all modelling systems.

Empirically (statistically) speaking, a bigger criticism of the current NRC (2001) procedure to estimate microbial protein flow could be that its evaluation method was biased, leading readers to have a false conclusion regarding its accuracy. When residuals (predicted minus measured) were regressed against measured microbial protein flows to the duodenum, a negative slope bias was detected. Similarly, a negative slope bias was detected for non-ammonia non-microbial N (NANMN) flows. In both cases, this would mean that microbial N and NANMN are being underpredicted with increasing measured values. Yet, a much smaller response was noted for total NAN (the sum of microbial N and NANMN fractions, which should logically accumulate negative slope bias). St-Pierre (2003) explained this apparent discrepancy as being caused by a biased evaluation procedure; when residuals were properly plotted against predicted values, the actual equation was considerably less biased than presented by the NRC (2001). Therefore, even though the prediction ignored effects of experiment and did not weight treatment means for variation among experiments (St-Pierre, 2001), which both have highly significant effects on regressions and interpretation of microbial protein production (Oldick *et al.*, 1999), the prediction actually appears to be relatively robust over a wide range of conditions, even if it lacks precision.

Microbial protein production is predicted based on a more mechanistic approach than NRC (2001) using the Cornell Net Carbohydrate and Protein System (or its derivative models), which has been evaluated by Alderman *et al.* (2001a,b). An early version of this model predicted average daily gain reasonably well over a wide range (0.7 to 1.5 kg/day) of predictions as assessed by an r^2

of 0.70 for a linear regression of predictions vs. measured data (Ainslie *et al.*, 1993). O'Connor *et al.* (1993) similarly concluded that the model predicted supply of individual amino acids to the duodenum well based on a high r^2 (0.81 to 0.90 for predicted vs. observed) over even larger ranges of approximately tenfold. Yet, a range in the data approaching 100% of the mean prediction can typically be visualized in their graphs, and Alderman *et al.* (2001b) noted that their data set was actually composed of two clusters, which could bias the interpretation. Besides the limitation in using r^2 (coefficient of determination) or R^2 (multiple coefficient of determination, including effects of trial as in Figs 5.1 and 5.2) from a *sample* to extend toward accuracy of a prediction for a *population*, extending the range of X measurements will inflate coefficient of determination as a measure of goodness of model fit for clustered data (Neter *et al.*, 1996). We note that the evaluation also would have been improved by appropriate residuals analysis for fit (see St-Pierre, 2003).

Cotta and Russell (1997) elaborated on the importance of synchronous N and carbohydrate supplies for microbial cell synthesis (see also Chapter 4). Mechanistic prediction of microbial protein flow to the duodenum has been well reviewed by Dijkstra *et al.* (1998). These models tend to emphasize the importance of synchronization of energy from carbohydrate fermentation with availability of RDP, however, which tend not to have been substantiated by direct *in vivo* experimentation (Castillo *et al.*, 2000; Dewhurst *et al.*, 2000; Bateman *et al.*, 2001b) and tended to cause overprediction of microbial protein flow in one evaluation (Bateman *et al.*, 2001a). With regard to stimulation of microbial protein production by increasing amino N, the yield of microbial growth was increased by 19% to 77%, depending on the model used (Dijkstra *et al.*, 2002). Such a large range emphasizes the predictive limitations for mechanistic models until further research is available. A sensitivity analysis (Bannink and De Visser, 1997) of the elaborate system described by Dijkstra (1994) indicated that more quantitative data are needed to improve the accuracy of parameters (coefficients) describing protozoal physiology and ecology for model robustness. We note that, although these systems might not be suspect to the errors associated with measuring microbial protein *in vivo*, they still are suspect to errors in measurement (and therefore prediction) of ruminal passage rate, which are also significant (Firkins *et al.*, 1998). Comparative accuracy and precision of virtually all models that are more mechanistic than the NRC (2001) model are difficult to assess at the present time, although mechanistic models probably hold more promise in the future to explain interactions among various dietary factors.

5.4 Methodological Issues Contributing to Variability in Estimation of Supply

5.4.1 Microbial markers

Markers to estimate microbial protein flow to the duodenum have been reviewed (Broderick and Merchen, 1992; Firkins *et al.*, 1998; Shingfield, 2000) and this topic is beyond the scope of this review. However, we note two current potential errors that could promote excessive variation among studies, contributing to the high significance of experiment in regression-based empirical approaches to predict metabolizable amino acid supply. Purines, the most common microbial marker, might have incomplete recovery or contain inhibitors when hydrolysed using the originally published conditions (Klopfenstein *et al.*, 2001). However, comparisons with ^{15}N (Broderick and Merchen, 1992; Shingfield, 2000) either do not support such large potential responses or indicate that recoveries are similar in both harvested bacteria and in duodenal samples, factoring out the error. Routine recovery checks in the first author's laboratory have documented the concentration of perchloric acid to have minor, if any, impact on purine recovery or concentration. As a result of the large importance of microbial N for capturing BUN as well as its importance in supply/requirement models, we recommend that researchers carefully evaluate marker procedures in their own laboratory conditions prior to continuing further research. Shingfield (2000) documented other sources of error for estimation of microbial N flow using excretion of purine derivatives and also potential escape of purines to the duodenum.

5.4.2 Protein degradability

Forage protein degradability probably adds considerable variation to prediction of metabolizable amino acid supply. Despite advancements in knowledge gained (Broderick, 1995), protein degradability still is highly variable (Kohn and Allen, 1995). Klopfenstein *et al.* (2001) outlined an improved methodology to estimate RUP of forages. More kinetics studies evaluating rates of degradation of protein fractions using ^{15}N-fertilized forages will help (Hristov *et al.*, 2001), but questions still remain regarding which fractions pass rapidly with ruminal fluid (Hvelplund *et al.*, 2001).

A fundamental principle of all kinetics studies is that dosing the tracer does not perturb the steady state of the tracee. We note a disturbing trend in current research to simply provide a large, potentially unphysiological bolus dose of some nitrogenous compound(s) into an unadapted rumen. Some published escape values for nitrogenous compounds likely have been inflated using such procedures. Investigators need to remember that: (i) a bolus dose must be shown not to affect the true metabolism/dilution of the tracer; or else (ii) a bolus dose of labelled tracer should replace an equal amount of unlabelled tracee that has been fed long enough to adapt rumen microbes. Interpretation of a log-linear elimination of tracer to document first-order kinetics is insufficient proof of the first assumption (as some authors have claimed). First-order kinetics can include multi-exponential dilutions or can aggregate a mix of heterogeneous rates.

5.5 Balancing Supply to Reduce Nitrogen Excretion

Although more limited for beef cattle, there are several reports of supply models being used to reduce N excretion for lactating dairy cows. Wu *et al.* (1997) summarized five experiments with respect to the Cornell model's ability to predict limiting amino acids and responses in milk production. The authors concluded that the model explained differences in milk yield, particularly for studies in which protein sources were manipulated compared with the use of rumen-protected methionine and/or lysine. Dietary protein could be reduced and milk N efficiency increased without a loss in milk production in one study (Kalscheur *et al.*, 1999). However, a constructive example (Dinn *et al.*, 1998) can demonstrate potential problems when this model is used to balance rations (rather than to evaluate them) to improve N efficiency. Diets were balanced to meet metabolizable lysine and methionine requirements estimated by the Cornell model while progressively decreasing dietary CP concentration and concomitantly increasing inputs of rumen-protected lysine and methionine. The partitioning of digestible protein toward milk N and away from urinary N increased progressively, as expected. The authors reported no change in milk N secretion, although it numerically decreased by 8.5%. DM intake and milk production both decreased significantly. St-Pierre and Thraen (1999) used the data of Dinn *et al.* (1998) to estimate that balancing diets for metabolizable amino acids actually would have cost US$1.10/kg reduction of N excretion. The Cornell model did not predict retained N well in another study (Haig *et al.*, 2002) and was marginally less effective than a procedure in which diets were balanced to meet predicted requirements of 15% and 5% of essential amino acid flow to the duodenum for lysine and methionine, respectively (Piepenbrink *et al.*, 1998). We note that the researchers' objective was to continue updating this model for field usage (Boston *et al.*, 2000), and ongoing efforts should increase its accuracy.

At the University of Reading, a series of studies have statistically evaluated dietary factors influencing N excretion. Castillo *et al.* (2000) compiled a database from 580 individual cows fed 90 treatments. They noted that, as N intake exceeded 400 g/day (corresponding to about 15% CP in the dietary DM), excretion of N in the urine increased exponentially. However, the authors noted that these data were from cows producing moderate amounts of milk (most <35 kg/day). For higher yielding cows under US conditions, Satter *et al.* (2002) recommended upper limits of about 17.5% CP. Still, both reviews note that the major response in CP intake above those amounts would be to increase urinary N output substantially.

The Reading group specifically investigated various managerial and dietary factors potentially influencing N excretion in urine. Kebreab *et al.* (2000) determined that cows fed early-cut grass silage had lower urinary N excretion but higher faecal N excretion when the grass was fertilized with a lower amount of N. Feeding a fibrous vs. starchy concentrate decreased faecal N loss but increased urinary N. Their data can be used to calculate that the starch-based concentrate increased the ratio of milk N:manure N excretion by 13% but only increased the ratio of (milk N plus retained N):manure N excretion by 5%. In another study (Castillo *et al.*, 2001a), cows that were fed highly degradable starch (mostly barley) had much higher N excretions in urine than those fed fibrous concentrate, low degradable starch (mostly ground maize) or soluble sugars (molasses). Numerically, the cows fed highly degradable starch had at least a 20% lower ratio of (milk N plus retained N):manure N excretion than the other groups. However, the group fed fibrous concentrates had an average of 48 g/day of N balance, which would equate to about 1.2 kg/day of body weight gain (Spanghero and Kowalski, 1997), which is probably high even for cows producing <21 kg/day of milk. In this study, the effects of feeding ground maize on tissue N balance support observations in late lactation cows receiving abomasal starch infusions at this location (Reynolds *et al.*, 2001), which we discussed previously. In another study (Castillo *et al.*, 2001b), concentrates with low or high percentages of CP were factorialized with high, medium or low RDP (soybean meal replaced by formaldehyde-treated soybean meal). Decreasing degradability greatly decreased urinary N while increasing tissue N balance. The RUP supply was always in excess of estimated requirements, but RDP became progressively limiting as degradability decreased, which likely would have progressively limited microbial N production.

After constructing a whole-body model to explain the preceding data, Kebreab *et al.* (2002) concluded that the efficiency of conversion of rumen-degradable protein into microbial protein 'had a major effect on N excretion especially by way of urine'. Similarly, the model predicted that increasing energy concentration (using the UK system) in the diet should decrease N losses, particularly in the urine. However, at an average N intake, N excretion in the urine still had a range of measured data about as large as the prediction. The authors stated that the model is a first step toward a mechanistic approach for nutrient modelling. This model, like others that have been reviewed, should be valuable for simulating N emissions from dairy systems, but predictive ability should improve with further development and adaptation to higher producing situations.

In the next few years, more studies should be available to evaluate the NRC (2001). Recently, Noftsger and St-Pierre (2003) balanced diets using the Cornell–Penn–Miner (CPM) model for metabolizable lysine and methionine based on feed samples screened before the study to have either low or high predicted intestinal digestibilities of the RUP. Only the high CP, high digestible RUP treatment was predicted to have a positive metabolizable protein balance (requirement < supply; Table 5.2). Therefore, selection for highly digestible RUP sources increased milk production, as expected, during the 12-week study. Despite a predicted negative metabolizable protein balance for cows fed both low protein diets, milk production for cows fed the blend of rumen-protected and -unprotected methionine was similar to those fed the diet with high CP/high digestible RUP. The diet with methionine increased efficiency of dietary N conversion into milk N and decreased N excretion (calculated by assuming zero N balance; i.e. dietary N intake − N secretion in milk) relative to N intake. Interestingly, during a 5-day digestibility experiment at the end of the production measurements, methionine addition did not increase N efficiency, perhaps because it might have ceased to be limiting by 16 weeks in lactation. Although demonstrating the difficulty of integrating N balance data with production data among published research, this report does highlight how emerging technology will likely be adapted in the future to improve efficiency of dietary conversion into milk protein.

Models have been developed to integrate dairy production and agronomic practice (Klausner *et al.*, 1998; Rotz *et al.*, 1999). We refer readers to Chapter 9 for a more extensive review of whole-farm implications. However, we note here that most, if not all, models ignore variation among cows within groups, nutrients in feeds and other factors that inflate 'safety factors' for protein intake on working farms. Table 5.3 estimates how uncertainty drives up CP percentage in dairy rations in practical situations (St-Pierre and Thraen, 1999). As can be seen, N efficiency was maximized at 14.9% CP, which agrees well with results from models based on individual cows in the UK (Castillo *et al.*, 2000; Kebreab *et al.*, 2002). Yet, such a strategy does not include effects of uncertainty, which would likely increase CP needed to maximise income over feed costs (Table 5.3). St-Pierre and Thraen (1999) argued that a strategy to maximize N efficiency while decreasing CP concentration of the diet would decrease N excretion by

Table 5.2. Least square means for performance measures for diets that vary in crude protein and digestibility of rumen-undegraded protein. From Noftsger and St-Pierre (2003).

	High CP[1]		Low CP[2]		
	DRUP	HDRUP	HDRUP	HDRUP + Met	SEM
Experiment 1 (n = 60; 12 weeks)					
DMI (kg/day)	21.7[a]	23.3[b]	23.2[b]	23.6[b]	0.49
Milk yield (kg/day)	40.8[a]	46.2[b]	42.9[a]	46.6[b]	0.72
N intake (g/day)	641[a]	690[b]	645[a]	651[a]	14.2
MP balance (g/day)[3]	−84	20	−58	−257	–
Milk N production (g/day)	188	214	203	228	3.9
Gross N efficiency[4]	29.5[a]	31.1[b]	31.7[b]	35.0[c]	0.60
Environmental efficiency[5]	2.47[a]	2.25[b]	2.19[b]	1.89[c]	0.06
Experiment 2 (n = 24; 5 days)					
N intake (g/day)	770[a]	735[a,b,c]	682[b,c]	679[b]	27.9
Faecal N (g/day)	279	271	257	263	10.9
Urine N (g/day)	268[a]	259[a,c]	216[b,d]	224[c,d]	19.3
Apparent N retention (g/day)	−1	−13	−16	−23	18.3
Productive N, % of N intake[6]	29.1	28.0	30.8	28.5	1.9
Environmental efficiency[7]	2.43[a]	2.44[a]	2.09[b]	2.24[a,b]	0.10

[1]High CP diets contained 18.3% CP. Diets had protein sources with low (LDRUP) or high (HDRUP) digestible rumen-undegraded protein.
[2]Low CP diets had 16.9% (LDRUP) or 17.0% (HDRUP) CP. The latter diet had extra supplemental methionine (Met) that was partially protected from ruminal degradation.
[3]Metabolizable protein (MP) balance (requirement−supply) from actual data using the NRC (2001) model.
[4]Calculated as milk N/N intake × 100.
[5]Calculated as kg N excreted/kg N in milk; N excreted calculated as N intake−milk N, assuming zero N balance.
[6]Productive N = milk N + retained N.
[7]Calculated as kg N excreted/kg N in milk; actual N excretion data were used.
[a,b,c,d]Treatment means in the same row with different superscripts are different ($P < 0.05$).

Table 5.3. Crude protein percentages required to optimize milk production, N efficiency, or income over feed costs (IOFC) with or without uncertainty of model parameters[a].

	Crude protein % required		
Scenario	Milk	N efficiency	IOFC
No uncertainty	18.5	17.0	17.7
With uncertainty	18.6	14.9	18.0

[a]Adapted from St-Pierre and Thraen (1999). Simulations are for a herd with high milk production potential (11,350 kg/year). N efficiency = kg milk/kg N excreted.

24% but would decrease milk production by 10.4%. Thus, this strategy was estimated to cost US$1.35 billion for a 'national' dairy herd, or up to US$9.55/kg of reduction of N excretion. Despite this uncertainty, the authors' simulations demonstrated that tighter grouping strategy would improve efficiency of N utilization. Yet, Jonker *et al.* (2002) noted that dairy farmers surveyed were not effectively grouping herds to reduce N loss. Clearly, confidence in ration balancing/modelling software needs to increase, including adaptability away from the 'average cow' toward group-feeding dynamics, before efficiency of N usage will be optimized.

5.6 Conclusions

The exchange of BUN with the gut is extensive and is probably an adaptive mechanism to enhance ruminal degradation of poor quality fibre even when N intake is low. When ruminal N increases, then BUN transfer to the urine becomes increasingly dominant and wasteful and potentially harmful to the environment. As much as 20% of the N lost into the environment, particularly from urine, is recoverable in cattle feeding operations. Methods are being refined to measure and predict the amount of microbial protein production in the rumen. Yet, despite the importance of the rumen, its metabolism interacts with the splanchnic and peripheral tissues. Research has documented how shifts in site of digestion and the metabolism by the splanchnic tissues influence whole-body N metabolism and excretion of urinary N. Improved integration of the rumen, gut and splanchnic tissues will advance the development of various models. Stage and level of production clearly influence metabolism of energy and amino acids, thereby affecting tissue N retention. Combined with better protein supplementation to meet metabolizable amino acid requirements, these systems will allow reduced inputs of dietary N and greater capture of BUN. As the prediction error in models is reduced and environmental regulations toughened, nutritional advisors should be able to use this information to decrease the amount of protein overfed with less risk of significant losses in animal productivity or loss of clientele.

References

Ainslie, S.J., Fox, D.G., Perry, T.C., Ketchen, D.J. and Barry, M.C. (1993) Predicting amino acid adequacy of diets fed to Holstein steers. *Journal of Animal Science* 71, 1312–1319.

Al-Dehneh, A., Huber, J.T., Wanderley, R., Theurer, C.B., Pessarakli, M. and DeYoung, D. (1997) Incorporation of recycled urea-N into ruminal bacteria flowing to the small intestine of dairy cows fed a high-grain or high-forage diet. *Animal Feed Science and Technology* 68, 327–338.

Alderman, G., France, J. and Kebreab, E. (2001a) A critique of the Cornell Net Carbohydrate and Protein System with emphasis on dairy cattle. 1. The rumen model. *Journal of Animal and Feed Sciences* 10, 1–24.

Alderman, G., Blake, J.S., France, J. and Kebreab, E. (2001b) A critique of the Cornell Net Carbohydrate and Protein System with emphasis on dairy cattle. 2. The post-rumen digestion model. *Journal of Animal and Feed Sciences* 10, 203–221.

Allen, M.S. (1997) Relationship between fermentation acid production in the rumen and the requirement for physically effective fiber. *Journal of Dairy Science* 80, 1447–1462.

Bannink, A. and De Visser, H. (1997) Comparison of mechanistic rumen models on mathematical formulation of extramicrobial and microbial processes. *Journal of Dairy Science* 80, 1296–1314.

Bargo, F., Muller, L.D., Kolver, E.S. and Delahoy, J.E. (2003) Invited review: production and digestion of supplemented dairy cows on pasture. *Journal of Dairy Science* 86, 1–42.

Bateman, H.G., Clark, J.H., Patton, R.A., Peel, C.J. and Schwab, C.G. (2001a) Accuracy and precision of computer models to predict passage of crude protein and amino acids to the duodenum of lactating cows. *Journal of Dairy Science* 84, 649–664.

Bateman, H.G., Clark, J.H., Patton, R.A., Peel, C.J. and Schwab, C.G. (2001b) Prediction of crude protein and amino acid passage to the duodenum of lactating cows by models compared with in vivo data. *Journal of Dairy Science* 84, 665–679.

Bell, A.W., Burhans, W.S. and Overton, T.R. (2000) Protein nutrition in late pregnancy, maternal protein reserves and lactation performance in dairy cows. *Proceedings of the Nutrition Society* 59, 119–126.

Bergen, W.R. (1978) Nutrient–nutrient, nutrient–hormonal and nutrient–genetic interactions. In: Church, D.C. (ed.) *Digestive Physiology and Nutrition of Ruminants.* Oxford Press, Portland, Oregon, pp. 430–441.

Boston, R.C., Fox, D.G., Sniffen, C., Janczewski, E., Munson, R. and Chalupa, W. (2000) The conversion of a scientific model describing dairy cow nutrition and production to an industry tool: the CPM Dairy Project. In: McNamara, J.P., France, J. and Beever, D.E. (eds) *Modeling Nutrient Utilization in Farm Animals.* CAB International, Wallingford, UK, pp. 361–377.

Broderick, G.A. (1995) Desirable characteristics of forage legumes for improving protein utilization in ruminants. *Journal of Animal Science* 73, 2760–2773.

Broderick, G.A. and Merchen, N.R. (1992) Markers for quantifying microbial protein synthesis in the rumen. *Journal of Dairy Science* 75, 2618–2632.

Castillo, A.R., Kebreab, E., Beever, D.E. and France, J. (2000) A review of efficiency of nitrogen utilisation in lactating dairy cows and its relationship with environmental pollution. *Journal of Animal and Feed Sciences* 9, 1–32.

Castillo, A.R., Kebreab, E., Beever, D.E., Barbi, J.H., Sutton, J.D., Kirby, H.C. and France, J. (2001a) The effect of energy supplementation on nitrogen utilization in lactating dairy cows fed grass silage diets. *Journal of Animal Science* 79, 240–246.

Castillo, A.R., Kebreab, E., Beever, D.E., Barbi, J.H., Sutton, J.D., Kirby, H.C. and France, J. (2001b) The effect of protein supplementation on nitrogen utilization in lactating dairy cows fed grass silage diets. *Journal of Animal Science* 79, 247–253.

Clark, J.H. and Davis, C.L. (1980) Some aspects of feeding high producing dairy cows. *Journal of Dairy Science* 63, 873–885.

Coleman, S.W. and Wyatt, R.D. (1982) Cottonseed meal or small grains forages as protein supplements fed at different intervals. *Journal of Animal Science* 55, 11–17.

Cotta, M.A. and Russell, J.B. (1997) Digestion of nitrogen in the rumen: a model for metabolism of nitrogen compounds in gastrointestinal environments. In: Mackie, R.I. and White, B.A. (eds) *Gastrointestinal Ecosystems and Fermentations*, Vol. 1. Chapman & Hall, New York, pp. 380–423.

Delgado-Elorduy, A., Theurer, C.B., Huber, J.T., Alio, A., Lozano, O., Sadik, M., Cuneo, P., De Young, H.D., Simas, I.J., Santos, J.E.P., Nussio, L., Webb, K.E. and Tagari, H. (2002a) Splanchnic and mammary nitrogen metabolism by dairy cows fed steam-rolled or steam-flaked sorghum grain. *Journal of Dairy Science* 85, 148–159.

Delgado-Elorduy, A., Theurer, C.B., Huber, J.T., Alio, A., Lozano, O., Sadik, M., Cuneo, P., De Young, H.D., Simas, I.J., Santos, J.E.P., Nussio, L., Webb, K.E. and Tagari, H. (2002b) Splanchnic and mammary nitrogen metabolism by dairy cows fed steam-rolled or steam-flaked corn. *Journal of Dairy Science* 85, 160–168.

Dewhurst, R.J., Davies, D.R. and Merry, R.J. (2000) Microbial protein supply from the rumen. *Animal Feed Science and Technology* 85, 1–21.

Dijkstra, J. (1994) Simulation of the dynamics of protozoa in the rumen. *British Journal of Nutrition* 72, 679–699.

Dijkstra, J., France, J. and Davies, D.R. (1998) Different mathematical approaches to estimating microbial protein supply in ruminants. *Journal of Dairy Science* 81, 3370–3384.

Dijkstra, J., Mills, J.A.N. and France, J. (2002) The role of dynamic modeling in understanding the microbial contribution to rumen function. *Nutrition Research Reviews* 15, 67–90.

Dinn, N.E., Shelford, J.A. and Fisher, L.J. (1998) Use of the Cornell Net Carbohydrate and Protein System and rumen-protected lysine and methionine to reduce nitrogen excretion from lactating dairy cows. *Journal of Dairy Science* 81, 229–237.

Doreau, M. and Ferlay, A. (1995) Effect of dietary lipids on nitrogen metabolism in the rumen: a review. *Livestock Production Science* 43, 97–110.

Drackley, J.K., Overton, T.R. and Douglas, G.N. (2001) Adaptations of glucose and long-chain fatty acid metabolism in liver of dairy cows during the periparturient period. *Journal of Dairy Science* 84 (E. Suppl.), E100–E112.

Faichney, G.J., Poncet, C., Lassalas, B., Jouany, J.P., Millet, L., Dore, J. and Brownlee, A.G. (1997) Effect of concentrates in a hay diet on the contribution of anaerobic fungi, protozoa and bacteria to nitrogen in rumen and duodenal digesta in sheep. *Animal Feed Science and Technology* 64, 193–213.

Firkins, J.L. (1996) Maximizing microbial protein synthesis in the rumen. *Journal of Nutrition* 126, 1347S–1354S.

Firkins, J.L. and Fluharty, F.L. (2000) Soy products as protein sources for beef and dairy cattle. In: Drackley, J.K. (ed.) *Soy in Animal Nutrition.* Federation of Animal Science Societies, Savoy, IL, pp. 182–214.

Firkins, J.L., Allen, M.S., Oldick, B.S. and St-Pierre, N.R. (1998) Modeling ruminal digestibility of carbohydrates and microbial protein flow to the duodenum. *Journal of Dairy Science* 81, 3350–3369.

Firkins, J.L., Eastridge, M.L., St-Pierre, N.R. and Noftsger, S.M. (2001) Effects of grain variability and processing on starch utilization by lactating dairy cattle. *Journal of Animal Science* 79 (E. Suppl.), E218–E238.

Gabler, M.T. and Heinrichs, A.J. (2003) Dietary protein to metabolizable energy ratios on feed efficiency and structural growth of prepubertal Holstein heifers. *Journal of Dairy Science* 86, 268–274.

Gibb, M.J., Ivings, W.E., Dhanoa, M.S. and Sutton, J.D. (1992) Changes in body components of autumn-calving Holstein-Friesian cows over the first 29 weeks of lactation. *Animal Production* 55, 339–360.

Greenfield, R.B., Cecava, M.J. and Donkin, S.S. (2000) Changes in mRNA expression for gluconeogenic enzymes in liver of dairy cattle during the transition to lactation. *Journal of Dairy Science* 83, 1228–1236.

Greenwood, R.H. and Titgemeyer, E.C. (2000) Limiting amino acids for growing Holstein steers limit-fed soyabean hull-based diets. *Journal of Animal Science* 78, 1997–2004.

Grummer, R.R. (1995) Impact of changes in organic nutrient metabolism on feeding the transition dairy cow. *Journal of Animal Science* 73, 2820–2833.

Haig, P.A., Mutsvangwa, T., Spratt, R. and McBride, B.W. (2002) Effects of dietary protein solubility on nitrogen losses from lactating dairy cows and comparison with predictions from the Cornell Net Carbohydrate and Protein System. *Journal of Dairy Science* 85, 1208–1217.

Harmon, D.L. and McLeod, K.R. (2001) Glucose uptake and regulation by intestinal tissues: implications and whole-body energetics. *Journal of Animal Science* 79 (E. Suppl.), E59–E72.

Hartwell, J.R., Cecava, M.J. and Donkin, S.S. (2000) Impact of dietary rumen undegradable protein and rumen-protected choline on intake, peripartum liver triacylglyceride, plasma metabolites and milk production in transition dairy cows. *Journal of Dairy Science* 83, 2907–2917.

Harvatine, D.I., Firkins, J.L. and Eastridge, M.L. (2002) Whole linted cottonseed as a forage substitute fed with ground or steam-flaked corn: digestibility and performance. *Journal of Dairy Science* 85, 1976–1987.

Hristov, A.N., Huhtanen, P., Rode, L.M., Acharya, S.N. and McAllister, T.A. (2001) Comparison of the ruminal metabolism of nitrogen from ^{15}N-labeled alfalfa preserved as hay or as silage. *Journal of Dairy Science* 84, 2738–2750.

Huntington, G.B. (1989) Hepatic urea synthesis and site and rate of urea removal from blood of beef steers fed alfalfa hay or a high concentrate diet. *Canadian Journal of Animal Science* 69, 215–223.

Hvelplund, T., Misciatelli, L. and Weisbjerg, M.R. (2001) Supply of the dairy cow with amino acids from dietary protein. *Journal of Animal and Feed Sciences* 10 (Suppl. 1), 69–85.

Iburg, M. and Lebzien, P. (2000) Requirements of lactating dairy cows for leucine and methionine at the duodenum. *Livestock Production Science* 62, 155–168.

James, T., Meyer, D., Esparza, E., DePeters, E.J. and Perez-Monti, H. (1999) Effects of dietary nitrogen manipulation on ammonia volatilization from manure from Holstein heifers. *Journal of Dairy Science* 82, 2430–2439.

Johnson, D.E. (1986) Fundamentals of whole animal calorimetry: use in monitoring body tissue deposition. *Journal of Animal Science* 63 (Suppl. 2), 111–114.

Jonker, J.S., Kohn, R.A. and High, J. (2002) Dairy herd management practices that impact nitrogen utilization efficiency. *Journal of Dairy Science* 85, 1218–1226.

Jouany, J.P. (1996) Effect of rumen protozoa on nitrogen utilization by ruminants. *Journal of Nutrition* 126, 1335S–1346S.

Kalscheur, K.F., Vandersall, J.H., Erdman, R.A., Kohn, R.A. and Russek-Cohen, E. (1999) Effects of dietary crude protein concentration and degradability on milk production responses of early, mid, and late lactation dairy cows. *Journal of Dairy Science* 82, 545–554.

Kebreab, E., Castillo, A.R., Beever, D.E., Humphries, D.J. and France, J. (2000) Effects of management practices prior to and during ensiling and concentrate type on nitrogen utilization in dairy cows. *Journal of Dairy Science* 83, 1274–1285.

Kebreab, E., France, J., Mills, J.A.N., Allison, R. and Dijkstra, J. (2002) A dynamic model of N metabolism in the lactating dairy cow and an assessment of impact on N excretion on the environment. *Journal of Animal Science* 80, 248–259.

Kennedy, P.M. and Milligan, L.P. (1980) The degradation and utilization of endogenous urea in the gastrointestinal tract of ruminants: a review. *Canadian Journal of Animal Science* 60, 205–221.

Klausner, S.D., Fox, D.G., Rasmussen, C.N., Pitt, R.E., Tylutki, T.P., Wright, P.E., Chase, L.E. and Stone, W.C. (1998) Improving dairy farm sustainability. I. An approach to animal and crop nutrient management planning. *Journal of Production Agriculture* 11, 225–233.

Klopfenstein, T.J. and Erickson, G.E. (2002) Effects of manipulating protein and phosphorus nutrition of feedlot cattle on nutrient management and the environment. *Journal of Animal Science* 80 (E. Suppl. 2), E106–E114.

Klopfenstein, T.J., Mass, R.A., Creighton, K.W. and Patterson, H.H. (2001) Estimating forage protein degradation in the rumen. *Journal of Animal Science* 79 (E. Suppl.), E208–E217.

Kohn, R.A. and Allen, M.S. (1995) Effect of plant maturity and preservation method on in vitro protein degradation of forages. *Journal of Dairy Science* 78, 1544–1551.

Kohn, R.A., Dou, Z., Ferguson, J.D. and Boston, R.C. (1997) A sensitivity analysis of nitrogen losses from dairy farms. *Journal of Environmental Management* 50, 417–428.

Kolver, E.S. and Muller, L.D. (1998) Performance and nutrient intake of high producing Holstein cows consuming pasture or a total mixed ration. *Journal of Dairy Science* 81, 1403–1411.

Komaragiri, M.V.S. and Erdman, R.A. (1997) Factors affecting body tissue mobilization in early lactation dairy cows. 1. Effect of dietary protein on mobilization of body fat and protein. *Journal of Dairy Science* 80, 929–937.

Krehbiel, C.R., Ferrell, C.L. and Freetly, H.C. (1998) Effects of frequency of supplementation on dry matter intake and net portal and hepatic flux of nutrients in mature ewes that consume low-quality forage. *Journal of Animal Science* 76, 2464–2473.

Kuipers, A., Mandersloot, F. and Zom, R.L.G. (1999) An approach to nutrient management on dairy farms. *Journal of Dairy Science* 82 (Suppl. 2), 84–89.

Lapierre, H. and Lobley, G.E. (2001) Nitrogen recycling in the ruminant: a review. *Journal of Dairy Science* 84 (E. Suppl.), E223–E236.

Leng, R.A. and Nolan, J.V. (1984) Nitrogen metabolism in the rumen. *Journal of Dairy Science* 67, 1072–1089.

Lobley, G.E., Connell, A., Lomax, M.A., Brown, D.S., Milne, E., Calder, A.G. and Farringham, D.A.H. (1995) Hepatic detoxification of ammonia in the ovine liver: possible consequences for amino acid catabolism. *British Journal of Nutrition* 73, 667–685.

Lobley, G.E., Weijs, P.J.M., Connell, A., Calder, A.G., Brown, D.S. and Milne, E. (1996) The fate of absorbed and exogenous ammonia as influenced by forage or forage–concentrate diets in growing sheep. *British Journal of Nutrition* 76, 231–248.

Lobley, G.E., Bremner, D.M. and Zuur, G. (2000) Effects of diet quality on urea fates in sheep as assessed by refined, non-invasive [^{15}N^{15}N] urea kinetics. *British Journal of Nutrition* 84, 459–468.

Maltby, S.A., Reynolds, C.K., Lomax, M.A. and Beever, D.E. (1993) The effect of increased absorption of ammonia and arginine on splanchnic metabolism of beef steers. *Animal Production* 56, 462 (Abstract).

Marini, J.C. and Van Amburgh, M.E. (2003) Nitrogen metabolism and recycling in Holstein heifers. *Journal of Animal Science* 81, 545–552.

Martin, A.K. (1966) Some errors in the determination of nitrogen retention of sheep by nitrogen balance studies. *British Journal of Nutrition* 20, 325–337.

McBride, B.W. and Kelly, J.M. (1990) Energy cost of absorption and metabolism in the ruminant gastrointestinal tract and liver: a review. *Journal of Animal Science* 68, 2997–3010.

McGuffey, R.K., Richardson, L.F. and Wilkinson, J.I.D. (2001) Ionophores for dairy cattle: current status and future outlook. *Journal of Dairy Science* 84 (E. Suppl.), E194–E203.

Meyer, D. and Mullinax, D.D. (1999) Livestock nutrient management concerns: regulatory and legislative overview. *Journal of Dairy Science* 82 (Suppl. 2), 51–62.

Milano, G.D. and Lobley, G.E. (2001) Liver nitrogen movements during short-term infusion of high levels of ammonia into the mesenteric vein of sheep. *British Journal of Nutrition* 86, 507–513.

Milano, G.D., Hotston-Moore, A. and Lobley, G.E. (2000) Influence of hepatic ammonia removal on ureagenesis, amino acid utilization and energy metabolism in the ovine liver. *British Journal of Nutrition* 83, 307–315.

Moorby, J.M. and Theobald, V.J. (1999) Short communication: the effect of duodenal ammonia infusions on milk production and nitrogen balance of the dairy cow. *Journal of Dairy Science* 82, 2440–2442.

Nelson, C.J. (1999) Managing nutrients across regions of the United States. *Journal of Dairy Science* 82 (Suppl. 2), 90–100.

Neter, J., Kutner, M.H., Nachtscheim, C.J. and Wasserman, W. (1996) *Applied Linear Regression Models*, 3rd edn. McGraw-Hill, Chicago, Illinois, pp. 82–84.

Noftsger, S. and St-Pierre, N.R. (2003) Supplementation of methionine and selection of highly digestible rumen undegradable protein to improve nitrogen efficiency for milk production. *Journal of Dairy Science* 86, 958–969.

Nolan, J.V. (1975) Quantitative models of nitrogen metabolism in sheep. In: McDonald, I.W. and Warner, A.C.I. (eds.) *Digestion and Metabolism in the Ruminant.* University of New England Publishing Unit, Armidale, Australia, pp. 416–431.

NRC [National Research Council] (1989) *Nutrient Requirements of Dairy Cattle*, 6th rev. edn. National Academy Press, Washington, DC.

NRC [National Research Council] (2000) *Nutrient Requirements of Beef Cattle*, 7th rev. edn. National Academy Press, Washington, DC, 248 pp.

NRC [National Research Council] (2001) *Nutrient Requirements of Dairy Cattle*, 7th rev. edn. National Academy Press, Washington, DC, 381 pp.

Oba, M. and Allen, M.S. (2003a) Effects of diet fermentability on efficiency of microbial nitrogen production in lactating dairy cows. *Journal of Dairy Science* 86, 195–207.

Oba, M. and Allen, M.S. (2003b) Hypophagic effects of ammonium are greater when infused with propionate compared with acetate in lactating dairy cows. *Journal of Nutrition* 133, 1100–1104.

Obitsu, T., Bremner, D., Milne, E. and Lobley, G.E. (2000) Effect of abomasal glucose infusion on alanine metabolism and urea production in sheep. *British Journal of Nutrition* 84, 157–163.

O'Connor, J.D., Sniffen, C.J., Fox, D.G. and Chalupa, W. (1993) A net carbohydrate and protein system for evaluating cattle diets: IV. Predicting amino acid adequacy. *Journal of Animal Science* 71, 1298–1311.

Oldham, J.D. and Smith, T. (1980) Protein–energy interrelationships for growing and for lactating cattle. In: Miller, E.L., Pike, I.H. and Van Es, A.J.H. (eds) *Protein Contribution of Feedstuffs for Ruminants: Application to Feed Formulation.* Butterworth Scientific, London, pp. 103–130.

Oldick, B.S., Firkins, J.L. and St-Pierre, N.R. (1999) Estimation of microbial nitrogen flow to the duodenum of cattle based on dry matter intake and diet composition. *Journal of Dairy Science* 82, 1497–1511.

Oldick, B.S., Firkins, J.L. and Kohn, R.A. (2000) Compartmental modelling with nitrogen-15 to determine effects of degree of fat saturation on intraruminal N recycling. *Journal of Animal Science* 78, 2421–2430.

Parker, D.S., Lomax, M.A., Seal, C.J. and Wilton, J.C. (1995) Metabolic implications of ammonia production in the ruminant. *Proceedings of the Nutrition Society* 54, 549–563.

Piepenbrink, M.S., Schingoethe, D.J., Brouk, M.J. and Stegeman, G.A. (1998) Systems to evaluate the protein quality of diets fed to lactating cows. *Journal of Dairy Science* 81, 1046–1061.

Plaizier, J.C., Martin, A., Duffield, T., Bagg, R., Dick, P. and McBride, B.W. (2000) Effect of a prepartum administration of monensin in a controlled-release capsule on apparent digestibilities and nitrogen utilization in transition dairy cows. *Journal of Dairy Science* 83, 2918–2925.

Putnam, D.E. and Varga, G.A. (1998) Protein density and its influence on metabolite concentration and nitrogen retention by Holstein cows in late gestation. *Journal of Dairy Science* 81, 1608–1618.

Reynolds, C.K. (1995) Quantitative aspects of liver metabolism in ruminants. In: Engelhardt, W.V., Leonhard-Marek, S., Breves, G. and Giesecke, D. (eds)

Ruminant Physiology: Digestion, Metabolism, Growth and Reproduction: Proceedings of the Eighth International Symposium on Ruminant Physiology. Ferdinand Enke Verlag, Stuttgart, Germany, pp. 351–371.

Reynolds, C.K. (2000) Forage evaluation using measurements of energy metabolism. In: Givens, D.I., Owen, E., Axford, R.F.E. and Omed, H.M. (eds) *Forage Evaluation in Ruminant Nutrition.* CAB International, Wallingford, UK, pp. 95–111.

Reynolds, C.K. (2002) Economics of visceral tissue metabolism in ruminants. Toll keeping or internal revenue service? *Journal of Animal Science* 80 (E. Suppl.), E74–E84.

Reynolds, C.K. (2005) Nitrogen metabolism by splanchnic tissues of ruminants. In: Burrin, D. and Merssman, H. (eds) *Biology of Metabolism of Growing Animals.* Elsevier Science, Oxford, UK, (in press).

Reynolds, C.K. and Huntington, G.B. (1988) Partition of portal-drained visceral net flux in beef steers. 1. Blood flow and net flux of oxygen, glucose and nitrogenous compounds across stomach and post-stomach tissues. *British Journal of Nutrition* 60, 539–551.

Reynolds, C.K., Humphries, D.J., Cammell, S.B., Benson, J., Sutton, J.D. and Beever, D.E. (1998) Effects of abomasal wheat starch infusion on splanchnic metabolism and energy balance of lactating dairy cows. In: McCracken, K.J., Unsworth, E.F. and Wylie, A.R.G. (eds) *Energy Metabolism of Farm Animals. Proceedings of the 14th Symposium on Energy Metabolism, Newcastle, Northern Ireland.* CAB International, Wallingford, UK, pp. 39–42.

Reynolds, C.K., Cammell, S.B., Humphries, D.J., Beever, D.E., Sutton, J.D. and Newbold, J.R. (2001) Effects of post-rumen starch infusion on milk production and energy metabolism in dairy cows. *Journal of Dairy Science* 84, 2250–2259.

Reynolds, C.K., Aikman, P.C., Lupoli, B., Humphries, D.J. and Beever, D.E. (2003) Splanchnic metabolism of dairy cows during the transition from late gestation through early lactation. *Journal of Dairy Science* 86, 1201–1217.

Rotz, C.A., Mertens, D.R., Buckmaster, D.R., Allen, M.S. and Harrison, J.H. (1999) A dairy herd model for use in whole farm simulations. *Journal of Dairy Science* 82, 2826–2840.

Russell, J.B. and Wilson, D.B. (1996) Why are ruminal cellulolytic bacteria unable to digest cellulose at low pH? *Journal of Dairy Science* 79, 1503–1509.

Santos, J.E.P., DePeters, E.J., Jardon, P.W. and Huber, J.T. (2001) Effect of prepartum dietary protein level on performance of primigravid and multiparous Holstein dairy cows. *Journal of Dairy Science* 84, 213–224.

Satter, L.D., Klopfenstein, T.J. and Erickson, G.E. (2002) The role of nutrition in reducing nutrient output from ruminants. *Journal of Animal Science* 80 (E. Suppl. 2), E143–E156.

Schneider, B.H. and Flatt, W.P. (1975) *The Evaluation of Feeds Through Digestibility Experiments.* University of Georgia Press, Athens, Georgia, pp. 168–178.

Sharp, R., Ziemer, C.J., Stern, M.D. and Stahl, D.A. (1998) Taxon-specific associations between protozoal and methanogen populations in the rumen and a model rumen system. *Federation of European Microbiological Societies Microbiology Ecology* 26, 71–78.

Shingfield, K.J. (2000) Estimation of microbial protein supply in ruminant animals based on renal and mammary purine metabolite excretion. A review. *Journal of Animal and Feed Sciences* 9, 169–212.

Sloan, B.K. (1997) Developments in amino acid nutrition of dairy cows. In: Garnsworthy, P.C. and Wiseman, J. (eds) *Recent Advances in Animal Nutrition.* Nottingham University Press, Nottingham, UK, pp. 167–198.

Spanghero, M. and Kowalski, Z.M. (1997) Critical analysis of N balance experiments with lactating cows. *Livestock Production Science* 52, 113–122.

St-Pierre, N.R. (2001) Integrating quantitative findings from multiple studies using mixed model methodology. *Journal of Dairy Science* 84, 741–755.

St-Pierre, N.R. (2003) Reassessment of biases in predicted nitrogen flows to the duodenum by NRC 2001. *Journal of Dairy Science* 86, 344–350.

St-Pierre, N.R. and Glamocic, D. (2000) Estimating unit costs of nutrients from market prices of feedstuffs. *Journal of Dairy Science* 83, 1402–1411.

St-Pierre, N.R. and Thraen, C.S. (1999) Animal grouping strategies, sources of variation, and economic factors affecting nutrient balance on dairy farms. *Journal of Dairy Science* 82 (Suppl. 2), 72–83.

Stern, M.D., Bach, A. and Calsamiglia, S. (1997) Alternative techniques for measuring nutrient digestion in ruminants. *Journal of Animal Science* 75, 2256–2276.

Stockdale, C.R. and Roche, J.R. (2002) A review of the energy and protein nutrition of dairy cows through their dry period and its impact on early lactation performance. *Australian Journal of Agricultural Research* 53, 737–753.

Sutter, F. and Beever, D.E. (2002) Energy and nitrogen metabolism in Holstein-Friesian cows during early lactation. *Animal Science* 70, 503–514.

Vanhatalo, A., Huhtanen, P., Toivonen, V. and Varvikko, T. (1999) Response of dairy cows fed grass silage diets to abomasal infusions of histidine alone or in combinations with methionine and lysine. *Journal of Dairy Science* 82, 2674–2685.

Van Horn, H.H. and Hall, M.B. (1997) Agricultural and environmental issues in the management of cattle manure. In: Recheigl, J.E. and MacKinnon, H.C. (eds) *Agricultural Uses of By-products and Wastes.* American Chemical Society, Washington, DC, pp. 91–109.

Van Horn, H.H., Newton, G.L. and Kunkle, W.E. (1996) Ruminant nutrition from an environmental perspective: factors affecting whole-farm nutrient balance. *Journal of Animal Science* 74, 3082–3102.

Van Saun, R.J., Idleman, S.C. and Sniffen, C.J. (1993) Effect of undegradable protein amount fed prepartum on postpartum production in first lactation Holstein cows. *Journal of Dairy Science* 76, 236–244.

Wallace, R.J., Onodera, R. and Cotta, M.A. (1997) Metabolism of nitrogen-containing compounds. In: Hobson, P.N. and Stewart, C.S. (eds) *The Rumen Microbial Ecosystem.* Chapman & Hall, New York, pp. 283–328.

Waterlow, J.C. (1999) The mysteries of nitrogen balance. *Nutrition Research Reviews* 12, 25–54.

Wells, J. E. and Russell, J.B. (1996) Why do many ruminal bacteria die and lyse so quickly? *Journal of Dairy Science* 79, 1487–1495.

Wessels, R.H. and Titgemeyer, E.C. (1997) Protein requirements of growing steers limit-fed corn-based diets. *Journal of Animal Science* 75, 3278–3286.

Whitelaw, F.G., Milne, J.S. and Wright, S.A. (1991) Urease (*EC* 3.5.1.5) inhibition in the sheep rumen and its effect on urea and nitrogen metabolism. *British Journal of Nutrition* 66, 209–225.

Whitlock, B.K., VandeHaar, M.J., Silva, L.F.P. and Tucker, H.A. (2002) Effect of dietary protein on prepubertal mammary development in rapidly growing dairy heifers. *Journal of Dairy Science* 85, 1516–1525.

Wu, Z. and Huber, J.T. (1994) Relationship between dietary fat supplementation and milk protein concentration in lactating cows: a review. *Livestock Production Science* 39, 141–155.

Wu, Z., Polan, C.E. and Fisher, R.J. (1997) Adequacy of amino acids in diets fed to lactating dairy cows. *Journal of Dairy Science* 80, 1713–1721.

6 Phosphorus Metabolism in the Rumen

R.L. Kincaid[1] and M. Rodehutscord[2]
[1]*Animal Sciences Department, Washington State University, Pullman, Washington, USA*
[2]*Institut für Ernährungswissenschaften, Martin-Luther-Universität Halle-Wittenberg, Halle, Germany*

6.1 Introduction

Among the signs of low phosphorus (P) intakes in cows are a general unthriftiness, body weight loss, reduced feed consumption, reluctance to move, abnormal stance, spontaneous bone fractures and finally, impaired reproductive performance (Call *et al.*, 1986). Read *et al.* (1986) suggested that feed intake depression during late lactation and early pregnancy was the most serious effect of P deficiency. The P needs of the rumen microbes may be greater than for the host animal and take priority for P. If the P requirements for growth of ruminal microbes are not met, fermentation and protein synthesis rates are reduced and digestibility of dietary organic matter (OM) may be lowered. The role of P for processes such as formation of phospholipids and nucleotides or ATP synthesis and regulation of enzyme activities in rumen microbes is similar to that in higher organisms. Thus, supplying the ruminal microbes with P is needed to maximize microbial growth and feed intake. Salivary inorganic phosphate (P_i) and dietary P are the two sources of P entering the rumen.

6.2 Estimates of P Requirement of Rumen Microbes

Durand and Kawashima (1980) reported a range in total P content of rumen microorganisms from 2% to 6% in dry mass. In the RUSITEC system, values were lower for both liquid (1.2%) and solid-associated bacteria (0.9% of dry mass; Komisarczuk *et al.*, 1987a), and there was no effect of P depletion from artificial saliva. Komisarczuk-Bony and Durand (1991) reviewed from the literature that the nitrogen (N):P ratio in different fractions of rumen microbes may vary from 4.0:1 to 8.4:1, with most data between 6.0:1 and 7.3:1. Attempts were made to quantify the P requirement of rumen microbes on the basis of net protein synthesis or degradation of carbohydrate fractions. Such values were usually expressed as concentra-

©CAB International 2005. *Nitrogen and Phosphorus Nutrition of Cattle*
(eds E. Pfeffer and A.N. Hristov)

tions in rumen fluid. The suggested lower concentration of P to maintain normal microbial growth in the rumen is 100 mg P/l (3.2 mmol/l) of ruminal fluid (Durand and Kawashima, 1980). In several studies in which low P diets were fed to cattle, P concentrations in ruminal fluid were well above this level (Witt and Owens, 1983; De Waal and Koekemoer, 1997). Concentrations of P in ruminal fluid can distinguish between P supplemented and non-supplemented cattle, but cannot distinguish between levels of P supplementation (De Waal and Koekemoer, 1997).

Other authors tried to estimate P requirements on the basis of microbial yield. Depending on what N:P ratio was assumed, these estimates range from 2.8 to 4.0 g P/kg digestible OM (Durand and Kawashima, 1980; Smith, 1984). The authors assume that the microbial protein yield is 30 g N/kg fermented OM and that the fraction fermented in the forestomach is 65% of digestible OM. The P requirement expressed in such a way may, however, be misleading because it ignores the high contribution of salivary P flow. In adult non-lactating animals, the P requirements are sufficiently low such that high plasma P_i concentrations and subsequent salivary P flow can compensate for dietary P concentrations below the aforementioned values.

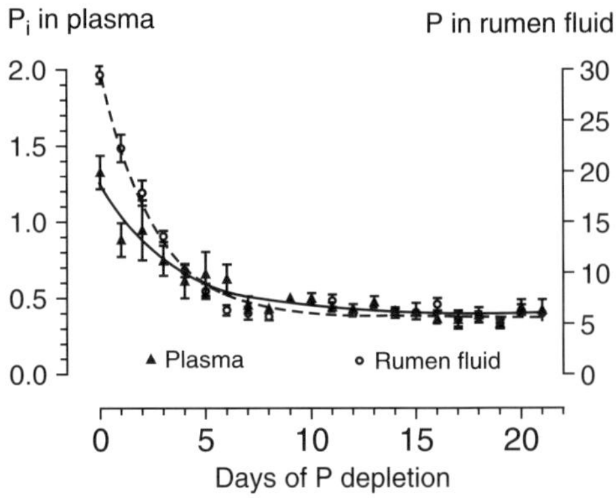

Fig. 6.1. Response of lactating dairy goats in P concentrations (mmol/l) during a 3-week P depletion period (adapted from Rodehutscord *et al.*, 1994).

6.3 Phosphorus Recycling to the Rumen via Saliva and Consequences of Phosphorus Deficiency

Recycling of endogenous P via salivary secretions into the rumen supplies much of the P to ruminal microbes. In small ruminants, the high correlation between P concentrations in blood plasma, saliva and ruminal fluid is well investigated (Breves, 1991). Similarly in cows, lower saliva P concentrations are induced by lower P concentrations in plasma (Valk *et al.*, 2002). Because the plasma P_i concentration depends on the dietary P concentration (Rodehutscord *et al.*, 1994), the supply of P to the rumen is affected by changes in the dietary P concentration in two ways: the direct diet effect and the subsequent saliva effect. After onset of a severe P depletion in lactating goats, plasma P_i and ruminal P decreased immediately from about 1.5 and 30 mmol/l to less than 0.5 and 5 mmol/l within about 10 days (Fig. 6.1). With the onset of P repletion, concentrations immediately increased again. The salivary gland is able to concentrate P in comparison with the blood plasma by a factor of up to 16, and a daily secretion of P_i via saliva between 30 and 60 g in cows is achieved by a combination of the gland's ability to concentrate P and the high salivary flow rate (Breves and Schröder, 1991). Thus, salivary P secretion may account for more than 50% of P entering the rumen (see Chapter 7). As dietary P intake decreases, salivary P increases as a percentage of total P entering the rumen. Feed intake and salivary flow rate is another important factor, and salivary P volume is related to dry matter intake (Karn, 2001).

In cows, salivary P concentrations during lactation varied between 4.3 and 8.4 mmol/l, and during the non-lactating period, saliva P concentration varied between 8.2 and 12.1 mmol/l (Valk *et al.*, 2002). These values agree with an earlier report that the mean composition of mixed saliva and parotid saliva of cows is 26 and 23 mEq P_i/l (Bailey and Balch, 1961), which is equivalent to 12.9 and 11.4 mmol/l, respectively. Different glands show, however, great differences in salivary P_i concentration (Kay, 1960), and the way of saliva sampling is a critical factor when literature data are to be compared. The increase in endogenous P as a percentage of total P in the rumen with decreasing dietary P concentration means that there is a net movement of endogenous P to the rumen, hence supporting the hypothesis that ruminal microbial P demands are greater than the animal's P demand (Preston and Pfander, 1964). Salivary P may be the main P source for ruminal microbes especially when insoluble phosphates are the main P source consumed by ruminants (Durand and Kawashima, 1980).

Although reduced feed intake is a common response of ruminants to P deficiency the mechanisms for this are not yet fully understood. The high correlation between plasma P_i concentration and salivary P flow makes it difficult to experimentally separate effects in the metabolism of the host from those in the rumen. Reduced OM fermentation in the rumen likewise contributes to reduced feed intake. However, studies by Milton and Ternouth (1985) indicate that intake could also be affected directly by reduced plasma P_i concentrations. The drain of P via milk makes the consequences of a reduced dietary P concentration more severe in lactating animals than in the non-lactating ones.

Often, P deficiency leads to reduced OM digestibility with effects on fermentation both in the rumen and in the hindgut (Breves and Höller, 1987), suggesting that microbes in different sections of the gastrointestinal tract are similarly affected. Reduced fermentation leads to the well-known reduction in the supply of microbial protein to the duodenum. Furthermore, in studies with lactating goats the efficiency of net protein synthesis (microbial protein synthesis per unit of digested OM) was by 0.30 times lower during P depletion as compared to adequate P supply (Petri *et al.*, 1988). Because the N:P ratio in isolated rumen microbes is largely unaffected by dietary P supply (Breves, 1991; Komisarczuk-Bony and Durand, 1991) microbes were unable to synthesize essential phosphate-containing cell materials, making P in rumen fluid the limiting factor for growth.

The impact of ruminal P concentration on carbohydrate digestibility is unclear. Early work by Raun *et al.* (1956) found increased digestibility *in vitro* with added P from either inorganic P or phytate-P. Witt and Owens (1983) did not find an effect of P on microbial digestion when the ruminal P concentration increased from 6.7 to 12.8 mmol/l. At much lower P levels, phosphate seems to be required in higher concentrations specifically for cellulolysis as compared to hemicellulolysis and amylolysis *in vitro* (Komisarczuk *et al.*, 1987b). A continuous reduction in ATP concentration with progressive P depletion was also reported in this study. Pure culture systems indicate a minimal P requirement of 0.2 and 0.5 mmol P/l for *Ruminococcus flavefaciens* and *Fibrobacter succinogenes*, respectively (Komisarczuk-Bony and Durand, 1991).

6.4 Inevitable Losses of Phosphorus Caused by Microbes

In ruminal bacteria, 80% of total P is present in the nucleic acids and 10% in phospholipids. The P content of nucleic acids is about 10% (Komisarczuk-Bony and Durand, 1991). Fermentable OM supply mainly determines microbial growth in the rumen, and 27.8 g microbial N enters a cow's duodenum per kg digestible OM intake (Rohr *et al.*, 1986). This is equivalent to 4.3 g P assuming an average N:P ratio in mixed rumen bacteria of 6.5:1 (Komisarczuk *et al.*, 1987a). Armstrong and Hutton (1975) reported that between 0.77 and 0.82 of the nucleic acid-N entering the small intestine disappeared within it. Assuming that microbially bound P is as digestible as microbially bound N would consequently mean a P excretion originating from microbial origin of 0.9 g/kg digestible OM intake. This accounts for roughly two-thirds of P that is inevitably lost in faeces (Rodehutscord *et al.*, 2000). The combination of P in faecal microbial debris and endogenous faecal P is about half of the total faecal P (Conrad, 1999), but this proportion is variable depending on the degree of excess in P supply.

6.5 Availability of Phosphorus in Feeds – Phytate-P and Phytate Degradation

Although leaves and stems of plants contain only trace amounts of phytin (phytic acid [myo-inositol hexakisphosphate] and its salts), about two-thirds of P is present as phytin-P in cereal grains, oilseeds and grain by-products (Nelson *et al.*, 1976; Eeckhout and De Paepe, 1994). Presumably, non-phytate P in forages and other feeds is readily available for absorption in ruminants. For example, Lofgreen and Kleiber (1954) reported about 94% of P in lucerne were absorbed in sheep. However, much of the P in seed grains is present as phytin from which the P cannot be absorbed by animals unless the P is hydrolysed. Phytate in soybean is distributed uniformly in the protein matrix of soybeans and not localized in specific areas of protein bodies such as in wheat

(Tombs, 1967). Phytate in rapeseed is located in globoid crystals that remain associated with the denatured protein in processed rapeseed meal (Yiu *et al.*, 1983).

Phytase activity in the rumen is largely of bacterial origin and associated with the cell pellet, not the ruminal fluid supernatant (Yanke *et al.*, 1998). The highest phytase activity is produced by those strains of ruminal bacteria associated with starch fermentation and not with the protozoa and fungi. The implication is that for myo-inositol hexakisphosphate to be hydrolysed, the phytate must be consumed by the bacteria. Phytin hydrolysis might also be caused by intrinsic phytase contained in the diet. Cereal grains and the corresponding bran have intrinsic phytase activity, particularly wheat, rye, triticale and, to a lesser extent, barley grains (Eeckhout and De Paepe, 1994). In maize, phytase is below detection limit. In pigs, plant phytase causes phytin hydrolysis in the gut and improves digestibility of plant P. This effect is, however, lower than the one caused by *Aspergillus niger* phytase (Zimmermann *et al.*, 2002), possibly due to the differences in the pH optimum between phytases or by differences in the resistance to proteolytic activity (Simon and Igbasan, 2002). At present it is not possible to give an estimate of the quantitative contribution of plant phytase to ruminal phytate hydrolysis.

Using calcium (Ca) phytate and *in vitro* rumen fermentation, Raun *et al.* (1956) found the same response in cellulose digestion to increasing phytate-P in the medium as for P originating from a standard inorganic source. This was taken as an indication for a 100% breakdown of phytate-P with an optimal rumen pH of 5.5 for phytate hydrolysis. Similarly, there was greater than 90% disappearance of phytate-P from dietary solids between 6 and 8 h of ruminal incubation. In lactating dairy cows, there was 99% disappearance of phytate-P in the faeces (Morse *et al.*, 1992). In calves, phytate hydrolysis occurs with the establishment of a microbial population in the developing rumen. When calves (56 days old) and steers (9 months old) were fed concentrate type diets with natural phytate-P, there was 100% hydrolysis in the steers and over 99% hydrolysis in the calves (Nelson *et al.*, 1976). In the study by Morse *et al.* (1992) on phytate-P disappearance *in vitro* from feeds, cottonseed meal had the slowest disappearance of phytate-P. In a separate experiment, total faecal collection was done in 11 cows and, using Cr as an indigestible marker, between 94% and 98% phytate hydrolysis was found.

Net absorption of dietary P did not differ due to source of supplemental P (mono- and dicalcium phosphate or wheat bran), average 45% across diets for lactating Holstein cows (Knowlton *et al.*, 2001). Absorption can, however, be affected by the level of P supply above requirement due to adaptive mechanisms of the animal. Hence, a marginal level of P supply must be assured in studies dealing with the availability of dietary P (AFRC, 1991; Rodehutscord *et al.*, 2000). Such marginal level is easier to ensure in lactating animals due to the drain of P via milk, which makes P availability studies undertaken with adult wether sheep hard to interpret. In lactating goats fed below their net requirement for P, net absorption of supplemented P was about 90% and not significantly different between monocalcium phosphate, dicalcium phosphate, grass hay and wheat bran (Koddebusch and Pfeffer, 1988). When solvent-extracted rapeseed meal was supplemented to a low-P basal diet, faecal P excretion of lactating dairy goats remained unaffected (Fig. 6.2).

Using a nylon bag technique to determine P availability, Bravo *et al.* (2000) found that rumen availability was not uniform amongst feedstuffs, varying from 33% (formaldehyde-treated rapeseed meal) to 85% (wheat). Phosphorus release was less in maize distillers (66%) than wheat by-products (85%), probably because in wheat, P is concentrated in peripheral envelopes whereas P in maize is concentrated in the germ. Intrinsic phytase activity from wheat might have played a

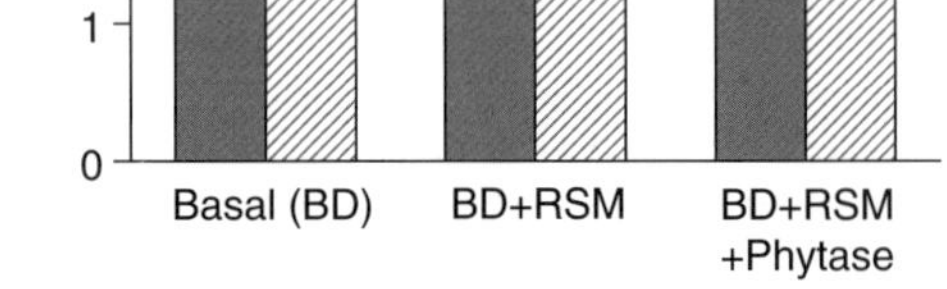

Fig. 6.2. Effect of a supplementation of solvent-extracted rapeseed meal (RSM) and an *Aspergillus niger* phytase on intake and faecal excretion of P in lactating dairy goats with a negative P balance (n = 4 per treatment) (Rodehutscord and Pfeffer, unpublished).

role as well. The remaining P in maize distillers may be integrated as a structural part of the cell walls of the peripheral part of the grain. More than 20% of phytate from rapeseed meal were recovered in the duodenum of sheep (Park *et al.*, 1999, 2000). Thus, phytate disappearance is not always complete in the rumen because of inaccessibility of the phytate to ruminal microorganisms. Hristov *et al.* (2004), however, reported high ruminal effective *in situ* degradability (mostly due to high solubility) of forage (from 93%, lucerne hay to 98%, maize silage) and concentrate (57%, barley grain to 92%, whole cottonseed) P.

Both processing of dietary ingredients and associative factors affect phytate hydrolysis. In non-ruminants, high Ca diets may reduce hydrolysis of phytate-P by forming insoluble Ca phytate that precipitates out of solution (Scheuermann *et al.*, 1988). Formaldehyde treatment of soybean meal and rapeseed meal reduced phytate degradation rate (Park *et al.*, 1999), and heat treatment did as well (Konishi *et al.*, 1999). Because phytate complexes with proteins, chemical and heat treatments that reduce the ruminal solubility of the protein–phytate complex reduce the rate and extent of phytate degradation. Thus, when rapeseed meal was heated at 143°C, between 31% and 58% of phytate-P escaped rumen degradation when the ruminal outflow rates were 0.02 to 0.08/h (Konishi *et al.*, 1999).

Although little phytate is normally found in ruminant faeces, complete hydrolysis of all dietary phytate in the rumen does not occur in all feeding conditions. Some ingested phytate escapes to the lower digestive tract where P absorption is reduced. Cattle fed barley varieties with reduced phytic acid-P excreted about 60% less phytic acid-P in their faeces (Taylor *et al.*, 2001). Utilization of P from inorganic and organic sources by ruminal bacteria in a semi-continuous culture system determined that effluent had a higher concentration of P_i from the inorganic than the organic P sources. This indicates that ruminal phytase activity does not hydrolyse all dietary phytate in some situations (Godoy and Meschy, 1999).

6.6 Conclusions

Microbial fermentation and growth in the rumen require P. The P needs of rumen microbes, expressed as dietary concentrations, may be higher than those of the host animal, especially in adult, non-lactating animals. Insufficient P supply to rumen microbes is often associated with reductions in feed intake, organic matter fermentation and efficiency of microbial protein synthesis. Recycling of endogenous P with saliva into the rumen supplies much P to the microbes and is an important conservation mechanism for P under low-dietary P conditions. The correlation between P concentrations in blood plasma, saliva and rumen fluid is high. P in undigested microbial mass is a major contributor to the inevitable P losses of the host. Dietary phytin can be efficiently hydrolysed in the rumen with presence of the enzyme phytase produced by rumen bacteria or supplied as intrinsic feed phytase. This contributes to the overall high level of P availability found in ruminants. However, depending on feed passage rate through the rumen and on accessibility of the molecules for microbes the ruminal phytin degradation may not be complete.

References

AFRC [Agricultural and Food Research Council] (1991) A reappraisal of the calcium and phosphorous requirements of sheep and cattle. *Nutrition Abstracts and Reviews (Series B)* 61, 573–612.

Armstrong, D.G. and Hutton, K. (1975) Fate of nitrogenous compounds entering the small intestine. In: McDonald, I.W. and Warner, A.C.I. (eds) *Digestion and Metabolism in the Ruminant.* The University of New England Publishing Unit, Armidale, Australia, pp. 432–447.

Bailey, C.B. and Balch, C.C. (1961) Saliva secretion and its relation to feeding in cattle. 2. The composition and rate of secretion of mixed saliva in the cow during rest. *British Journal of Nutrition* 15, 383–402.

Bravo, D., Meschy, F., Bogaert, C. and Sauvant, D. (2000) Ruminal phosphorus availability from several feedstuffs measured by the nylon bag technique. *Reproduction, Nutrition, Development* 40, 149–162.

Breves, G. (1991) Physiologische Grundlagen des gastrointestinalen P-Umsatzes und Bedeutung einer nicht bedarfsdeckenden P-Versorgung bei kleinen Wiederkäuern. *Übersichten zur Tierernährung* 19, 23–44.

Breves, G. and Höller, H. (1987) Effects of dietary phosphorus depletion in sheep on dry matter and organic matter digestibility. *Journal of Animal Physiology and Animal Nutrition* 58, 280–286.

Breves, G. and Schröder, B. (1991) Comparative aspects of gastrointestinal phosphorus metabolism. *Nutrition Research Reviews* 4, 125–140.

Call, J.W., Butcher, J. E., Shupe, J.L., Blake, J.T. and Olson, A.E. (1986) Dietary phosphorus for beef cows. *American Journal of Veterinary Research* 47, 475–481.

Conrad, H.R. (1999) Dietary phosphorus, excretory phosphorus and environmental concerns. In: *Proceedings of the Cornell Nutrition Conference*, Rochester, New York.

De Waal, H.O. and Koekemoer, G.J. (1997) Blood, rib bone and rumen fluid as indicators of phosphorus status of grazing beef cows supplemented with different levels of phosphorus at Armoedsvlakte. *South African Tydskr. Veek* 27, 76–84.

Durand, M. and Kawashima, R. (1980) Influence of minerals in rumen microbial digestion. In: Ruckebusch, Y. and Thivend, P. (eds) *Digestive Physiology and Metabolism in Ruminants.* MTP Press, Lancaster, UK, pp. 375–408.

Eeckhout, W. and De Paepe, M. (1994) Total phosphorus, phytate-phosphorus and phytase activity in plant feedstuffs. *Animal Feed Science and Technology* 47, 19–29.

Godoy, S. and Meschy, F. (1999) Utilisation of phytic phosphorus by rumen bacteria in a semi-continuous culture system (rusitec). *South African Journal of Animal Science* 29 (ISRP), 148–149.

Hristov, A.N., Melgar, A., Foley, A. and Kincaid, R. (2004) Ruminal degradability of nitrogen, phosphorus, and potassium from dairy feedstuffs. *Journal of Dairy Science* 87 (Suppl. 1), 49.

Karn, J.F. (2001) Phosphorus nutrition of grazing cattle: a review. *Animal Feed Science and Technology* 89, 133–153.

Kay, R.N.B. (1960) The rate of flow and composition of various salivary secretions in sheep and calves. *Journal of Physiology, London* 150, 515–537.

Knowlton, K.F., Herbein, J. H., Meister-Weisbarth, M.A. and Wark, W.A. (2001) Nitrogen and phosphorus partitioning in lactating Holstein cows fed different sources of dietary protein and phosphorus. *Journal of Dairy Science* 84, 1210–1217.

Koddebusch, L. and Pfeffer, E. (1988) Untersuchungen zur Verwertbarkeit von Phosphor verschiedener Herkünfte an laktierenden Ziegen. *Journal of Animal Physiology and Animal Nutrition* 60, 269–275.

Komisarczuk, S., Durand, M., Beaumatin, Ph. and Hannequart, G. (1987a) Effects of phosphorus deficiency on rumen microbial activity associated with the solid and the liquid phase of a fermentor (Rusitec). *Reproduction, Nutrition, Development* 27, 907–919.

Komisarczuk, S., Merry, R.J. and Mc Allan, A.B. (1987b) Effect of different levels of phosphorus on rumen microbial fermentation and synthesis determined using a continuous culture technique. *British Journal of Nutrition* 57, 279–290.

Komisarczuk-Bony, S. and Durand, M. (1991) Effects of minerals on microbial metabolism. In: Jouany, J.P. (ed.) *Rumen Microbial Metabolism and Ruminant Digestion.* Institut National de la Recherche Agronomique, Paris, pp. 179–198.

Konishi, C., Matsui, T., Park, W., Yano, H. and Yano, F. (1999) Heat treatment of soybean meal and rapeseed meal suppresses rumen degradation of phytate phosphorus in sheep. *Animal Feed Science and Technology* 80, 115–122.

Lofgreen, G.P. and Kleiber, M. (1954) Further studies on the availability of phosphorus in alfalfa hay. *Journal of Animal Science* 13, 258–264.

Milton, J.T.B. and Ternouth, J.H. (1985) Phosphorus metabolism in ruminants. II. Effects of inorganic phosphorus concentration upon food intake and digestibility. *Australian Journal of Agricultural Research* 36, 647–654.

Morse, D., Head, H. H. and Wilcox, C.J. (1992) Disappearance of phosphorus in phytate from concentrates in vitro and from rations fed to lactating dairy cows. *Journal of Dairy Science* 75, 1979–1986.

Nelson, T.S., Daniels, L.B., Hall, J.R. and Shields, L.G. (1976) Hydrolysis of natural phytate phosphorus in the digestive tract of calves. *Journal of Animal Science* 42, 1509–1512.

Park, W.-Y., Matsui, T., Konishi, C., Kim, S.-W., Yano, F. and Yano, H. (1999) Formaldehyde treatment suppresses ruminal degradation of phytate in soyabean meal and rapeseed meal. *British Journal of Nutrition* 81, 467–471.

Park, W.-Y., Matsui, T., Yano, F. and Yano, H. (2000) Heat treatment of rapeseed meal increases phytate flow into the duodenum of sheep. *Animal Feed Science and Technology* 88, 31–37.

Petri, A., Müschen, H., Breves, G., Richter, O. and Pfeffer, E. (1988) Response of lactating goats to low phosphorus intake. 2. Nitrogen transfer from rumen ammonia to rumen microbes and proportion of milk protein derived from microbial amino acids. *Journal of Agricultural Science, Cambridge* 111, 265–271.

Preston, R.L. and Pfander, W.H. (1964) Phosphorus metabolism in lambs fed varying phosphorus intakes. *Journal of Nutrition* 83, 369–378.

Raun, A., Cheng, E. and Burroughs, W. (1956) Phytate phosphorus hydrolysis and availability to rumen microorganisms. *Agricultural Food Chemistry* 4, 869–871.

Read, M.V.P., Engels, E.A.N. and Smith, W.A. (1986) Phosphorus and the grazing ruminant. 2. The effects of supplementary P on cattle at Glen and Armoedsvlakte. *South African Journal of Animal Science* 16, 7–12.

Rodehutscord, M., Pauen, A., Windhausen, P., Brintrup, R. and Pfeffer, E. (1994) Effects of drastic changes in P intake on P concentrations in blood and rumen fluid of lactating ruminants. *Journal of Veterinary Medicine A* 41, 611–619.

Rodehutscord, M., Heuvers, H. and Pfeffer, E. (2000) Effect of organic matter digestibility on obligatory faecal phosphorus loss in lactating goats, determined from balance data. *Animal Science* 70, 561–568.

Rohr, K., Lebzien, P., Schafft, H. and Schulz, E. (1986) Prediction of duodenal flow of non-ammonia nitrogen and amino acid nitrogen in dairy cows. *Livestock Production Science* 14, 29–40.

Scheuermann, S.E., Lantzsch, H.-J. and Menke, K.-H. (1988) In vitro und in vivo Untersuchungen zur Hydrolyse von Phytat. I. Löslichkeit von Phytat. *Journal of Animal Physiology and Animal Nutrition* 60, 55–63.

Simon, O. and Igbasan, F. (2002) In vitro properties of phytases from various microbial origins. *International Journal of Food Science and Technology* 37, 813–822.

Smith, R.H. (1984) Minerals and rumen function. In: *Nuclear Techniques in Tropical Animal Diseases and Nutritional Disorders.* IAEA, Vienna, Austria, pp. 79–96.

Taylor, N., Bowman, G.P., Anderson, K.A., Surber, L.M.M., Blake, T.K., Raboy, V., Dorsch, J.A. and Broyles, J.G. (2001) Phytic acid levels in barley for beef cattle. *Proceedings of the Western Section of the American Society of Animal Science* 52, 492–495.

Tombs, M.P. (1967) Protein bodies of the soybean. *Plant Physiology* 42, 797–813.

Valk, H., Sebek, L.B.J. and Beynen, A.C. (2002) Influence of phosphorus intake on excretion and blood plasma and saliva concentrations of phosphorus in dairy cows. *Journal of Dairy Science* 85, 2642–2649.

Witt, K.E. and Owens, F.N. (1983) Phosphorus: ruminal availability and effects on digestion. *Journal of Animal Science* 56, 930–937.

Yanke, L.J., Bae, H.D., Senlinger, L.B. and Cheng, K.J. (1998) Phytase activity of anaerobic ruminal bacteria. *Microbiology* 144, 1565–1573.

Yiu, S.H., Altosaar, I. and Fulcher, R.G. (1983) The effects of commercial processing on the structure and microchemical organization of rapeseed. *Food Microstructure* 2, 165–173.

Zimmermann, B., Lantzsch, H.-J., Mosenthin, R., Schöner, F.-J., Biesalski, H.K. and Drochner, W. (2002) Comparative evaluation of the efficacy of cereal and microbial phytases in growing pigs fed diets with marginal phosphorus supply. *Journal of the Science of Food and Agriculture* 82, 1298–1304.

7 Phosphorus Metabolism in Ruminants and Requirements of Cattle

E. Pfeffer[1], D.K. Beede[2] and H. Valk[3]
[1]*Institut für Tierernährung der Universität Bonn, Bonn, Germany*
[2]*Department of Animal Science, Michigan State University, East Lansing, Michigan, USA*
[3]*Animal Sciences Group, Lelystad, The Netherlands*

7.1 Introduction

Phosphorus (P), linked into phosphate (PO_4) or substances containing phosphate, plays an essential role in metabolism of plants, animals and microorganisms. P-containing nucleic acids form the basis of genetics, nucleotides enable organisms to utilize chemically bound energy, phospholipids

©CAB International 2005. *Nitrogen and Phosphorus Nutrition of Cattle* (eds E. Pfeffer and A.N. Hristov)

are essential compounds of cellular membranes and, in vertebrates, calcium phosphates, as hydroxyl apatite [$Ca_5(PO4)_3OH$], give strength to the skeleton. Inorganic phosphate participates in buffering the pH of biological fluids.

The essentiality of P for animals is beyond question and this element may be limiting animal performance in some areas of the world. Until recently, scientifically based recommendations concerning the supply of P to farm animals appear to have been guided largely by the goal to avoid any risk arising from potential shortages in P intake.

With increasing demands of environmental sustainability in all sectors of agriculture, however, the conscience has grown that excess enrichment of P in the soil may be just as dangerous to the environment as its emaciation. Sustainable farm units must be managed with the objective of zero P balance, i.e. during some defined period of time the amount of P imported into the unit should equal the amount exported from the unit (see also Chapter 9).

In cattle farms, importation of P may occur in the form of fertilizer and feeds whereas P is exported in the form of milk and animal bodies. If import of P exceeds export, reducing or even ceasing the use of P-containing fertilizer may be seen as a first step towards avoiding the imbalance, and this step has the advantage of lowering the cost of production. But, in many areas this may not be sufficient for achieving a zero P balance. Exporting manure from the farm unit may be an adequate means under certain conditions, but as a general strategy it is unlikely to be the best solution in the majority of situations. A more promising strategy is to restrict dietary P supply from purchased feeds and supplements, and to use lower P-containing basal ration ingredients (Beede and Davidson, 1999; Beede, 2003; Sutton and Beede, 2003).

The need for differentiating between feeds grown on the farm and purchased feeds is addressed in more detail in Chapter 9. If the amount of P imported into the farm with purchased feeds does not exceed the amount of P exported in milk and animal bodies then P balance of zero or less can be achieved. As long as the amount of P excreted by animals does not exceed the amount consumed in the form of feeds produced on-farm, zero P balance in the farm unit on the whole is possible.

In animal nutrition research it is now accepted that recommendations of P supply to farm animals must aim at preventing not only dietary insufficiencies but also unnecessary surpluses. The present state of knowledge relating to P requirements of the microbes in the forestomach of ruminants is presented in Chapter 6. In the current chapter we try to first review the relevant scientific literature on P metabolism in ruminants and especially point out differences compared with non-ruminants. Secondly, this information is used as a basis for deriving P requirements of cattle, and thirdly, a critical synopsis of recommendations in four countries is presented. The effect of dietary P intake on fertility of dairy cows, often discussed among extension workers and cattle farmers, is addressed separately in Chapter 8.

7.2 Phosphorus Metabolism in Ruminants

7.2.1 Balance studies up to the middle of the 20th century

The role of P in the metabolism of animals has attracted the interest of scientists practically from the start of animal nutrition as a scientific discipline. In one of the very early metabolism studies it was found that in adult wethers faecal P excretion practically equalled the quantity of P ingested whereas the urine contained only negligible quantities of this element (Henneberg, 1870, p. 230). Using silica as an inert marker in the digesta, Wildt (1874) concluded that substantial secretion of P into the forestomach of sheep must take place, almost amounting to the quantity ingested, and that absorption of a corresponding amount must take place in the small intestine. Thus, the essential characteristics of P metabolism in ruminants were published before the end of the 19th century; namely that the gut is the dominant pathway for excretion of P and that the difference between P intake and faecal P excretion reflects the difference between quantities of P absorbed from and quantities secreted into the digestive tract.

If the quantity of P excreted in faeces is less than P intake, the difference indicates the extent of net absorption. If P excreted exceeds P intake, the difference is identical to the extent of net secretion. Net secretion by definition is equivalent to negative net absorption. Thus, the following equation holds:

$$\text{P net absorption (g/day)} = \text{P intake (g/day)} - \text{faecal P excretion (g/day)} \quad (1)$$

Net absorption also can be expressed as a fraction of P intake for which the term net absorption efficiency (from the lumen of the digestive tract) will be used here:

$$\text{P net absorption efficiency} = \frac{\text{P net absorption (g/day)}}{\text{P intake (g/day)}} \quad (2)$$

Total secretion or absorption of P, respectively, cannot be quantified by faecal collection alone.

Duncan (1958) very carefully reviewed the then available literature concerning calcium (Ca) and P balances in ruminants and hypothesized that amounts of these elements provided to dairy cows were often far in excess of their needs and that no benefit was obtained from such excesses. With respect to sheep and goats she stated:

> Metabolism trials on the small ruminants, sheep and goats, are few and have not been well planned. More and better studies are needed.

Comparing cumulative estimates of Ca and P retention from metabolism studies vs. actual changes in body composition in cattle and sheep showed that metabolism studies grossly over-estimated actual retention. Duncan's examination of published metabolism studies revealed no probable explanation for this over-estimation and she concluded:

> The use of radioactive isotopes to study the details of mineral metabolism, which at the moment has scarcely begun, may detect a cause for the discrepancy.

7.2.2 Secretion and absorption of P studied with ^{32}P in intact animals

The use of the radioisotope ^{32}P as tracer of P in higher organisms was first reported from the Institute of Theoretical Physics at the University of Copenhagen. Chiewitz and Hevesy (1935) fed a single dose of labelled sodium phosphate to rats and found that after 1 month about 58% of the isotope was excreted in faeces and urine whereas about 25% was recovered from the skeleton of the animals. Hevesy *et al.* (1939) gave single doses to human patients and used the ratio between the respective specific activities in urine and stools as a basis for identifying the origin of faecal P.

The use of ^{32}P for studying kinetics of P in ruminants was initiated at the University of California–Davis, and the first of a series of publications concerning this subject (Kleiber *et al.*, 1951) begins as follows:

> Fecal phosphorus may be partitioned into exogenous and endogenous phosphorus. The exogenous fecal phosphorus is undigested phosphorus from food. The endogenous phosphorus reaches the intestinal contents either by diffusion from blood or intestinal tissue fluid, as part of secretions such as saliva, as component of cells or cell fragments sloughed off from the intestinal lining, or contained in phagocytes. The phosphorus which comes to the feces in microorganisms is either exogenous, namely that part which the microorganisms have taken up from undigested compounds in the intestinal tract, or endogenous, namely that part which the intestinal microorganisms get from the digestive juices or other sources of endogenous phosphorus.

This original definition of endogenous and exogenous P will be used throughout this chapter, irrespective of various changes in semantics in the scientific literature that have occurred repeatedly in the course of the years. The term 'endogenous', therefore, only indicates that P originating from the animal's body which has been transferred from the blood into the lumen of the digestive tract – it does not indicate a necessity of excretion of this P.

A key principle of the isotope dilution method is that body P is labelled by subcutaneous, intramuscular or intravenous administration of the isotope. The term 'specific activity' (SA) is used to characterize the ratio between radioactive and total P. 'Disintegrations in time/g P' or '% of a standard radioactivity/g P' can be used as the dimension for SA. Because all endogenous P is derived from blood plasma, the SA of P must be identical in secreted P and in inorganic phosphate (P_i) of blood plasma, because plasma P_i is the sole source of secreted P. When endogenous and exogenous P are mixed in the digestive tract, the degree of 'dilution' of SA of secreted P is a function of the ratio between exogenous and endogenous P. If the lag between time of secretion into the digestive tract and time of excretion in faeces is taken into account properly, then the ratio between the respective SA in faecal P and P_i of blood plasma or a comparable reference must indicate that proportion of faecal P which is of endogenous origin. Using disintegrations per minute (DIpM) as one of the potential units for radioactivity, this can be written as:

$$\frac{\text{SA of faecal P (DIpM/g)}}{\text{SA of plasma P}_i\text{ (DIpM/g)}} = \frac{\text{Endogenous faecal P (g/day)}}{\text{Total faecal P (g/day)}} \quad (3)$$

which also can be written as:

$$\text{Endogenous faecal P (g/day)} = \frac{\text{SA of faecal P (DIpM/g)}}{\text{SA of plasma P}_i\text{ (g/day)} \times \text{Total faecal P (g/day)}} \quad (4)$$

Kleiber *et al.* (1951) found in two Jersey cows producing about 10 kg of milk daily that endogenous faecal P made up 43% and 70%, respectively, of total faecal P. Subsequently, a large number of studies were carried out using the radioisotope ^{32}P for labelling endogenous P. Table 7.1 shows the endogenous faecal P as a percentage of total faecal

Table 7.1. Ranges of endogenous P as percentage of total P in ruminant faeces and absorption efficiencies (given in the original papers) as well as recalculated relative quantities of P secretion and P absorption.

Animals	Faecal P endogenous (%)	P absorption efficiency[a]	As % of P intake		Reference
			Secretion[b]	Absorption[b]	
Cows	43–70	0.50–0.64	75–130	88–111	Kleiber *et al.* (1951)
Lambs	75–95	0.81–0.96	97–104	115–131	Lofgreen and Kleiber (1953)
Wethers	90–93	0.93–0.96	920–1462	948–1500	Lofgreen and Kleiber (1954)
Steers	58–63	0.75–0.76	139–166	180–202	Tillman and Brethour (1958a)
Wethers	67–73	0.65–0.80	203–274	191–261	Tillman and Brethour (1958b)
Wethers	82–85	0.87–0.90	357–462	379–498	Tillman and Brethour (1958c)
Steers	55–56	0.68–0.78	120–126	150–176	Tillman *et al.* (1959)
Bulls	48–89	0.71–0.88	94–369	122–423	Brüggemann *et al.* (1959)
Wethers	46–82	0.48–0.82	87–390	89–395	Brüggemann *et al.* (1959)
Wethers	36–66	0.43–0.69	56–193	67–200	Lofgreen (1960)
Lambs	54–63	0.20–0.55	117–396	100–408	Lueker and Lofgreen (1961)
Wethers	57–62	0.51–0.58	134–161	122–146	Guéguen (1962)
Calves	45	0.80	83	146	Guéguen (1963)
Lambs	70–80	0.81–0.86	244–392	254–423	Preston and Pfander (1964)
Wethers	60–71	0.53–0.79	169–266	167–290	Young *et al.* (1966c)
Wethers	52–69	0.42–0.68	118–192	97–181	Potthast *et al.* (1976)
Lambs	48–63	0.57–0.82	92–171	110–201	Field *et al.* (1982)
Lambs	15–39	−0.33	−57	−54	Valdivia *et al.* (1982)
Lambs	68–78	0.77–0.84	219–345	245–374	Boxebeld *et al.* (1983)
Sheep	45–56	0.44–0.55	100–122	100–122	Braithewaite (1984)
Lambs	58–67	0.17–0.77	139–206	42–235	Braithewaite (1985)
Lambs	60–74	0.80–0.87	149–289	192–335	Field *et al.* (1985)
Ewes	53–64	0.55–0.69	114–155	129–171	Braithewaite (1986)
Calves	52–69	0.56–0.84	107–210	150–260	Challa and Braithwaite (1988a,b)
Lambs	24–43	0.30–0.54	38–76	41–95	Garcia-Bojalil *et al.* (1988)
Sheep	75–78	0.55–0.69	114–155	129–173	Ternouth (1989)
Cows	75–78	0.68–0.72	292–360	282–313	Martz *et al.* (1990)
Heifers	64–87	0.66–0.92	204–933	185–950	Coates and Ternouth (1992)
Sheep	81–83	0.83–0.85	425–503	448–503	Rajaratne *et al.* (1994)
Steers	66–84	0.50–0.82	200–505	205–393	Bortolussi *et al.* (1996)

[a]See Eq. (7) in text for calculation of absorption efficiency.
[b]See Eqs (9) and (10) in text for explanation of derivation and calculations of secretion and absorption.

P of cattle and sheep found in 29 papers published between 1951 and 1996. One of the studies was conducted with weaned calves weighing less than 100 kg (Guéguen, 1963) and seven were with growing cattle weighing between 100 and 250 kg (Tillman and Brethour, 1958a; Brüggemann *et al.*, 1959; Tillman *et al.*, 1959; Challa and Braithwaite, 1988a,b; Coates and Ternouth, 1992; Bortolussi *et al.*, 1996). One study was performed with Holstein cows weighing 650 kg (Martz *et al.*, 1990). In two of the studies using lambs (Valdivia *et al.* 1982; Garcia-Bojalil *et al.*, 1988), the endogenous proportions of faecal P ranged below 43% found in one of the two cows of Kleiber *et al.* (1951). An exceedingly low proportion of only 36% also was found for one of five treatments studied by Lofgreen (1960). No reason can be given to explain why these results do not fit into the general range of the majority of the published work. On the other hand, the minimum endogenous proportion of faecal P was more than 50% in 21 of the 29 publications, and the respective maximum was equal to or exceeded the 70% of the second cow of Kleiber *et al.* (1951) in 17 of the studies. About one-third of all results ranged between 70% and 95%. Ignoring the apparent outlier results, it can be assumed generally that about two-thirds to three-quarters of faecal P in ruminants are of endogenous origin. In milk-fed calves, endogenous faecal P amounted to 26% of total faecal P and this proportion was further reduced by supplementation of phosphate in milk (Challa and Braithwaite, 1989).

Exogenous faecal P, i.e. that part of faecal P which is unabsorbed dietary P, is calculated as:

$$\begin{aligned}\text{Exogenous faecal P (g/day)} &\\ = \text{Total faecal P (g/day)} &\\ - \text{endogenous faecal P (g/day)} & \quad (5)\end{aligned}$$

Knowledge of exogenous faecal P enables the calculation of the absorption of dietary P as:

$$\begin{aligned}\text{Absorbed dietary P (g/day)} &\\ = \text{P intake (g/day)} &\\ - \text{exogenous faecal P (g/day)} & \quad (6)\end{aligned}$$

Efficiency of P absorption, which indicates absorption as a fraction of dietary P consumed (P intake) is calculated as:

$$\text{Dietary P absorption efficiency} = \frac{\text{Absorbed dietary P (g/day)}}{\text{P intake (g/day)}} \quad (7)$$

Table 7.1 also shows efficiencies of P absorption calculated by the respective authors of the cited papers. In 23 of the publications, mostly using non-lactating animals, these values range above 0.50, the majority of values are found between 0.60 and 0.80.

Figure 7.1 resembles the illustration given by Lofgreen and Kleiber (1953) for interpreting results of their ^{32}P studies. In that paper the term 'metabolic P' is used for endogenous P whereas the term 'undigested P' is used for exogenous P. The illustration gives the impression that digesta first pass through a section of the digestive tract in which absorption of P takes place and then through a section in which endogenous P is added. This, however, does not correctly illustrate the actual movements of P in the digestive tract.

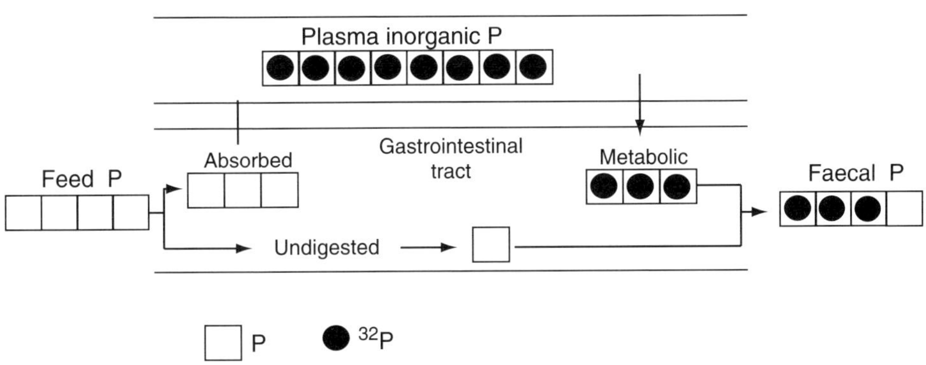

Fig. 7.1. Scheme used by Lofgreen and Kleiber (1953) to illustrate the isotope dilution method as 'method for determining the proportion of the faecal phosphorus which is of metabolic origin'.

Endogenous faecal P excretion is not a direct measure of total P secretion into the digestive tract. It only represents that part of the secreted quantity that is not reabsorbed during the passage along the tract. Further derivations are possible if it is legitimate to assume that secretion of P takes place mostly proximal to the site of absorption and that complete mixing between endogenous and exogenous P can be taken as certain. Young *et al.* (1966c) reported identical SA of P in the solid and the liquid phases of faeces after sheep were given parenteral doses of ^{32}P-labelled orthophosphate. They concluded that mixing of endogenous and exogenous P in the digestive tract must be complete. Practically identical patterns were found of the respective SA of P in ruminal contents and in faeces after correction for the time lag due to passage through the digestive tract when sheep were given an intravenous dose of ^{32}P (Potthast *et al.*, 1976). This not only confirmed complete mixing of P in the forestomach, but also indicated that quantities of P secreted into the postruminal part of the digestive tract must have been insignificant, relative to the P in salivary secretions. Théwis and Francois (1985) found practically identical SA of total P in duodenal and ileal digesta of sheep after intravenous dosing of ^{32}P, which also indicates that secretion of P into the small intestine after the entrance of pancreas and bile must have been relatively small. Therefore, Fig. 7.2 as an extension of Fig. 7.1 is preferred to illustrate P movements in the digestive tract of ruminants. The underlining dominant role of salivary P secretion into the reticulorumen is more clearly emphasized. Because complete mixing of P of the different origins is assumed (see also Chapter 6), it is concluded that the absorption efficiency for dietary P, as shown in Eq. (7), must be applicable also for endogenous P secreted into the digestive tract, as shown in Eq. (8):

$$\frac{\text{Exogenous faecal P (g/day)}}{\text{P intake (g/day)}} = \frac{\text{Endogenous faecal P (g/day)}}{\text{P secretion (g/day)}} \qquad (8)$$

This equation easily can be transformed into:

$$\text{P secretion (g/day)} = \frac{\text{P intake (g/day)}}{\text{Exogenous faecal P (g/day)}} \times \text{Endogenous faecal P (g/day)} \qquad (9)$$

When an animal is fed the same diet over time and is in steady state, then the sum of P entering the digestive tract from feed and from secretions must be equal to the sum leaving, either by absorption or by excretion in faeces, which allows the following calculation:

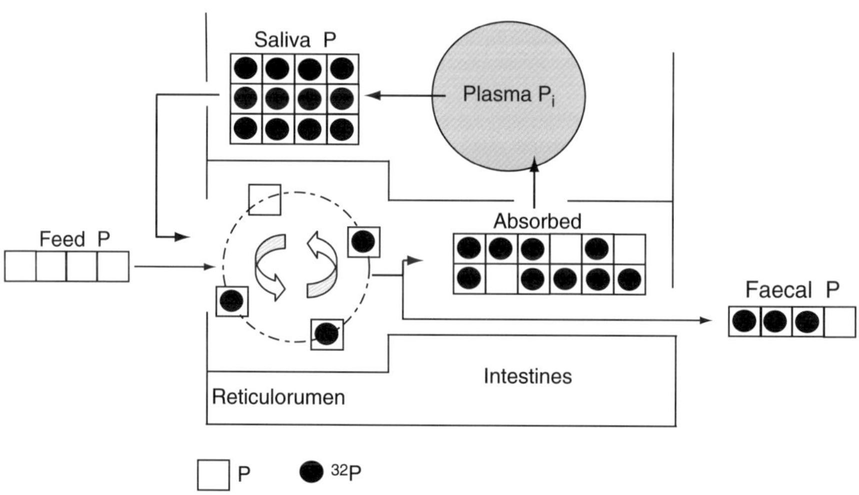

Fig. 7.2. Extended version of Fig. 7.1 paying tribute to the importance of saliva for P flows in the digestive tract of ruminants.

$$\text{P absorption (g/day)} = \text{P intake (g/day)} + \text{P secretion (g/day)} - \text{total faecal P (g/day)} \quad (10)$$

The efficiency of P absorption as a fraction of the sum of P entering the digestive tract is defined as:

$$\text{P absorption efficiency} = \frac{\text{P absorption (g/day)}}{\text{P intake (g/day)} + \text{P secretion (g/day)}} \quad (11)$$

Secretion and absorption of P were recalculated using Eqs (9) and (10), respectively, for the 29 cited publications. The ranges for both, relative to the respective P intake, are shown in Table 7.1. It is obvious that in most of the experiments with cattle and sheep secretion as well as absorption of P exceeded P intake, in some cases by several fold. Thus, it can be stated generally that P absorption largely exceeds P net absorption from the digestive tract of ruminants. Consequently, identification of sites of P secretion into and absorption from the digestive tract was regarded as a consideration of major scientific relevance.

7.2.3 Phosphorus in secretions into the digestive tract

McDougall (1948) showed that saliva of sheep contained large quantities of phosphate and thus proved that at least a major part, if not all, of the P secretion into the reticulorumen mentioned by Wildt (1874) took place via salivary secretion. Kay (1960b) showed that secretion of the parotid glands in kids develops in the first 3 months of life. Rumination is the most effective stimulus of secretion of parotid saliva in cattle (Bailey and Balch, 1961; Kaufmann and Orth, 1966). Inorganic P concentrations in ruminal fluid are determined principally by the extent of salivary P secretion (Tomas *et al.*, 1967; Tomas, 1973).

Table 7.2 shows ranges of phosphate concentrations in different types of saliva. Differences among secretions from different glands may be one source of variation in P concentration of mixed saliva. Yet, there are substantial ranges in P concentration of specific types of saliva.

Perge *et al.* (1982) found the phosphate concentration of mixed saliva to be influenced not only by intake of P and of Ca but also by the time of sampling in sheep fed once daily. Varying plasma P_i concentrations within wide ranges by dietary depletion or phosphate loading via infusions directly influenced salivary P concentration but not salivary flow in cattle (Riad *et al.*, 1987) or sheep (Scott, 1978; Wright *et al.*, 1984; Breves *et al.*, 1987). Bailey (1961) estimated that salivary flow in cows consuming about 10 kg dry matter (DM) daily ranged between 100 and 190 l/day.

It was noted for sheep that chloride is the predominant anion contained in bile, pancreatic juice and secretions into the upper jejunum, and that about equal concentrations of chloride and bicarbonate are found in secretions into lower jejunum, ileum, caecum and spiral colon, whereas only very small concentrations of phosphate are found in all of these secretions (Kay and Pfeffer, 1970). It is, therefore, concluded that the quantity of P secreted into the intestine of ruminants is almost irrelevant in comparison to salivary P secretion.

7.2.4 Measurements of phosphorus flow in the digestive tract with cannulated animals

The method of surgically fitting cannulas into well-defined sites of the intestine led to a better understanding of net movements of mineral elements through the major segments of the digestive tract of ruminants. Table 7.3 shows flow rates of total P through the proximal duodenum as well as faecal excretion reported in 15 publications involving sheep and six with cattle.

Flow of P through the proximal duodenum exceeded P intake in each of the experiments reported with the exception of the work of Leibholz (1974). It is not possible to explain why net secretion of P prior to the duodenum was not observed in all experiments of that publication. The fact that contrary to the other papers a very substantial net absorption of Ca prior to the duodenum was noted suggests doubts about the reliability of digesta flows measured by Leibholz (1974); the use of a reference marker for correcting measured flow of digesta is not mentioned in the paper.

In each experiment of the other references cited in Table 7.3, the flow of P at the proximal duodenum substantially exceeded P intake, which is consistent with the previously mentioned findings

Table 7.2. Ranges of P concentration in saliva of ruminants.

Animals	Type of saliva	Salivary P (mmol/l)	Plasma P_i (mmol/l)	S/P[a]	Reference
Sheep	Mixed	17–28			McDougall (1948)
	Parotid	6–42			
Sheep	Parotid	12–40			Kay (1960a)
	Submaxillary	1–88			
	Sublingual	0.3–2.0			
	Labial	1–5			
	Inferior molar	22–26			
	Palatine	12			
	Residual	22–35			
Calves	Parotid	8–24			
	Submaxillary	0.2–2.0			
	Inferior molar	9–27			
	Residual	10–17			
Goats	Parotid	15–40			Kay (1960b)
Steer	Parotid	16–32	1.4	5	Bailey and Balch (1961)
Cows	Mixed	15–66			Bailey (1961)
Lambs	Parotid	16–20			Tribe and Peel (1963)
	Residual	9–16			
	Mixed	11–23			
Wethers	Parotid	14–20			
	Residual	12–16			
	Mixed	21–35			
Sheep	Parotid	13–23	1.4–2.7	8.4–10.2	Tomas *et al.* (1967)
Sheep	Mixed	16–32	2.1–3.4	9–12	Perge *et al.* (1982)
Sheep	Parotid	18–75	1.3–8.0	8–23	Manas-Almendros *et al.* (1982)
Heifers	Mixed	7–14	1.0–1.8	6–9	Gartner *et al.* (1982)
Sheep	Mixed	11.7 ± 2.6	1.7 ± 0.2	7.5 ± 1.7	Breves *et al.* (1987)
		4.7 ± 1.0	0.7 ± 0.1	6.8 ± 1.2	
Cows	Mixed	0.9–2.6	4.3–12.1	3.5–6.8	Valk *et al.* (2002)

[a]S/P = Salivary P/Plasma P_i.

about secretion of P in saliva. Faecal excretion of P was lower than P intake in most of the experiments reported, but P net absorption efficiency from the total tract (see Eq. (2)) varied widely among different experiments; this phenomenon will be addressed subsequently in this chapter.

Results cited in Table 7.3 allow calculation of P net absorption as well as the efficiency of P net absorption from the intestines, analogous to Eqs (1) and (2), respectively:

$$\text{Intestinal P net absorption (g/day)} = \text{P flow at proximal duodenum (g/day)} - \text{faecal P (g/day)}, \tag{12}$$

and

$$\text{Intestinal P net absorption efficiency} = \frac{\text{Intestinal P net absorption (g/day)}}{\text{P flow at duodenum (g/day)}} \tag{13}$$

Corresponding calculations can be done for defined sections of the intestine:

$$\text{P net absorption}_{\text{small intestine}} \text{ (g/day)} = \text{P flow at duodenum (g/day)} - \text{P flow at ileum (g/day)}, \tag{14}$$

$$\text{P net absorption efficiency}_{\text{small intestine}} = \frac{\text{P net absorption}_{\text{small intestine}} \text{ (g/day)}}{\text{P flow at duodenum (g/day)}} \tag{15}$$

Table 7.3. Relative flows of P into the small intestine and net absorption efficiencies of P from the whole tract and from intestinal sections of sheep and cattle.

		P net absorption efficiency			
			From the intestines[b]		
	Duodenal flow, % of P intake	From the total tract[a]	Small[c]	Large[d]	Reference
Sheep	500	0.17	0.82	0.10	Bruce *et al.* (1966)
	192–352	−0.06–0.20	0.34–0.64	0.02–0.30	Pfeffer *et al.* (1970)
	118–248	0.07–0.26	0.22–0.58	0.00–0.20	Grace *et al.* (1974)
	85–119	0.25–0.61		0.29–0.57	Leibholz (1974)
	315	0.16	0.71	0.08	Ben-Ghedalia *et al.* (1975)
	180	0.68	0.84	0.00	Dillon and Scott (1979)
	238–314	0.30–0.40	0.71–0.75	0.00–0.12	Greene *et al.* (1983a)
	337	<0.01	0.60	0.25	Théwis and Francois (1985)
	218–425	0.09–0.26	0.68–0.80	−0.04–0.18	Wylie *et al.* (1985)
	160–296	−0.24–0.62	0.42–0.55	−0.11–0.06	Breves *et al.* (1985)
	169–227	0.07–0.47		0.59–0.69	Scott and Buchan (1985)
	323–428	−0.06–0.00		0.65–0.75	Grings and Males (1987)
	215–243	0.23–0.36		0.67–0.70	Scott and Buchan (1988)
	159–174	0.48–0.56	0.68–0.75	−0.12–0.00	Khorasani and Armstrong (1990)
	306–316	0.32–0.34	0.75–0.77	0.04–0.15	Kirk *et al.* (1994)
Cattle	131–195	0.04–0.40		0.48–0.68	Pfeffer and Kaufmann (1972)
	137–216	0.24–0.48	0.35–0.72	0.00–0.24	Bertoni *et al.* (1976)
	167–201	0.36–0.52	0.68–0.74	−0.07–0.05	Greene *et al.* (1983b)
	157–268	0.31–0.53	0.58–0.78	−0.18–0.18	Khorasani and Armstrong (1992)
	127–148	0.32–0.42		0.50–0.54	Rahnema *et al.* (1994)
	132–184	0.20–0.39		0.51–0.59	Khorasani *et al.* (1997)

[a]See Eq. (2) for calculating P net absorption efficiency from the total tract.
[b]See Eq. (13) for calculating P net absorption efficiency from the intestines.
[c] See Eq. (15) for calculating P net absorption from the small intestine.
[d]See Eq. (17) for calculating P net absorption from the large intestine.

$$\text{P net absorption}_{\text{large intestine}}\ (\text{g/day}) = \text{P flow at ileum (g/day)} - \text{faecal P (g/day)} \qquad (16)$$

and

$$\text{P net absorption efficiency}_{\text{large intestine}} = \frac{\text{P net absorption}_{\text{large intestine}}\ (\text{g/day})}{\text{P flow at ileum (g/day)}} \qquad (17)$$

It can be concluded that P net absorption from the intestine of ruminants is much larger than P net absorption from the whole digestive tract, and that intestinal P net absorption efficiency may exceed 0.80 under given circumstances. The coincidence of high rates of salivary P secretion and of extensive absorption of P in the intestine results in the endogenous cycle of P between blood and the lumen of the digestive tract typical for all ruminant species.

Flow of P through the terminal ileum was measured in most experiments with sheep and in three of the six publications with cattle (Table 7.3). From these results it must be concluded that intestinal net absorption of P takes place mainly between proximal duodenum and terminal ileum. In most, though not all experiments, P flow through the terminal ileum was slightly higher than faecal P excretion. Therefore, it can be concluded that a small proportion to total net absorption of P occurs in the large intestine. In a series of perfusion studies into the colon and rectum of sheep some P net secretion into the gut was observed as long as phosphate-free buffers were used. However, when buffer solutions contained phosphate, there always was net absorption of P_i, which increased with increasing phosphate concentrations (Höller *et al.*, 1988).

Further specific definition of the site of P absorption in ruminants is based on studies using sheep with cannulas in different sites of the small intestine (Kay, 1969; Ben-Ghedalia *et al.*, 1975). As shown in Table 7.4, elevation of pH of the digesta is comparably modest during passage through the first third of the small intestine. Correspondingly low pH was found in duodenal contents of cows in which re-entrant cannulas were positioned well distal to the addition of bile and pancreatic secretions (Kaufmann *et al.*, 1972). These values can be compared with pH measured in defined sites of the small intestine of piglets (Eidelsburger *et al.*, 1992; Risley *et al.*, 1992), humans (Fallingborg *et al.*, 1994), horses (Meyer *et al.*, 1997) or mice (Delcenserie *et al.*, 2001). As summarized in Fig. 7.3, pH of digesta of the proximal small intestine is lower in ruminants than in non-ruminants. This means that for a considerable part of the intestinal passage, P_i of digesta is present mainly as the monovalent dihydrogen phosphate, $H_2PO_4^-$ and no precipitation of calcium phosphates would be expected. As in blood plasma most of the P_i is present as the bivalent monohydrogen phosphate, HPO_4^{2-}, diffusion of the primary phosphate through the gut wall is favoured by the very high concentration gradient of the primary phosphate from the mucosal to the serosal side.

About 92% of the P in duodenal digesta was found in the liquid phase when sheep were fed hay and this was reduced to 87% when concentrates were fed (Scott and Buchan, 1985). In sheep fed diets adequate in P, inorganic P made up 92% of total P flowing through the proximal duodenum and 32% of that flowing through the terminal ileum; during P depletion this proportion was reduced to 46% at the duodenum and not changed at the terminal ileum (Breves *et al.*, 1985).

Théwis *et al.* (1978) analysed total P and phospholipid P in digesta in the major sections of the digestive tract of sheep. The respective portion of total P present in the form of phospholipids was 2.9% in the diet, more than 6% in contents of the reticulorumen, more than 15% in contents of the upper small intestine and about 8% in contents of the large intestine. Whereas the difference between diet and ruminal contents indicates microbial synthesis of phospholipids, the peak proportion in the upper small intestine probably is a resultant of bile secretion and of preferential absorption of P_i in the upper part and increased absorption of phospholipids in the lower sections of the small intestine. These results, however, do not allow conclusions regarding the quantity of bile phospholipids secreted per unit of time.

When the flow of P at the proximal duodenum was measured, it was found in almost all of the experiments to exceed P intake by an order of magnitude that may be attributed to salivary flow. This occasionally has led to the interpretation of P net secretion prior to the duodenum as a direct measurement of salivary P secretion (Scott and Buchan, 1985, 1988; Challa and Braithwaite,

Table 7.4. Luminal pH and phosphate concentrations in digesta at different sites of the small intestine of sheep.

	Reference				
	Kay (1969)		Ben Ghedalia *et al.* (1975)		
Distance from pylorus (m)	pH	Inorganic P (mmol/l)	pH	Soluble P (mmol/kg)	Ca/P (g/g)
0.05	2.4–2.7	16–34	2.60–3.00	21.3	1.5
0.5–0.6	3.6–4.3	10–14			1.6
2–2.5	4.8–5.2	6–7			
3			4.11–5.15	5.46	2.9
7	7.4	1.2	5.95–7.02	3.20	3.5
9	7.5	1.8			
Terminal ileum	7.8–8.0	1.1–6.6	7.80–8.22	5.06–5.66	5.0–5.7

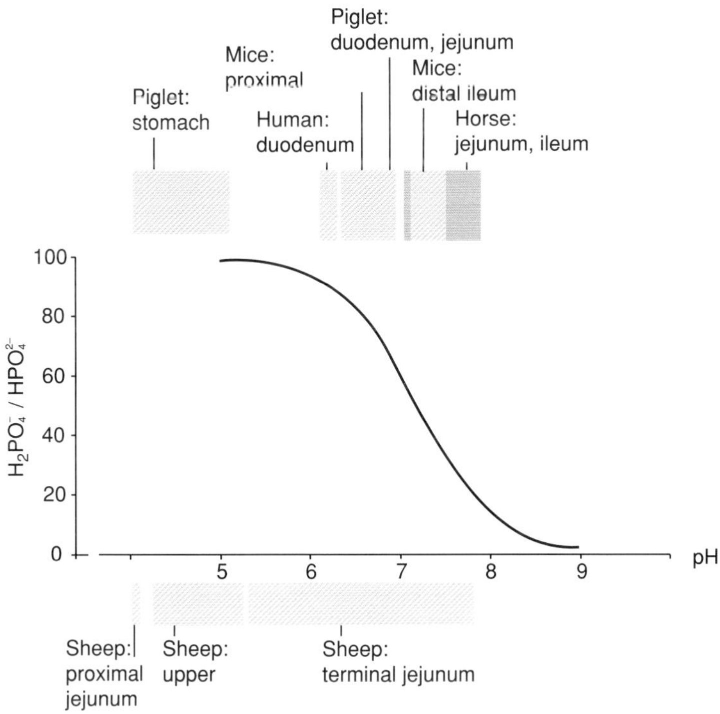

Fig. 7.3. Ranges of pH found in intestinal digesta of non-ruminants and ruminants as well as corresponding ratio between $H_2PO_4^-$ and HPO_4^{2-}.

1988a,b,c; Challa *et al.*, 1989). Yet, the ruminal mucosa is not to be regarded as impermeable to phosphate. Scarisbrick and Ewer (1951) injected ^{32}P into the rumen and after 10 min found a higher SA of P in the ruminal vein than in the carotid artery. Also, they found arteriovenous differences after short-term changes in phosphate concentration in rumen contents. They concluded:

> The net absorption of inorganic phosphate from the rumen over a long period of time, therefore, seems to be at most only small in amount, but at any particular instant there may be a substantial movement of inorganic phosphate into or out of the blood traversing the ruminal wall.

Net secretion into the rumen was found in sheep fed diets either adequate or deficient in P when P_i concentrations on the mucosal side did not exceed 2.2 mmol/l, whereas net absorption was found when concentrations exceeded 4.1 mmol/l. These findings were recorded irrespective of the P status of the sheep (Breves *et al.*, 1988; Beardsworth *et al.*, 1989).

There is one paper which indicates that substantial net absorption of P may take place in the omasum of ruminating calves (Edrise and Smith, 1986). This finding, however, has not been confirmed by other researchers, presumably due to the specific complications for research work concerned with that organ.

7.2.5 Regulation of phosphorus absorption

Depletion of Ca increases plasma concentrations of calcitriol (also known as 1,25-$(OH)_2$ vitamin D_3 or vitamin D hormone) and increases Ca absorption efficiency in non-ruminant and ruminant animals (Braithwaite, 1974; Fox and Care, 1978; Abdel-Hafeez *et al.*, 1982; Fox and Ross, 1985).

Phosphorus depletion increased plasma calcitriol concentrations and absorption efficiencies of both Ca and P in rats (Hughes *et al.*, 1975; Ribovich and DeLuca, 1978) and pigs (Fox and Care, 1978; Fox *et al.*, 1978; Fox and Ross, 1985).

Contrary to this finding, however, P depletion increased absorption efficiency of P in ruminants without raising that of Ca (Young *et al.*, 1966b; Abdel-Hafeez *et al.*, 1982).

Differences in vitamin D metabolism of different species are of relevance in this context. During P depletion no increase in plasma concentration of calcitriol was found in sheep (Abdel-Hafeez *et al.*, 1982; Breves *et al.*, 1985; Maunder *et al.*, 1986) or in lactating goats (Müschen *et al.*, 1988). In addition, Maunder *et al.* (1986) found that neither metabolic clearance rate nor production rate of calcitriol were altered by dietary P depletion. However, the finding of Riad *et al.* (1987) is not in accord with these results, namely that intramuscular injections of 1α-OH-vitamin D_3 in heifers increased plasma P_i concentrations and concurrently decreased salivary P_i concentrations and thus secretions.

Neither the substantial net secretion of P into the digestive tract prior to the duodenum nor the net absorption of P from the intestines were ever accompanied by comparable respective net movements of Ca, as shown in Table 7.5. This explains why in tracer studies using isotopes of both elements, absorption of P in most cases exceeded that of Ca (see Table 7.6). Uncoupling of absorptive processes for P and Ca, respectively, allows maximum absorption of P during periods of P deficiency and at the same time decreased absorption of Ca (Young *et al.*, 1966a; Breves *et al.*, 1985). This fact marks a substantial biological difference between ruminants and non-ruminants.

The independence between absorption of P and Ca explains why P-depleted dairy goats increased faecal excretion of Ca (Müschen *et al.*, 1988) and why additional dietary supplementation of Ca did not increase the negative consequences of P deficiency in lactating goats (Deitert and Pfeffer, 1993) and in weaned kids (Pfeffer *et al.*, 1996). Table 7.6 summarizes results of studies in which absorption of both P and Ca were determined using the respective radioisotopes. In 14 of the 15 treatments P absorption exceeded Ca absorption irrespective of the Ca:P ratio in the diet. Obviously, excessive dietary Ca concentrations do not negatively affect P absorption and dietary Ca:P ratio is not regarded as having impact on the amount of P required by ruminants.

Table 7.7 summarizes results of a study in which *in vivo* and *in vitro* measurements were combined for growing kids fed diets varying in dietary P and/or Ca. Reducing only P intake drastically lowered retention of both elements and caused hypophosphataemia in combination with hypercalcaemia without affecting plasma concentrations of parathyroid hormone (PTH) or calcitriol and drastically increased unidirectional as well as net P_i fluxes from the mucosal to the serosal side of duodenal and jejunal preparations. Reducing only Ca intake caused increased plasma P_i concentrations and gave rise to greater plasma concentrations of PTH and calcitriol with no significant changes in P_i fluxes through the mucosa of duodenum or jejunum, measured *in vitro*. Kids fed diets low in both elements produced results very similar to

Table 7.5. Net movements of Ca and P in stomachs and intestines of ruminants (g/day).

	Intake		Stomachs[a]		Intestines[a]		
Species	Ca	P	Ca	P	Ca	P	Reference
Sheep	4.2	2.9	+0.5	+4.8	+0.2	−5.0	Pfeffer *et al.* (1970)
	8.6	3.2	−2.0	+2.7	+0.2	−3.2	Grace *et al.* (1974)
	8.4	5.8	−1.2	+3.8	±0.0	−6.7	Khorasani and Armstrong (1990)
	17.8	4.1	+1.5	+8.9	−3.7	−9.6	Ben Ghedalia *et al.* (1975)
	3.0	1.7	+0.4	+4.7	−0.9	−5.1	Wylie *et al.* (1985)
	6.4	1.0	−0.5	+1.9	±0.0	−0.7	Breves *et al.* (1985)
	6.2	4.2	±0.0	+5.7	−1.3	−3.3	
Cattle	32	40	+5	+27	−4	−40	Pfeffer and Kaufmann (1972)
	74	57	−10	+25	−10	−49	Bertoni *et al.* (1976)
	37	17	−8	+17	±0.0	−24	Khorasani and Armstrong (1992)
	195	92	−1	+27	−17	−65	Rahnema *et al.* (1994)
	154	94	−19	+48	−30	−79	Khorasani *et al.* (1997)

[a]+ indicates net secretion and − indicates net absorption.

Table 7.6. Dietary Ca:P ratio and absorption of dietary Ca and total P in ruminants.

	In the diet		Absorption (g/day)			
Animals	P (g/kg DM)	Ca:P	Dietary Ca[a]	Total P[b]	Absorbed Ca / Absorbed P	Reference
Lambs	0.76	10.4:1	6.6	7.8	0.85:1	Young *et al.* (1966b,c)
	2.48	1.9:1	7.0	23	0.31:1	
	2.64	9.9:1	25	29	0.86:1	
Lambs	3.6	2.1:1	2.2	12.9	0.17:1	Boxebeld *et al.* (1983)
	1.5	2.1:1	1.3	5.8	0.22:1	
	1.5	8.2:1	2.8	3.8	0.73:1	
Lambs	4.3	0.4:1	1.2	9.0	0.13:1	Field *et al.* (1985)
		0.8:1	2.0	9.2	0.22:1	
		1.1:1	2.1	11.2	0.20:1	
Lambs	0.7	7.0:1	1.2	0.2	6.1:1	Braithwaite (1985)
	1.7	3.0:1	1.2	1.4	0.83:1	
	2.6	1.9:1	1.7	4.5	0.37:1	
Ewes	DMI not reported	2.1:1	0.8	5.9	0.13:1	Braithwaite (1986)
		1.7:1	2.1	8.5	0.24:1	
		2.4:1	1.4	3.3	0.44:1	
Calves	1.3	5.2:1	3.6	5.4	0.66:1	Challa and Braithwaite (1988a)
	3.3	2.2:1	6.8	16.3	0.41:1	
	4.8	1.4:1	6.4	23.3	0.28:1	
Dairy cows	1.5	3.3:1	26	17	1.5:1	Martz *et al.* (1990)
	2.1	2.4:1	50	65	0.77:1	

[a]Absorption of dietary Ca calculated in analogy to Eq. (6).
[b]Absorption of total P calculated according to Eq. (10).

Table 7.7. Balances of P and Ca, plasma concentrations of P_i, Ca, PTH and calcitriol as well as P_i flux rates through the walls of duodenum and jejunum of ruminating kids fed diets varying in their P and Ca concentrations (Pfeffer *et al.*, 1995; Schröder *et al.*, 1995).

Dietary P (g/kg DM)	4.6	2.0	4.6	2.1
Dietary Ca (g/kg DM)	10.9	10.9	3.9	3.9
P intake (g/day)	2.94	1.08	3.20	1.18
P in faeces (g/day)	1.75	1.11	1.69	1.03
P in urine (g/day)	0.04	0.01	0.34	0.02
P retained (g/day)	1.15	−0.03	1.16	0.13
Ca intake (g/day)	7.22	5.76	3.07	2.39
Ca in faeces (g/day)	5.09	5.35	1.51	1.97
Ca in urine (g/day)	0.14	0.20	0.06	0.22
Ca retained (g/day)	1.98	0.22	1.50	0.21
Plasma concentrations				
P_i (mmol/l)	2.22	0.61	2.78	0.87
Ca (mmol/l)	2.73	3.02	2.70	2.95
PTH (pmol/l)	95	85	148	99
Calcitriol (pmol/l)	107	102	233	224
P_i fluxes[a]				
At duodenum				
J_{ms}	44.2	95.4	34.7	86.5
J_{sm}	8.7	10.1	7.7	11.3
J_{net}	35.5	85.4	27.0	75.3
At jejunum				
J_{ms}	86.1	171.6	122.2	156.2
J_{sm}	34.5	39.0	30.8	32.1
J_{net}	51.6	132.6	91.5	124.1

[a]Fluxes of P_i through intestinal tissue were measured *in vitro* using ^{32}P: J_{ms} = from mucosal to serosal side; J_{sm} = from serosal side to mucosal side and $J_{net} = J_{ms} - J_{sm}$ (calculated).

those fed the diet deficient only in P with the exception of plasma calcitriol which was elevated as in kids fed the diet low in Ca and adequate in P. These results fully confirm that P is absorbed independent of plasma calcitriol concentration.

7.2.6 Faecal excretion of phosphorus

Animals excrete P in the faeces for one of three potential reasons.

1. A fraction of the P contained in feeds may be present in a chemical binding that cannot be absorbed. As a consequence, due to the nature of the feed, this fraction has to be excreted. This fraction may be substantial in non-ruminants, but is practically negligible in ruminants due to microbial breakdown of phytates in the rumen (see Chapter 6).

2. Some P inevitably is lost in faeces independent of P intake. These inevitable losses result from metabolism either of the host animal or of the microorganisms in the digestive tract and are obligatory to normal basal functions of the animal.

3. If intake of absorbable P exceeds that needed for inevitable losses and requirements for growth, reproduction or lactation, then this surplus P is excreted to maintain homoeostasis. Normally in ruminants most of this surplus is excreted with faeces, whereas in non-ruminants renal excretion predominates.

There is no possibility of separately quantifying faecal P excretion among these three possible reasons for excretion. This is a challenge for nutritionists in developing more complete understanding of P metabolism and excretion of ruminants.

7.2.6.1 Absorbability of phosphorus from different dietary sources

If animals would, under all circumstances, absorb the maximum possible fraction of dietary P, this would mean that exogenous faecal P excretion always would be minimal and a function of the respective sources of dietary P. As a consequence, any homoeostatic regulation of total faecal P excretion could be achieved only by changes of endogenous faecal P excretion. Based on information in Section 7.2.3, it appears unlikely that changes in P secretion alone could be an efficient means of P homoeostasis. Therefore, it would be consequent to assume different efficiencies of absorption for dietary and secreted P, respectively. This differentiation between dietary P and secreted P is, however, hardly justifiable based on the cited evidence for complete mixing of exogenous and endogenous P in the forestomach (Young *et al.*, 1966c; Potthast *et al.*, 1976).

Therefore, as a clear alternative of interpreting isotope studies it is assumed that mixing of P in the rumen is complete. If this is the case, then there is no logic in differentiating between absorption efficiencies of endogenous and exogenous P, respectively, they are identical. As a consequence it has to be accepted that endogenous P as a fraction of total faecal P indicates only the ratio at which dietary and secreted P are being mixed in the digestive tract, i.e. practically the ratio between dietary and salivary P.

Isotope dilution studies in animals provided with sufficient or even excessive amounts of P, therefore, are not an adequate method for differentiating between availabilities of varying sources of dietary P. Such a differentiation would only be correct under conditions that force animals to maximize P absorption.

Koddebusch and Pfeffer (1988) fed rations to dairy goats containing either less than 1 g P or about 2 g P/kg DM. This was achieved by feeding a P-deficient basal diet and supplementing with one of four sources of P (dried grass, wheat bran, monocalcium phosphate or dicalcium phosphate). No isotope of phosphorus was used in the study, but the important point was that even the supplemented diets supplied P only at a level where goats were forced to maximize absorption. As shown in Table 7.8 net absorption efficiency of P, as defined in Eq. (2), was 0.23 for the basal diet and 0.46 for the supplemented diets. Calculating the differences between the respective supplemented and unsupplemented diets proves that net absorption efficiency of supplemented P under these conditions exceeded 0.9. No differences due to the source of supplemental P were detected.

As such extremely high net absorption efficiencies of P from supplements are verified, it must be concluded that faecal P excretion during P deficiency is primarily caused not by a low ‘availability’ depending on the specific dietary source of P, but rather because of other reasons. As these losses

Table 7.8. Intake, faecal excretion and net absorption efficiency of P in dairy goats fed diets very low or moderately low in P (Koddebusch and Pfeffer, 1988).

	Level of dietary P supply		
Item	Very low	Moderately low	Difference
Daily P intake (mg/kg BW)	33.6 ± 8.6	53.7 ± 12.4	20.1
Daily faecal P (mg/kg BW)	27.2 ± 6.0	28.7 ± 8.8	1.5
Net absorption efficiency of P	0.23 ± 0.11	0.46 ± 0.05	0.93

are not determined by the nature of the feed components fed, they must be caused by factors associated with the animal. These factors, causing inevitable losses of faecal P, are discussed in Section 7.2.6.2.

During the pre-ruminant phase in early life net absorption efficiency of P contained in milk exceeded 0.95 in calves (Guéguen, 1963; Challa and Braithwaite, 1989), in lambs (Dillon and Scott, 1979) and in kids (Boeser *et al.*, 2003).

The lack of influence of the source of dietary P on P absorption efficiency must be seen as a consequence of microbial metabolism in the forestomach, whereas in non-ruminants such differences in availability of dietary P from varying sources are of great relevance.

7.2.6.2 Inevitable faecal losses of phosphorus

When animals are fed diets very low in P they may not be able to reduce faecal P excretion to such an extent as would be necessary to achieve an equilibrium P balance (zero balance). These losses are defined as inevitable or obligatory losses and it is assumed that they are caused neither by the 'quality' or absorbability of dietary P sources nor by the level of P intake, but rather by the physiology of the host animal and/or by microbial metabolism.

As inevitable faecal P losses may be caused by or a result of either the host animal or the microbes in the digestive tract, it is worth comparing P excretion in the pre-ruminant and ruminant phases.

Walker and Al-Ali (1987) fed milk replacers low in P to lambs in their first 3 weeks after birth and observed daily faecal P excretions in the order of 4 mg/kg of body weight (BW). Kids raised on goat's milk up to about 17 kg body weight retained 88% of the P ingested with milk, but separate determinations in faeces and urine were not carried out in that work (Pfeffer and Rodehutscord, 1998). In recent balance studies with kids in their first 3 weeks of life it was found that daily P excretions per kg body weight were 1.4 mg in faeces and 25 mg in urine, respectively. This comparably high urinary P excretion obviously results from Ca being the limiting factor in milk for accretion of both elements in the bodies of kids. When Ca citrate was added to goat's milk, faecal P excretion increased to 2.8 mg and renal P excretion decreased to only 0.5 mg/kg of body weight per day, respectively (Boeser, unpublished). It is concluded that only a very small fraction of the urinary P of milk-fed pre-ruminants is inevitable loss of P from the body, as most of the loss could be avoided by increasing Ca supply.

The technique of intragastric infusion excludes any influences of microbial metabolism on balance results, because nutrients are supplied solely in the form of solutions into the rumen and the abomasum. Using this technique, Rajaratne *et al.* (1996) lowered the daily P supply of three adult sheep weighing about 40 kg from 1.29 to 0.13 g. At this marginal P supply sheep reduced their daily faecal P excretion to rates ranging between 6 and 30 mg/kg. The rates of these losses agree fairly closely with the 7 mg/kg daily calculated for pigs fed diets insufficient in P (Rodehutscord *et al.*, 1998). It must be concluded that the isolated metabolism of the ruminant host animal can conserve P under conditions of low P supply as efficiently as non-ruminant mammals.

Whereas regressing inevitable faecal P losses on body weight generally is accepted for non-ruminants, the validity of such a regression must be doubted for ruminants. In a series of investigations in Bonn, negative P balances were found in 138 trials carried out with lactating goats, 24 trials with pregnant and non-lactating goats and eight trials with non-pregnant and non-lactating goats. Renal P excretion was negligible in each of these goats.

No significant correlation was found between faecal P excretion and live weight (LW), whereas a highly significant correlation existed between faecal P excretion and dry matter intake (DMI) indicating that per kg DMI, 0.88 g P were inevitably lost in faeces of these goats (Pfeffer, 1989). From a reanalysis of 158 data points of growing cattle, Ternouth *et al.* (1996) concluded that excretions of endogenous faecal P in growing cattle consuming forage diets were correlated most closely with DMI and on average amounted to 0.505 g/kg DMI. Because on average 77% of the faecal P of cattle on the respective diets is endogenous (Coates and Ternouth, 1992; Bortolussi *et al.*, 1996; see Table 7.1), it may be speculated that regressing total faecal P excretion to DMI on these low P diets might result in roughly 0.66 g P/kg DMI (computed as 0.505 g/kg DMI divided by 0.77).

Results of Challa and Braithwaite (1988a,b) can be used for estimating inevitable faecal P losses of calves as a function of dietary P concentration. These authors fed a diet low in P at constant rates to ruminating calves and varied P supply to the animals by either supplementing P to the diet or infusing into the abomasums varying amounts of orthophosphate. We have taken P retention as the dependent variable (y) plotted against P concentration in dietary DM as the independent variable (x) in Fig. 7.4 and fitted the data to the following equation:

$$y = b_0(1 - e^{-k(x-c)}) \qquad (18)$$

In the present case the following constants were computed: $b_0 = 28.57$; $k = 0.483$ and $c = 1.17$. As c is that point on the abscissa at which y is 0, it is concluded that zero balance for P can only be achieved at a dietary P concentration of not less than about 1.2 g/kg DM.

Spiekers *et al.* (1993) fed a low P diet of constant composition at rates of either 16.9 or 10.0 kg DM daily to dairy cows. Irrespective of the treatment, cows excreted about 1.2 g P/kg DM ingested, which agrees remarkably well with the results shown in Fig. 7.4 for calves.

The Technical Committee on Responses to Nutrients (AFRC, 1991) expressed theoretical considerations starting its reappraisal of maintenance requirements for Ca and P:

> The predominant constituent of maintenance requirements for both Ca and P is the obligatory endogenous faecal loss (E), i.e. the amount of Ca and P secreted into the gastrointestinal tract which is not reabsorbed further down the tract.... Most information on E is from direct measurement using radioisotope and is most easily interpreted when E is independent of the dietary intake of the element. This independence holds less strongly for P than for Ca. E(P) is held to consist of two fractions, obligatory and excretory (the latter arising from dietary excess), and only experiments in which the absorption of P from the diet was equal to or less than net requirement will give valid information on the obligatory component...).

Deviating from the view of AFRC (1991), we prefer to strictly differentiate between 'endogen-

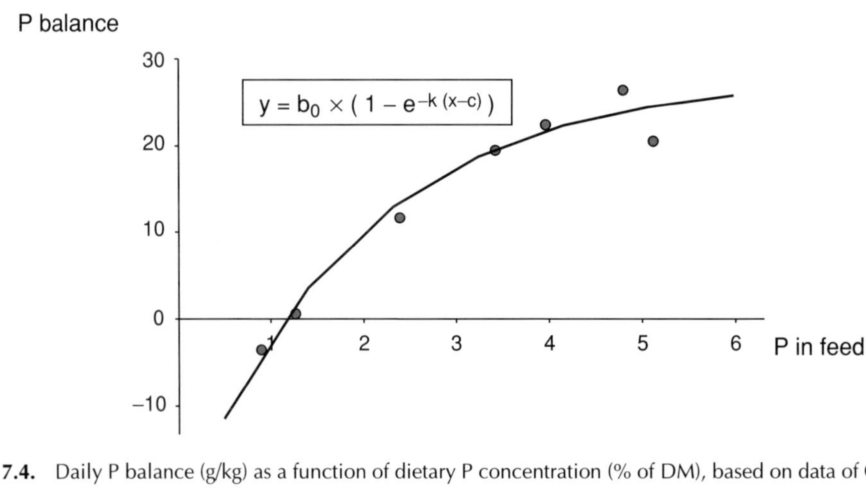

Fig. 7.4. Daily P balance (g/kg) as a function of dietary P concentration (% of DM), based on data of Challa and Braithwaite (1988a,b).

ous faecal P' as defined in Section 7.2.2 on one side and 'inevitable' or 'obligatory' faecal loss on the other side. In ruminant animals, the latter is seen primarily as a consequence of microbial metabolism and may have been captured from either endogenous P, or from potentially available P from exogenous sources.

It is assumed by AFRC (1991) that sloughing of intestinal epithelial cells is the major reason for inevitable faecal P losses in ruminants and that this phenomenon is caused mainly by fibrous plant material of low digestibility. It is hypothesized that, therefore, a negative correlation may exist between metabolizability of energy (= *q*) and inevitable faecal P losses. It is concluded: 'There is a marked increase in daily need for P, though not for Ca, when *q* value of the diet falls below 0.7 . . . '

The AFRC (1991) assumption and hypothesis are tenuous based on more recent research results. Rodehutscord *et al.* (2000) challenged this hypothesis by adding either sawdust as indigestible organic matter (OM) or starch as completely digestible OM to a diet for dairy goats with low basal P. Supplementing sawdust did not affect the amount of faecal P significantly, whereas supplementing starch caused a significant increase in faecal P excretion. Therefore, it is concluded that the digestible and not the indigestible part of OM intake has the most influence on inevitable faecal P losses in ruminants. On the basis of this finding the difference between the regression of Ternouth *et al.* (1996) for growing cattle fed forage diets and the results of Spiekers *et al.* (1993) for dairy cows fed mixed rations containing 58% concentrates appears less dramatic. Dry matter digestibility in most of the diets fed to the growing cattle was well below 0.6 (Coates and Ternouth, 1992; Bortolussi *et al.*, 1996), whereas that of the dairy cows was 0.75 (Spiekers *et al.*, 1993).

More research is needed for fully understanding factors influencing inevitable faecal P losses in ruminants.

7.2.6.3 Surplus phosphorus excretion in faeces

When P supply to sheep was increased by infusion of phosphate solutions either into the blood or into the rumen, daily quantities of P net secretion prior to the duodenum and P net absorption from the intestine were not influenced significantly, but faecal P excretion was increased by lowering intestinal P net absorption efficiency (Scott *et al.*, 1984a,b).

This shows that adaptation of intestinal absorption efficiency to the level of P supply contributes substantially to homoeostatic regulation of P in ruminants. The increase in faecal P excretion did not affect digestibility of DM. However, it did significantly decrease faecal DM concentration and thus increase faecal water excretion.

Adding solutions of monosodium dihydrogen orthophosphate to milk fed to pre-ruminant calves caused faecal P and urinary P excretion to increase at a ratio of about 1:2 and P retention to decrease concurrently (Challa and Braithwaite, 1989). This result is not in agreement with current studies carried out in Bonn in which faecal P excretion of kids was not affected as a consequence of supplementation of phosphate to milk and all extra P was excreted in the urine (Boeser, 2004).

7.2.7 Renal excretion of phosphorus

Phosphorus balances were reported in several of the papers cited in Tables 7.1 and 7.2. Well above 90% of the total P excretion was via the faeces in most of these studies, leaving only marginal importance to excretion in urine (Grace *et al.*, 1974; Bertoni *et al.*, 1976; Boxebeld *et al.*, 1983; Braithwaite, 1984, 1985, 1986; Wylie *et al.*, 1985; Martz *et al.*, 1990; Khorasani and Armstrong, 1992; Bortolussi *et al.*, 1996). Comparable balances were reported in dairy cow studies (Morse *et al.*, 1992; Delaquis and Block, 1995; Knowlton *et al.*, 2001; Knowlton and Herbein, 2002; Valk *et al.*, 2002). Ignoring renal P excretion completely would, therefore, not appreciably bias the calculated P balances in these experiments. This situation must, however, not be generalized because there are situations in which P excretion may be elevated, even in ruminants.

In the 1960s beef production turned to feeding very high concentrate diets. Reed *et al.* (1965) reported that in steers fed such diets P excreted in urine could exceed the quantity of P excreted in faeces. This surprising finding gave reason to several investigations. Table 7.9 summarizes papers in which partitioning of P excretion by ruminants was found to deviate from the predominant faecal route with a substantial contribution by the kidneys.

Table 7.9. Partitioning of P excretion in ruminants between faeces and urine as influenced by the diet.

Animals (BW)	Diet	Dietary P % of DM	DMI (kg/day)	P excretion (g/day) Faeces	Urine	Reference
Steers 250 kg	Concentrate	0.47	3.2	6.3	7.2	Reed *et al.* (1965)
	+10% roughage	0.44	4.4	9.6	7.1	
	+20% roughage	0.41	4.8	11.9	5.0	
	+30% roughage	0.38	4.8	11.3	3.9	
Steers 660 kg	Concentrate	0.47	6.7	Not given	6.7	Topps *et al.* (1966)
	+10% roughage	0.44	9.0		2.7	
	+20% roughage	0.41	10.9		2.1	
	+30% roughage	0.38	10.3		1.4	
Calves 105 kg	Roughage	0.43	2.6	4.8	4.1	Scott *et al.* (1971)
	Concentrates	0.64	2.0	4.5	4.8	
Sheep 53 kg	Roughage	0.50	1.4	5.6	0.3	Scott (1972)
	Soya meal concentrates	0.63	1.4	4.8	2.1	
	Fishmeal concentrates	1.00	1.4	6.6	5.5	
Calves 80 kg	Roughage	0.50	2.0	Not given	0.57	
	Soya meal concentrates	0.63	2.0		4.69	
	Fishmeal concentrates	1.00	2.0		5.68	
Lambs BW not given	Milk substitute	0.76	0.13	0.01	0.32	Dillon and Scott (1979)
	Milk substrate + concentrate	0.62/0.56	0.3/0.4	0.16	0.44	
	Concentrates	0.52	0.45	0.60	0.97	
Sheep 29 kg	Concentrate/	0.27	0.7	0.96	0.03	Wylie *et al.* (1985)
	roughage +	0.27	0.7	0.88	0.15	
	$KHCO_3$ into	0.27	0.7	1.02	0.01	
	rumen into	0.27	0.7	0.92	0.06	
	abomasum into ileum					
Sheep 50 kg	Concentrates	0.80	0.8	3.39	2.12	Scott and Buchan (1985)
	Concentrates: hay 50:50	0.80	0.8	4.84	1.03	
	Hay	0.80	0.8	5.97	1.80	
Sheep 46 kg	Coarse hay	0.80	0.8	4.86	1.24	Scott and Buchan (1988)
	Fine hay	0.79	0.8	4.05	1.93	
Calves 140 kg	Concentrates,	0.13	2.0	2.5	0.04	Challa and Braithwaite (1988a)
	low-P+	0.33	2.0	3.6	0.4	
	$CaHPO_4$	0.48	2.1	5.0	0.4	
	$+Na_2PO_4$					
Sheep 58 kg	Semipurified	0.587	1.0	2.85	2.11	Khorasani and Armstrong (1990)
Cattle 300 kg	Hay or silages	0.34–0.41	4.0–5.7	7.2–10.4	0.3–5.8	Khorasani and Armstrong (1992)

It appears that renal P excretion is not changed immediately at weaning from urine to faeces (Dillon and Scott, 1979) and this phenomenon is being further investigated at present in Bonn (Loof *et al.*, 2004).

7.2.8 Whole-body phosphorus kinetics

Many kinetic studies using radioactive or stable isotopes are based on the scheme of metabolism as a system of connected pools with corresponding fluxes between or among pools. To determine details about kinetics an inert marker may be used which is distributed evenly in a pool without interfering with any biological kinetics. If an isolated pool is imagined which is of constant size it must be concluded that fluxes into and out of this pool are equal. If a set amount of a marker is added to this pool and evenly distributed within it, the rate at which the concentration of the marker decreases in time is an indicator of the turnover rate and after extrapolating the marker concentration back to the time of dosing, the quotient of marker dose to marker concentration at time zero indicates the pool size. In a system of pools marker concentrations may be analysed not only in the primary pool into which the marker is administered, but also in potential secondary pools into which marker is transferred over time from the primary pool.

When a mineral element is absorbed from the digestive tract, it enters into a pool, which is made up mainly by the quantity of that element in blood plasma. Biological half-lives of mineral elements in this pool are short, especially for Ca and P_i, as there exists a rapid transfer of both elements into skeletal and soft tissues of the body. As a consequence, the SA of Ca or P in plasma must fall according to an exponential function after a single dose of ^{45}Ca or ^{32}P into blood as the primary pool. Specific activity after a single dosing of the isotope does not, however, follow a single exponential function. Resorption of both elements from the bone and return from soft tissues into blood take place concurrently with accretion into the skeleton, and some of the isotope will thus return into the primary pool. When kinetics of P and Ca are investigated, P_i and Ca in blood plasma commonly are used as the respective primary pools.

Grace (1981) offered 24 meals per day at hourly intervals in order to produce near steady-state conditions in metabolism of sheep. Following a single dose of ^{32}P as orthophosphate into the jugular vein on one side, blood was sampled from the alternate side after 1, 2.5, 5, 10, 20, 40 and 60 min and following at increasing intervals for 2 weeks. In total he found a model to fit measured values which is shown in Fig. 7.5. In this model, the pool of total exchangeable P consists of four compartments and total flow of P into this pool (P absorbed from the digestive tract + P removed from bone and soft tissues) is equal to P outflow (urinary P + faecal endogenous P + P deposited into non-exchangeable bone and uptake by soft tissues). The model does not take into account total secretion and total absorption as calculated by Eqs (9) and (10), respectively.

The author commented about the model shown in Fig. 7.5 as follows:

> The P exchangeable pool (M_T) is represented by four interchanging compartments (M_1-M_4) which reflect a dilution of the ^{32}P in progressively larger masses of P. The compartments cannot be defined in terms of anatomical structures but are more likely to be related to metabolic P processes having similar turnover rates. Further it seems most likely that compartment M_1 includes the plasma inorganic P and the P in rapid equilibrium with it while compartment M_4, because of its size and relatively slow turnover rate, would include some skeletal P.

This example shows that although modelling is highly attractive from a purely mathematical point of view, its perfection may be of limited value with regard to practical application. If compartments result purely from the time needed for the marker to become distributed evenly, this hardly identifies biologically meaningful information.

Schneider *et al.* (1987) extended the model to sheep in which cannulas were placed in different sites of the digestive tract. Also, these authors used two isotopes (^{32}P and ^{33}P) for concurrently labelling two different primary pools. Their model comprised nine compartments and in total they identified 15 specified P fluxes. The approach of using increasingly more sophisticated modelling software will allow identification of more pools and fluxes. Future development will have to show to what extent this can increase understanding of biological processes.

Figure 7.6 demonstrates the changes in metabolism caused by varying P intake. For this purpose, the findings of Challa and Braithwaite (1988a) were

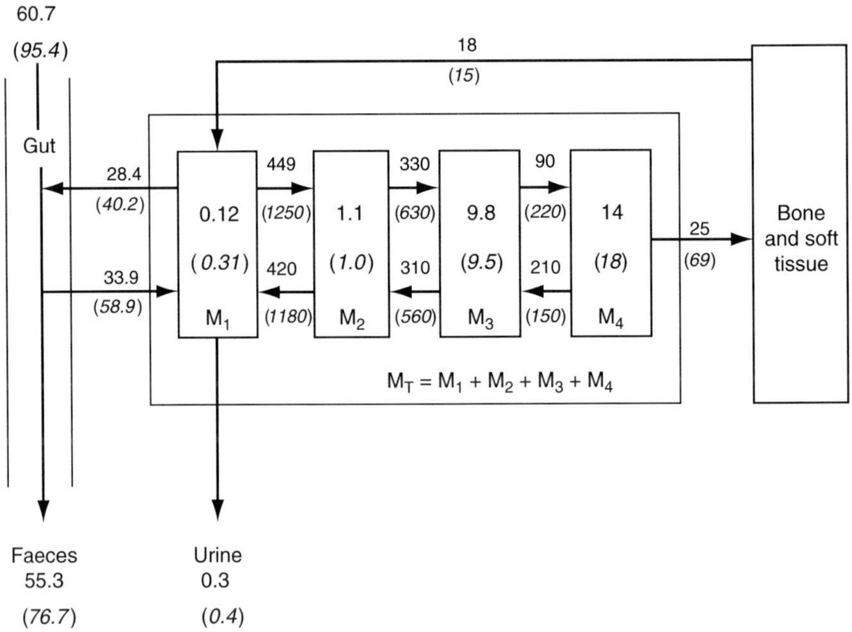

Fig. 7.5. Model of P kinetics in sheep fed 850 g DM daily of lucerne or perennial ryegrass (Grace, 1981). M_1, M_2, M_3 and M_4 are compartments of the exchangeable P pool M_T. Numerical values attached to arrows indicate rates of P intake and P excretion, rates of P secretion, P absorption, P accretion and P resorption as well as rates of P transport between compartments of the exchangeable P pool, respectively (mg/kg per day), numerical values within rectangles indicate the respective masses of P within each compartment (g) in sheep fed 850 DM of lucerne chaff (or perennial ryegrass).

recalculated using Eqs (9) and (10), respectively. Friesian calves weighing about 140 kg were fed diets of concentrates plus straw calculated to be adequate with the exception of P for maintenance plus 250 g growth per day. Phosphorus supply from each of the three diets was either deficient, or adequate or excessive, respectively. When P intake was increased from deficient to adequate and to excessive, respectively, P secretion increased at comparable rates and P absorption increased at higher rates. Phosphorus accretion into the skeleton increased from the deficient to the adequate supply and showed only a very small response to further increase in P intake (excessive), whereas P resorption from the skeleton remained practically unaffected by P intake. The net changes in kinetics due to increasing provision of dietary P were increases in faecal P excretion, renal P excretion and P retention.

Figure 7.7 shows comparable schemes of P kinetics in pregnant and lactating ewes based on the results of Braithwaite (1986). In both cases, P supply was insufficient to maintain the P equilibrium (zero balance) of the maternal organism. Although in the case of the pregnant ewe the combined balance of maternal plus conceptus units (P intake − faecal P − urinary P) appeared mildly positive ($53.3 - 48.1 - 0.4 = 4.8$ mg/kg LW daily), P transfer to foetal metabolism amounted to almost one-third of P intake, which exceeded the maternal plus conceptus balance by a factor of 3.5. In the case of the lactating ewe, the balance was clearly negative (-34.5 mg/kg LW daily) due to the drain of P into milk. Resorption of P from the skeleton exceeded P accretion by 2- and 3.6-fold in pregnant and lactating ewes, respectively.

Substantial losses of ash from the skeletons of ewes recorded between mid-gestation and mid-lactation were replaced 2 months after the end of lactation, if dietary P concentration was 4.5 g/kg DM; however, rations containing only 1.5 g P/kg DM were unable to cause the respective replacement of P lost during late pregnancy and lactation

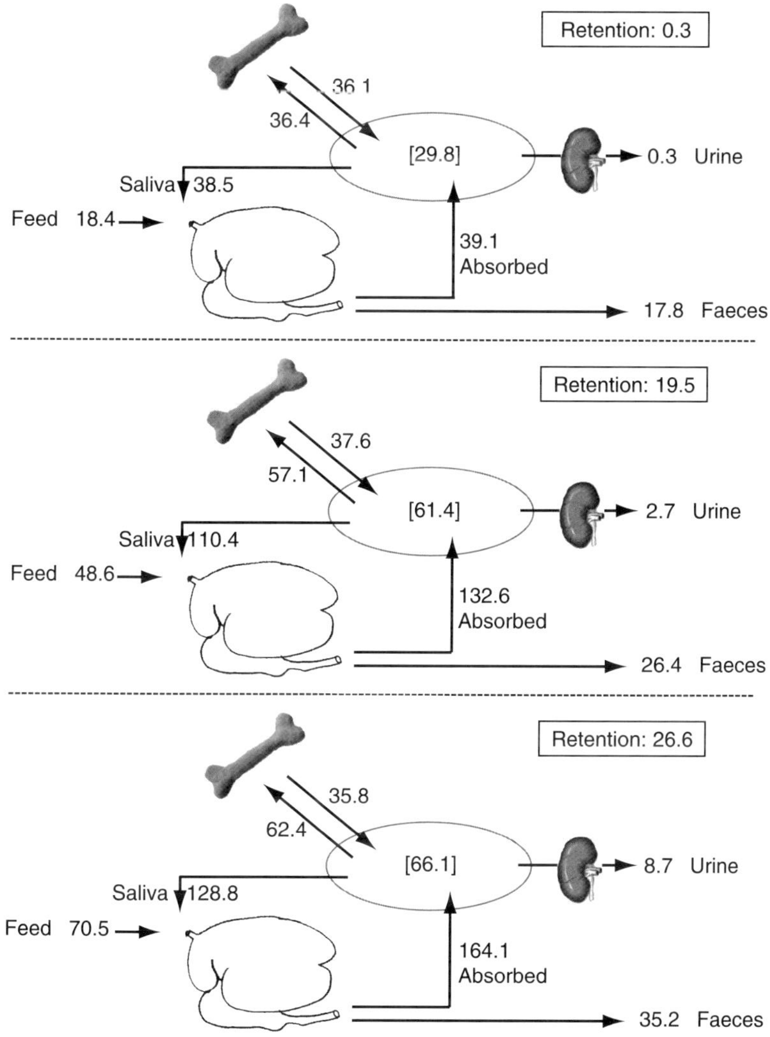

Fig. 7.6. Phosphorus kinetics in calves (140 kg LW) fed diets varying in P concentration (recalculated from Challa and Braithwaite, 1988a). Numerical values attached to arrows indicate the respective rates of P intake and excretion, P secretion, P absorption, P accretion and P resorption. Numbers in brackets indicate sizes of rapidly exchangeable P (mg/kg).

(Benzie *et al.*, 1959). It is assumed that comparable phases of loss and replacement of bone minerals also occur in other ruminant species.

Inevitable faecal P losses caused continuous negative P balances and a steady draining of P reserves from the body of mature wether sheep fed a diet containing only 1 g P/kg DM (Breves *et al.*, 1985). During this experimentally induced P depletion, cortical thickness and bone density decreased and the trabecular structure of the distal radius became coarser and less dense with reduced cross-linking between trabeculae (Breves and Prokopp, 1990). More dramatic losses of P as well as Ca from the skeleton due to insufficient dietary supply were provoked experimentally in lactating goats which were unable to sufficiently reduce faecal P excretion (Müschen *et al.*, 1988; Pfeffer *et al.*, 1993, 1994; Rodehutscord *et al.*, 1994b). In growing kids gaining 123 g body weight daily the inevitable faecal P losses prevented any net

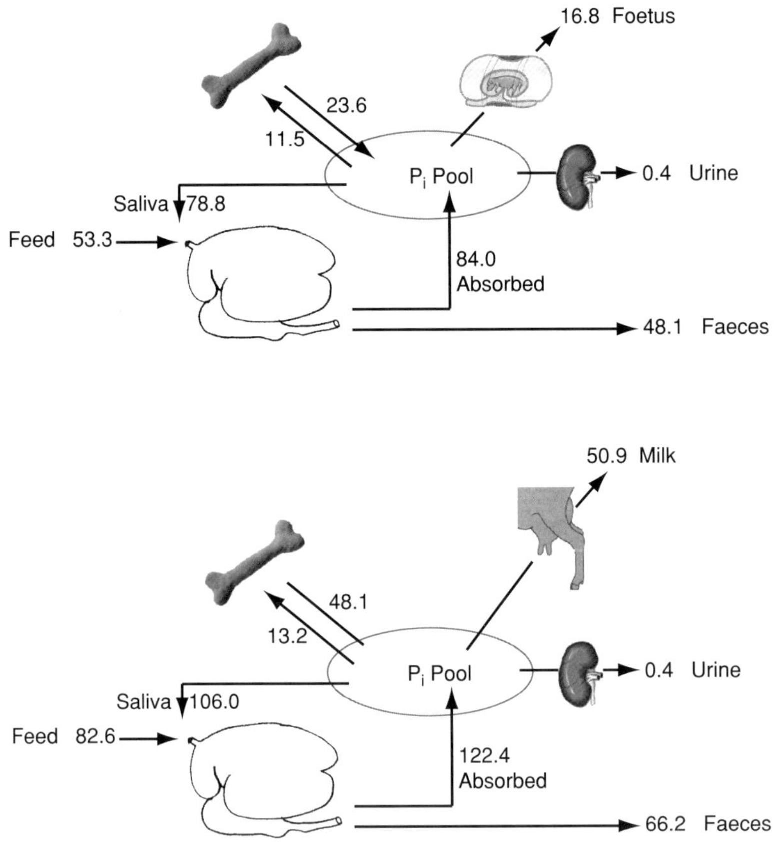

Fig. 7.7. Phosphorus kinetics in pregnant or lactating ewes (recalculated from Braithwaite, 1986). Numerical values attached to arrows indicate the respective rates of P intake and excretion, P secretion, P absorption, P accretion and P resorption as well as rates of transfer to the fetus and secretion with milk.

accretion of P or Ca (Pfeffer *et al.*, 1995) when diets contained only 2 g P/kg DM.

When lactating goats or cows were changed abruptly from diets containing at least 3.4 g P/kg DM to diets containing not more than 1.4 g P/kg DM, plasma P_i concentration dropped within 10 days from 1.8 to 0.2 mmol/l without causing dramatic effects on health or appearance of the animals. Concurrently, phosphate concentration in ruminal fluid decreased from about 30 to about 6 mmol/l (Rodehutscord *et al.*, 1994a). When P depleted animals were switched back to adequate P intake, P_i concentrations in saliva and ruminal fluid returned to normal values within similarly short periods, long before P deficits were restored. These latter results are the basis of a hypothetical explanation of how a cow of 650 kg LW may react to a sudden change from sufficient to highly insufficient P intake. In the following approximation example, all numbers are only rough estimates, the precision of which must not be over interpreted.

The skeleton of the cow is thought to weigh 60 kg and contain 3.8 kg P. Soft tissues and organs are thought to weigh 500 kg (wet basis) and contain 600 g P. Further, gut fill of the cow may weigh 90 kg and contain 100 g P. Extracellular body fluid amounts to about 100 l and has the P_i concentration found in blood plasma. Most of the roughly 70 l of fluid of gut contents is located in the rumen and P_i concentration of ruminal fluid is, therefore, used to calculate total P_i in fluid of digesta. As long as P intake is adequate, P_i concentrations in plasma and ruminal fluid are assumed to be 1.8 and 30 mmol/l, respectively. This means that the respective amounts of P_i are 180 mmol in extracellular body fluid and 2100 mmol in digesta fluid, which adds up to a pool of rapidly exchangeable P of about 2.3 mol which is equivalent to about 70 g. If after a sudden reduction in P intake, volumes of extracellular and digesta fluids do not change but the respective P_i concentrations are reduced by up to 90% within 10 days, this means that more than 60 g P may be mobilized from this rapidly exchangeable pool of P within the first phase of depletion. Afterwards, a continuous drain of P from the skeleton will occur which finally may cause severe damage to the animal.

Lactating goats fed diets containing only about 1 g P/kg DM lost about a fourth of the P and Ca from their skeletal tissues without being able to stop this drain (Pfeffer *et al.*, 1994). When cows are switched back to adequate P intake after a period of P depletion, the immediate increases in P_i concentrations of plasma, saliva and ruminal fluid indicate that the rapidly exchangeable pool of P is replenished immediately, whereas it takes a much longer time to deposit in the skeleton amounts of P and Ca equivalent to the preceding losses. If this hypothesis is accepted then plasma P_i concentration cannot be regarded as a very useful indicator for monitoring the P status of ruminants.

Lowered phosphate concentrations in ruminal fluid may impair microbial activity, as discussed in Chapter 6. Phosphorus depletion not only decreased digestibility of OM prior to the duodenum and posterior to the ileum of sheep (Breves and Höller, 1987), but it also reduced the contribution of amino acids of microbial origin to milk protein in dairy goats (Petri *et al.*, 1988).

7.3 Factorial Derivation of Phosphorus Requirements in Cattle

The ARC (1965) employed a twofold approach in assessment of requirements of ruminant animals for major mineral elements. This approach continues to be used in establishment of the most recent requirement estimates (AFRC, 1991; GfE, 1993; NRC, 2001; Valk and Beynen, 2002). This approach is described as follows (ARC, 1965, p. 14):

> We have employed a two-fold approach in the assessment of requirements of ruminant animals for the major mineral elements. First, by a factorial method, the minimal requirements of animals of different classes, producing at different rates, have been calculated and secondly, these factorial estimates have been compared with the results of experiments and feeding trials in which the element has been given in two or more different amounts and the resultant performance of the animals measured. The factorial method assesses requirements of animals in two stages. In the first, the net mineral requirement of the animal is obtained from estimates of storage and excretion of the element made during growth, pregnancy and lactation and any inevitable losses of the element from the body (the endogenous losses). Secondly, the availability of dietary sources of the element is assessed from metabolism experiments. The net requirement divided by the availability of the element gives the requirement in terms of a dietary amount.

Amounts retained daily during particular phases of growth and at specified stages of pregnancy as well as amounts contained in the milk yielded as parts of 'net requirements' of an element are accepted unanimously. There is, however, some controversy with regard to the fourth part of net requirements, the inevitable or obligatory losses (maintenance). ARC (1965) also used the term 'net endogenous requirement' for the latter. Judged on the basis of knowledge of today the twofold use of the term 'endogenous' is regrettable because it contributes to misunderstanding.

In this chapter 'endogenous' is used consequently according to the definition given previously (Section 7.2.2):

> P that originates from the animal's body and has been transferred from the blood into the lumen of the digestive tract – it does not indicate a necessity of excretion of this P and is not a synonym to inevitable.

7.3.1 Deposition of phosphorus in the body during growth

Table 7.10 shows P concentrations in empty bodies of Friesian bulls fed to gain on average 1 kg LW daily and slaughtered at one of five target weights between about 150 and 575 kg, respectively. Throughout, P concentrations were close to 7.0 g/kg empty body weight which corresponded to about 5.8 g/kg LW (Schulz *et al.*, 1974).

Hoey *et al.* (1982) analysed body composition of Hereford heifers fed a basal diet containing only

Table 7.10. Phosphorus concentration in empty body and skeleton of Friesian bulls at different LW (Schulz *et al.*, 1974).

		Empty body		Skeleton		
LW (kg)	EBW (kg)	*n*	P concentration (g/kg)	*n*	Weight (% of EBW)	P concentration (g/kg)
152±3	123±3	6	7.2±0.3	6	16.0±1.1	38.1±2.2
267±2	218±2	6	6.9±0.4	6	13.8±0.7	41.7±2.7
370±2	309±2	6	7.0±0.4	8	12.8±0.4	46.2±1.7
480±1	416±2	6	7.0±0.5	12	11.6±0.5	51.5±3.1
576±2	509±2	6	6.6±0.3	12	10.7±0.6	53.2±2.0

Table 7.11. Accretion of P in fetus and conceptus of beef heifers and dairy cows.

Reference	Ferrell *et al.* (1982)		House and Bell (1993)	
Animals	81 Angus, Hereford and Red Poll crossbred heifers (332±6 kg BW), mated to Brown Swiss bulls		18 Multiparous Holstein cows (714±14 kg BW), artificially bred to one Holstein bull	
Day of gestation	Weight of fetus (kg)	P in conceptus (g)	Weight of fetus (kg)	P in conceptus (g)
100	0.35	1.00	0.41	3.2
130	1.23	3.96	1.49	10.2
160	3.55	12.9	4.39	27.8
190	8.54	34.6	10.6	66.5
220	17.1	76.2	21.2	139
250	28.6	138	34.5	235
280	39.7	205	46.2	403

0.9 g P/kg DM. This diet was fed either without P supplement or an additional 12 g P per day as monoammonium phosphate. A third group of heifers received the supplement, but the quantity of diet was restricted to intake of the group fed basal diet without supplement. Empty whole-body P concentrations ranged from 6.0 g/kg in the group without supplemental P to 6.8 g/kg in the group receiving the supplement but at restricted feed intake. These results show that in growing cattle empty body P concentrations may vary to some degree depending upon P intake.

7.3.2 Deposition of phosphorus in products of conception

It is generally accepted that growth of products of conception follows an exponential curve in mammalian species. By far, the greatest proportion of conceptus tissue accretion occurs in the last third of pregnancy.

Mineral element accretion in fetuses and conceptus of single- and twin-bearing ewes was studied by Grace *et al.* (1986). Table 7.11 summarizes findings of P accretion in fetuses of beef heifers (Ferrell *et al.*, 1982) and of Holstein cows (House and Bell, 1993). Although the difference in P deposition between these two cattle populations appears remarkable, it is emphasized that quantities deposited are relatively small compared with inevitable losses, with P accretion during growth, or with transfer of P into milk during lactation. From results presented in Table 7.11 it is concluded that daily P deposition in the conceptus even near the very end of gestation will not exceed 4 g in the beef breeds and 7 g in Holstein cows, respectively.

7.3.3 Phosphorus concentration in milk

Table 7.12 shows P concentrations in milk found by some authors within the last 15 years. Values for colostrum generally are higher than for milk,

Table 7.12. Phosphorus concentration in bovine milk.

Breed	*n*	Stage of lactation	P in milk (g/kg)	Reference
Brahman	36	Parturition	1.27	Salih *et al.* (1987)
		After 3 months	0.90	
British Friesian	134	'Indoor period'	0.81–0.95	Brodison *et al.* (1989)
	217	'Pasture period'	0.85–1.10	
Friesian and Friesian–Jersey cross bred	34	Day of parturition	1.73–1.46	Law *et al.* (1993)
		Day 1 of lactation	1.32	
		Days 6–9 of lactation	1.01	
Holstein–Friesian	52	Complete lactation	0.90	Brintrup *et al.* (1993)
Holstein–Friesian	26	Complete lactation	0.85–0.88	Wu *et al.* (2000)
Holstein–Friesian	36	Early and mid lactation	0.68–0.78	Knowlton *et al.* (2001)
Holstein–Friesian	24	Complete lactation	0.87–0.94	Valk *et al.* (2002)
Holstein–Friesian	13	11 weeks of lactation	0.89	Knowlton and Herbein (2002)

but they fall within the first week of lactation. No influence of the amount of P intake was found on P concentration in milk. Influences of the stage of lactation were not evident in any of the cited studies. A single subcutaneous injection of slow-release bovine somatotropin 7 days before the expected calving date significantly reduced P concentration in colostrum, presumably due to increased colostrum volume, but this difference was no longer evident by 6 to 9 days of lactation (Law *et al.*, 1994). It is reasonable to assume generally a P concentration in milk of 0.9 g/kg over the course of a lactation. Especially in high yielding dairy cows the quantity of P secreted into milk must be regarded as the dominating factor establishing P requirements.

7.3.4 Inevitable losses of phosphorus

Urinary P excretion obviously can be reduced to negligible amounts in ruminants. For this reason it is reasonable to ignore inevitable urinary losses as a factor in determining P requirements of cattle. From the discussion in Section 7.2.6.2 it is clear that inevitable losses of P from the body of cattle are difficult to predict, because effects of low P intake on the host animal's metabolism must be differentiated from those taking place in microbial metabolism (see Chapter 6).

Inevitable P losses as defined in this chapter are not considered as part of 'net requirements' in the sense of ARC (1965), as they are measured as faecal excretion and, therefore, need not be corrected by 'availability' to 'give the requirement in terms of dietary amount'. For dairy herds it is generalized that mixed rations based on silages of maize or grass and comparatively high proportions of concentrates will cause inevitable faecal P losses of 1.2 g/kg DMI.

7.3.5 Availability of dietary phosphorus

From what was discussed in the foregoing sections, especially in Sections 7.2.2, 7.2.4 and 7.2.6.1, respectively, it has to be concluded that quite differing values for 'availability of dietary P' may be derived from experimental data, depending on the definition of 'availability' chosen. In a factorial approach of deriving recommendations for mineral supply, however, this factor very much dominates the result because all data concerning net requirements have to be transformed into total dietary P requirement by dividing by the 'availability' or absorption efficiency.

Findings of Lofgreen (1960) are often regarded as evidence that differences in absorption efficiency of P from different inorganic sources exist, as well as between dicalcium phosphate and calcium phytate. On the other hand, no differences in overall utilization of P could be established between monocalcium phosphate and calcium phytate (Tillman and Brethour, 1958b) or between monosodium phosphate and acid sodium pyrophosphate (Tillman and Brethour, 1958c). From the work of Koddebusch and Pfeffer (1988) it is concluded that ruminants will utilize P from the different relevant sources with very high efficiency (greater than 0.9), if P intake is limiting performance.

Currently, for example, the availability values or absorption efficiencies for dairy cattle diets used by the various working parties are 0.58, 0.70 and 0.70 for AFRC (1991), GfE (1993) and NRC (2001), respectively. Valk and Beynen (2002) calculated maximum absorption efficiencies of 0.72 for dry cows and 0.77 for lactating cows, but reduced it to 0.70 to allow a safety margin in practical feeding.

Uncertainty concerning the proper method of deriving availabilities as criteria typical for feed components may have caused working parties to decide for comparably low values of availability in order to include safety margins in the recommendations. Critical interpretation of the literature cited in Section 6.5 of Chapter 6 and Section 7.2.6.1 of this chapter, respectively, appears to justify the general assumption of an availability of not less than 0.9 for P contained in the feeds commonly fed to cattle. However, the relevant experiments had to be carried out, for methodological reasons, under conditions of P limiting performance. As cattle are not fed to limit performance by P intake, availability of marginal P may be less than 0.9 in practical situations, according to the principle of diminishing returns. It has to be stated that data with high-yielding dairy cows are insufficient to really prove whether assuming a general availability of 0.9 is appropriate and will not cause practical problems. More research is needed for this problem.

7.4 Evaluation of Derived Recommendations from Results of Feeding Trials

7.4.1 Dairy cows

According to the approach of ARC (1965), factorial estimates of requirements have to be compared with results of feeding trials in which the element has been given in two or more amounts and the resultant performance of the animals measured.

Table 7.13 summarizes the results of nine investigations published in the last two decades. Although based on diverse basal rations, dietary P supply was the variable within each study. These studies were carried out with dairy cows yielding between about 5000 and more than 12,000 kg of milk/cow per year and were sufficiently long to allow general conclusions. Concentrations in dietary DM of 0.24% (Call *et al.*, 1987) or 0.23% (Valk and Sebek, 1999) produced symptoms of P deficiency, mainly reduced feed intake and, consequently, reduced milk yield. Neither milk yield nor health status of the cows was affected as long as the P concentration was not less than 0.27% DM.

Milk yields were comparably low in the work of Brodison *et al.* (1989) and numbers of cows were comparably low in most of the other papers cited. This reservation, however, is not to be held with regard to the work of Lopez *et al.* (2004a) who used more than 100 cows per treatment for comparing P concentrations of 0.57% and 0.37% in the DM of total mixed rations fed for over the first half of lactation. These numbers were sufficiently high to allow the additional statement that reproductive performance was not impaired by the lower concentration (Lopez *et al.*, 2004b).

It is concluded from Table 7.13 that rations for dairy cows need not contain more than 0.37% P in their DM if feed intake is adequate to meet the needs of animals for energy and other nutrients.

We decided to base the recommendation on P quantities rather than on dietary P concentration because of the vast differences in diet composition within and among countries, and because of often observed differences in amount of DMI as influenced by animal and dietary factors. Further, we believe that the approach of using P quantities (i.e. g/cow per day) is more accurate practically to achieve the scientifically determined requirement because the P flow within the animal is taken into account, than using dietary P concentration.

Based on the findings discussed above, recommendation standards for P supply of dairy cows have been revised in Germany (GfE, 1993), the USA (NRC, 2001) and the Netherlands (Valk and Beynen, 2002). Table 7.14 shows the amounts of P recommended by each of the working parties and AFRC (1991) over a range of milk yields. The P concentration in rations at set DMI rates are also shown. Recommendations in Germany and the USA are lower than those of AFRC (1991) and do not differ greatly, whereas the Dutch standards are notably lower. In each of the systems, dietary concentration of P has to increase with increasing milk yield. At very high yields, the 0.37% of dietary DM, which, according to Table 7.13 are sufficient, are remarkably exceeded in the British

Table 7.13. Lactational performance of dairy cows fed diets varying in P concentrations over extended periods of time.

No. of cows	Duration	Dietary P (% of DM)	DMI (kg/day)	Milk yield (kg/day)	Reference
13	1 lactation	0.42		21.2[a]	Call *et al.* (1987)
8		0.32		22.2[a]	
13		0.24		17.3[a]	
39	2 years	0.42–0.46		17.1	Brodison *et al.* (1989)
39		0.34–0.36		16.7	
26	1 lactation	0.39	17.4	24.5	Brintrup *et al.* (1993)
26		18.1	25.4		
8	3 lactations	0.33	20.2–22.6	24.5–33.0	Valk and Sebek (1999)
8	4 lactations	0.27	19.9–22.5	24.1–34.2	
8	1 lactation	0.23	20.4	23.2	
9	44 weeks of lactation	0.49	23.4	36.2	Wu *et al.* (2000)
9		0.40	22.4	36.5	
8		0.31	23.0	35.0	
21	1st year	0.48 (0.44)[b]	20.4	28.8	Wu and Satter (2000)
21		0.38 (0.31)[b]	20.7	29.6	
27	2nd year	0.48 (0.44)[b]	23.4	32.1	
26		0.38 (0.31)[b]	23.2	32.0	
13	3 years	0.47	24.6	39.3	Wu *et al.* (2001)
14		0.39	25.0	38.7	
10		0.31	25.0	42.3	
5	11 weeks	0.67	24.1	45.8	Knowlton and Herbein (2002)
5		0.51	26.6	48.4	
4		0.34	25.3	49.5	
124	165 days of lactation	0.57		34.9	Lopez *et al.* (2004a)
123		0.37		35.1	

[a]Milk yield in the work of Call *et al.* (1987) is presented as fat-corrected milk.

[b]P concentrations in TMR fed from September/October to May (numbers in brackets indicate dietary P concentration during the grazing period, May to August).

Table 7.14. Four recommendations for P supply to dairy cows and relative total P excretions to be expected from cows fed according to each of the recommendations.

		Recommended P supply[a] (g/day)				Total P excretion[b] (g/kg milk)			
Milk (kg/day)	DMI (kg/day)	UK	D	USA	NL	UK	D	USA	NL
15	17.0	56	46	51	40	2.8	2.2	2.5	1.8
25	20.3	77	65	65	55	2.2	1.7	1.7	1.3
35	23.6	99	84	83	69	1.9	1.5	1.5	1.1
45	26.9	121	103	96	83	1.8	1.4	1.2	0.9
55	30.0	142	121	114	97	1.7	1.3	1.2	0.9

[a]Recommendations calculated according to: UK, AFRC (1991); D, GfE (1993); USA, NRC (2001); NL, Valk and Beynen (2002).

[b]$\frac{\text{Recommended P supply (g/day)} - 0.9 \text{ (g/kg)} \times \text{milk yield (kg/day)}}{\text{Milk yield (kg/day)}}$ (assumes animals are not pregnant and not growing).

system. On the other hand, dietary P concentration recommended for the lower yielding cows in the Dutch system is close to that P concentration which was found insufficient for higher yielding cows in the work of Valk and Sebek (1999). More research is needed to clarify whether microbial needs for P (see Chapter 6) may in low yielding cows exceed P requirements of their host.

Table 7.14 shows total P excretions to be expected in dairy cows fed according to each of the country's recommended P feeding rates (g/cow per day). This excretion value has been calculated on the assumption of a 'zero-balance' of P which may not be correct for each of the specific phases of lactation, because P mobilization from the body occurs during early and peak lactation and deposition into the body occurs during advanced phases. It is, however, assumed that mobilization and deposition of P occurring in the course of lactation rather compensate each other over the entire lactation.

If fed according to the respective recommendations, P excretion per kg milk produced will decrease with increasing milk yield in each of the four systems. Feeding according to the Dutch standards will cause least P excretion and potential loading to the land and feeding according to AFRC (1991) will cause greatest excretion and P loading. The fact that the German system causes comparably high P excretion results mainly from the assumption of a P concentration of 1.0 g/kg milk which is not justified by data on milk composition shown in Table 7.11 and from correcting the assumed inevitable losses of 1 g P/kg of DMI by an availability of 0.7 which includes an unnecessary 'safety margin'.

7.4.2 Beef cattle

Beef production in the USA uses two different systems. One is based on high-forage/roughage, primarily for beef cow herds and growing young cattle prior to harvest or before they enter the finishing phase. The other system is more intensive (typically in feedlots) and highly dependent on high-grain feeding to 'finish' cattle for harvest at higher body condition [greater intramuscular fat (marbling) deposition]. Definition of the two systems is an important distinction because the high-forage/roughage system typically involves lower (marginal) P inputs, whereas the feedlot system often results in P inputs from feedstuff concentrates in excess of requirements of the beef cattle being fed. Also, often the forage/roughage-based system exists in geographical areas much less likely to have high P soils. Therefore, spreading manure on these soils naturally via grazing is possible with minimal risk to the environment (as long as spreading is not near or in water bodies). Also, the forages/roughages available in this system typically have low P contents compared with the needs of the cattle.

FORAGE/ROUGHAGE SYSTEM. Karn (2001) provided a comprehensive review of P nutrition of grazing cattle. Compared with most modern dairy systems P needs and inputs into beef grazing systems are quite low. In beef grazing systems the animals' needs for P during the year as well as the amount supplied by the forage/roughage can vary widely. Actively growing forages typically provide more dietary P on a DM basis than dead standing forage or stockpiled feed. Whether or not the brood beef cow is gestating and/or lactating influences the requirement for P; but its needs are never very great compared with that of a high-yielding dairy cow (NRC, 1996, 2001).

Studies with range cows in the western USA generally showed sub-par performance (e.g. lower calf weaning weights) of cows without supplemental P for extended periods of time (Judkins *et al.*, 1985). However, the detriment was most pronounced in years of drought or after cows had been in the low-P treatment for more than 1 year; reproductive performance also was lower during that subsequent year. Little (1980) showed that reproductive performance was not compromised in beef cows fed about two-thirds the NRC (1996) P recommendation. Assurance of adequate dietary P during the breeding season of beef cattle is critical, although it appears that relatively small amounts of supplementation are needed. Reference is often made to a series of experiments conducted in South Africa more than 70 years ago in which maintenance, reproduction and calf weaning weights of grazing cattle on poor-P soils were improved with P supplementation during the dry season (Theiler and Green, 1932; see also Chapter 8). Grazed plant material in these studies was extremely low in P. Overall, supplementation with free-choice mixtures of P mineral (e.g. dicalcium phosphate) and NaCl in forage/roughage systems are practically efficacious especially during breeding season and peak lactation (Karn, 2001).

FEEDLOT SYSTEM. Feedlot systems have been a primary focus of P management in the USA in recent years. The NRC (1996) specifies a P requirement for growth of 3.9/100 g of retained body protein; typically US feedlot cattle retain 150 to 200 g of protein per day (average daily gain of 1.5 to 2.2 kg). In addition, some P is required for maintenance. Based on an assumed absorption coefficient for P of 0.68 for feedlot diets, between 16 and 30 g of total dietary P is estimated as the requirement for maintenance and growth depending upon growth weight and body size between 300 and 600 kg (NRC, 1996). At typical daily DMI of 8 to 12 kg this amounts to dietary P concentrations between 0.2% and 0.3%, DM basis.

With the NRC (1996) recommendations as a reference point, studies were done to evaluate the minimum dietary P input for normal animal growth that also would result in less manure P. Research from the University of Nebraska–Lincoln showed that the NRC (1996) recommendation for ration P for feedlot cattle may be greater than necessary. Erickson *et al.* (1999) fed yearling steers feedlot rations with less than 0.14% P (DM basis) or about 70% of the NRC (1996) P requirement and observed normal performance, similar to that of cattle fed higher concentrations of P. With younger finishing calves, the P requirement was about 14 g/day; this amount was provided with 0.16% P given the feed intake rates in the experiment (Erickson *et al.*, 2002). This amount, less than the recommendation of NRC (1996), resulted in normal feedlot performance and bone characteristics, similar to that of contemporary calves consuming more P.

Because most feedlot rations, composed primarily of maize or other grain concentrates and by-product feeds, provide P in excess of that required for normal feedlot performance, the Nebraska researchers suggested that determining the P requirement of feedlot cattle *per se* is not very important practically. However, they emphasized the importance of removal of all supplemental P from feedlot cattle rations because the typical basal ingredients provide an overage of P. Supplementation (excess input) of P in feedlot finishing rations is not necessary nutritionally, and is costly financially and has potential environmental implications. Furthermore, even with no supplemental P, reducing dietary P further to more closely meet animals' requirements would require use of some very low-P feedstuffs.

In Germany, GfE (1995) derived recommendations for P supply of growing cattle which are based on the same factors that had been used for dairy cattle GfE (1993). These recommendations depend mainly on LW and on daily weight gain, ranging from 14 to 21 and from 22 to 26 g daily per head at growth rates of 800 or 1400 g/day, respectively. It is believed that no supplementation of inorganic P is required for optimizing performance of beef cattle under practical conditions in Germany.

7.5 Optimizing Phosphorus in Cattle Rations

With zero P balance in cattle farms gaining importance as one criterion of environmentally sustainable farming, dairy and beef units have to pay more attention to the amount of P imported in the form of purchased feeds. This will have to be taken care of by feed compounders as well as by farmers and farm managers.

Where grass and grass silage are the major forages, concentrates comparably low in crude protein (CP) are used for increasing dietary energy concentration, which is often achieved by either grain or by-products like sugarbeet or citrus pulp. Per unit of energy (irrespective of the question whether metabolizable energy or net energy is used for evaluating feeds), grains contain about twice as much P as the pulps do. Thus, the choice of purchased sources of feed energy may notably influence the P balance of the farm.

Where, on the other hand, maize silage and comparable crops are the dominating forages, there will be a need for supplementary CP. This need is most often met by by-products of the food industry that may be available at attractive market prices. These products vary quite widely in the amount of P associated with a given amount of CP, as shown in Table 7.15.

This shows that there are possibilities of reducing the amounts of P imported into cattle units with the aim of achieving zero P balance.

7.6 Conclusions

Salivary secretion of P in ruminants often exceeds intake of dietary P. Exogenous P from feed and endogenous P from saliva are mixed completely in the forestomach due to microbial metabolism. Pas-

Table 7.15. Phosphorus and CP concentrations in by-products of food industry that may be used as protein supplements in dairy cow rations (Beede, 2003).

By-product feed	P (% of DM)	CP (% of DM)	P/CP (g/kg)
Maize gluten meal	0.60	65.0	9.2
Soybean meal (44% CP; expellers)	0.70	49.9	14.0
Brewer's grain (dried)	0.67	29.2	22.9
Cottonseed, whole	0.60	23.5	25.5
Distiller's grain, maize (with solubles, dried)	0.83	29.7	27.9
Rape meal (mechanically extracted)	1.10	37.8	29.1
Maize gluten feed	1.00	23.8	42.0
Maize steep liquors	1.70	33.0	51.5
Wheat middlings	1.00	18.5	54.0
Wheat bran	1.18	17.3	68.2

sage of P into the intestine always greatly exceeds P intake and P absorption occurs mostly from the small intestine without differentiation between endogenous and exogenous origin. In ruminants, absorption of phosphate is independent of calcitriol [1,25-$(OH)_2$-vitamin D_3] and, therefore, excessive Ca intake has no negative impact on efficiency of utilization of dietary P. Contrary to non-ruminants, the extent of P absorption is hardly, if at all, influenced by the nature of feed components and absorbability is always very high. Inevitable faecal losses of P are greater in ruminants than in non-ruminants and appear correlated to feed intake rather than to body weight. Surplus P is excreted in faeces on rations containing forage; only when pure concentrate rations are fed, urinary excretion of P may become substantial.

P concentration of milk averages 0.09% and is not influenced by the level of P intake. P concentration of body mass can vary depending on P intake and is about 0.7% of empty body weight in well fed cattle. When the factorial approach is used for deriving recommendations, discrepancies exist between different working groups with respect to assumed inevitable losses and availabilities.

Long-term feeding trials indicate that P concentration of dietary dry matter need not exceed 0.37% in herds producing more than 10,000 kg milk per cow annually. In most rations there is no necessity to supplement inorganic phosphates.

References

Abdel-Hafeez, H.M., Manas-Almendros, M., Ross, R. and Care, A.D. (1982) Effects of dietary phosphorus and calcium on the intestinal absorption of Ca in sheep. *British Journal of Nutrition* 47, 69–77.

AFRC [Agricultural and Food Research Council – Technical Committee on Responses to Nutrients] (1991) A reappraisal of the calcium and phosphorus requirements of sheep and cattle, report 6. *Nutrition Abstracts and Reviews (Series B)* 61, 573–608.

ARC [Agricultural Research Council] (1965) *The Nutrient Requirements of Farm Livestock, No. 2, Ruminants.* HMSO, London, 264 pp.

Bailey, C.B. (1961) Saliva secretion and its relation to feeding cattle. 3. The rate of secretion of mixed saliva in the cow during eating, with an estimate of the magnitude of total daily secretion of mixed saliva. *British Journal of Nutrition* 15, 443–451.

Bailey, C.B. and Balch, C.C. (1961) Saliva secretion and its relation to feeding cattle. 1. The composition and rate of secretion of parotid saliva in a small steer. *British Journal of Nutrition* 15, 371–382.

Beardsworth, L.J., Beardsworth, P.M. and Care, A.D. (1989) The effect of ruminal phosphate concentration on the absorption of calcium, phosphorus and magnesium from the reticulo-rumen of the sheep. *British Journal of Nutrition* 61, 715–723.

Beede, D.K. (2003) Ration phosphorus management: requirements and excretion. *Proceedings of the Four State Applied Nutrition and Management Conference*, Midwest Plan Service, Iowa State University, Ames, Iowa, pp. 145–151.

Beede, D.K. and Davidson, J.A. (1999) Phosphorus: nutritional management for Y2K and beyond. *Proceedings of the Tri-State Dairy Nutrition Conference*, The Ohio State University, Columbus, Ohio, pp. 51–99.

Ben-Ghedalia, D., Tagari, H., Zamwel, S. and Bondi, A. (1975) Solubility and net exchange of calcium, magnesium and phosphorus in digesta flowing along the gut of sheep. *British Journal of Nutrition* 33, 87–94.

Benzie, D., Boyne, A.W., Dalgarno, A.C., Duckworth, J. and Hill, R. (1959) Studies of the skeleton of the sheep. III. The relationship between phosphorus intake and resorption and repair of the skeleton in pregnancy and lactation. *Journal of Agricultural Science, Cambridge* 52, 1–12.

Bertoni, G., Watson, M.J., Savage, G.P. and Armstrong, D.G. (1976) The movements of minerals in the digestive tract of dry and lactating Jersey cows. 1. Net movements of Ca, P, Mg, Na, K and Cl. *Zootecnica e Nutrizione Animale* 2, 107–118.

Boeser, U. (2004) Untersuchungen zu den Ausscheidungswegen für Calcium und Phosphor bei jungen Ziegenlämmern. Dissertation, Agricultural Faculty, University of Bonn, Germany, 139 pp.

Boeser, U., Hovenjürgen, M., Huber, K., Hattendorf, J., Breves, G. and Pfeffer, E. (2003) Effects of liquid and solid diets on P-excretion by Saanen goat kids during weaning. *Proceedings of the Society of Nutrition Physiology* 12, 44.

Bortolussi, G., Ternouth, J.H. and McMenniman, N.P. (1996) Dietary nitrogen and phosphorus depletion in cattle and their effects on live weight gain, blood metabolite concentrations and phosphorus kinetics. *Journal of Agricultural Science, Cambridge* 126, 493–501.

Boxebeld, A., Gueguen, L., Hannequart, G. and Durand, M. (1983) Utilization of phosphorus and calcium and minimal maintenance requirement for phosphorus in growing sheep fed a low-phosphorus diet. *Reproduction, Nutrition, Developement* 23, 1043–1053.

Braithwaite, G.D. (1974) The effect of changes of dietary calcium concentration on calcium metabolism in sheep. *British Journal of Nutrition* 31, 319–331.

Braithwaite, G.D. (1984) Changes in phosphorus metabolism of sheep in response to the increased demands for P associated with an intravenous infusion of calcium. *Journal of Agricultural Science, Cambridge* 102, 135–139.

Braithwaite, G.D. (1985) Endogenous faecal loss of phosphorus in growing lambs and the calculation of phosphorus requirements. *Journal of Agricultural Science, Cambridge* 105, 67–72.

Braithwaite, G.D. (1986) Phosphorus requirements of ewes in pregnancy and lactation. *Journal of Agricultural Science, Cambridge* 106, 271–278.

Breves, G. and Höller, H. (1987) Effects of dietary phosphorus depletion in sheep on dry matter and organic matter digestibility. *Journal of Animal Physiology and Animal Nutrition* 58, 280–286.

Breves, G. and Prokopp, M. (1990) Dietary phosphorus depletion in sheep: long-term effects on bone structure. *Journal of Animal Physiology and Animal Nutrition* 64, 74–79.

Breves, G., Ross, R. and Höller, H. (1985) Dietary phosphorus depletion in sheep: effects on plasma inorganic phosphorus, calcium, 1,25-$(OH)_2$-Vit.D_3 and alkaline phosphatase and on gastrointestinal P and Ca balances. *Journal of Agricultural Science, Cambridge* 105, 623–629.

Breves, G., Rosenhagen, C. and Höller, H. (1987) Die Sekretion von anorganischem Phosphor mit dem Speichel bei P-depletierten Schafen. *Journal of Veterinary Medicine* A 34, 442–47.

Breves, G., Höller, H., Packheiser, P., Gäbel, G. and Martens, H. (1988) Flux of inorganic phosphate across the sheep rumen wall *in vivo* and *in vitro*. *Quarterly Journal of Experimental Physiology* 73, 343–351.

Brintrup, R., Mooren, T., Meyer, U., Spiekers, H. and Pfeffer, E. (1993) Effects of two levels of phosphorus intake on performance and faecal phosphorus excretion of dairy cows. *Journal of Animal Physiology and Animal Nutrition* 69, 29–36.

Brodison, J.A., Goodall, E.A., Armstrong, J.D., Givens, D.I., Gordon, F.J., McCaughey, W.J. and Todd, J.R. (1989) Influence of dietary phosphorus on performance of lactating dairy cattle. *Journal of Agricultural Science, Cambridge* 112, 303–311.

Bruce, J., Goodall, E.D., Kay, R.N.B., Phillipson, A.T. and Vowles, L.E. (1966) The flow of organic and inorganic materials through the alimentary tract of the sheep. *Proceedings of the Royal Society* B 166, 46–62.

Brüggemann, J., Bronsch, K., Lörcher, K. and Seuss, H. (1959) Resorptionsstudien an Nutztieren. III. Die Bestimmung der Phosphorresorption aus Mineralfuttern. *Zeitschrift für Tierphysiologie, Tierernährung und Futtermittelkunde* 14, 224–241.

Call, J.W., Butcher, J.E., Shupe, J.L., Lamb, R.C., Boman, R.L. and Olson, A.E. (1987) Clinical effects of low dietary phosphorus concentrations in feed given to lactating dairy cows. *American Journal of Veterinary Science* 48, 133–136.

Challa, J. and Braithwaite, G.D. (1988a) Phosphorus and calcium metabolism in growing calves with special emphasis on phosphorus homoeostasis. 1. Studies of the effect of changes in the dietary phosphorus intake on phosphorus and calcium metabolism. *Journal of Agricultural Science, Cambridge* 110, 573–581.

Challa, J. and Braithwaite, G.D. (1988b) Phosphorus and calcium metabolism in growing calves with special emphasis on phosphorus homoeostasis. 2. Studies of the effect of different levels of phosphorus, infused abomasally, on phosphorus metabolism. *Journal of Agricultural Science, Cambridge* 110, 583–589.

Challa, J. and Braithwaite, G.D. (1988c) Phosphorus and calcium metabolism in growing calves with special emphasis on phosphorus homoeostasis. 3. Studies of the effect of continuous intravenous infusion of different levels of phosphorus in ruminating calves receiving adequate dietary phosphorus. *Journal of Agricultural Science, Cambridge* 110, 591–595.

Challa, J. and Braithwaite, G.D. (1989) Phosphorus and calcium metabolism in growing calves with special emphasis on phosphorus homoeostasis. 4. Studies on milk-fed calves given different amounts of dietary phosphorus but a constant intake of calcium. *Journal of Agricultural Science, Cambridge* 113, 285–289.

Challa, J., Braithwaite, G.D. and Dhanoa, M.S. (1989) Phosphorus homoeostasis in growing calves. *Journal of Agricultural Science, Cambridge* 112, 217–226.

Chiewitz, O. and Hevesy, G. (1935) Radioactive indicators in the study of phosphorus metabolism in rats. *Nature* 136, 754–755.

Coates, D.B. and Ternouth, J.H. (1992) Phosphorus kinetics of cattle grazing tropical pastures and implications for the estimation of their phosphorus requirements. *Journal of Agricultural Science, Cambridge* 119, 401–409.

Deitert, C. and Pfeffer, E. (1993) Effects of a reduced P supply with adequate or high Ca intake on performance and mineral balances in dairy goats during pregnancy and lactation. *Journal of Animal Physiology and Animal Nutrition* 69, 12–21.

Delaquis, A.M. and Block, E. (1995) Acid–base status, renal function, water, and macromineral metabolism of dry cows fed diets differing in cation–anion difference. *Journal of Dairy Science* 78, 604–619.

Delcenserie, R., Menudier, A., Yzet, T., Sevestre, H. and Dupas, J.L. (2001) Intestinal pH and *Staphylococcus* contamination of pancreas, liver and spleen during experimental acute pancreatitis in mice. *Pancreatology* 1, 129–199.

Dillon, J. and Scott, D. (1979) Digesta flow and mineral absorption in lambs before and after weaning. *Journal of Agricultural Science, Cambridge* 92, 289–297.

Duncan, D.L. (1958) The interpretation of studies of calcium and phosphorus balance in ruminants. *Nutrition Abstracts and Reviews* 28, 695–715.

Edrise, B.M. and Smith, R.H. (1986) Exchanges of magnesium and phosphorus at different sites in the ruminant stomach. *Archives of Animal Nutrition* 36, 1019–1027.

Eidelsburger, U., Kirchgessner, M. and Roth, F.X. (1992) Zum Einfluss von Ameisensäure, Calciumformiat und Natriumhydrogencarbonat auf pH-Wert, Trockenmassegehalt, Konzentration an Carbonsäuren und Ammoniak in verschiedenen Segmenten des Gastrointestinaltraktes. 8. Untersuchung zur nutritiven Wirksamkeit von organischen Säuren in der Ferkelaufzucht. *Journal of Animal Physiology and Animal Nutrition* 68, 20–32.

Erickson, G.E., Klopfenstein, T.J., Milton, C.T., Hanson, D. and Calkins, C. (1999) Effect of dietary phosphorus on finishing steer performance, bone status and carcass maturity. *Journal of Animal Science* 77, 2832–2836.

Erickson, G.E., Klopfenstein, T.J., Milton, C.T., Brink, D., Orth, M.W. and Whittet, K.M. (2002) Phosphorus requirement of finishing feedlot calves. *Journal of Animal Science* 80, 1690–1695.

Fallingborg, F., Christensen, L.A., Jacobsen, B.A., Ingeman-Nielsen, M., Rasmussen, H.H., Abildgaard, K. and Rasmussen, S.N. (1994) Effect of olsazine and mesalazine on intraluminal pH of the duodenum and proximal jejunum in healthy humans. *Scandinavian Journal of Gastroenterology* 29, 498–500.

Ferrell, C.L., Laster, D.B. and Prior, R.L. (1982) Mineral accretion during prenatal growth of cattle. *Journal of Animal Science* 54, 618–624.

Field, A.C., Coop, R.L., Dingwall, R.A. and Munro, C.S. (1982) The phosphorus requirements for growth and maintenance of sheep. *Journal of Agricultural Science, Cambridge* 99, 311–317.

Field, A.C., Woolliams, J.A. and Dingwall, R.A. (1985) The effect of dietary intake of calcium and dry matter on the absorption and excretion of calcium and phosphorus by growing lambs. *Journal of Agricultural Science, Cambridge* 115, 237–243.

Fox, J. and Care, A.D. (1978) Effect of low calcium and low phosphorus diets on the intestinal absorption of phosphate in intact and parathyroidectomized pigs. *Journal of Endocrinology* 82, 417–424.

Fox, J. and Ross, R. (1985) Effects of low phosphorus and low calcium diets on the production and metabolic clearance rates of 1,25-dihydroxycholecalciferol in pigs. *Journal of Endocrinology* 105, 169–173.

Fox, J., Pickard, D.W., Care, A.D. and Murray, T.M. (1978) Effect of low phosphorus diets on intestinal calcium absorption and the concentration of calcium-binding protein in intact and parathyroidectomized pigs. *Journal of Endocrinology* 78, 379–387.

Garcia-Bojalil, C.M., Ammerman, C.B., Henry, P.R., Littell, R.C. and Blue, W.G. (1988) Effects of dietary phosphorus, soil ingestion and dietary intake level on performance, phosphorus utilization and serum and alimentary tract mineral concentrations in lambs. *Journal of Animal Science* 66, 1508–1519.

Gartner, J.W., Murphy, G.M. and Hoey, W.A. (1982) Effects of induced, subclinical phosphorus deficiency on feed intake and growth of beef heifers. *Journal of Agricultural Science, Cambridge* 98, 23–29.

GfE (Ausschuss für Bedarfsnormen der Gesellschaft für Ernährungsphysiologie) (1993) Überarbeitete Empfehlungen zur Versorgung von Milchkühen mit Calcium und Phosphor. *Proceedings of the Society of Nutrition Physiology* 1, 108–113.

GfE (Ausschuss für Bedarfsnormen der Gesellschaft für Ernährungsphysiologie) (1995) *Energie- und Nährstoffbedarf landwirtschaftlicher Nutztiere. Nr. 6. Empfehlungen zur Energie- und Nährstoffversorgung der*

Mastrinder. DLG-Verlag, Frankfurt am Main, Germany, 85 pp.

Grace, N.D. (1981) Phosphorus kinetics in the sheep. *British Journal of Nutrition* 45, 367–374.

Grace, N.D., Ulyatt, M.J. and Macrae, J.C. (1974) Quantitative digestion of fresh herbage by sheep. III. The movement of Mg, Ca, P, K and Na in the digestive tract. *Journal of Agricultural Science, Cambridge* 82, 321–330.

Grace, N.D., Watkinson, J.H. and Martinson, P.L. (1986) Accumulation of minerals by the fetus(es) and conceptus of single- and twin-bearing ewes. *New Zealand Journal of Agricultural Research* 29, 207–222.

Greene, L.W., Fontenot, J.P. and Webb, K.E. (1983a) Site of magnesium and other macromineral absorption in steers fed high levels of potassium. *Journal of Animal Science* 57, 503–510.

Greene, L.W., Webb, K.E. and Fontenot, J.P. (1983b) Effect of potassium level on site of absorption of magnesium and other macroelements in sheep. *Journal of Animal Science* 56, 1214–1221.

Grings, E.E. and Males, R. (1987) Effects of potassium on macromineral absorption in sheep fed wheat straw-based diets. *Journal of Animal Science* 64, 872–879.

Guéguen, L. (1962) L'utilisation digestive réelle du phosphore du foin de luzerne par le mouton, mesurée a l'aide de ^{32}P. *Annales de Biologie animale, Biochimie, Biophysique* 2, 143–149.

Guéguen, L. (1963) Influence de la nature du régime alimentaire sur l'excretion fécale de phosphore endogène chez le veau. *Annales de Biologie animale, Biochimie, Biophysique* 3, 243–253.

Henneberg, W. (1870) *Neue Beiträge zur rationellen Fütterung der Wiederkäuer.* Deuerlich'sche Buchhandlung, Göttingen, 457 pp.

Hevesy, G., Hahn, L. and Rebbe, O. (1939) Excretion of phosphorus. *Det Kgl. Danske Videnskabernes Selskab. Biologiske Meddelelser* XIV, 3–23.

Hoey, W.A., Murphy, G.M. and Gartner, R.J.W. (1982) Whole body composition of heifers in relation to phosphorus status with particular reference to the skeleton. *Journal of Agricultural Science, Cambridge* 98, 31–37.

Höller, H., Figge, A., Richter, J. and Breves, G. (1988) Nettoresorption von Calcium und von anorganischem Phosphat aus dem perfundierten Colon und Rectum von Schafen. *Journal of Animal Physiology and Animal Nutrition* 59, 9–15.

House, W.A. and Bell, A.W. (1993) Mineral accretion in the fetus and adnexa during late gestation in Holstein cows. *Journal of Dairy Science* 76, 2999–3010.

Hughes, M.R., Brumbaugh, P.F., Haussler, M.R., Wergedal, J.E. and Baylink, D.J. (1975) Regulation of serum 1α, 25-dihydroxyvitamin D_3 by calcium and phosphate in the rat. *Science* 190, 578–580.

Judkins, M.B., Wallace, J.D., Parker, E.E. and Wright, J.D. (1985) Performance and phosphorus status of range cows with and without phosphorus supplementation. *Journal of Range Management* 38, 139–143.

Karn, J.F. (2001) Phosphorus nutrition of grazing cattle: a review. *Animal Feed Science and Technology* 38, 133–153.

Kaufmann, W. and Orth, A. (1966) Untersuchungen über Einflüsse des Futters und der Pansenfermentation auf die Speichelsekretion. *Zeitschrift für Tierphysiologie, Tierernährung und Futtermittelkunde* 21, 110–120.

Kaufmann, W., Pfeffer, E. and Dirksen, G. (1972) Untersuchungen zur Verdauungs-physiologie der Milchkuh mit der Umleitungstechnik am Duodenum. II. Messtechnik, Flussrate am Duodenum und Verdaulichkeit der Energie in Mägen und Darm. *Fortschritte in der Tierphysiologie und Tierernährung* 1, 33–37.

Kay, R.N.B. (1960a) The rate and composition of various salivary secretions in sheep and calves. *Journal of Physiology* 150, 515–537.

Kay, R.N.B. (1960b) The development of parotid salivary secretion in young goats. *Journal of Physiology* 150, 538–545.

Kay, R.N.B. (1969) Digestion of protein in the intestines of adult ruminants. *Proceedings of the Nutrition Society* 28, 140–151.

Kay, R.N.B. and Pfeffer, E. (1970) Movements of water and electrolytes into and from the intestine of the sheep. In: Phillipson, A.T. (ed.) *Physiology of Digestion and Metabolism in the Ruminant.* Oriel Press, Newcastle upon Tyne, UK, pp. 390–402.

Khorasani, G.R. and Armstrong, D.G. (1990) Effect of sodium and potassium level on the absorption of magnesium and other macro-minerals in sheep. *Livestock Production Science* 24, 223–235.

Khorasani, G.R. and Armstrong, D.G. (1992) Calcium, phosphorus, and magnesium absorption and secretion in the bovine digestive tract as influenced by dietary concentration of these elements. *Livestock Production Science* 31, 271–243.

Khorasani, G.R., Janzen, R.A., McGill, W.B. and Kennelly, J.J. (1997) Site and extent of mineral absorption in lactating cows fed whole-crop cereal grain silage or alfalfa silage. *Journal of Animal Science* 75, 239–248.

Kirk, D.J., Fontenot, J.P. and Rahnema, S. (1994) Effects of feeding lasalocid and monensin on digestive flow and partial absorption of minerals in sheep. *Journal of Animal Science* 72, 1029–1037.

Kleiber, M., Smith, A.H., Ralston, N.P. and Black, A.L. (1951) Radiophosphorus (^{32}P) as tracer for measuring endogenous phosphorus in cows feces. *Journal of Nutrition* 45, 253–263.

Knowlton, K.F. and Herbein, J.H. (2002) Phosphorus partitioning during early lactation in dairy cows fed diets varying in phosphorus content. *Journal of Dairy Science* 85, 1227–1236.

Knowlton, K.F., Herbein, J.H., Meister-Weisbarth, M.A. and Wark, W.A. (2001) Nitrogen and phosphorus partitioning in lactating Holstein cows fed different sources of dietary protein and phosphorus. *Journal of Dairy Science* 84, 1210–1217.

Koddebusch, L. und Pfeffer, E. (1988) Untersuchungen zur Verwertbarkeit von Phosphor verschiedener Herkünfte an laktierenden Ziegen. *Journal of Animal Physiology and Animal Nutrition* 60, 269–275.

Law, F.M.K., Leaver, D.D., Martin, T.J., Selleck, K., Clarke, I.J. and Moate, P.J. (1994) Effect of treatment of dairy cows with slow release bovine somatotropin during the periparturient period on minerals in plasma and milk and on parathyroid hormone-released protein in milk. *Journal of Dairy Science* 77, 2242–2248.

Leibholz, J. (1974) The flow of calcium and phosphorus in the digestive tract of the sheep. *Australian Journal of Agricultural Research* 1974, 147–154.

Little, D.A. (1980) Observations on the phosphorus requirement of cattle for growth. *Research in Veterinary Science* 28, 258–268.

Lofgreen, G.P. (1960) The availability of the phosphorus in dicalcium phosphate, bonemeal, soft phosphate and calcium phytate for mature wethers. *Journal of Nutrition* 70, 58–62.

Lofgreen, G.P. and Kleiber, M. (1953) The availability of the phosphorus in alfalfa hay. *Journal of Animal Science* 12, 366–371.

Lofgreen, G.P. and Kleiber, M. (1954) Further studies on the availability of the phosphorus in alfalfa hay. *Journal of Animal Science* 13, 258–264.

Loof, M., Huber, K., Hattendorf, J., Breves, G. and Pfeffer, E. (2004) Alteration of regulative P excretions from urine to faeces by growing Saanen goat kids fed either with a solid or a liquid diet. *Proceedings of the Society of Nutrition Physiology* 13, 24.

Lopez, H., Kanitz, F.D., Moreira, V.R., Wiltbank, M.C. and Satter, L.D. (2004a) Effect of dietary phosphorus on performance of lactating dairy cows: milk production and cow health. *Journal of Dairy Science* 87, 139–145.

Lopez, H., Kanitz, F.D., Moreira, V.R., Satter, L.D. and Wiltbank, M.C. (2004b) Reproductive performance of dairy cows fed two concentrations of phosphorus. *Journal of Dairy Science* 87, 146–157.

Lueker, C.E. and Lofgreen, G.P. (1961) Effects of intake and calcium to phosphorus ratio on absorption of these elements by sheep. *Journal of Nutrition* 74, 233–238.

Manas-Almendros, M., Ross, R. and Care, A.D. (1982) Factors affecting the secretion of phosphate in parotid saliva in the sheep and goat. *Quarterly Journal of Experimental Physiology* 67, 269–280.

Martz, F.A., Belo, A.T., Weiss, M.F., Belyea, R.L. and Goff, J.P. (1990) True absorption of calcium and phosphorus from alfalfa and corn silage when fed to lactating cows. *Journal of Dairy Science* 73, 1288–1295.

Maunder, E.M.W., Pillay, A.V. and Care, A.D. (1986) Hypophosphataemia and vitamin D metabolism in sheep. *Quarterly Journal of Experimental Physiology* 71, 391–399.

McDougall, E.L. (1948) Studies on ruminant saliva. I. The composition and output of sheep's saliva. *Biochemical Journal* 43, 99–108.

Meyer, H., Flothow, C. and Radicke, S. (1997) Preileal digestibility of coconut fat and soybean oil in horses and their influence on metabolites of microbial origin of the proximal digestive tract. *Archives of Animal Nutrition* 50, 63–74.

Morse, D., Head, H.H., Wilcox, C.J., van Horn, H.H., Hissem, C.D. and Harris, B. (1992) Effects of concentration of dietary phosphorus on amount and route of excretion. *Journal of Dairy Science* 75, 3039–3049.

Müschen, H., Petri, P., Breves, G. and Pfeffer, E. (1988) Response of lactating goats to low phosphorus intake. 1. Milk yield and faecal excretion of P and Ca. *Journal of Agricultural Science, Cambridge* 111, 255–263.

NRC (National Research Council) (1996) *Nutrient Requirements for Beef Cattle*, 7th rev. edn. National Academy Press, Washington, DC, 242 pp.

NRC (National Research Council) (2001) *Nutrient Requirements of Dairy Cattle*, 7th rev. edn. National Academy Press, Washington, DC, 381 pp.

Perge, P., Hardebeck, H., Sommer, H. and Pfeffer, E. (1982) Untersuchungen zur Beeinflussung der Calcium- und Phosphorgehalte in Blutserum und Speichel von Hammeln durch die Versorgung. *Zeitschrift für Tierphysiologie, Tierernährung und Futtermittelkunde* 48, 113–121.

Petri, A., Müschen, H., Breves. G., Richter, O. and Pfeffer, E. (1988) Response of lactating goats to low phosphorus intake. 2. Nitrogen transfer from rumen ammonia to rumen microbes and proportion of milk protein derived from microbial amino acids. *Journal of Agricultural Science, Cambridge* 111, 265–271.

Pfeffer, E. (1989) Phosphorus requirements in goats. In: Recent progress on mineral nutrition and mineral requirements in ruminants. *Proceedings of the International Meeting on Mineral Nutrition and Mineral Requirements in Ruminants*, Kyoto, Japan, 3–4 September 1989, pp. 42–46.

Pfeffer, E. and Kaufmann, W. (1972) Untersuchungen zur Verdauungsphysiologie der Milchkuh mit der Umleitungstechnik am Duodenum. 4. Netto-Bewegung einiger mineralischer Mengenelemente durch die Wand der Mägen und des Darms. *Fortschritte in der Tierphysiologie und Tierernährung* 1, 33–37.

Pfeffer, E. and Rodehutscord, M. (1998) Body chemical composition and utilization of dietary energy by male Saanen kids fed either milk to satiation or solid complete feeds with two proportions of straw. *Journal of Agricultural Science, Cambridge* 131, 487–495.

Pfeffer, E., Thompson, A. and Armstrong, D.G. (1970) Studies on intestinal digestion in the sheep. 3. Net movement of certain inorganic elements in the digestive tract on rations containing different proportions of hay and rolled barley. *British Journal of Nutrition* 24, 197–204.

Pfeffer, E., Pauen, A. and Haverkamp, R. (1993) Changes in retention of P and Ca and courses of blood plasma concentrations of inorganic phosphate and calcium in dairy goats following a change in P supply from reduced to adequate in combination with adequate or high Ca intake. *Journal of Animal Physiology and Animal Nutrition* 69, 22–28.

Pfeffer, E., Windhausen, P. and Rodehutscord, M. (1994) Changes in body composition of dairy goats caused by phosphorus depletion. *Journal of Animal Physiology and Animal Nutrition* 72, 65–72.

Pfeffer, E., Rodehutscord, M. and Breves, G. (1995) Effects of reducing dietary calcium and/or phosphorus on performance and body composition in male kids. *Journal of Animal Physiology and Animal Nutrition* 74, 243–252.

Pfeffer, E., Rodehutscord, M. and Breves, G. (1996) Effect of varying dietary calcium and phosphorus on growth and on composition of empty bodies and isolated bones of male kids. *Archives of Animal Nutrition* 49, 243–252.

Potthast, V., Moctar, M., Abdel Rahman, K. and Pfeffer, E. (1976) Untersuchungen an Hammeln über Sekretion und Absorption von Speichel-Phosphor. *Zeitschrift für Tierphysiologie, Tierernährung und Futtermittelkunde* 36, 334–340.

Preston, R.L. and Pfander, W.H. (1964) Phosphorus metabolism in lambs fed varying phosphorus intakes. *Journal of Nutrition* 83, 369–378.

Rahnema, S., Wu, Z., Ohajuruka, O.A., Weiss, W.P. and Palmquist, D.L. (1994) Site of mineral absorption in lactating cows fed high-fat diets. *Journal of Animal Science* 72, 229–235.

Rajaratne, A.A.J., Scott, D. and Buchan, W. (1994) Effects of a change in phosphorus requirement on phosphorus kinetics in the sheep. *Research in Veterinary Science* 56, 262–264.

Rajaratne, A.A.J., Buchan, W., Kyle, D. and Scott, D. (1996) Use of sheep maintained by intragastric infusion as a model for studying the endogenous loss of phosphorus in ruminants. *Research in Veterinary Science* 60, 92–93.

Reed, W.D.C., Elliott, R.C. and Topps, J.H. (1965) Phosphorus excretion of cattle fed on high-energy diets. *Nature* 208, 953–954.

Riad, F., Lefaivre, J. and Barlet, J.-P. (1987) 1,25-Dihydroxycholecalciferol regulates salivary phosphate secretion in cattle. *Journal of Endocrinology* 112, 427–430.

Ribovich, M.L. and DeLuca, H.F. (1978) Effect of dietary calcium and phosphorus on intestinal calcium absorption and vitamin D metabolism. *Archives of Biochemistry and Biophysics* 188, 145–156.

Risley, C.R., Kornegay, E.T., Lindermann, M.D., Wood, C.M. and Eigel, W.N. (1992) Effect of feeding organic acids on selected intestinal content measurements at varying times postweaning in pigs. *Journal of Animal Science* 70, 196–206.

Rodehutscord, M., Pauen, A., Windhausen, P., Brintrup, R. and Pfeffer, E. (1994a) Effects of drastic changes in P intake on P concentrations in blood and rumen fluid of lactating ruminants. *Journal of Veterinary Medicine* A 41, 611–619.

Rodehutscord, M., Pauen, A., Windhausen, P. and Pfeffer, E. (1994b) Balances of phosphorus and calcium in dairy goats during periods of phosphorus depletion and subsequent phosphorus repletion. *Journal of Animal Physiology and Animal Nutrition* 72, 57–64.

Rodehutscord, M., Haverkamp, R. and Pfeffer, E. (1998) Inevitable losses of phosphorus in pigs, estimated from balance data using diets deficient in phosphorus. *Archives of Animal Nutrition* 51, 27–38.

Rodehutscord, M., Heuvers, H. and Pfeffer, E. (2000) Effect of organic matter digestibility on obligatory faecal phosphorus loss in lactating goats, determined from balance data. *Animal Science* 70, 561–568.

Salih, Y., McDowell, L.R., Hentges, J.F., Mason, R.M. and Wilcox, C.J. (1987) Mineral content of milk, colostrums and serum as affected by physiological state and mineral supplementation. *Journal of Dairy Science* 70, 608–612.

Scarisbrick, R. and Ewer, T.K. (1951) The absorption of inorganic phosphate from the rumen of sheep. *Biochemical Journal* 49, lxxix (abstract).

Schneider, K.M., Boston, R.C. and Leaver, D.D. (1987) Quantitation of phosphorus excretion in sheep by compartmental analysis. *American Journal of Physiology* 252, R720–R731.

Schröder, B., Käppner, H., Failing, K., Pfeffer, E. and Breves, G. (1995) Mechanisms of intestinal phosphate transport in small ruminants. *British Journal of Nutrition* 74, 635–648.

Schulz, E., Oslage, H.J. and Daenicke, R. (1974) Untersuchungen über die Zusammensetzung der Körpersubstanz sowie den Stoff- und Energieansatz bei wachsenden Mastbullen. *Fortschritte in der Tierphysiologie und Tierernährung* 4, 5–70.

Scott, D. (1972) Excretion of phosphorus and acid in the urine of sheep and calves fed either roughage or concentrate diets. *Quarterly Journal of Experimental Physiology* 57, 379–392.

Scott, D. (1978) The effects of intravenous phosphate loading on salivary phosphate secretion and plasma parathyroid hormone levels in the sheep. *Quarterly Journal of Experimental Physiology* 63, 147–156.

Scott, D. and Buchan, W. (1985) The effects of feeding either roughage or concentrate diets on salivary phosphorus absorption and urinary phosphorus excretion in the sheep. *Quarterly Journal of Experimental Physiology* 70, 365–375.

Scott, D. and Buchan, W. (1988) The effects of feeding pelleted diets made from either coarsely or finely ground hay on phosphorus balance and on the partition of phosphorus excretion between urine and faeces in sheep. *Quarterly Journal of Experimental Physiology* 73, 315–322.

Scott, D., Whitelaw, F.G. and Kay, M. (1971) Renal excretion of acid in calves fed either roughage or concentrate diets. *Quarterly Journal of Experimental Physiology* 56, 18–32.

Scott, D., McLean, A.F. and Buchan, W. (1984a) The effect of variation in phosphorus intake on net intestinal phosphorus absorption, salivary phosphorus secretion and pathway of excretion in sheep fed roughage diets. *Quarterly Journal of Experimental Physiology* 69, 439–452.

Scott, D., McLean, A.F. and Buchan, W. (1984b) The effects of intravenous phosphate loading on salivary phosphorus secretion, net intestinal phosphorus absorption and pathway of excretion in sheep fed roughage diets. *Quarterly Journal of Experimental Physiology* 69, 453–461.

Spiekers, H., Brintrup, R., Balmelli, M. and Pfeffer, E. (1993) Influence of dry matter intake on faecal phosphorus losses in dairy cows fed rations low in phosphorus. *Journal of Animal Physiology and Animal Nutrition* 69, 37–43.

Sutton, A. and Beede, D.K. (2003). Feeding strategies to lower nitrogen and phosphorus in manure. In: *Best Environmental Management Practices – Farm Animal Production.* Publication E-2822. Michigan State University Extension, East Lansing, Michigan.

Ternouth, J.H. (1989) Endogenous losses of phosphorus by sheep. *Journal of Agricultural Science, Cambridge* 113, 291–297.

Ternouth, J.H., Bortolussi, G., Coates, D.B., Hendricksen, R.E. and McLean, R.W. (1996) The phosphorus requirements of growing cattle consuming forage diets. *Journal of Agricultural Science, Cambridge* 126, 503–510.

Theiler, A. and Green, H.H. (1932) Aphosphorosis in ruminants. *Nutrition Abstracts and Reviews* 1, 359–385.

Théwis, A. and Francois, E. (1985) Intestinal absorption and secretion of total and lipid phosphorus in adult sheep fed chopped meadow hay. *Reproduction, Nutrition, Developement* 25, 389–397.

Théwis, A., Francois, E. and Thielemans, M.-F. (1978) Etude quantitative de l'absorption et de la sécrétion du phosphore total et du phosphore phospholipidique dans le tube digestif du mouton. *Annales de Biologie animale, Biochimie, Biophysique* 18, 1181–1195.

Tillman, A.D. and Brethour, J.R. (1958a) Dicalcium phosphate and phosphoric acid as phosphorus sources for beef cattle. *Journal of Animal Science* 17, 100–103.

Tillman, A.D. and Brethour, J.R. (1958b) Utilization of phytin phosphorus by sheep. *Journal of Animal Science* 17, 104–112.

Tillman, A.D. and Brethour, J.R. (1958c) Ruminant utilization of sodium meta-, ortho- and pyrophosphates. *Journal of Animal Science* 17, 792–796.

Tillman, A.D., Brethour, J.R. and Hansard, S.L. (1959) Comparative procedures for measuring the phosphorus requirement of cattle. *Journal of Animal Science* 18, 249–255.

Tomas, F.M. (1973) Parotid salivary secretion in sheep: its measurement and influence on phosphorus in rumen fluid. *Quarterly Journal of Experimental Physiology* 58, 131–138.

Tomas, F.M., Moir, R.J. and Somers, M. (1967) Phosphorus turnover in sheep. *Australian Journal of Agricultural Research* 18, 635–645.

Topps, J.H., Reed, W.D.C. and Elliott, R.C. (1966) Studies of the metabolism of cattle given high concentrate diets. *Journal of Agricultural Science, Cambridge* 66, 233–240.

Tribe, D.E. and Peel, L. (1963) Total salivation in grazing lambs. *Australian Journal of Agricultural Research* 14, 330–339.

Valdivia, R., Ammerman, C.B., Henry, P.R., Feaster, J.P. and Wilcox, C.J. (1982) Effect of dietary aluminium and phosphorus utilization and tissue mineral composition in sheep. *Journal of Animal Science* 55, 402–410.

Valk, H. and Beynen, A.C. (2002) Proposal for the assessment of phosphorus requirements of dairy cows. *Livestock Production Science* 79, 267–272.

Valk, H. and Sebek, L.B.J. (1999) Influence of long-term feeding of limited amounts of phosphorus on dry matter intake, milk production, and body weight of dairy cows. *Journal of Dairy Science* 82, 2157–2163.

Valk, H., Sebek, L.B.J. and Beynen, A.C. (2002) Influence of phosphorus intake on excretion and blood plasma and saliva concentrations of phosphorus in dairy cows. *Journal of Dairy Science* 85, 2642–2649.

Walker, D.M. and Al-Ali, S.J. (1987) The endogenous phosphorus excretion of preruminant lambs. *Australian Journal of Agricultural Research* 38, 1061–1069.

Wildt, E. (1874) Ueber die Resorption und Secretion der Nahrungsbestandtheile im Verdauungskanal des Schafes. *Journal für Landwirthschaft* 22, 1–34.

Wright, R.D., Blair-West, J.R., Nelson, J.F. and Tregear, G.W. (1984) Handling of phosphate by a parotid gland (ovine). *American Journal of Physiology* 246, F916–F926.

Wu, Z. and Satter, L.D. (2000) Milk production and reproductive performance of dairy cows fed two concentrations of phosphorus for two years. *Journal of Dairy Science* 83, 1052–1063.

Wu, Z., Satter, L.D. and Sojo, R. (2000) Milk production, reproductive performance, and fecal excretion of phosphorus by dairy cows fed three amounts of phosphorus. *Journal of Dairy Science* 83, 1028–1041.

Wu, Z., Satter, L.D., Biohowiak, A.J., Stauffacher, R.H. and Wilson, J.H. (2001) Milk production, estimated phosphorus excretion, and bone characteristics of dairy cows fed different amounts of phosphorus for two or three years. *Journal of dairy Science* 84, 1738–1748.

Wylie, M.J., Fontenot, J.P. and Greene, L.W. (1985) Absorption of magnesium and other macrominerals in sheep infused with potassium in different parts of the digestive tract. *Journal of Animal Science* 61, 1219–1229.

Young, V.R., Luick, J.R. and Lofgreen, G.P. (1966a) The influence of dietary phosphorus intake on the rate of bone metabolism in sheep. *British Journal of Nutrition* 20, 727–732.

Young, V.R., Richards, W.P.C., Lofgreen, G.P. and Luick, J.R. (1966b) Phosphorus depletion in sheep and the ratio of calcium to phosphorus in the diet with reference to calcium and phosphorus absorption. *British Journal of Nutrition* 20, 783–794.

Young, V.R., Lofgreen, G.P. and Luick, J.R. (1966c) The effects of phosphorus depletion, and of calcium and phosphorus intake, on the endogenous excretion of these elements by sheep. *British Journal of Nutrition* 20, 795–805.

8 Effects of Dietary Phosphorus and Nitrogen on Cattle Reproduction

J.D. Ferguson[1] and D. Sklan[2]
[1]*School of Veterinary Medicine, University of Pennsylvania, Kennett Square, Pennsylvania, USA*
[2]*Faculty of Agriculture, Hebrew University, Rehovot, Israel*

8.1 Introduction

Concerns about eutrophication of streams and estuaries from nonpoint source pollution of nitrogen (N) and phosphorus (P) from animal waste has stimulated a re-examination of feeding standards for crude protein (CP) and P in dairy rations. Excretion of N and P is directly correlated with intake (Belonje and van den Berg, 1980; Cohen, 1980; Sanson *et al.*, 1990; Jonker *et al.*, 1998) and thus manure content of N and P is directly influenced by dietary inputs, so there is an environmental incentive to reduce dietary inputs. On one hand excessive reductions in N and P content of dairy rations may impair productivity and reproductive performance whereas increasing N content of dairy rations from 1.92% dry matter (DM) up to 3.2% of DM has been associated with increases in milk production, although higher rates of N feeding may reduce fertility (Ferguson and Chalupa, 1989). Exact N requirements are challenging to predict due to rumen metabolism of feed consumed. These nutrients are often overfed either to provide a safety margin or stemming from the view that N or P are limiting farm performance. Better understanding of the requirements for these nutrients should allow more confidence in feeding lower dietary concentrations without compromising performance while reducing manure nutrient concentrations and potentially reducing environmental harm.

8.2 Phosphorus in the Body

Phosphorus is not found in elemental form in biological systems but is present in phosphate salts or as phosphate covalently bound to organic

©CAB International 2005. *Nitrogen and Phosphorus Nutrition of Cattle* (eds E. Pfeffer and A.N. Hristov)

compounds. Due to the multiple biological forms of P, it is less ambiguous to describe the P requirement as the elemental form. Phosphorus is important in many biological functions as a structural element, a key element in energy transactions, a buffer, a critical component in regulation of cell function and a stabilizer of DNA and RNA polymers.

In the body, roughly 85% of P is found in bone and teeth, 14% in soft tissue and 1% in extracellular fluid (Blair-West *et al.*, 1992). In dairy cows, 80% to 90% of body P is contained in skeletal tissue and teeth (NRC, 2001). Phosphorus in bone is contained in calcium (Ca) phosphate and hydroxyl-apatite crystals ($Ca_3(PO_4)_2$ and $Ca_{10}(PO_4)_6(OH)_2$). The bone content of Ca and P is dynamic and a function of age, physiological state and skeletal ash. Bone typically has a Ca:P ratio of 2.2:1 by weight (Ganong, 1999). Ash content in normal bone is above 40% of which 17% to 18% is P (Ternouth, 1989; Ternouth *et al.*, 1996). A 650 kg cow may have 7 to 8 kg of P in bone. Bone P and Ca are available to support deficits when demand is high and intake is low, such as occurring in early lactation. A total cumulative deficit of 600 to 1000 g of P may be supplied from bone stores (NRC, 2001).

Phosphorus performs other important functions in the body. The central role of P in energy transactions and regulation of ion transport through ATP and cAMP has been extensively described (Ganong, 1999). In ruminants, salivary P, as phosphate, has an important role in contributing to rumen buffering and supplying P to rumen microorganisms (see Chapters 6 and 7). Phosphorus has an important role in stabilizing DNA and RNA and phosphorylation of serine residues in casein help to form casein complexes in milk.

8.3 Influence of Phosphorus on Reproduction

The influence of P on reproductive efficiency cannot be discussed without a review of the history of the association of dietary P with the clinical syndrome of aphosphorosis in cattle. The association of the intake of forage from P-deficient soils with clinical syndromes was described by Tuff (1923), Eckles *et al.* (1926), Meador (1926, 1927), Theiler *et al.* (1927), Henderson and Weakly (1930), Theiler and Green (1932), Eckles *et al.* (1932), Becker *et al.* (1933), Mitchell (1936), Theiler *et al.* (1937), Green (1939), Schmidt (1940), Rose (1954), Barnes and Jephcott (1955), Little (1970) and Cohen (1980). Descriptions of problems observed in cattle grazing P-deficient pastures have not been significantly different from observations of Tuff (1923) to Cohen (1980). The earliest report of malnutrition in grazing cattle was in 1785 by Le Vaillant (cited by Mitchell, 1936 and Theiler and Green, 1932). This author observed cattle scavenging his camp for bones during travels through Mamaqualand. He astutely noted that he only observed this behaviour when pastures were dry, and did not observe this behaviour when pastures were lush. In 1835, Don Azara observed that wild cattle in Paraguay ate dried bones of carcasses decomposing on pastures. In the late 1800s bone chewing by cattle was associated with failure to thrive, reduced calving rates, fractures or softening of skeleton, downer animals and sudden death. By the 1920s, bone chewing and accompanying conditions had been described in Norway, South Africa, Australia, New Zealand and the USA (Texas, Minnesota, Florida and other US regions).

The syndrome was characterized in cattle grazing poor pastures, often in drought years. Affected animals developed ill-thrift associated with lameness and softening or fracturing of bones. Animals lost condition, assumed an abnormal posture and gait, developed a depraved appetite for old bones and carrion, had poor reproduction and eventually became recumbent and died (Atkinson *et al.*, 1916; Tuff, 1923). Cows nursing calves in these regions grew poorly. Milk cows were more often affected than non-lactating cows and the condition was more pronounced following a drought than during wet seasons. Less than 60% of cows nursing calves became pregnant compared with over 80% of dry cows on these pastures. Thus, the average calving interval was over 2 years. The abnormal gait and creaking sounds made as the animals walked gave name to the condition as Styfsiekte (stiff-sickness) in South Africa, Peg Leg in Australia and Sweeny or Creeps in the USA. Histologic examination of bones and articular surfaces led to a more formal classification of the condition as osteomalacia in adult cows and ricketts in young animals. Scientists in South Africa (Theiler *et al.*, 1924, 1927, 1928, 1937; Theiler and Green, 1932), Minnesota (Eckles *et al.*, 1926, 1932) and Norway (Tuff, 1923) were instrumental in describing the condition and identifying the causative factor as low P intake on poor quality pastures and termed the condition aphosphorosis.

Worldwide, the condition was seen predominantly in cattle grazing pastures and rangeland in geographic regions with low soil P (Theiler *et al.*, 1924; Eckles *et al.*, 1926; Theiler and Green, 1932). Forages produced from these soils typically had P contents below 0.20% on DM basis and often lower than 0.10% (Tuff, 1923; Theiler *et al.*, 1924; Eckles *et al.*, 1926). However, Meigs and Woodward (1921), Haag *et al.* (1929) and Huffman *et al.* (1930) identified aphosphorosis as a risk in confined dairy cattle when consuming rations with lucerne hay as the primary forage source. If the lucerne had low P content (<0.20% DM) and no supplemental protein sources were fed, milk cows could not consume sufficient P despite receiving supplemental maize, oats or barley grain and aphosphorosis could occur in these situations.

As recognition of the syndrome developed, rapid methods of diagnosis were sought. Palmer and Eckles (1927), Theiler *et al.* (1927), Malan *et al.* (1928) and Green (1939) found that plasma inorganic phosphate (P_i) concentrations were typically below 0.6 mmol/l in affected cattle, and plasma P_i concentrations in cows from unaffected regions appeared to correlate well with dietary intake (Palmer *et al.*, 1930) and responded rapidly to changes in dietary P intake. Theiler and Green (1932) and other workers proposed that a blood plasma sample was a reliable diagnostic tool.

A typical finding in herds experiencing osteomalacia was low calf crops, particularly following drought years when P content of forages was markedly depressed (Tuff, 1923; Theiler *et al.*, 1924, 1928; Eckles *et al.*, 1926). It was observed that lactating cows were less likely to become pregnant than dry cows, and clinical signs were more likely to be seen in lactating cows than in dry cows. Lactation was recognized as a critical drain of P and a risk factor for disease development. Dry cows often showed no symptoms yet milk cows may be severely affected (Eckles *et al.*, 1926). Once lactation was terminated, body stores of P were repleted over an extended open period and clinical signs abated. Often cows affected with clinical signs were lactating and had atrophic ovaries. The proportion which became pregnant was low and calf crops were around 50%.

Eckles *et al.* (1926) described the cause of this infertility as cessation of visible oestrus associated with small, inactive ovaries, which was based on examination of the reproductive tract *per rectum*. Following calving, cows seemed to have one or two oestrous cycles; if bred they would become pregnant; but if they were not bred, oestrous cycles would cease. These cows would not resume ovarian activity for a year, well after the cessation of milk production (Eckles *et al.*, 1926). As a result, pregnancy rates in cows nursing a calf were significantly reduced (50–60%) compared with cows with no calf (70–90%) at side (Theiler *et al.*, 1928; O'Moore, 1950; Rose, 1954; Hart and Mitchell, 1965). This led to the association of aphosphorosis with reduced fertility caused by anoestrous in cattle. This view was supported by the work of Theiler *et al.* (1928). Striking differences were noted in the calf crop between cattle grazed on P-deficient rangeland, which received bonemeal supplementation vs. cattle that did not receive a supplement, which grazed on the same cropland. Average P content of pastures in pica regions was 0.05% compared to 0.135% P in nonpica regions (Theiler and Green, 1932). Typically, problems were severe in winter, as P content in lush spring pastures was often around 0.24%, but declined to 0.03% P when the pasture was mature and extended on into winter. Thus, by winter, cows would have been consuming low P forage for an extended time. Lactating cows were more affected than non-lactating cows due to the higher P requirement. Problems seen on these pastures included stunted growth, reduced milk production, reduced weaning weights of calves, reduced calf crops and lameness. Supplementation with bonemeal increased P intake by over 0.15% and alleviated problems observed in cattle on these pastures.

Irregular breeding was thought to be induced by the drain of P during lactation in cattle on low P rangeland. The decline in serum P_i was also thought to interfere with metabolism at the cellular level. O'Moore (1950) expressed the prevailing opinion in a letter to *Nature* as follows:

> sub-clinical aphosphorosis is often associated with unthriftiness, low milk-yield in dairy cows and anoestrous or oestrus with repeated failures to conceive after service. First-calf heifers usually exhibit the severest symptoms.

A widespread opinion developed that the primary cause of infertility associated with a deficiency of dietary P was anoestrous and/or irregular and abnormal oestrous cycles.

Theiler *et al.* (1928) concluded that P alone was the cause of infertility in range cattle on P-deficient pastures, because bone meal supplementation or P supplementation restored normal calving rates. Merely supplementing P increased calving rates from below 60% to over 80%. Not only was fertility enhanced, but also signs of bone eating were eliminated. The authors recognized that this experiment was not a definitive proof that P was the sole cause of the infertility, since P supplemented animals ate more than non-supplemented animals. However, because solely supplementing P led to dramatic improvements in animal performance, they felt P deficiency was the main cause of the reproductive inefficiency.

Eckles *et al.* (1935) recognized that low P forages often had low CP content. Furthermore, they recognized that cattle suffering from aphosphorosis ate less than unaffected cattle, thus reducing total nutrient intake. Therefore, they felt that the infertility associated with low P forages might be confounded with low CP and low nutrient intakes. To examine some of the confounding issues, these workers purchased cows from P deficient farms in 1928. The majority of these cows were not pregnant and had no evidence of ovarian cycling. They placed these cows on rations which supplied required amounts of CP and energy yet were still low in P. Over a 3-month period, a majority of these cows resumed normal oestrous cycles despite the low P rations. To test this further, Eckles *et al.* (1935) set out to examine uncomplicated P deficiency as a cause of infertility over a 3-year period in dairy cows consuming low P rations but adequate in CP and energy. Eight mature cows were assigned to receive a low P diet, with P supplement given to maintain plasma P_i concentrations around 0.78 mmol/l. Phosphorus was provided daily at 10 to 31 g/cow to maintain plasma concentrations while lactating or non-lactating. Diets ranged from 0.11% to 0.19% P content of total DM. The cow study was confounded by three animals becoming positive for brucellosis, one animal apparently dying from fatty liver shortly after calving, one animal aborting twins and one animal aborting from *Arcanobacterium pyogenes* (*Corynebacterium pyogenes*), which made it impossible to continue all cows across a 3-year period of time. However, despite these confounding factors, cows had regular oestrous cycles and good conception rates (CR). Based on dates of observed oestrus behaviour and inseminations, oestrus detection was 73%. CR was 60% to inseminations after at least 4 months of low P feeding, well after plasma P_i concentrations had declined to 0.78 mmol/l. The mean plasma P_i concentration was 0.71 mmol/l (range from 0.54 to 0.89 mmol/l). The authors concluded that P caused low breeding efficiency but had no effect on ovarian activity or oestrus expression when CP and energy were not limiting. A closer inspection of their data suggests that the low P diet had no effect on reproduction; the low breeding efficiency was associated with the confounding problems they encountered. Milk production was low by today's standards, average milk per day across lactating and dry days ranged from 3.2 to 11.1 kg/day, but the low P intake in this study seemed to have no effect on fertility. Subsequent to this work, Palmer *et al.* (1941) fed P and protein deficient diets to 11 Holstein heifers for 24 to 59 months. Nine animals were 6 to 7 months of age at the start of the trial, and two were 22 to 24 months of age. Daily P intake was limited to 5–8 g/head. They were interested in testing the effects of low protein and P intake on age at sexual maturation, oestrous cycling and fertility. Sexual maturity was delayed, cows reaching puberty between 15.5 and 21 months of age. Animals cycled normally once oestrous cycles were initiated. Eight of the nine younger animals became pregnant and six of the eight conceived on a first insemination. Three animals completed two lactations before being slaughtered for bone analysis. Although animals were stunted, some quite severely, reproduction was not impaired. Plasma P_i concentrations ranged from 1.0 to 1.6 mmol/l prior to calving, after which they then declined to less than 1 mmol/l. Ash content of bone was reduced compared with reported normal ranges.

Although lacking large numbers of animals and contemporary controls, early observational studies suggested that rations that contained 0.15% to 0.20% P on a DM basis did not impair fertility as long as CP and energy were adequate. Although control animals were not employed in the studies, these levels of P were typically observed in forages on farms experiencing osteomalacia and attending conditions in lactating cows. It was expected that these concentrations of dietary P would result in animals with less than 60% pregnancy status and reduced oestrous cycling, but they did not. By 1940, several studies suggested that infertility associated with aphosphorosis may be due to confounding factors associated with deficiencies with multiple nutrients, but problems in the field were responsive to P supplementation from bone meal or other supplements such as dicalcium phosphate or monosodium phosphate, thus most opinions were that P deficiency influenced reproduction through ovarian dysfunction. Due to continued infertility problems observed in herds consuming low P pastures, the view in the industry was that low dietary P was a primary cause of infertility.

After World War II, a series of papers was published in England which continued to contribute to this viewpoint and implicated P, the Ca:P ratio and vitamin D as a cause of infertility in dairy herds. Hignett (1950) in a paper, which was innovative in presenting fertility as an integrated process of nutrient intake, infectious disease and management, proposed that P was a major contributor to infertility through luteal insufficiency and cystic ovaries. He observed that CR was reduced in herds grazing heavily limed pastures during the summer. Fertility improved on these farms in the winter, when higher P supplements were fed, Ca intakes were lower and cows were not consuming limed pasture. Hignett (1950) felt that high Ca intake and low P intake was reducing fertility on these farms. As evidence, Hignett (1950) presented data from a survey of herds categorized by the use of lime on pastures. Herds fed on limed pastures were suggested to have narrower Ca:P ratios in the feed intake, and had lower CR than herds which did not use lime on pastures. When several of these herds were recommended to reduce liming, CR improved. The high intake of Ca was proposed to be causing a conditional P deficiency. However, serum P_i was within normal concentrations on these farms, but this was explained as mobilization of bone mineral preventing a decline in serum P_i in cows consuming the limed pastures. This work was fairly anecdotal, but supported by survey studies in New Zealand (Webster, 1932).

Subsequently, three papers (Hignett and Hignett, 1951, 1952, 1953) were published to support these observations. These publications examined the relationship of P, Ca and vitamin D to fertility in dairy cows. Breeding and feeding records of 802 cows from 39 herds were collected during the autumn and winter housing periods in England. At the time of insemination, feed intake and milk production were recorded, and samples of feed were collected for analysis of Ca and P. CR to first service was then examined in categories created by ranking Ca and P intake relative to requirement (P intake − P requirement). The ARC recommendations at that time were that milk cows receive 10 g of P daily for maintenance plus 2 g P per kg of milk (ARC, 1965). Categories were created for Ca and P differences. Increasing Ca difference appeared to increase CR; increasing P difference also appeared to increase CR. In fact, there was a major mistake in the calculations of CR across Ca categories that dramatically influences the conclusions. The authors concluded that fertility was improved with higher intakes of P than recommended by the ARC (Hignett and Hignett, 1951). No attempt was made to control for confounding across the herd. In addition, higher categories of Ca or P intake may have been due to higher concentrate intake.

The conclusions from the survey (Hignett, 1950) are worth re-examining, because this study is one of the more cited papers on P influencing fertility. The authors had calculated an overall chi-square statistic for four categories of Ca intake relative to requirement. The overall chi-square appears significant. However, a careful reading of the paper and calculating the number of animals in each category and the pregnant animals in each category reveals that a major mistake was made in calculating the CR in the highest Ca category. The authors report a CR of 94% in the highest category; the actual CR has to be 72.7% based on the information described in the paper. There is no significant Ca effect when that correction is made. Across the seven categories of P differences reported, CR ranges from 50% in the lowest category to 77% in the highest category. Overall the authors report a chi-square for the seven categories of 18.576. When the table is examined in a logistic model, the only significant effects are for the lowest and the next lowest P categories relative to the highest P category. However, the highest P category were cows from one farm which had very good CR. The authors admitted that these cows influenced the conclusions referring to Ca. Because herd and concentrate intake were not controlled in the method of analysis, the results are greatly confounded and should have been interpreted cautiously.

In addition to the reports cited above, P as a cause of infertility in dairy cows was further promoted by other reports, which found that fertility was enhanced when P supplementation was increased or implicated low P in feed with sterile cows. These include reports by Morrow (1969) and others (Carson *et al.*, 1978; Cates and Christensen, 1983; Brooks *et al.*, 1984). Jakovac *et al.* (1967), in a survey of 250 cows that were considered infertile, found that serum levels of P_i and total protein were below values established for normal conditions of feeding and that P concentration in hay was below normal values. These reports were anecdotal in that herd situations of low fertility were attributed to low dietary P or imbalances in Ca:P and were responsive to P supplementation. However, other possible confounders were not controlled for and often serum P_i values were not below 1.3 mmol/l, a concentration often considered the minimum normal concentration.

The prevailing attitude developed is reflected in a quote (DeBoer *et al.*, 1981) that 'Essentiality for efficient reproduction in ruminants has been demonstrated for P by Steevens *et al.* (1971)'. Yet, Steevens *et al.* (1971) had demonstrated no such effect. They reported that a few cows consuming low P diets in their study had cystic ovaries, but the proportion was not significant and was based on rectal palpation. The confounding effects associated with P supplementation in these reports from farm situations may easily have been confounded by other management changes and not have been related directly to P intake. But, the mindset that P was associated primarily with infertility and ovarian function appeared to be firmly in place.

Studies in lactating dairy cows and heifers that examined P content and reproduction in controlled studies that included varying concentrations of P or controlled for confounding variables include the work by Lindsey and Archibald (1929), Eckles *et al.* (1935), Palmer *et al.* (1941), Littlejohn and Lewis (1960), Steevens *et al.* (1971), Carstairs *et al.* (1980, 1981), DeBoer *et al.* (1981), Kincaid *et al.* (1981), Hurley *et al.* (1982), Call *et al.* (1987), Brodison *et al.* (1989), Brintrup *et al.* (1993), Wu and Satter (2000a) and Wu *et al.* (2001) (Table 8.1). Carstairs *et al.* (1980, 1981) included two P groups and two energy groups to examine this interaction. Phosphorus content of diets ranged from 0.16% to 0.56%. Across all the studies, there is no effect of P content on CR, percentage of oestruses detected or days to first insemination (Fig. 8.1).

Figure 8.1 presents the change in CR across dietary P concentration within each study in Table 8.1. Each study is represented by one continuous line plotted across dietary P content. There is no apparent trend in CR within studies as dietary P increases or decreases. In order to examine if a relationship with dietary P may be obscured by the sample sizes in Table 8.1, we grouped dietary P into categories. Categories were <0.3%, 0.30% to 0.349%, 0.35% to 0.399%, 0.40% to 0.449%, 0.45% to 0.499% and ≥0.50% DM, respectively. Grouping dietary P into categories based on DM concentration and examining CR across studies again revealed no significant influence of P on CR or other reproductive variables (Fig. 8.2, CR). The mean CR values are presented as points in Fig. 8.2. The lines represent the 95% confidence limits around the point estimates in Fig. 8.2. There was no statistically significant trend across P dietary categories across studies.

The studies indicated in Table 8.1 were combined to examine pregnancy outcome for a breeding period (usually an annual basis) from dairy and beef cattle and heifers (beef and dairy) against dietary P concentration (Fig. 8.3). Each line represents an individual study. Points connect across P concentrations in the diets. A general pattern seems to be that pregnancy increases as P increases up to 0.4% of DM, and then appears to decline at high dietary P levels.

To examine this analytically, two logistic models were explored. One model was fit across all data, controlling for study (Fig. 8.4). A second-order polynomial of P concentration in DM significantly described pregnancy proportion. Below 0.2% dietary P and above 0.5% dietary P, pregnancy proportion declined. However, the overall model fit across studies with individual dietary P concentration was poor and the results of this quadratic fit for dietary P should be viewed with caution. Possibly the poor fit was due to large differences in animal numbers across P content in different studies.

Table 8.1. Controlled studies reporting phosphorus content and fertility in dairy cows.

n	Dietary P (% of DM)	CR[a]	% Oestrus detection	DFI[b]	Pregnant (%)	Reference
10	0.34	0.769	45		100	Lindsey and Archibald (1929)
10	0.54	0.769	38		100	
8	0.16	0.600	79	105	86	Eckles *et al.* (1935)
16	0.37	0.385	43		100	Steevens *et al.* (1971)
16	0.55	0.476	36		100	
16	0.56	0.385	37		100	
16	0.37	0.227	32		100	
16	0.55	0.526	37		100	
16	0.56	0.357	39		100	
12	0.50	0.391	25		75	Carstairs *et al.* (1980, 1981)
12	0.40	0.538	25		58	
12	0.50	0.410	29		75	
12	0.40	0.375	29		100	
13	0.24	0.769	45	77	92	Call *et al.* (1987)
8	0.32	0.526	66	91	87	
13	0.42	0.667	50	72	77	
46	0.355	0.520	36	74	87	Brodison *et al.* (1989)
37	0.425	0.590	36	74	86	
36	0.364	0.600	33	75	86	
26	0.458	0.570	44	80	77	
40	0.342	0.630	34	79	80	
32	0.435	0.630	29	83	97	
52	0.39	0.476		55	79	Brintrup *et al.* (1993)
52	0.33	0.434		47	90	
8	0.31	0.714		70	100	Wu and Satter (2000a)
9	0.40	0.625		92	89	
9	0.49	0.435		67	89	
21	0.38	0.400		76	95	Wu and Satter (2000b)
21	0.48	0.385		77	90	
26	0.38	0.625		66	96	
27	0.48	0.476		72	86	
10	0.31	0.588		90	75	Wu *et al.* (2001)
14	0.39	0.625		77	80	
13	0.47	0.833		94	85	

[a]Conception rate, calculated as 1/(services per conception).
[b]Days to first insemination.

To more uniformly account for animal numbers, data were categorized by P concentration in the DM and a second model was run. A fourth-order polynomial fit the data, adjusting for study. The model had an overall fit of 0.388, a significant improvement over the first model (Fig. 8.5). This model also suggested a quadratic relationship with dietary P and pregnancy proportion. Pregnancy proportion was reduced when dietary P was below 0.20% to 0.15% and it tended to decline when dietary P approached 0.5% of DM in rations (Fig. 8.5). However, the number of cows in the studies where P was greater than 0.5% of DM were small. In addition, few studies contained dietary P concentrations that ranged from <0.20% to >0.50% of DM. Thus, the quadratic effect may be confounded with study distribution of dietary P concentrations, which were examined within each study.

Based on the studies reported and summarized here, dietary P does not seem to have a major impact on reproduction until dietary

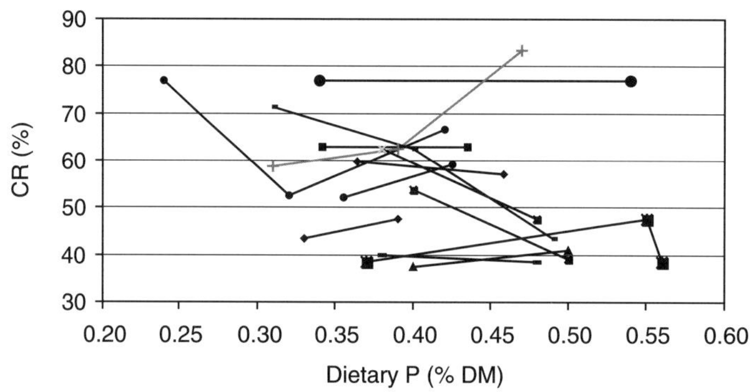

Fig. 8.1. Conception rate (CR) plotted against dietary P content (% DM) in studies with dairy cows. Each plot line represents a study (presented in Table 8.1).

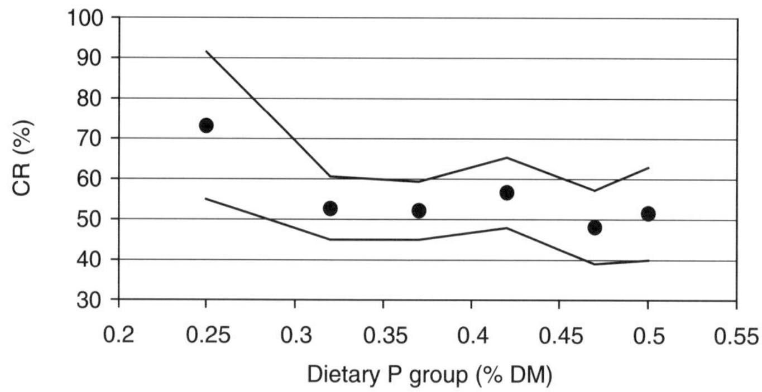

Fig. 8.2. Conception rate (CR) plotted against dietary P content categories (% DM) in studies with dairy cows. Point estimates represent mid ranges for each category (six categories). Lines represent 95% confidence limits. There is no significant difference across P categories.

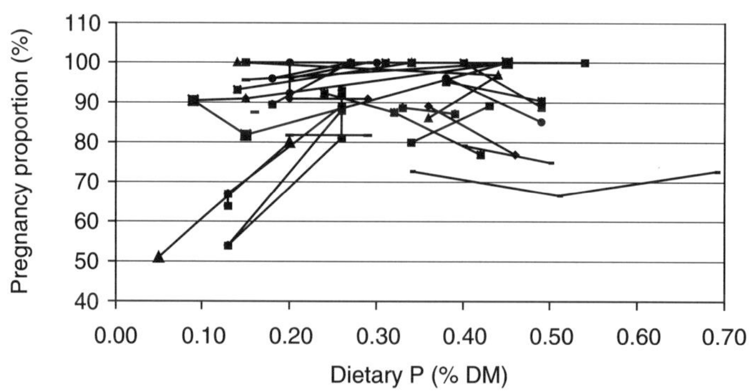

Fig. 8.3. Studies with beef and dairy cattle reporting the proportion of cows which became pregnant over a breeding season or on an annual basis against P content in dietary DM. Each plot line represents a study. It appears pregnancy proportion is reduced when dietary P is below 0.15% DM (data from Theiler *et al.*, 1928; Lindsey and Archibald, 1929; Eckles *et al.*, 1935; Call *et al.*, 1978, 1987; Carstairs *et al.*, 1980, 1981; DeBoer *et al.*, 1981; Cates and Christensen, 1983; Brodison *et al.*, 1989; Karn, 1992; Brintrup *et al.*, 1993; Wu and Satter 2000a; Wu *et al.*, 2001).

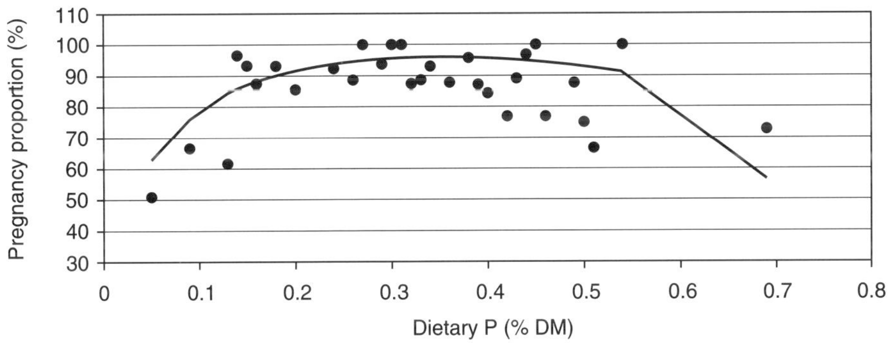

Item	beta	SEM	χ^2	$P <$
Intercept	−0.37	0.69	0.28	0.590
Phosphorus (%DM)	4.95	10.79	0.21	0.646
Phosphorus^2 (%DM)	−124.0	56.67	4.39	0.028
Phosphorus^3 (%DM)	331.4	121.1	7.49	0.006
Phosphorus^4 (%DM)	−247.4	88.41	7.83	0.005
Study				
Likelihood ratio	18		41.09	0.008

Fig. 8.4. Quadratic fit (logistic regression) of pregnancy proportion and P content of dietary DM (data from Theiler *et al.*, 1928; Lindsey and Archibald, 1929; Eckles *et al.*, 1935; Call *et al.*, 1978, 1987; Carstairs *et al.*, 1980, 1981; DeBoer *et al.*, 1981; Cates and Christensen 1983; Brodison *et al.*, 1989; Karn, 1992; Brintrup *et al.*, 1993; Wu and Satter, 2000a; Wu *et al.*, 2001).

concentrations are below 0.10%. In fact, many pasture studies in beef cattle find no benefit with supplementation when pasture concentrations are 0.15% to 0.20%. Marginal pastures will be more of a problem in drought years when P concentration will decline. In addition, lactating cows will experience a higher probability of deficiency than non-lactating cattle due to the P drain into milk. Dairy cattle tolerate dietary P concentrations between 0.20% and 0.30%, however, at these concentrations milk production has typically been lower compared to cows receiving rations containing more than 0.30% P. It seems that dairy cattle can safely be fed dietary P concentrations of 0.33% to 0.40% with no negative effects on reproduction or milk production. A recent report by Lopez *et al.* (2004) found no difference in fertility and milk production in cows fed a 0.37% P diet vs. a 0.57% P diet.

Since manure P content directly increases with dietary P concentration, from an environmental perspective dietary P concentration should be as low as possible to maintain health and productivity. Again, dietary concentrations from 0.33% to 0.40% of DM appear more than adequate to sustain dairy production.

8.4 Protein and Fertility

8.4.1 An overview of N metabolism in the cow

Increasing the CP content of dairy rations from 13% to 20% of DM has been found to be associated with increasing milk production at a decreasing rate reaching a plateau (Oldham, 1984; Wu and Satter, 2000b; Broderick, 2003). However, it has become clear that CP in the diet may not directly predict the amount of metabolizable protein (MP) absorbed by the cow (NRC, 2001). The variable extent of rumen degradation of CP sources, the changeable production of microbial protein as a function of fermentable organic matter and the total amount of feed consumed combine to influence the flow of true protein (from feed and microbial sources) to the small intestine.

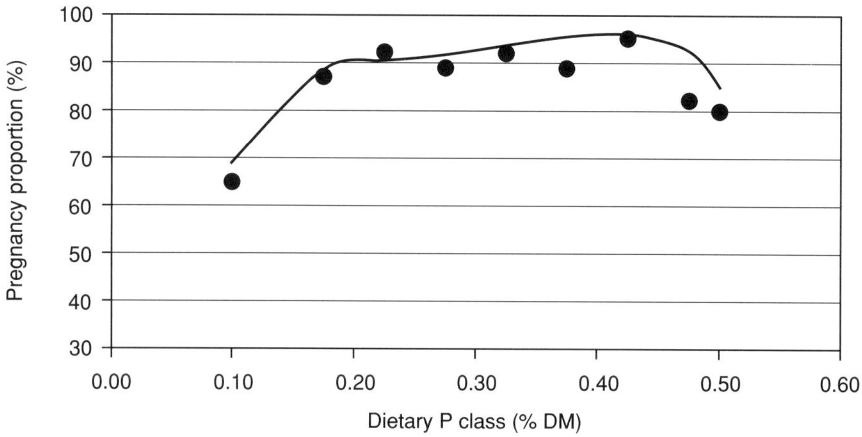

Item	DF	beta	SEM	*P*<
Intercept	1	6.45	3.11	0.0381
phos	1	−127.1	56.58	0.0247
phos*phos	1	702.8	345.6	0.0420
phos*phos*phos	1	−1714.1	862.7	0.0469
phos*phos*phos*phos	1	1502.7	753.6	0.0461
study	12			<0.0001
Likelihood ratio	15		chi-sq 15.90	*P* <0.3888

Dietary P	Num	Range DM%
0.100	1534	<0.160
0.175	225	0.160–0.209
0.225	13	0.210–0.249
0.275	2905	0.250–0.309
0.325	226	0.310–0.349
0.375	208	0.350–0.409
0.425	167	0.400–0.449
0.475	107	0.450–0.499
0.500	30	≥0.500

Fig. 8.5. Quadratic fit (logistic regression) or pregnancy proportion against P content of dietary DM grouped into categories: <0.15%; 0.15%–0.19%; 0.20–0.24%; 0.25%–0.29%; 0.30%–0.34%; 0.35%–0.39%; 0.40%–0.44%; 0.45%–0.49% and ≥0.50% (data from Theiler *et al.*, 1928; Lindsey and Archibald, 1929; Eckles *et al.*, 1935; Call *et al.*, 1978, 1987; Carstairs *et al.*, 1980, 1981; DeBoer *et al.*, 1981; Cates and Christensen 1983; Brodison *et al.*, 1989; Karn, 1992; Brintrup *et al.*, 1993; Wu and Satter, 2000a; Wu *et al.*, 2001).

The potential conversion of ammonia released from the rumen degradation of feed protein (RDP) into bacterial protein is a function of the amount of carbohydrate fermented in the rumen. The NRC (2001) estimates that 130 g of microbial protein will be produced per 1 kg of total digestible nutrients adjusted for the level of feed intake. If RDP is fed in insufficient amounts, rumen production of microbial protein will be reduced, and if RDP exceeds the amount that can be incorporated into bacterial protein, ammonia released from RDP will be absorbed and transported to the liver to be converted to urea. Urea is passed in the urine and lost to the animal as waste, thus excessive RDP results in excretion of a significant amount of N, reducing the overall efficiency of CP utilization for milk production. Underfeeding RDP can result in a reduction in DM intake, microbial protein synthesis and milk production (Santos *et al.*, 1998).

At high levels of milk production, sufficient microbial protein cannot be produced in the rumen to meet the total animal requirement. Supplemental feed protein must be provided to meet the additional MP required for higher milk production. Feed protein which escapes rumen degradation is referred to as rumen undegradable protein (RUP). Rumen undegradable protein must supplement in amount and complement in amino acid content the MP from bacteria to supply sufficient amino acids for milk protein synthesis. As milk production increases, a higher proportion of feed protein must escape rumen degradation to supplement microbial protein to supply sufficient amino acids to meet animal requirements. The proportion of feed protein which is not degraded in the rumen is a characteristic of individual feed ingredients and rumen passage rates. Therefore, the best protein sources to complement microbial protein may vary with increasing milk production to provide the appropriate balance of amino acids to meet productive requirements.

Care must be taken in balancing RDP and RUP in cattle rations to meet the needs of rumen bacteria for ammonia N and ruminant tissues for MP supply. If increasing RUP content of dairy rations reduces RDP content, insufficient ammonia N and peptides may be supplied to rumen bacteria, reducing microbial protein synthesis. The amount of RDP needed in the ration is a function of rumen fermentable carbohydrate. It has been estimated that RDP should comprise 10% to 13% of DM (Hoover and Miller, 1991) in dairy rations. Rumen degradable protein supplied in quantities exceeding bacterial requirements, supplies N above the amount that can be captured into microbial protein. Excess rumen N appears primarily as ammonia. Ammonia will be absorbed across the rumen wall into the portal circulation. It must be detoxified to urea by the liver. The liver releases the urea into the systemic circulation where it equilibrates rapidly in body water. A proportion of body urea is recycled to the digestive tract or is excreted from the body in milk and urine each day.

MP supply may be higher than required, particularly at lower levels of milk production. Excess amino acids are catabolized as an energy source, releasing N as ammonia. There is no large body storage pool of amino acids as exists for energy in the form of adipose tissue. Thus, excess MP ultimately is catabolized. The energy released from catabolism of MP may be used for tissue needs or stored in adipose tissue if in excess of requirement. The ammonia produced must be detoxified by conversion to urea to prevent toxicity to cells. The N released from metabolism of MP also contributes to the tissue urea pool.

The challenge in formulating dairy rations is to balance the supply of RDP, rumen fermentable carbohydrate and RUP to meet the needs for milk production and minimize excretion of urea. As rumen ammonia increases due to increasing supply of RDP above that needed in the rumen, the urea content of milk increases in a linear fashion (Ropstad *et al.*, 1989). Therefore, balancing the supply of RDP to ensure minimal ammonia excess in the rumen can enhance the efficiency of N utilization. Likewise, provision of optimal blends of amino acids in MP can enhance the efficiency of tissue metabolism, further improving the efficiency of N utilization in the cow (Rulquin *et al.*, 1993). Sophisticated animal models have been developed which utilize more specific analysis of feeds to predict the microbial production of protein in the rumen from degradation of feed protein and carbohydrate. The amount of MP from RUP feed protein and bacterial protein is calculated along with the supply of amino acids in the MP. Application of these models in ration formulation can be used to improve the efficiency of N utilization on farms. Often, however, producers desire to err on the side of slightly overfeeding CP to ensure that milk production is not limited. In addition, forage preferences of producers and cattle grouping may limit the extent to which total N efficiency may be optimized within individual farms.

Efficient supply of RDP and RUP to the cow may be assessed by examining the efficiency of capture of feed N into milk N. Efficiency improves as milk production increases and as proper balances of RDP and RUP are incorporated into the diet. This concept suggests that urea lost in urine each day will reflect this efficiency. That is, urea excretion should be minimized while capture in milk N is maximized. Taking this concept further, there should be an optimal range in plasma urea associated with the balance of RUP and RDP. Indeed, Hof *et al.* (1994) and Jonker *et al.* (1998) have suggested that plasma urea will reflect balances of RUP and RDP. Since urea freely diffuses across body water compartments, milk urea concentration is highly correlated with plasma urea concentrations and provides a convenient sampling medium (Roseler *et al.*, 1993). Urinary excretion of urea should correlate with milk urea. Therefore, milk urea concentration can provide a useful monitoring tool to assess protein feeding efficiency and estimate urinary losses.

8.4.2 Protein and reproduction

The effect of concentration of dietary CP on fertility of dairy cows has been investigated (Folman *et al.*, 1973, 1981; Jordan and Swanson, 1979a,b; Kaufmann, 1982; Kaim *et al.*, 1983; Carroll *et al.*, 1988; Canfield *et al.*, 1990) (Table 8.2). In most

Table 8.2. Ratio of services per conception. Summary of literature results by crude protein (CP) concentration in the diet.

Dietary CP concentration (% of DM)			
12 to 13	15 to 17	17 to 20	Reference
2.37	2.10	–	Chandler *et al.* (1976)
1.50	1.90	2.60	Jordan and Swanson (1979a,b)
2.29	2.60	2.70	Edwards *et al.* (1980)
–	1.80	2.25	Folman *et al.* (1981)
–	2.00	2.80	Piatkowski *et al.* (1981)
–	1.80	2.36	Kaim *et al.* (1983)
2.13	1.90	1.92	Huber (1983)
–	1.76	1.72	Howard *et al.* (1987)
1.50	–	1.80	Carroll *et al.* (1988)
–	2.00	2.86	Canfield *et al.* (1990)
–	1.22	1.65	Elrod and Butler (1993)

studies, increasing CP concentration has resulted in increased services: conception ratio and days open, but responses have not been consistent across all the studies. McCormick *et al.* (1999) found first service CR reduced in cows grazing a ryegrass pasture consuming a high CP supplement (21% CP, 75% RDP, total diet; first service CR 24.1%) compared with two groups of cows consuming the ryegrass pasture supplemented with moderate CP concentrations and different RUP content (17.7% and 17.2% CP, 72% and 64% RUP, first service CR, 41.0% and 38.7%, respectively). Despite the apparent trend in Table 8.2 which suggests that higher CP diets are associated with higher services per conception (SPC) than medium or low CP diets, a meta-analysis did not identify any significant association of increasing CP with higher SPC, or reduced CR (Ferguson and Chalupa, 1989).

Recognizing that CP does not adequately describe MP status, the RDP and RUP concentrations of the diets in the studies in Table 8.2 were calculated based on NRC (Ferguson and Chalupa, 1989; NRC, 2001) values. A meta-analysis

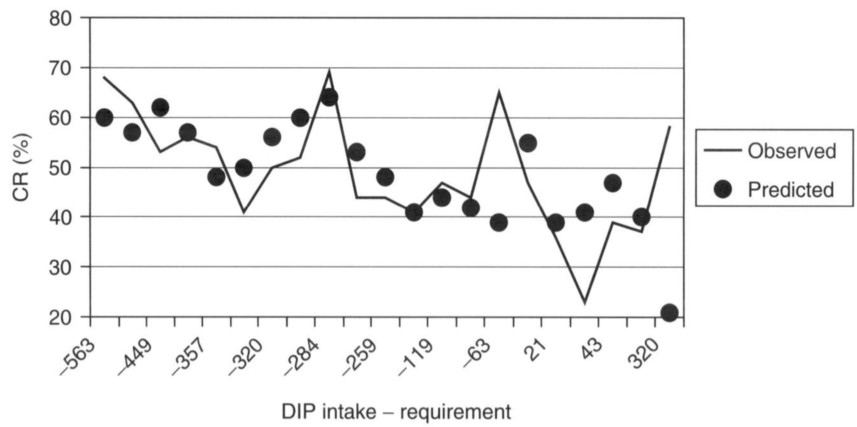

- Degradable intake protein (DIP) (NRC, 2001) relative to requirement (DIRP) influenced conception rate (CR) in studies.
  - –1.0845 × DIPR influenced CR.
- Increasing DIP was associated with lower conception rate.
- Crude protein content was not associated with conception rate.

Fig. 8.6. Observed and predicted fit of conception rate against the difference in rumen degradable protein intake and rumen degradable protein requirement calculated based on NRC (2001).

identified that the difference between the requirement for RDP and the dietary supply of RDP ($RDP_{requirement} - RDP_{supply}$ = DIPR, kg) was significantly associated with a reduction in CR (Fig. 8.6; Ferguson and Chalupa, 1989). This review paper was the first to identify RDP as a risk factor influencing fertility in dairy cows. Two possible models were identified in the paper by Ferguson and Chalupa (1989). The first model identified a linear relationship between increasing DIPR and CR. This model suggests that an increasing production of rumen ammonia produced from degradation of increasing dietary RDP is a factor contributing to declines in CR. A second model in that paper identified an interaction between supply of RDP and RUP relative to requirement as an important factor influencing CR. Increasing the positive balance of RUP improved fertility. Increasing RDP above requirement reduced fertility when RUP was in positive supply, but improved fertility when RUP was deficit. What each model identified was that protein supply relative to requirement, whether RDP or a combination of RDP and RUP, influenced fertility. From a meta-analysis of literature reports it is not possible to identify mechanisms or biological effects, but such an analysis may suggest research directions.

The next logical step, based on the observations that increasing RDP above rumen requirement is associated with increased rumen ammonia and plasma and milk urea N (MUN), would be to examine the relationship between MUN and fertility. Indeed, studies have identified that increases in blood plasma urea N (PUN) and/or MUN are associated with reductions in fertility in dairy cattle (Call *et al.*, 1978; Ferguson *et al.*, 1988, 1993; Armstrong *et al.*, 1990; Canfield *et al.*, 1990; Elrod and Butler, 1993; Butler *et al.*, 1996; McEvoy *et al.*, 1997; Sinclair *et al.*, 2000; Rajala-Schultz *et al.*, 2001). Subsequent research has examined effects of increasing RDP on fertility (Canfield *et al.*, 1990; Elrod and Butler, 1993; Sinclair *et al.*, 2000), but little research has examined the interaction of RDP and RUP on fertility.

8.4.3 Mechanisms and interactions

Ferguson and Blanchard employed nine herds, divided into two feeding groups. Five herds were fed a 16.5% CP diet with half the cows receiving a 62% RDP and half the cows receiving a 70% RDP from calving through 120 to 150 days post-calving (Ferguson and Blanchard, unpublished). The remaining four herds were split into two groups, one of which was fed a 16.5% CP diet with 62% RDP and the other group fed a 19% CP diet with 70% RDP from calving through 120 to 150 days post-calving. Independent of CP level, the 70% RDP diet was associated with reduced CR to first insemination in cows with early first service insemination (< 75 days post-calving), with body condition loss, and in cows which had a history of postpartum metritis (Ferguson and Blanchard, unpublished; Table 8.3, metritis interactions). Furthermore, cows which had metritis failed to recover fertility with time when fed the 70% RDP compared with cows fed the 62% RDP diet. This suggested that increasing RDP negatively affects the uterine environment. The researchers did not observe a difference in rate of uterine involution across the diets used in the nine-herd study. This study suggested that time post-calving, metritis and body condition loss interact with excess RDP to affect fertility.

The interactions of RDP with metritis on reduction in fertility were also suggested by Carroll *et al.* (1988). These workers observed that cows with reproductive health problems were more negatively affected by higher CP diets, which in their study resulted in increases in RDP. In addition these workers found higher urea

Table 8.3. Metritis effects on fertility in cows fed diets with two levels of rumen degradable protein (RDP) and inseminated less than 121 days postpartum; all services (Ferguson and Blanchard, unpublished observations).[a]

	Number of inseminations	Observed CR (%)[b]
Diet 1 (16.5% CP, 71% RDP)	16	26.0[c]
Diet 2 (16.5% CP, 63% RDP)	28	54.0[d]

[a]Cows with metritis post-calving; odds ratio (90% confidence interval) = 4.71 (1.44, 15.42).
[b]Observed CR = conception rate observed in the respective groups for all inseminations.
[c,d]Means differ significantly ($P < 0.05$).

concentrations in uterine fluid and lower CR associated with higher CP diets, particularly in cows which had metritis.

The effects of body condition (Ferguson *et al.*, 1994) loss (scale 1 to 5) are similar to the effects of metritis. Cows with negative changes in body condition from calving to first breeding on the high RDP diet had low first service CR compared with cows on the lower RDP diet (24% vs. 46%, respectively). Cows with stable or increasing body condition score from calving to first breeding had no difference in CR. There was no overall difference in body condition loss between the diets in the study (Folman *et al.*, 1973, 1981; Ferguson *et al.*, 1988). Folman *et al.* (1981) had found CP to affect fertility in older cows with more weight loss. It has long been recognized that weight change affects fertility (Youdan and King, 1977). Westwood *et al.* (2000) observed that cows fed a diet with high RDP sources had significantly lower CR at first service, particularly when body tissue mobilization was high.

It was apparent from determining days to first ovulation, overall body condition change and production that the effects of increased RDP were not on reproductive function or through more negative energy balance. Data from Sinclair *et al.* (2000) and McEvoy *et al.* (1997) also suggest that reproductive function as measured by luteinizing hormone release and time to ovulation from initiation of oestrus was not influenced by RDP. Overall, the associations with metritis, body condition loss and time post-calving suggest that RDP may influence fertility through effects on uterine environment, the oocyte or the early developing embryo.

In a herd fed the 16.5% CP diet and divided into feeding groups receiving 62% RDP and 70% RDP, interbreeding intervals, determined from milk progesterone analysis and from days between inseminations, suggested that 70% RDP influenced uterine environment and pregnancy to first insemination (Ferguson and Blanchard, unpublished). Interoestrous intervals may be considered to be abnormally short (<17 days between progesterone declines) or abnormally long (>24 days between progesterone declines). Prior to first insemination, 35% of interoestrus intervals were irregular for cows consuming the 70% and 62% RDP diets (CP 16.5%). After insemination for first service, 60% of the interoestrus intervals were irregular in cows consuming the 70% RDP compared with 19% of intervals in cows consuming the 62% RDP diet. Of the interoestrus cycles, 38% were >24 days in cows consuming the 70% RDP diet compared with 14% in cows consuming the 62% RDP diet. The short oestrous intervals (<17 days) suggest uterine inflammation caused these cows to have short interoestrus intervals; whereas the higher proportion of long interoestrus intervals (>24 days) suggest early embryonic death. Interoestrus intervals post-insemination were significantly different between the dietary groups. Cows consuming the 70% RDP diet had more

Table 8.4. Percentage of interoestrus intervals in days between consecutive low (<1 ng/ml) milk progesterone profiles and interbreeding intervals in days between first and second service in cows fed diets with two levels of RDP.

	Pre-breeding[a]		Post-breeding[b]		Breeding interval between first and second service[c]	
	71% RDP (n = 110)	63% RDP (n = 100)	71% RDP (n = 35)	63% RDP (n = 36)	71% RDP (n = 37)	63% RDP (n = 41)
<10	–	–	3	–	–	2
10–17	20	16	20	6	11	–
18–24	65	65	39	80	49	62
25–35	13	12	26	11	19	7
36–48	2	3	6	3	8	24
>48	–	4	6	–	13	–

[a]Pre-breeding: the percentage of cows with an interval in days between milk progesterone concentrations <1.0 ng/ml prior to first insemination.

[b]Post-breeding: the percentage of cows with an interval in days between milk progesterone concentrations <1.0 ng/ml in cows which failed to become pregnant at first insemination.

[c]The percentage of cows with an interval in days between second and first insemination.

irregular intervals compared with cows consuming the 62% RDP based on milk progesterone profiles (Table 8.4). Observed interbreeding intervals corresponded with these changes in progesterone profile. The increase in irregular interoestrus intervals post-insemination in cows consuming the 70% RDP diet suggested the uterine environment was influenced by diet and affected fertility. The heat detection rate on the farm used in the study in Table 8.3 was 65% and 90% accurate.

Elrod and Butler (1993), in a very elegant experiment, demonstrated that diets which contained higher RDP were associated with reduced CR in heifers, longer oestrous intervals following breeding if conception failed and lower uterine pH on day seven post-oestrus. Data suggested that lower fertility was associated with higher RDP, higher serum urea and early embryonic death associated with changes in the uterine environment.

These effects in the uterine environment may be on the early embryo. Blanchard *et al.* (1990) found that embryo quality was reduced in cows consuming a 16.5% CP diet that contained 70% RDP compared with 62% RDP (Table 8.5). The effect was not apparent in all cows, but found in a higher proportion of cows consuming the high RDP diet, particularly in cows in their 4th parity and older. Approximately one-third of cows consuming the higher RDP diet failed to yield any fertilized embryos. Subsequent to this study we fed 12 cows a 21% CP diet with 70% RDP. Four of the 12 cows failed to yield fertilized embryos. In the cows that failed to yield fertilized embryos in both studies, unfertilized zygotes were collected. Numbers of corpora lutea (CL) and zygotes were not different across treatments. These studies suggest that the problem is in fertilization failure. Larson *et al.* (1997) found that cows with higher MUN had more failed pregnancies that were associated with regular interoestrus interval, based on sequential milk progesterone testing, compared with cows that became pregnant. Unlike data in Table 8.4, which suggest irregular interoestrus intervals with higher intakes of RDP and MUN, these data suggest interoestrus intervals should be regular. Possibly interaction with a prior history of metritis influences the mechanism of lowered fertility.

Several studies have examined RDP effects on oocyte maturation *in vitro* and early embryo development. Sinclair *et al.* (2000) have linked RDP with spikes in serum ammonia and effects of oocyte maturation and early blastocyst development. McCormick *et al.* (1999) observed that plasma ammonia concentrations measured at or near insemination were negatively correlated ($r = -0.27$, $P < 0.0002$) with pregnancy at that insemination, suggesting ammonia may play a role in reducing reproductive performance in cows fed high RDP diets. McEvoy *et al.* (1997) observed similar effects in sheep when embryos were harvested from ewes fed increasing dietary concentrations of urea. DeWit *et al.* (2001) and Ocon and Hansen (2003) found that oocytes incubated in increasing concentrations of urea had reduced proportions of fertilized oocytes that developed to blastocysts. DeWit *et al.* (2001) found that

Table 8.5. Effects of diet on embryo yield and quality (Blanchard *et al.*, 1990); Ferguson and Blanchard, unpublished, Group three).

	Diets		
	Group one	Group two	Group three
Crude protein (CP) (%)	16.5	16.4	21.0
Degradable intake protein (% CP)	71	63	71
Cows (*n*)	19	19	12
Failure (*n*)	7	1	4
Lactation			
Second parity	7	7	–
Failure[a]	2	1	–
Fourth parity+	11	11	12
Failure	5[b]	0[c]	4[b]

[a]Failure = no fertilized embryos collected.
[b,c]Means differ significantly ($P < 0.05$).

increasing urea was associated with reduced fertilization and cleavage rate, but had no effect on embryos after fertilization. However, Ocon and Hansen (2003) reported that fewer oocytes developed to blastocysts due to decreased developmental competence. The urea concentrations used in these studies were similar to concentrations found in cows fed excess RDP.

Few studies have examined the relationship between RUP and fertility. Westwood *et al.* (2000, 2002) concluded that increasing RUP in isonitrogenous diets improved feed intake, reduced serum non-esterified fatty acids postpartum and improved reproductive performance particularly in cows of high genetic merit. Triplett *et al.* (1995) fed a basal diet to postpartum beef cows with three supplements, increasing in RUP (low RUP, 38.1%; moderate RUP, 56.3% and high RUP, 75.6%). Cows receiving the low RUP supplement had lower first service CR than cows receiving the moderate and high RUP supplements (29.2% vs. 57.6% and 54.6%, respectively). Overall pregnancy proportion tended to be lower for the cows receiving the low RUP supplement than the moderate and high supplements (43.2%, 61.5% and 56.4%, respectively). It is difficult to separate the effects of increasing RUP on fertility from the simultaneous reduction in RDP which occurred in these studies.

8.4.5 Urea and conception rate

Ferguson *et al.* (1988) observed that herd fertility was sensitive to elevated urea levels, associated with higher RDP. The data suggested that PUN concentrations above 3.3 mmol/l were detrimental to fertility. Canfield *et al.* (1990) associated elevated PUN with reduced CR in an experiment with higher dietary RDP. In subsequent work, data from 323 cows suggested that increasing PUN was associated with reduced fertility in a stepwise fashion (Ferguson *et al.*, 1993).

Together, the results suggest that fertility declines in cows with PUN above 2.5 to 2.7 mmol/l (Rajala-Schultz *et al.*, 2001) and it appears to be further reduced when values are above 3.3 mmol/l. Cows on well-balanced diets for RDP and RUP will have PUN concentrations between 1.7 and 2.7 mmol/l. Thus, high production can be supported with adequate protein and minimal urea concentrations. Table 8.6 presents data summarized from Ferguson *et al.* (1993), Butler *et al.* (1996) and Rajala-Schultz *et al.* (2001) for MUN category and pregnancy. In all three studies, increasing MUN is associated with a lower likelihood ratio (LR) test for pregnancy or a lower risk ratio for pregnancy (Table 8.5).

8.4.6 Application to farm situations

Many factors influence fertility in a dairy herd. These include management, cow, bull-inseminator and environmental factors. Within cow, factors are specific: infectious diseases, opportunistic disease such as metritis, metabolic conditions, calving difficulties, lameness, mastitis, body condition loss and milk production. Milk urea N and protein feeding are one component of cow factors, which may influence fertility. Often herds may have low or high mean MUN values and have excellent, good or poor reproductive performance due to other factors. An appropriate way to interpret how change in MUN would impact performance is to only apply the comparison within a herd and not across herds. This could be done by calculating the

Table 8.6. Likelihood ratio and risk ratio for multiple categories of plasma urea N (PUN, mg/100 ml) or milk urea N (MUN, mg/100 ml)[a].

Ferguson *et al.* (1993)		Butler *et al.* (1996)		Rajala-Schultz *et al.* (2001)		
PUN	LR	PUN	LR	MUN	RR	LR
<10	1.43	<16	2.65	<10	2.4	2.50
10–14.9	1.01	16–18.9	1.61	10–12.7	1.4	0.90
15–19.9	0.90	19–21.9	0.81	12.7–15.4	1.2	0.71
20–24.9	0.92	22–24.9	0.80	>15.4	1.0	0.56
>25	0.53	>25	0.73			

[a]LR= likelihood ratio; RR = risk ratio; LR calculated in Rajala-Schultz based on the RR and a pre-test CR of 37.5%.

impact of lowering MUN on CR using LR to calculate expected changes on CR or alternatively use an adjusted CR calculated from the odds ratio of increasing MUN on fertility (Sackett *et al.*, 1991).

Using LR, a post-test odds may be calculated based on MUN. For example, a group of cows have an MUN of 19.0. The herd CR is 35%. The pre-test odds can be calculated as $(0.35)/(1 - 0.35) = 0.538$. Post-test odds are equal to the pre-test odds times the LR. The LR for an MUN of 19.0 is 0.90 for Ferguson *et al.* (1993), 0.81 for Butler *et al.* (1996) and 0.56 for Rajala-Schultz *et al.* (2001) (Table 8.5). The post-test odds would be 0.484, 0.436 and 0.301 for the three studies, respectively. The post-test CR in this group of cows would be (post-test odds)/(1+post-test odds) 0.326, 0.304 and 0.231, respectively, for the three studies. The change in CR for this group of cows could be calculated for the concentration of MUN, and the economic cost of semen assessed. The value of changing MUN on fertility could be calculated. In addition, the excretion of N based on MUN could be evaluated (Jonker *et al.*, 1998).

8.5 Conclusions

Protein effects on fertility appear to be associated with elevations in urea or non-protein N in plasma and uterine fluids. Cows under different types of stress, such as body condition loss, metritis or early insemination have more detrimental effects than those without these conditions. Well-balanced diets should not lead to high urea concentrations or reduction in fertility. Feeding rumen degradable protein above the amount needed for rumen synthesis of microbial protein is associated with reduced fertility. The effects may be more pronounced in stressed cows. Effects appear to be due to changes in the uterine environment, increasing early embryonic mortality and are associated with milk or blood urea nitrogen concentrations above 3.3 mmol/l. The effect on fertility, although significant, is not of a large magnitude in most circumstances.

References

ARC [Agricultural Research Council] (1965) *The Nutrient Requirements of Farm Livestock, No. 2, Ruminants.* HMSO, London, 264 pp.

Armstrong, J.D., Goodall, E.A., Gordon, F.J., Rice, D.A. and McCaughey, W.J. (1990) The effects of levels of concentrate offered and inclusion of maize gluten or fish meal in the concentrate on reproductive performance and blood parameters of dairy cows. *Animal Production* 50, 1–10.

Atkinson, D., Harbaugh, Hickman, Law, Lowe, Mohler, Murray, Pearson, Ransom, and Trumbower (1916) *Special Report on Diseases of Cattle. U.S. Department of Agricultural Bureau of Animal Industry.* Government Printing Office, Washington, DC, pp. 264–267.

Barnes, J.E. and Jephcott, B.R. (1955) Phosphorus deficiency in cattle in the Northern Territory and its control. *Australian Veterinary Journal* 31, 302–316.

Becker, R.B., Neal, W.M. and Shealy, A.L. (1933) Stiffs or Sweeny (phosphorus deficiency) in cattle. *University of Florida Agricultural Experimental Station Bulletin* 264, 1–27.

Belonje, P.C. and van den Berg, A. (1980) The use of faecal analyses to estimate the phosphorus intake by grazing sheep. II. The repeatability of the technique and the influence of varying the phosphorus intake. *Onderstepoort Journal of Veterinary Research* 47, 169–172.

Blair-West, J.R., Denton, D.A., McKinley, M.J., Radden, B.G., Ramshaw, E.H. and Wark, J.D. (1992) Behavioral and tissue responses to severe phosphorus depletion in cattle. *American Journal of Physiology* 263, R656–R663.

Blanchard, T., Ferguson, J., Love, L., Takeda, T., Henderson, B., Hasler, J. and Chalupa, W. (1990) Effect of dietary crude-protein type on fertilization and embryo quality in dairy cattle. *American Journal of Veterinary Research* 51, 905–908.

Brintrup, R., Mooren, T., Meyer, U., Spiekers, H. and Pfeffer, E. (1993) Effects of two levels of phosphorus intake on performance and fecal phosphorus excretion of dairy cows. *Journal of Animal Physiology and Animal Nutrition* 69, 29–36.

Broderick, G.A. (2003) Effects of varying dietary protein and energy levels on the production of lactating dairy cows. *Journal of Dairy Science* 86, 1370–1381.

Brodison, J.A., Goodall, E.A., Armstrong, J.D., Givens, D.I., Gordon, F.J., McCaughey, W.J. and Todd, J.R. (1989) Influence of dietary phosphorus on the performance of lactating dairy cattle. *Journal of Agricultural Science, Cambridge* 112, 303–311.

Brooks, H.V., Cook, T.G., Mansell, G.P. and Walker, G.A. (1984) Phosphorus deficiency in a dairy herd. *New Zealand Veterinary Journal* 32, 174–176.

Butler, W.R., Calaman, J.J. and Beam, S.W. (1996) Plasma and milk urea nitrogen in relation to pregnancy rate in lactating dairy cattle. *Journal of Animal Science* 74, 858–865.

Call, J.W., Butcher, J.E., Blake, J.T., Smart, R.A. and Shupe, J.L. (1978) Phosphorus influence on growth and reproduction of beef cattle. *Journal of Animal Science* 47, 216–225.

Call, J.W., Butcher, J.E., Shupe, J.L., Lamb, R.C., Boman, R.L. and Olson, A.E. (1987) Clinical effects of low dietary phosphorus concentrations in feed given to lactating dairy cows. *American Journal of Veterinary Research* 48, 133–136.

Canfield, R.W., Sniffen, C.J. and Butler, W.R. (1990) Effects of excess degradable protein on postpartum reproduction and energy balance in dairy cattle. *Journal of Dairy Science* 73, 2342–2349.

Carroll, D.J., Barton, B.A., Anderson, G.W. and Smith, R.D. (1988) Influence of protein intake and feeding strategy on reproductive performance of dairy cows. *Journal of Dairy Science* 71, 3470–3481.

Carson, R.L., Caudle, A.B. and Riddle, H.E. (1978) The relationship between narrow calcium–phosphorus ratio and reproductive problems in a dairy herd: a case report. *Theriogenology* 9, 505–507.

Carstairs, J.A., Morrow, D.A. and Emery, R.S. (1980) Postpartum reproductive function of dairy cows as influenced by energy and phosphorus status. *Journal of Animal Science* 51, 1122–1130.

Carstairs, J.A., Neitzel, R.R. and Emery, R.S. (1981) Energy and phosphorus status as factors affecting postpartum performance and health of dairy cows. *Journal of Dairy Science* 64, 34–41.

Cates, W.F. and Christensen, D.A. (1983) The effect of nutrition on conception rate in beef cows at Cumberland House, Saskatchewan. *Canadian Veterinary Journal* 24, 145–147.

Chandler, P.T., Brown, C.A., Johnston, R.P., MacLeod, G.K., McCarthy, R.D., Moss, B.R., Rakes, A.H. and Satter, L.D. (1976) Protein and methionine hydroxy analog for lactating cows. *Journal of Dairy Science* 59, 1897–1909.

Cohen, R.D.H. (1980) Phosphorus in rangeland ruminant nutrition: a review. *Livestock Production Science* 7, 25–37.

DeBoer, G., Buchanan-Smith, J.G., MacLeod, G.K. and Walton, J.S. (1981) Responses of dairy cows fed alfalfa silage supplemented with phosphorus, copper, zinc, and manganese. *Journal of Dairy Science* 64, 2370–2377.

DeWit, A.A.C., Cesar, M.L.F. and Kruip, T.A.M. (2001) Effect of urea during in vitro maturation on nuclear maturation and embryo development of bovine cumulus–oocyte complexes. *Journal of Dairy Science* 84, 1800–1804.

Eckles, C.H., Becker, R.B. and Palmer, L.S. (1926) A mineral deficiency in the rations of cattle. *Minnesota Agricultural Experimental Station Bulletin* 229, 1–86.

Eckles, C.H., Gullickson, T.W. and Palmer, L.S. (1932) Phosphorus deficiency in the rations of cattle. *Minnesota Agricultural Experimental Station Technical Bulletin* 91, 1–118.

Eckles, C.H., Palmer, L.S., Gullickson, T.W., Fitch, C.P., Boyd, W.L., Bishop, L. and Nelson, J.W. (1935) Effect of uncomplicated phosphorus deficiency on estrous cycle, reproduction, and composition of tissues on mature dry cows. *Cornell Veterinarian* 25, 22–43.

Edwards, J.S., Bartley, E.E. and Dayton, A.D. (1980) Effects of dietary protein concentration on lactating cows. *Journal of Dairy Science* 63, 243–248.

Elrod, C.C. and Butler, W.R. (1993) Reduction of fertility and alteration of uterine pH in heifers fed excess ruminally degradable protein. *Journal of Animal Science* 71, 694–701.

Ferguson, J.D. and Chalupa, W. (1989) Impact of protein nutrition on reproduction in dairy cows. *Journal of Dairy Science* 72, 746–766.

Ferguson, J.D., Blanchard, T., Galligan, D.T., Hoshall, D.C. and Chalupa, W. (1988) Infertility in dairy cattle fed a high percentage of protein degradable in the rumen. *Journal of the American Veterinary Association* 192, 659–662.

Ferguson, J.D., Galligan, D.T., Blanchard, T. and Reeves, M. (1993) Serum urea nitrogen and conception rate: the usefulness of test information. *Journal of Dairy Science* 76, 3742–3746.

Ferguson, J.D., Galligan, D.T. and Thomsen, N. (1994) Principal descriptors of body condition score in Holstein cows. *Journal of Dairy Science* 77, 2695–2703.

Folman, Y., Rosenberg, M., Herz, Z. and Davidson, M. (1973) The relationship between plasma progesterone concentration and conception in postpartum dairy cows maintained on two levels of nutrition. *Journal of Reproduction and Fertility* 34, 267–278.

Folman, Y., Neumark, H., Kaim, M. and Kaufmann, W. (1981) Performance, rumen and blood metabolites in high-yielding cows fed varying protein percents and protected soybean. *Journal of Dairy Science* 64, 759–768.

Ganong, W.F. (1999) *Review of Medical Physiology*. Appleton and Lange, Stamford, Connecticut.

Green, H.H. (1939) Recent advances in our knowledge of diseases associated with mineral balance in the blood of ruminants. *Veterinary Record* 51, 1179–1198.

Haag, J.R., Jones, J.S., Jones, I.R. and Brandt, P.M. (1929) The physiological effect of rations restricted principally or solely to the alfalfa plant. 1. The calcium, phosphorus and nitrogen metabolism of dairy cattle. *Journal of Dairy Science* 12, 445–455.

Hart, B. and Mitchell, G.L. (1965) Effect of phosphate supplementation on the fertility of an open range beef cattle herd on the Barkly Tableland. *Australian Veterinary Journal* 41, 305–309.

Henderson, H.O. and Weakly, C.E.J. (1930) The effect of feeding different amounts of calcium and phosphorus upon growth and development of dairy animals. *West Virginia Agriculture Experimental Station Bulletin* 231, 25–55.

Hignett, S.L. (1950) Factors influencing herd fertility in cattle. *Veterinary Record* 62, 652–674.

Hignett, S.L. and Hignett, P.G. (1951) The influence of nutrition on reproductive efficiency in cattle. I. The effect of calcium and phosphorus intake on the fertility of cows and heifers. *Veterinary Record* 63, 603–609.

Hignett, S.l. and Hignett, P.G. (1952) The influence of nutrition on reproductive efficiency in cattle. II. The effect of the phosphorus intake on ovarian activity and fertility of heifers. *Veterinary Record* 64, 203–206.

Hignett, S.l. and Hignett, P.G. (1953) The influence of nutrition on reproductive efficiency in cattle. III. The influence of vitamin D status on the effect of calcium and phosphorus intake on the fertility of cows and heifers. *Veterinary Record* 65, 21–24.

Hof, G., Tamminga, S. and Lenaers, P.J. (1994) Efficiency of protein utilization in dairy cows. *Livestock Production Science* 38, 169–178.

Hoover, W.H. and Miller, T.K. (1991) Rumen digestive physiology and microbial ecology. *Veterinary Clinics of North America – Food Animal Practice* 7, 311–325.

Howard, H.J., Aalseth, E.P., Adams, G.D., Bush, L.J., McNew, R.W. and Dawson, L.J. (1987) Influence of dietary protein on reproductive performance of dairy cows. *Journal of Dairy Science* 70, 1563–1571.

Huber, J.T. (1983) Influence of ration protein percent and feeding of nonprotein nitrogen during early lactation on reproductive efficiency of cows. Michigan State University, East Lansing, Michigan.

Huffman, C.F., Robinson, C.S. and Winter, O.B. (1930) The calcium and phosphorus metabolism of heavily milking cows. *Journal of Dairy Science* 13, 432–448.

Hurley, W.L., Edgerton, L.A., Olds, D. and Hemken, R.W. (1982) Estrous behavior and endocrine status of dairy heifers with varied intakes of phosphorus. *Journal of Dairy Science* 65, 1979–1986.

Jakovac, M., Supe, S. and Mikulec, K. (1967) Deficiency of phosphorus as one of the causes of disturbances in the reproduction of cattle. *Veterinarski Arhir* 37, 223–229.

Jonker, J.S., Kohn, R.A. and Erdman, R.A. (1998) Using milk urea nitrogen to predict nitrogen excretion and utilization efficiency in lactating dairy cows. *Journal of Dairy Science* 81, 2681–2692.

Jordan, E.R. and Swanson, L.V. (1979a) Effect of crude protein on reproductive efficiency, serum total protein, and albumin in the high-producing dairy cow. *Journal of Dairy Science* 62, 58–63.

Jordan, E.R. and Swanson, L.V. (1979b) Serum progesterone and luteinizing hormone in dairy cattle fed varying levels of crude protein. *Journal of Animal Science* 48, 1154–1158.

Kaim, M., Folman, Y. and Neumark, H. (1983) The effect of protein intake and lactation number on post-partum body weight loss and reproductive performance of dairy cows. *Journal of Dairy Science* 37, 229.

Karn, J. (1992) Ad libitum phosphorus supplementation of range cows in the northern great plains. *Journal of Production Agriculture* 5, 409–413.

Kaufmann, W. (1982) Variation in der Zusammensetzung des Rohstoffes Milch unter besonderer Berücksichtigung des Harnstoffgehaltes. *Milchwissenschaft* 37, 6–9.

Kincaid, R.L., Hillers, J.K. and Cronrath, J.D. (1981) Calcium and phosphorus supplementation of rations for lactating cows. *Journal of Dairy Science* 64, 754–758.

Larson, S.F., Butler, W.R. and Currie, W.B. (1997) Reduced fertility associated with low progesterone postbreeding and increased milk urea nitrogen in lactating cows. *Journal of Dairy Science* 80, 1288–1295.

Lindsey, J.B. and Archibald, J.G. (1929) Mineral supplements for dairy cows. *Journal of Dairy Science* 12, 102–116.

Little, D.A. (1970) Factors of importance in the phosphorus nutrition of beef cattle in Northern Australia. *Australian Veterinary Journal* 46, 241–248.

Littlejohn, A.I. and Lewis, G. (1960) Experimental studies of the relationship between calcium–phosphorus ratio of the diet and fertility in heifers: a preliminary report. *Veterinary Record* 72, 1137–1144.

Lopez, H., Kanitz, F.D., Moreira, V.R., Satter, L.D. and Wiltbank, M.C. (2004) Reproductive performance of dairy cows fed two concentrations of phosphorus. *Journal of Dairy Science* 87, 146–157.

McCormich, M.E., French, D.D., Brown, T.F., Cuomo, G.J., Chapa, A.M., Fernandez, J.M., Beatty, J.F. and Blouin, D.C. (1999) Crude protein and rumen undegradable protein effects on reproduction and lactation performance of Holstein cows. *Journal of Dairy Science* 82, 2697–2708.

Malan, A.L., Green, H.H. and duToit, P.J. (1928) Studies in mineral metabolism. V. Composition of bovine blood on phosphorus deficient pasture. *Journal of Agricultural Science, Cambridge* 18, 376–383.

McEvoy, T.G., Robinson, J.J., Aitken, R.P., Findlay, P.A. and Robertson, I.S. (1997) Dietary excesses of urea influence the viability and metabolism of preimplantation sheep embryos and may affect fetal growth among survivors. *Animal Reproduction Science* 47, 71–90.

Meador, D.J. (1926/1927) Osteomalacia in cattle. *Cornell Veterinarian*, 16–17.

Meigs, E.B. and Woodward, T.E. (1921) The influence of calcium and phosphorus in the feed on the milk yield of dairy cows. *Journal of Dairy Science* 4, 185–217.

Mitchell, H.H. (1936) Phosphorus deficiencies and disease. *Veterinary Medicine* 31, 66–71.

Morrow, D.A. (1969) Phosphorus deficiency and infertility in dairy heifers. *Journal of the American Veterinary Association* 154, 761–768.

NRC [National Research Council] (2001) *Nutrient Requirements of Dairy Cattle*, 7th rev. edn. National Academy Press, Washington, DC, 381 pp.

Ocon, O.M. and Hansen, P.J. (2003) Disruption of bovine oocytes and preimplantation embryos by urea and acidic pH. *Journal of Dairy Science* 86, 1194–1200.

Oldham, J.D. (1984) Protein–energy interrelationships in dairy cows. *Journal of Dairy Science* 67, 1090–1114.

O'Moore, L.B. (1950) Aphosphorosis in Ireland. *Nature* 165, 192.

Palmer, L.S. and Eckles, C.H. (1927) Effect of phosphorus deficient rations on blood composition. *Proceedings of the Society of Experimental and Biological Medicine* 24, 307–309.

Palmer, L.S., Cunningham, W.S. and Eckles, C.H. (1930) Normal variation in the inorganic phosphorus of the blood of dairy cattle. *Journal of Dairy Science* 13, 174–195.

Palmer, L.S., Gullickson, T.W., Boyd, W.L., Fitch, C.P. and Nelson, J.W. (1941) The effect of rations deficient in phosphorus and protein on ovulation, estrous and reproduction of dairy heifers. *Journal of Dairy Science* 24, 199–210.

Piatkowski, B., Voigt, J. and Girschewski, H. (1981) Einfluss des Rohproteinniveaus auf die Fruchtbarkeit und den Harnstoffgehalt in Körperflussigkeiten bei Hochleistungskühen. *Archiv für Tierernährung* 31, 497–504.

Rajala-Schultz, P.J., Saville, W.J.A., Frazer, G.S. and Wittum, T.E. (2001) Association between milk urea nitrogen and fertility in Ohio dairy cows. *Journal of Dairy Science* 84, 482–489.

Ropstad, E., Vik-Mo, L. and Refsdal, A.O. (1989) Levels of milk urea, plasma constituents and rumen liquid ammonia in relation to the feeding of dairy cows during early lactation. *Acta Veterinaria Scandinavica* 30, 199–208.

Rose, A.L. (1954) Osteomalacia in the Northern Territory. *Australian Veterinary Journal* 30, 172–177.

Roseler, D.K., Ferguson, J.D., Sniffen, C.J. and Herrema, J. (1993) Dietary protein degradability effects on plasma and milk urea nitrogen and milk nonprotein nitrogen in Holstein cows. *Journal of Dairy Science* 76, 525–534.

Rulquin, H., Pisulewski, P.M., Verite, R. and Guinard, J. (1993) Milk production and composition as a function of postruminal lysine and methionine supply: a nutrient–response approach. *Livestock Production Science* 37, 69–90.

Sackett, D.L., Haynes, R.B., Guyatt, G.H. and Tugwell, P. (1991) *Clinical Epidemiology. A Basic Science for Clinical Medicine*, 2nd edn. Little, Brown and Company, Boston, MA.

Sanson, D.W., Walker, G.L., Clanton, D.C. and Eskridge, K.M. (1990) Relationship between phosphorus intake and blood or fecal phosphorus in gestating cows. *Journal of Range Management* 43, 238–241.

Santos, F.A.P., Santos, J.E.P., Theurer, C.B. and Huber, J.T. (1998) Effects of rumen-undegradable protein on dairy cow performance: a 12-year literature review. *Journal of Dairy Science* 81, 3182–3213.

Schmidt, H. (1940) Calcium and phosphorus deficiencies in cattle and horses: clinical picture, treatment and prevention. *Journal of the American Veterinary Association* 96, 441–458.

Sinclair, K.D., Kuran, M., Gebbie, F.E., Webb, R. and McEvoy, T.G. (2000) Nitrogen metabolism and fertility in cattle. II. Development of oocytes recovered from heifers offered diets differing in their rate of nitrogen release in the rumen. *Journal of Animal Science* 78, 2670–2680.

Steevens, B.J., Bush, L.J., Stout, J.D. and Williams, E.I. (1971) Effect of varying amounts of calcium and phosphorus in rations for dairy cows. *Journal of Dairy Science* 54, 655–661.

Ternouth, J.H. (1989) Endogenous losses of phosphorus by sheep. *Journal of Agricultural Science, Cambridge* 113, 291–297.

Ternouth, J.H., Bortolussi, G., Coates, D.B., Hendricksen, R.E. and McLean, R.W. (1996) The phosphorus requirements of growing cattle consuming forage diets. *Journal of Agricultural Science, Cambridge* 126, 503–510.

Theiler, A. and Green, H.H. (1932) Aphosphorosis in ruminants. *Nutrition Abstracts and Reviews* 1, 359–385.

Theiler, A., Green, H.H. and DuToit, P.J. (1924) Phosphorus in the livestock industry. *Journal of the Department of Agriculture, South Africa* 8, 460.

Theiler, A., Green, H.H. and duToit, P.J. (1927) Minimum mineral requirements in cattle. *Journal of Agricultural Science, Cambridge* 17, 291–314.

Theiler, A., Green, H.H. and duToit, P.J. (1928) Studies in mineral metabolism. III. Breeding of cattle on phosphorus deficient pasture. *Journal of Agricultural Science, Cambridge* 18, 369–371.

Theiler, A., Du Toit, P.J. and Malan, A.L. (1937) Studies in mineral metabolism XXXVII. The influence of variations in the dietary phosphorus and in the Ca:P ratio on the production of ricketts in cattle. *Onderstepoort Journal of Veterinary Science* 8, 375–414.

Triplett, B.L., Neuendorff, D.A. and Randel, R.D. (1995) Influence of undegraded intake protein supplementation on milk production, weight gain, and reproductive performance in postpartum Brahman cows. *Journal of Animal Science* 73, 3223–3229.

Tuff, P. (1923) Osteomalacia and its occurrences in cattle in Norway. *Journal of Comparative Pathology* 36, 143–155.

Webster, W.M. (1932) Bovine sterility in New Zealand. *Australian Veterinary Journal* 8, 199–222.

Westwood, C.T., Lean, I.J., Garvin, J.K. and Wynn, P.C. (2000) Effects of genetic merit and varying dietary protein degradability on lactating dairy cows. *Journal of Dairy Science* 83, 2926–2940.

Westwood, C.T., Lean, I.J. and Garvin, J.K. (2002) Factors influencing fertility of Holstein dairy cows: a multivariate description. *Journal of Dairy Science* 85, 3225–3237.

Wu, Z. and Satter, L.D. (2000a) Milk production and reproductive performance of dairy cows fed two concentrations of phosphorus for two years. *Journal of Dairy Science* 83, 1052–1063.

Wu, Z. and Satter, L.D. (2000b) Milk production during the complete lactation of dairy cows fed diets containing different amounts of protein. *Journal of Dairy Science* 83, 1042–1051.

Wu, Z., Satter, L.D., Blohowiak, A.J., Stauffacher, R.H. and Wilson, J.H. (2001) Milk production, estimated phosphorus excretion, and bone characteristics of dairy cows fed different amounts of phosphorus for two or three years. *Journal of Dairy Science* 84, 1738–1748.

Youdan, P.G. and King, J.O.L. (1977) The effects of body weight changes on fertility during the post-partum period in dairy cows. *British Veterinary Journal* 133, 635.

9 Improving the Efficiency of Nutrient Use on Cattle Operations

J. Schröder[1], A. Bannink[2] and R. Kohn[3]
[1]Wageningen University and Research Centre, Plant Sciences Group, Wageningen, The Netherlands
[2]Wageningen University and Research Centre, Animal Sciences Group, Lelystad, The Netherlands
[3]Department of Animal and Avian Sciences, University of Maryland, College Park, Maryland, USA

9.1 Introduction

Inputs of nitrogen (N) and phosphorus (P) are indispensable to sustain agriculture. These inputs have increased drastically along with agricultural production. Concomitantly, losses of N and P to the environment have also increased, as N in particular cannot be fully utilized in any production system (Isermann, 1993; Galloway, 1998; Carton and Jarvis, 2001; De Clercq *et al.*, 2001). Nutrient losses can negatively affect the quality of soils, groundwater, surface water and air. Losses may put drinking water quality at risk, as well as the functioning of ecosystems including the Earth as a whole, via direct effects (concentration of N and P compounds) and via indirect effects (acidification, denitrification-induced solution of sulphate and metals, destruction of ozone, global warming, toxic algae blooming, loss of biodiversity) (Sharpley *et al.*, 1987; Cartwright *et al.*, 1991; Menzel, 1991; Tunney *et al.*, 1997; Galloway, 1998; Novotny, 1999; Pierzynski *et al.*, 2000; Carton and Jarvis, 2001). So, in addition to its self-evident beneficial effects, agriculture can also have detrimental effects on the health and welfare of present and future generations.

Nutrients can be lost from agriculture via numerous pathways. For N, one of these pathways

©CAB International 2005. *Nitrogen and Phosphorus Nutrition of Cattle* (eds E. Pfeffer and A.N. Hristov)

is the volatilization of ammonia (NH_3). Manure, and urine in particular, is the main source of NH_3, although mineral fertilizers, standing and wilting crops and silage heaps can emit minor amounts as well (Jarvis and Pain, 1990; Bussink and Oenema, 1998; Dueck *et al.*, 2001). Manures emit NH_3 from housing and storage, during grazing and after spreading. N can also be lost through denitrification, producing harmful nitrous oxide (N_2O) and nitric oxide (NO) along with innocuous elementary N_2. Denitrification occurs in manure heaps, urine and dung patches and, most importantly, from soils (Velthof and Oenema, 1997; Velthof *et al.*, 1998). Nitrogen can also leach through the soil profile into ground- and surface water, mainly as NO_3 (Addiscott and Powlson, 1992). On sloping, poorly drained or frozen soils, N can also be lost via runoff. As for P, runoff, and to a lesser extent leaching, are important pathways for losses. P is less mobile than N in soil, and therefore P losses are easier to control. Nutrients can be agriculturally unavailable, and from that point of view lost, without being instantly transferred to the wider environment. This occurs when N and P are biologically or physically sequestered in the soil. However, it must be emphasized that soils do not have an infinite capacity to immobilize nutrients. Sooner or later immobilization and mineralization will reach an equilibrium (Jarvis *et al.*, 1996a). To some extent losses are inevitable. Therefore, supplementation of nutrients is a prerequisite for sustainable agriculture. However, a goal of modern agriculture is to reduce nutrient losses to the environment as much as possible within economic and social constraints (Jarvis *et al.*, 1996b; Jarvis and Aarts, 2000). This chapter intends to evaluate how nutrient losses from cattle operations can be reduced and efficiency can be increased.

9.2 A Simple Model

9.2.1 Model description

Nutrient inputs (I) such as fertilizer and legume-fixed N exceed productive nutrient outputs (O) such as food and fibre, in many agricultural production systems around the world. The balance ($I - O$) can be regarded as an indicator of losses (L), although great caution is required to use L judiciously for this purpose. A positive balance *per se* does not reveal the nature or irreversibility of the losses. Besides, the absence of a positive balance does not necessarily imply the absence of losses (Jarvis *et al.*, 1996b; Jarvis and Aarts, 2000; Schröder *et al.*, 2003a).

The efficiency of nutrient use can be expressed as the ratio of O and I (O/I). To evaluate the environmental impact of complete production systems, it is relevant to determine the balance both per unit area (L) and per unit of produce (L/O) (De Wit, 1992; Kohn *et al.*, 1997; Jarvis and Aarts, 2000; Schröder *et al.*, 2003a). Thus, nutrient utilization efficiency (O/I) and nutrient losses per unit of produce (L/O) are intricately related, $L/O = I/O - 1$, i.e. the losses per unit of produce equal the inverse of the farm efficiency minus 1. The same relationships that were demonstrated for whole farm systems also apply to the efficiencies of the farm subsystems: herd, manure handling, soil, cropping and feeding. From an animal nutrition point of view, nutrient losses and use efficiency can be expressed in terms of feed and bedding material inputs (F) and output of animal products such as milk and meat (A). The term 'meat' in this chapter means the total body of slaughter animals for the sake of simplicity. Further differentiation between edible and inedible parts would not change nutrient balances of the farm. Faeces, urine and worn bedding constitute the loss from this subsystem (M). This loss from the herd (M) is an input in the manure handling subsystem. When exported to another (part of the) farm M becomes available to be applied to soils but a fraction of it is inevitably lost from housing and storage as gaseous NH_3-N (Fig. 9.1). The mathematical description of this limited system equals:

$$F = \text{feed} + \text{bedding}\left(\frac{\text{kg nutrient}}{\text{ha} \times \text{year}}\right)$$

and

$$A = \text{milk} + \text{meat}\left(\frac{\text{kg nutrient}}{\text{ha} \times \text{year}}\right),$$

$$\begin{aligned} M &= F - A \\ &= \text{manure removal from housing and storage} \\ &\quad \times \left(\frac{\text{kg nutrient}}{\text{ha} \times \text{year}}\right), \\ FP &= \frac{A}{F} \\ &= \text{efficiency of nutrient utilization for milk} \\ &\quad \text{and/or meat} \\ &\quad \left(\frac{\text{kg nutrient}}{\text{ha} \times \text{year}}\right), \end{aligned}$$

$$M = F \times (1 - FP)$$

and

$$\frac{M}{A} = \frac{1}{FP} - 1.$$

The efficiency of the system within the herd and farm building boundaries is mainly determined by the ability of the herd to convert nutrients in feed (represented by F in the equations above) and bedding material into nutrients in animal products (represented by P in the equations above) and can be expressed in terms of the conversion coefficient FP. The larger the FP, the smaller the relative amount of nutrients produced as manure (M).

When a system is restricted at the level of the herd and the farm building, and manure is considered a valuable fertilizer resource, ruminant production systems can appear quite efficient. However, after the components herd and manure, the components soil and crop can be discerned in most ruminant production systems. Nutrient transfers between these components are also inevitably associated with losses and, therefore, losses and efficiencies are not merely determined by animals and their nutrition, i.e. by FP (Jarvis *et al.*, 1996b). To complicate the analysis even further, nutrient inputs and outputs across the farm boundaries could be included as well. At the whole farm level, nutrients in meat and milk constitute the output (O) together with nutrients exported in crops when the farm is mixed. Inputs (I) consist of nutrients in feeds, bedding and manures when imported to the farm and of nutrients introduced via fertilizers, deposition and biological fixation. Agriculture can thus be seen as a chain of activities transferring nutrients in a cyclic way from: (i) feed; via (ii) animals and men; (iii) manure and by-products from society; (iv) soil; (v) crops utilized within the farm or exported; to (i) feed again. Hence, the spatial and temporal scale of efficient nutrient management is not confined to the level of herd and ration but extended to the farm, regional, national or even global scale (Fig. 9.2).

The overall functioning of ruminant production systems in terms of L, L/O and O/I within farm boundaries is, therefore, also determined by a set of conversion coefficients for the transfer of manure nutrients to soil (MS), from soil nutrients to nutrients in harvested crops (SC) and from nutrients in crops to what appears of them as feed (CF), in addition to the coefficient for conversion from nutrients in feed to nutrients animal products (FP) (Kohn *et al.*, 1997; Jarvis and Aarts, 2000; Schröder *et al.*, 2003a). All coefficients are dimensionless:

$$\frac{\text{kg}}{\text{kg}} \times \frac{\text{ha} \times \text{year}}{\text{ha} \times \text{year}}$$

Table 9.1 indicates the range of values observed for these four coefficients. The ranges reflect that it is easier to properly recycle P than the mobile element N. Differences in the coefficients FP, MS,

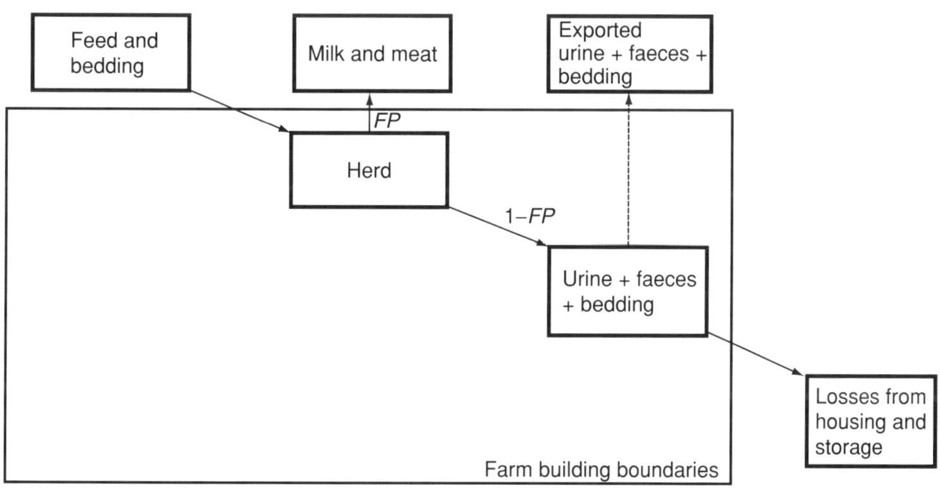

Fig. 9.1. Nutrient fluxes at the herd level.

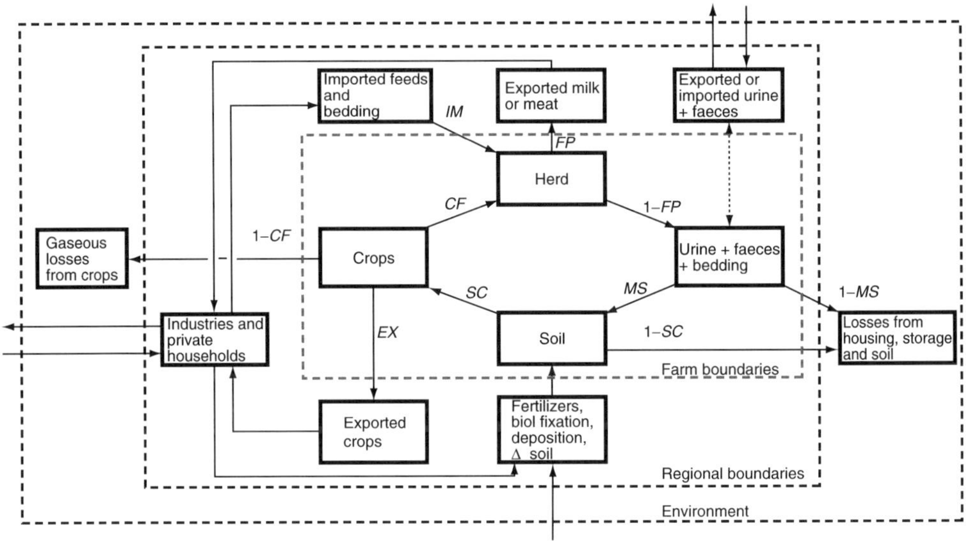

Fig. 9.2. Nutrient fluxes at the whole farm and regional levels.

SC and *CF* can be attributed to differences in the operational management skills of farmers and deliberately chosen tactics with regard to crop types, feed types, fertilizer types, animal types and grazing regimes, and the types of housing and manure handling equipment. In addition, factors that cannot be controlled on individual farms, and in some cases cannot be predicted, also contribute to the subsystem efficiencies (e.g. soil quality, weather).

The model depicted in Fig. 9.2 helps to evaluate the consequences of changes in the four key coefficients *FP*, *MS*, *SC* and *CF* for *O*/*I* and *L*/*O* (Kohn *et al.*, 1997; Schröder *et al.*, 2003a). The model further allows calculation of required inputs per unit area and losses per unit area (*L*, kg/ha) with additional information of either the targeted output per unit area or the attainable crop yield per unit area, or the other way around. The mathematical description of the model proposed by Schröder *et al.* (2003a) is also helpful in showing that *O*/*I*, *L*/*O* and *L* are not just determined by the operational skills of a farmer and his tactical decisions, but also by strategic decisions pertaining to the extent to which the farmer leaves the processing of crops to others (as in an arable farm) or the extent to which the livestock farmer opts for self-sufficiency concerning his feed and bedding supply. These aspects are characterized by the fraction of harvested nutrients being exported (*EX*) and the fraction of nutrients in feed and bedding material being imported (*IM*, Fig. 9.2). Note that losses taking place outside the farm remain concealed in the larger the values of *EX* and *IM*. Variation of the strategic decisions pertaining to *EX* and *IM* has considerable consequences for *O*/*I*, *L*/*O* and *L* as will be demonstrated. Hence, differences in these indices do not necessarily reflect differences in just the operational management skills.

Table 9.1. Indicative range of conversion coefficient values [(kg/ha/year)/(kg/ha/year)].

		Nitrogen		Phosphorus	
Coefficient	Step	Low	High	Low	High
FP	From feed and bedding to milk and meat	0.1	0.3	0.2	0.4
MS	From excreted manure to soil-applied manure	0.4	0.9	1.0	1.0
SC	From soil to harvestable crop	0.4	0.9	0.5	1.0
CF	From harvestable crop to available feed	0.7	0.9	0.9	1.0

The mathematical description of the whole farming system can be simplified by solving equations simultaneously to yield the following:

$$\frac{O}{I} = \frac{p}{q},$$

$$L = I \times \left(1 - \frac{p}{q}\right)$$

and

$$\frac{L}{O} = \frac{q}{p} - 1,$$

with

$$p = FP + \frac{EX \times (1 - IM)}{CF \times (1 - EX)}$$

and

$$q = IM + \frac{p - FP}{EX \times SC} - (1 - FP) \times MS$$

under the condition that

$$(1 - FP) \times MS\left(\frac{1 - IM}{CF \times (1 - EX) \times SC}\right)$$

assuring that the inputs through fertilizers + deposition + biological fixation cannot become less than zero in the diagram presented in Fig. 9.2. After some rewriting, the equations reveal that *O*/*I* equals a value of *SC* for a stockless arable farm (*EX* approaching a value of 1) using mineral fertilizer only (so $MS = 1$) and a value of

$$\frac{SC \times CF \times FP}{1 - SC \times CF \times (1 - FP) \times MS}$$

for a self-sufficient specialized livestock farm (*EX* and *IM* approaching values of 0).

To identify the scope for system improvement, analysis of the six underlying conversion coefficients is indispensable (Kohn *et al.*, 1997; Schiere and Van Keulen, 1999; Van Bruchem *et al.*, 1999a; Aarts *et al.*, 2000a,b; Schröder *et al.*, 2003a). Schröder *et al.* (2003a) also showed that the disintegration of mixed farms into specialized farms might lead to an apparent improvement of the nutrient use efficiency at the farm level, but this improvement may disappear when the efficiency is evaluated at a higher spatial scale. Van Noordwijk (1999) arrived at a similar conclusion. The major reason for this is that specialization leads to partial outboarding of subsystems and their associated losses.

Before making suggestions on how to stop leaks, i.e. how to maximize *FP*, *MS*, *SC* and *CF*, we first explore the consequences of their variation for *O*/*I*, *L* and *L*/*O* at the whole farm level and on a yearly basis.

9.2.2 Model explorations

For the sake of simplicity the explorations are first focused on self-sufficient specialized dairy farms (so *IM* and $EX = 0$) and initially address just N. Explorations start with a farm with moderately low conversion coefficients. Subsequently, we introduce improvements of these coefficients one by one and finally combine all improvements in an ideotypic farm. We have distinguished three situations, which may be valid for separate political entities. In Table 9.2 we fixed the crop N yield to an arbitrary level ('no policies'), in Table 9.3 we fixed the production ('milk quota system') and in Table 9.4 we fixed the loss per unit area ('environmental legislation').

Table 9.2 shows that especially improvement of the herd efficiency *FP* (from 0.15 to 0.25) promotes the output in milk and meat. Improvement of the crop efficiency *SC* (from 0.50 to 0.80) mainly reduces the loss per unit area. Simultaneous improvement of all four coefficients results in an increased production, a much higher output:input ratio and drastic reductions of both the loss per unit area and per unit output.

Table 9.3 shows that within a given output level (e.g. milk quota) improvement of the crop efficiency *SC* and, to a lesser extent, improvement of the herd efficiency *FP* are most effective in terms of loss and output:input ratio. It is the combination of improvements again which gives an even better performance of the system.

Table 9.4 shows that within given loss per unit area demands, improvements of separate coefficients hardly affect the N-input requirements. Improvement of the crop efficiency *SC* has a relatively strong positive effect on the output, i.e. the production of milk and meat. Combined improvements create room to apply considerable more N input, i.e. fertilizer, boost crop yield and maximize output, i.e. the production of milk and

Table 9.2. Nitrogen use efficiency, loss and milk + meat output at a whole farm level, as affected by improved conversion coefficients at a fixed crop yield level.

	Improvement directed at					
Coefficient	None	Herd	Manure	Crop	Harvest	All
FP (kg/kg)	0.15	0.25	0.15	0.15	0.15	0.25
MS (kg/kg)	0.75	0.75	0.95	0.75	0.75	0.95
SC (kg/kg)	0.50	0.50	0.50	0.80	0.50	0.80
CF (kg/kg)	0.80	0.80	0.80	0.80	0.90	0.90
IM (kg/kg)	0.0	0.0	0.0	0.0	0.0	0.0
EX (kg/kg)	0.0	0.0	0.0	0.0	0.0	0.0
N-crop (kg/ha)	250	250	250	250	250	250
N-input[a] (kg/ha)	373	388	339	185	357	152
N-milk and meat (kg/ha)	30	50	30	30	34	56
O/I[b] (kg/kg)	0.08	0.13	0.09	0.16	0.09	0.37
L[b] (kg/ha)	342	337	308	155	323	96
L/O (kg/kg)	11	7	10	5	10	2

[a]Fertilizer, deposition and biological fixation.
[b]*O*, nutrient outputs; *I*, nutrient inputs and *L*, nutrient losses.

Table 9.3. Nitrogen use efficiency, loss and required crop production level at a whole farm level, as affected by improved conversion coefficients at fixed production quota.

	Improvement directed at					
Coefficient	None	Herd	Manure	Crop	Harvest	All
FP (kg/kg)	0.15	0.25	0.15	0.15	0.15	0.25
MS (kg/kg)	0.75	0.75	0.95	0.75	0.75	0.95
SC (kg/kg)	0.50	0.50	0.50	0.80	0.50	0.80
CF (kg/kg)	0.80	0.80	0.80	0.80	0.90	0.90
IM (kg/kg)	0.0	0.0	0.0	0.0	0.0	0.0
EX (kg/kg)	0.0	0.0	0.0	0.0	0.0	0.0
N-milk and meat (kg/ha)	30	30	30	30	30	30
N-input[a] (kg/kg)	373	233	339	185	317	81
N-crop (kg/kg)	250	150	250	250	222	134
O/I[b] (kg/kg)	0.08	0.13	0.09	0.16	0.09	0.37
L[b] (kg/kg)	342	202	308	155	287	51
L/O (kg/kg)	11	7	10	5	10	2

[a]Fertilizer, deposition and biological fixation.
[b]*O*, nutrient outputs; *I*, nutrient inputs and *L*, nutrient losses.

meat. Note that O/I and L/O within scenarios do not differ in Tables 9.2–9.4, as logically follows from the formulation in Section 9.2.1. Whole farm N efficiency O/I increases from values around 0.10 to almost 0.40 kg N/kg N in the ideotypic farm, whereas the corresponding N loss per unit output (L/O) decreases from values around 5–10 to 2 kg N/kg N.

The extent to which a farm relies on feed produced outside the farm ($IM > 0$) has a notable impact on the loss and efficiency O/I, as illustrated in Fig. 9.3. Substitution of fertilizer-N by imported feed-N and bedding-N reduces the loss per unit area (L) and per unit output (L/O), and drastically improves the output:input ratio (O/I). In this example substitution took place up to a point where no other inputs than feed (including bedding) were used to compensate for the annual N outputs and losses: up to 62% of the feed-N was imported ($IM = 0.62$), given the adopted values of the con-

Table 9.4. Nitrogen use efficiency, milk + meat output and permitted input at a whole farm level, as affected by improved conversion coefficients at a fixed loss level.

	Improvement directed at					
Coefficient	None	Herd	Manure	Crop	Harvest	All
FP (kg/kg)	0.15	0.25	0.15	0.15	0.15	0.25
MS (kg/kg)	0.75	0.75	0.95	0.75	0.75	0.95
SC (kg/kg)	0.50	0.50	0.50	0.80	0.50	0.80
CF (kg/kg)	0.80	0.80	0.80	0.80	0.90	0.90
IM (kg/kg)	0.0	0.0	0.0	0.0	0.0	0.0
EX (kg/kg)	0.0	0.0	0.0	0.0	0.0	0.0
L[a] (kg/ha)	100	100	100	100	100	100
N-input[b] (kg/ha)	109	115	110	119	111	158
N-crop (kg/ha)	73	74	81	161	78	260
N-milk and meat (kg/ha)	9	15	10	19	11	59
O/*I*[a] (kg/kg)	0.08	0.13	0.09	0.16	0.09	0.37
L/*O* (kg/kg)	11	7	10	5	10	2

[a]*O*, nutrient outputs; *I*, nutrient inputs and *L*, nutrient losses.
[b]Fertilizer, deposition and biological fixation.

version coefficients. However, from a sustainability point of view there are limits to the substitution of fertilizer-N by feed-N. The reason for this is that the amount of feed-P associated with a certain amount of feed-N may exceed the herd and crop requirements substantially. This will lead to P accumulation and eventually P losses from the soil. To determine which *IM* level is permitted in order to prevent these P losses completely (so, $MS_P = SC_P = CF_P = 1$), the N/P ratio and attainable N yield of home-grown feed, the N:P ratio in exported milk and/or meat, and the surmised

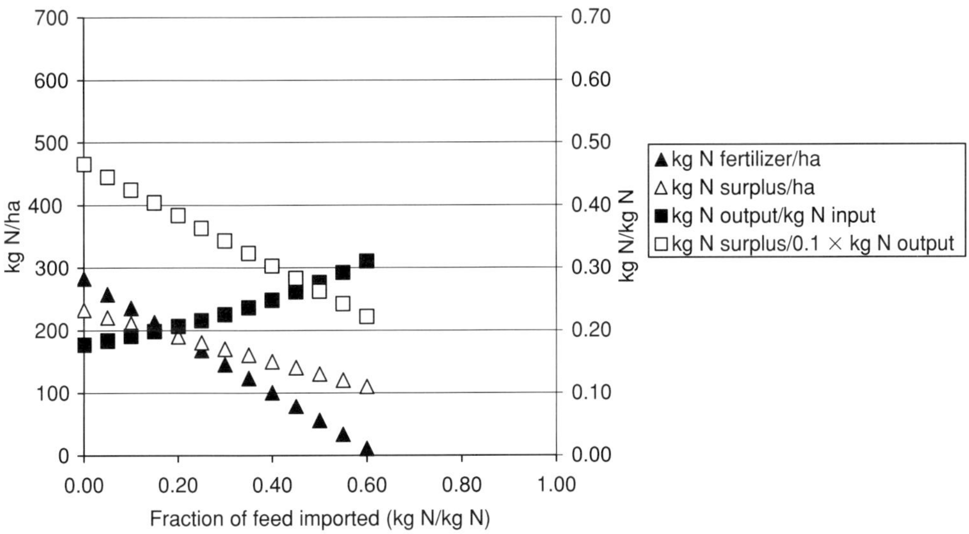

Fig. 9.3. Nitrogen (N)-fertilizer use per unit area, N loss per unit area, N loss per unit N-output and N-output per unit N-input, as affected by the fraction of the feed-N derived from imports (FP_N, MS_N, SC_N and CF_N were arbitrarily set at 0.20, 0.85, 0.65 and 0.85 kg N/kg N, respectively; N-output in milk and/or meat was fixed at 50 kg N/ha).

conversion coefficients of N and P must be taken into account when: (i) the permitted feed P imported (kg/kg milk and/or meat N output) = IMPU; and (ii) the permitted feed-N imported (kg/kg milk and/or meat N output) = IMNU, then

$$\text{IMPU} = \frac{1}{\text{N:P ratio in milk and/or meat output}}$$

and

$$\text{IMNU} = \frac{1}{FP_N} - \text{IMPU} \times \left(\frac{1}{FP_P} - 1\right) \times (\text{N:P ratio in home-grown feed}) \times CF_N$$

(with subscripts indicating whether the coefficient refers to N or P).

When: (i) the maximum attainable milk (and/or meat) N output (kg/ha) = MAON; (ii) the maximum attainable crop N (kg/ha) = MACN; (iii) the permitted feed P import (kg/ha) = IMP and the permitted feed-N import (kg/ha) = IMN, then

$$\text{MAON} = \frac{\text{MACN}}{\frac{1}{FP_N} - \text{IMNU}},$$

$$\text{IMP} = \text{IMPU} \times \text{MAON},$$

$$\text{IMN} = \text{IMNU} \times \text{MAON},$$

$$\frac{\text{IMN}}{\text{IMP}} = \text{required N:P ratio of imported feed},$$

$$IM_P = FP_P$$

and

$$IM_N = \text{IMNU} \times FP_N.$$

When the input of N via imported feed (including bedding) is thus restricted by the P import, the remaining N requirements can only be met by fertilizers or biological fixation in addition to the deposited N. The additional N requirement per unit milk and/or meat N output (ANU) through deposition, fertilizers or biological fixation, then equals:

$$\text{ANU} = \left(\frac{1}{FP_N} - \text{IMNU}\right) \times \frac{1}{SC_N \times CF_N} - \left(\frac{1}{FP_N} - 1\right) \times MS_N.$$

And so AN, the additional N requirement (kg/ha) equals:

$$\text{AN} = \text{ANU} \times \text{MAON}$$

This scenario's outcomes, based on these formulations, show that demands for self-sufficiency tighten considerably (i.e. lower *IM* values) in the presence of environmental P restrictions (Table 9.5).

Extending the fraction of crop-N exported ($EX > 0$) also has an apparent positive effect on the efficiency of farms with a fixed output of milk and/or meat nutrients per unit area home-grown feed. A smaller fodder area means fewer animals and manure. Fertilizer inputs hence increase. As

Table 9.5. Fraction of feed nutrients to be imported ('self-insufficiency'), associated phosphorus (P) loss and additional nitrogen (N) requirements through fertilizers, deposition or biological fixation, as affected by conversion coefficients of feed P into milk and/or meat P (N output in milk and/or meat fixed at 50 kg N/ha; FP_N, MS_N, SC_N and CF_N: 0.20, 0.85, 0.65 and 0.85 kg N/kg N, respectively; $MS_P = SC_P = CF_P =$ 1 kg P/kg P).

P-restrictions	No	No	Yes	Yes
FP_P (kg P/kg P)	0.25	0.35	0.25	0.35
N:P ratio in milk and/or meat output	5.5	5.5	5.5	5.5
N:P ratio in home-grown feed	9.0	9.0	9.0	9.0
Imported feed-N (kg/ha)	156	156	41	121
Imported feed-P (kg/ha)	24	14	9	9
N:P ratio in imported feed	6.5	11.4	4.6	13.3
IM_N (kg N/kg N)	0.62	0.62	0.17	0.48
IM_P (kg P/kg P)	0.66	0.53	0.25	0.35
P loss (kg P/ha)	15	5	0	0
Additional N requirements (kg N/ha)	0	0	208	64

the absolute losses associated with the processing of crops (*CF*) and manure (*MS*) decrease concomitantly, the loss drops and the output:input ratio increases. Note that the observed changes are not related to changes of conversion coefficients within the farm. The improvements can hence not be attributed to improved 'operational skills'. Effects are illustrated in Fig. 9.4 for a farm with an output of milk and/or meat of 50 kg N/ha of feed production.

9.3 Measures to Improve the Conversion of Nutrients Within Farms

9.3.1 From feed and bedding to milk, meat and manure

Our definition of *FP* in Section 9.2 included the conversion of bedding material. Bedding material has a negative effect on *FP* as it represents an input, which hardly leads to an output in milk or meat. However, it seems fair to say that the overall impact of bedding on *FP* is generally small, as the annual use of, for instance, 1 t of cereal straw per cow involves about 5 kg N and 1 kg P which is little compared to a typical annual feed input of 200 kg N and 30 kg P per cow. Therefore, no further attention will be paid to bedding in this section. Bedding can play a role, however, in the conversion of nutrients from manure to soil (*MS*) and from soil to crops (*SC*) and will therefore be treated in Sections 9.3.2 and 9.3.3.

The conversion of the nutrients N and P from feed to animal product can be improved, firstly, by increasing the level of production per animal, and secondly, by feeding N and P more closely to the requirements imposed by the level of production, thereby preventing the excretion of excess N and P with urine and faeces. Although most of the matter discussed in this section in principle holds for meat production as well, the remainder of this section focuses solely on dairy farming.

There are several measures to increase milk yield per animal and thereby achieve farm production levels with fewer cows. The most obvious measure is to improve the genetic potential for milk production, but also more technical measures are possible, such as the application of growth hormone, increasing the frequency of milking from twice to thrice a day or extending the photoperiod by artificial lightning. Simulation studies by Dunlap *et al.* (2000) demonstrated that all three technical measures together may increase the efficiency of N utilization of a whole herd (including growing heifers)

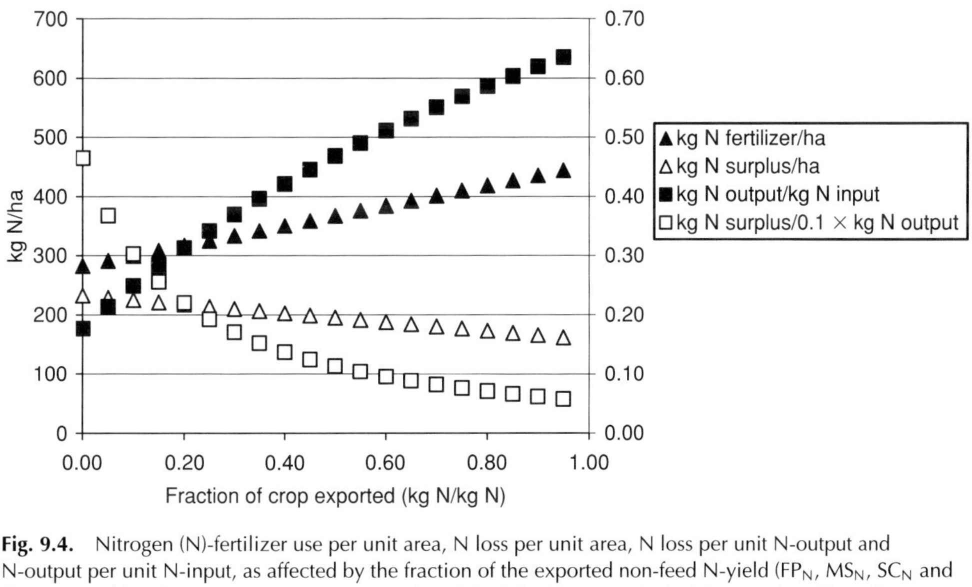

Fig. 9.4. Nitrogen (N)-fertilizer use per unit area, N loss per unit area, N loss per unit N-output and N-output per unit N-input, as affected by the fraction of the exported non-feed N-yield (FP_N, MS_N, SC_N and CF_N were arbitrarily set at 0.20, 0.85, 0.65 and 0.85 kg N/kg N, respectively; N-output in milk and/or meat was fixed at 50 kg N/ha per unit feed production area).

from a base value of 0.287 to 0.313, with increased calving interval having only a minor effect on these simulated figures. Nevertheless, if an increase in milk yield per cow is associated with a higher rate of replacement of lactating cows, and the consequent requirement of a larger number of young livestock, then the increase in efficiency of nutrient utilization may be partially compensated (Aarts *et al.*, 1999b). The quality of forage offered to non-producing offspring will often be near that offered to dairy cows for practical reasons of farm management. The quality offered to young livestock may thus well exceed the requirements and, therefore, the efficiency of retention with growth remains much lower than the retention with milk production by dairy cows. Nevertheless, there may be several reasons not to board out livestock and to maintain a high rate of replacement of dairy cows, such as the farmer's policy of transition to a herd with a higher genetic merit in order to reduce the number of cows, to compensate for the higher incidence of health and productive disorders, to minimize costs of veterinary services and so on. Whatever the precise argument of the farmer for a specific strategy of herd management (feeding, rearing, numbers, age and production groups) and total farm management (area available, forage production management, harvestable crops available, milk quota), it is without doubt that changes in *FP* of the whole herd cannot simply be equated to the change of the efficiency of milk production by dairy cows. If higher milk yields simultaneously require higher requirements of cow replacement, then there is a trade-off between the incremental effect of productive cows and the detrimental effect of higher numbers of non-productive offspring on *FP*. Although increasing milk yield per cow still seems to pay off, the effect may hence be smaller than indicated by an analysis of milking cow production data only.

Increased milk yields per cow require that a higher intake of metabolizable energy be achieved. To sustain higher levels of production with high-yielding cows, the possibilities to increase dry matter (DM) intake are limited. Normally it will be necessary to offer good quality diets with a higher energy density. This will only improve the conversion of N and P from feed to milk (*FP*), however, if these nutrients maintain to be fed close to requirement levels. In this respect the dietary content of N and P must be considered as the main determinant of the efficiency of N and P retention, even more than the production levels achieved per cow.

In the case of N, several measures can be taken to improve the conversion coefficient for N (FP_N). The first obvious measure that can be taken is changing the forage composition of the diet. Diets almost completely based on fresh herbage with a high CP content are highly digestible, and a relatively large fraction of absorbed N will not be utilized for animal production but excreted with urine as urea (Valk *et al.*, 2002). On the other hand, diets mainly composed of concentrates or low CP roughages (e.g. maize) may result in a much more efficient N retention (FP_N) up to levels just below 0.40.

In the case of P, a high phytase activity of rumen contents ensures that almost all P will flow into the small intestine in the inorganic form. Although feed-P is highly digestible in principle, the normal excretory route is through faecal excretion only. In contrast to N, almost no P is excreted with urine and, apart from P retained by the cow or in milk, all P absorbed from the gastrointestinal tract flows back to it via saliva production. Valk (2002) demonstrated how P excretion with faeces depends on P content in feed and the level of animal production. In lactating cows (in early-, mid- as well as late-lactation), efficiency of P retention (FP_P) in milk ranged from 0.42 to more than 0.55 on diets ranging from 3.4 to 2.3 g P/kg of feed DM. Most P in faeces will be in the inorganic form subject to P leaching after manure application. Once in the soil, this P will chemically interact with P in the soil matrix. For further considerations we refer to Chapter 7.

Another measure that can improve FP_N is to improve the balance between rumen digestible energy and N. Most current protein systems for ruminants calculate this type of balance in order to compare fermentable energy and N, as an indication of the efficiency of N utilization for microbial protein synthesis in the rumen. In addition to the high CP content of a fresh herbage diet, a relatively high quantity of rumen degradable protein (RDP) compared to the quantity of rumen digestible energy is another main cause of large quantities of feed-N being fermented to NH_3, absorbed from the rumen to blood and eliminated from the system as urea by urine production (Valk, 2002). Exchanging part of these diets by energy-rich compounds will capture more RDP in microbial protein which, after subsequent digestion, becomes available as MP. Relatively low quantities of RDP compared to rumen digestible energy strongly stimulate the reflux of urea from blood to the rumen (Dijkstra *et al.*, 1992), allowing micro-organisms to synthesize despite the low quantities of N delivered by the feed. On such diets the efficiency of N utilization by a lactating cow is maximized. Hence, it can be concluded that manipulation of the quantity of RDP that is not incorporated by microbial synthesis may improve *FP* drastically.

In order to demonstrate the importance of the nutritional measures discussed above, some results concerning N utilization by dairy cows from our own research will be presented and compared with the general relationships proposed by Kebreab *et al.* (2001). The latter authors established empirical relationships between N intake, secretion of N in milk and N excretion with urine and faeces derived from trials with lactating cows on 30 different diets in five trials. These empirical relationships have been depicted in Figs 9.5–9.7 as solid lines, and were evaluated against 90 observations from our own research obtained for 16 diets in seven feeding trials.

Figure 9.5 shows that our observations of faecal N excretion, as a function of daily N intake, are slightly underestimated by the empirical linear relationship (Faecal N $= 76.7 + 0.16 \times$ N intake, in g/day, $r^2 = 0.30$) of Kebreab *et al.* (2001). In congruence with the observations of Kebreab *et al.* (2001), the variation in faecal N excretion at a certain N intake can be extremely large. We estimated the upper boundary of faecal N excretion to be 250 g/day (as values up to 300 are considered unrealistically large and probably result from the measurement technique).

Kebreab *et al.* (2001) propose an exponential response of urine N excretion to daily N intake (urine N $= 0.003 \times$ N intake$^{1.8}$, in g/day, $r^2 = 0.67$). Figure 9.5 shows that this prediction slightly overestimates our data. Although the variation was very large again, similar to the large variation found by Kebreab *et al.* (2001), it can be concluded that higher intakes of N are associated with larger amounts of N being excreted with urine.

Our data of N retention in milk as a function of N intake could not confirm the positive linear relationship (Milk N $= 38.2 + 0.19 \times$ N intake, in g/day, $r^2 = 0.30$) of Kebreab *et al.* (2001) (Fig. 9.6). It must be noted that our data set included diets beyond intakes of 600 g N/day, whereas the prediction of Kebreab *et al.* (2001) was based on the range from 300 to 600 g/day. Within the latter range we also observed a positive linear response. Predicted and observed values of milk N production remained in the same range from just below 100 to 175 g N/day.

From a viewpoint of gaseous N losses, it is of considerable importance to distinguish between urine and faecal N excretion, because the former

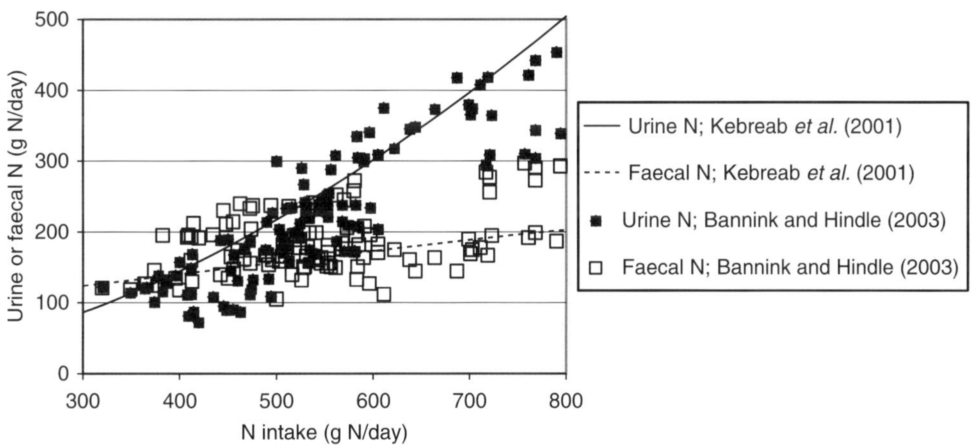

Fig. 9.5. Relationship between nitrogen (N) intake (in g N/day) and N excretion with urine and faeces (both in g N/day); curves indicate the relationships established by Kebreab *et al.* (2001), whereas individual data originate from an evaluation data set of Bannink and Hindle (2003).

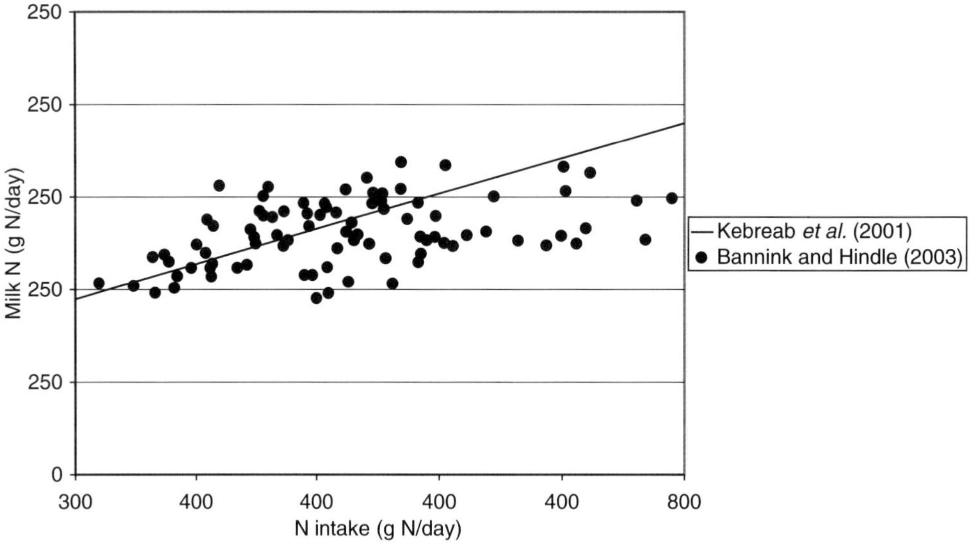

Fig. 9.6. Relationship between nitrogen (N) intake (in g N/day) and N excretion with milk (in g N/day); curve indicates the relationships established by Kebreab *et al.* (2001), whereas individual data originate from an evaluation data set of Bannink and Hindle (2003).

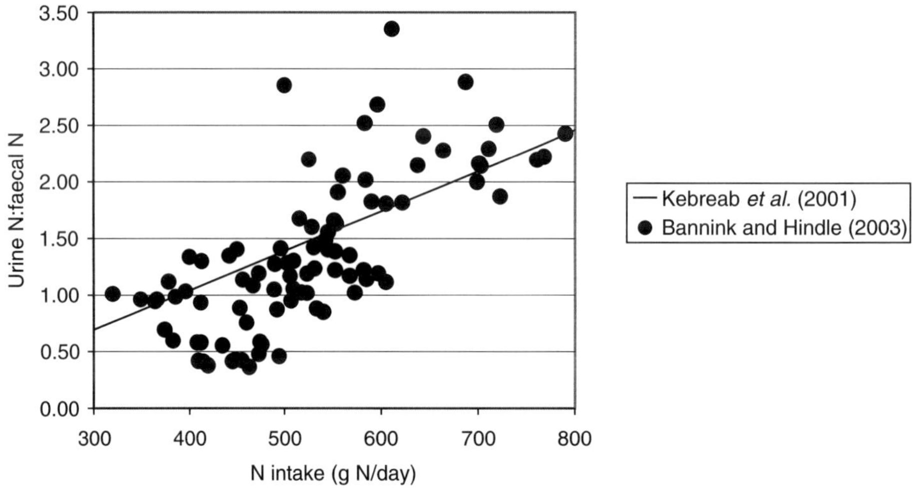

Fig. 9.7. Relationship between nitrogen (N) intake (in g N/day) and ratio of urine N:faecal N excretion; curve was calculated from the relationships established by Kebreab *et al.* (2001), whereas individual data originate from an evaluation data set of Bannink and Hindle (2003).

is much more susceptible to volatilization (Monteny and Erisman, 1998) and, therefore, the ratio of urine N to faecal N may be considered as a determinant for the coefficient of N loss from urine and faeces (1−*MS*). In Fig. 9.7 our data are compared with the ratio derived from Kebreab *et al.* (2001). Although this relationship overestimates the ratio of urine N:faecal N of our data, in particular at lower intake rates as expected from Figs 9.5 and 9.7, it confirms the positive response of the ratio to N intakes. Our data show that the variation of the ratio increases with intake level, ranging from values below 1.0 to values above 3.0.

Apparently, the relationships in Figs 9.5–9.7 are not very close and other nutritional factors, besides the rate of N intake, must have caused a large variation in the partitioning of N. Anyhow, excretion of N with faeces is much less affected by dietary changes than is excretion with urine. Besides, its dependence on the digestibility of the N-containing components in the diet will be determined largely by the quantity of feed DM ingested. Fermentation in the large intestine may become stimulated with the arrival of considerable quantities of fermentable matter (e.g. feeding large quantities of potato starch or beetpulp). In such extreme cases, considerable quantities of N may be incorporated into microbial matter and be excreted with faeces instead of urine. With practical diets, these effects remain rather small, however, and cannot explain the large variation in urine: faecal N ratios as shown in Fig. 9.7. Although positive effects of N intake on N excretion with urine and on urine:faecal N ratio are to be expected when the full range of N intakes is considered (300 to 800 g N/day), such effects may become completely overruled by other nutritional factors with a more narrow range of N intake (e.g. 400 to 500 g N/day). From the wide range of this ratio it also becomes clear that dietary manipulations may have an considerable impact, not only on *FP* (and $1 - FP$), but also on NH_3 emission rates and hence on the coefficient $1 - MS$ (and *MS*).

In addition to the nutritional measures, discussed above, other management measures may be undertaken to improve the efficiency of N utilization. First, the requirement for MP depends on the level of production that is achieved. Managing separate groups of cows which are in a different stage of lactation (groups for different levels of production) may allow for group-specific manipulation of the diet and for feeding closer to cow requirements.

A technical measure that can be implemented in farm management is the evaluation of milk urea data. Milk urea data together with other farm characteristics may be used to decide what measures can be taken to improve the N efficiency (Hof *et al.*, 1997; Jonker *et al.*, 1998; Schepers and Meijer, 1998). Milk urea content may be considered an indicator of average urea concentrations in blood, because urea diffuses from blood to milk, and hence of absorbed, non-retained N which needs to be excreted. Therefore, differences in rumen imbalance between energy and N, or an oversupply of absorbed N relative to cow requirements for the production of milk protein, will be reflected in elevated milk urea concentration. In this way, milk urea also may become a useful indicator of NH_3 emission rates that are to be expected with a certain nutritional management (Smits *et al.*, 2001). Milk urea was found to correlate closely to urine N excretion and NH_3 emission from cow houses, particularly caused by NH_3 emission immediately following urination on a wet urine and faeces fouled floor ($1 - MS$). These N losses with urination may account for a large part of all gaseous N losses at a dairy farm, its contribution very much depending on the type of farm management and the techniques adopted for storage and subsequent application of manure (Smits *et al.*, 2001).

In balance trials with lactating cows on a variety of diets mainly composed of fresh herbage, strongly varying in herbage-N content (Valk, 2002), the amount of N excreted with urine and urea-N concentrations in milk ranged from 132 to 393 g/day and 73 to 302 mg/l milk, respectively. Variation in urine N should be accounted by changes in the quantity of urea excreted, as the other nitrogenous compounds in urine remain more or less constant (Van Vuuren and Smits, 1997; Valk, 2002). Considering a minimum fraction of urea-N in total urine N of 50% (Valk, 2002), the approximately fivefold change in the quantity of urea excreted with urine corresponded reasonably well to the fourfold increase in milk urea-N.

Data from the trials presented in Figs 9.5–9.7 were also used by Bannink and Hindle (2003) to analyse the relationships between milk urea and other milk parameters and urine N. The following relationships could be established:

$$\text{N intake} = 129.4 + 1275 \times (\text{milk urea-N}) + 3.16 \times \text{milk N} + 8.19 \times \text{milk},$$

with milk urea-N in g/kg, both N intake and milk N in g/day and milk in kg/day; n=90; $r^2 = 0.846$; RSD=38.6.

$$\text{Urine N} = 44.7 + 1349 \times (\text{milk urea-N}) + 1.91 \times \text{milk N} - 10.87 \times \text{milk},$$

with milk urea-N in g/kg, both urine N and milk N in g/day and milk in kg/day; $n = 90$; $r^2 = 0.846$; RSD = 38.6.

$$\text{Urine N and faecal N} = 124.0 + 1320 \times (\text{milk urea-N}) + 1.87 \times \text{milk N} - 6.90 \times \text{milk},$$

with milk urea-N in g/kg, urine N, faecal N and milk N in g/day; milk in kg/day; $n = 90$; $r^2 = 0.800$; RSD=43.2.

These relationships suggest that the milk urea content may be a useful indicator of the quantities of N excreted with either urine, or with urine and faeces, and even of the quantity of N consumed. This implies that there is also potential for the use of data on milk urea content to analyse the situation at any dairy cattle operation with regard to the efficiency of N utilization (*FP*), the rates of N excretion by cows ($1 - FP$) and the potential impact on NH_3 emission rates ($1 - MS$). Although these results are not directly applicable for analysis on a whole-herd basis, they do give insight in the potential impact that nutrition has on N retention, and on N excretion and subsequent N emission.

In addition to the quantity of urea excreted, the volume of urine produced is the second determinant of urea concentration in urine, and hence may have an effect on rates of NH_3 emission. Urine volume was demonstrated to vary strongly with the quantities of sodium (Na), potassium (K) and N excreted with urine (Bannink *et al.*, 1999). Forage-based diets normally have a high K content well in excess of the requirement by a lactating cow, and because both Na and K are highly absorbable, large quantities of K and perhaps also Na will be excreted with urine (McDowell, 1992). This factor is a major determinant for the volume of urine produced daily, and generally seems to overrule the effect of other factors like ambient temperature and humidity, or luxury consumption of water. Therefore, a lower urea concentration because of a larger urine volume reduces the rate of NH_3 emission from a single urination. Nevertheless, increased urine volume will have a less proportional impact on NH_3 emission as more urine results in more frequent urination and more frequent refreshment of emitting puddles of urine on areas fouled with faeces (Monteny and Erisman, 1998; Smits *et al.*, 2001).

In principle, the cation anion difference in ruminant diets also may influence NH_3 emission from excreted urine ($1 - MS$) by its effect on urine pH (Oenema *et al.*, 2001). Urine pH changes from alkaline to more acidic values below a dietary cation anion difference (calculated as $K + Na - Cl - 2 \times S$) of $+100$ meq/kg feed DM; the change is strongest around a difference value of zero (Bannink and Van Vuuren, 1998). As discussed above, under practical conditions with forage-based diets with a high K content, the dietary cation anion difference will remain too positive and urine pH too alkaline to let dietary manipulation have any effect on urine acidity. Lower values are achievable though when forages high in K are replaced by maize or maize silage, for example, and when mineral salts high in sulphur or chloride relative to K and Na are included in the diet. Then, a strong reduction of NH_3 emission rates from urine may be expected, as it has already been demonstrated in numerous feeding trials with pigs (Canh *et al.*, 1998).

Summarizing, it may be concluded that dietary measures can have a large impact on rates of NH_3 emission ($1 - MS$), particularly immediately following urination which seems to be the main contribution to the total NH_3 emission from Dutch cattle operations (Monteny and Erisman, 1998; Smits *et al.*, 2001; conditions characterized by coverage of manure storages, manure pits, semi-closed cow housing). When considering the nutrient retention (*FP*) of the whole herd, not only the nutritional and production aspects on producing cows should be considered, but that of the relatively low-productive offspring as well.

9.3.2 From excreted to soil-incorporated manure

The efficiency by which nutrients in manure are returned into the soil pool (*MS*) depends on the magnitude of gaseous N losses from stables and storage facilities, and on losses from manure application and grazing. Loss in the form of NH_3 is the most important one. Ammonia losses may vary from less than 5% (so $MS = 0.95$) to 50% ($MS = 0.50$) of the amount of excreted N (Van der Molen *et al.*, 1990; Bussink and Oenema, 1998; Monteny and Erisman, 1998; Huijsmans and De Mol, 1999; Schils *et al.*, 1999; Aarts *et al.*, 2000b). Losses due to denitrification (N_2, N_2O, NO) appear limited in this section of a ruminant production system, relative to the losses associated with the soil component (Velthof and Oenema, 1997; Velthof *et al.*, 1998; Chadwick *et al.*, 1999).

Ammonia losses can be reduced by measures that lower the NH_3 content of manures in the first place. Examples of such measures are the addition of materials low in N/high in carbon (C) to either the animal diet (Smits *et al.*, 1995, 1997; Paul *et al.*, 1998; Van Bruchem *et al.*, 1999b) or to the excrements via bedding material (Berry *et al.*, 2002; Chadwick *et al.*, 2002).

Dietary measures aiming at an increase of *MS* from land-spread manure or manure excreted during grazing were extensively reviewed in Section 9.3.1. Note that for a sufficiently large reduction of the ammoniacal N content in manure to make any subsequent measure redundant, rations may have to become so poor in N that the farm income will be compromised. Data presented in Fig. 9.7 suggest that milking cows taking in as little as 300 g N/day may still produce excrements with urine N:faecal N ratio of 0.5–1.0, i.e. of which 33–50% of the excreted N is ammoniacal.

Maize silage is commonly used to reduce the unnecessary high N concentration of well-fertilized grass and/or legume-dominated rations. As a consequence, *FP* may increase, and the amount of excreted urine N and, hence, ammoniacal N will drop. However, the intended improvements of *FP* and *MS* may in this case have a trade-off in terms of *SC*, as the *SC* value of arable crops including maize is generally lower than that of permanent grassland (see Section 9.3.3 and Prins *et al.*, 1988; Schröder *et al.*, 1997a; Ten Berge *et al.*, 2000b). When materials low in N/high in C are added to the excrements instead of the diet (i.e. when solid manure instead of slurry is produced) the intended increase in *MS* in the field can be counteracted by a decrease of *MS* on the yard due to the losses resulting from regular turning of manure heaps. As much as 40% of the initial N content of solid manures can be lost in this way (Dewes, 1995; Bokhorst and Ten Berg, 2001; Berry *et al.*, 2002). Moreover, when manures are spring-applied, crop demand is generally better matched with supply by liquid manures than by solid manures, even if the long-term residual N effect of the solid manure is taken into account (Berry *et al.*, 2002; Schröder, 2002). Hence, even if *MS* would be improved by the production of solid instead of liquid manures, opting for solid manure may carry a price in terms of *SC*.

Slatted instead of solid floors represent another measure to reduce the NH_3 losses from stables because slats reduce the effective surface and allow urine and faeces to be removed quickly to a covered storage. Chemical treatment, especially acidification (Schils *et al.*, 1999), and mechanical treatment (e.g. the removal of urine and dung from stable floors with scrapers and sprinklers) have also been proposed as emission-reducing measures. Coverage of the outdoor manure storage also significantly contributes to the reduction of NH_3 volatilization. Obviously, a covered storage would be a questionable investment in cattle operations that recycle lagoon liquid for barn flushing or spread slurry through spray irrigation without incorporation; volatilization abatement needs to be attended throughout the whole cycle (Fig. 9.2).

The more the animals are allowed to graze (per day, per year), the smaller the fraction of the total excretion collected indoors. The relative importance of some of the previous measures to reduce NH_3 volatilization ($1 - MS$) therefore depends on grazing regimes. Grazing itself is also associated with NH_3 losses from urine and dung patches. It is the combination of factors such as housing type, handling method of collected manure and control over diet composition which determines whether the intended reduction of NH_3 losses is better served by unlimited grazing than by zero grazing. Effects of indoor and outdoor measures on the conversion coefficient *MS*, including the effects of housing, dietary changes and grazing, were extensively reviewed by Monteny and Erisman (1998) and Bussink and Oenema (1998). Recent studies in the Netherlands indicate that grazing generally reduces the overall NH_3 losses from dairy systems, relative to situations in which the herd is confined in housings (Smits *et al.*, 2001). However, grazing can have drawbacks in terms of the conversion coefficient *SC* (see Section 9.3.3.1).

Substitution of surface application of manure by injection is another measure to reduce NH_3 losses and thus increase *MS*. This is especially relevant for manures rich in ammoniacal N such as slurries and urine (Van der Meer *et al.*, 1987; Bussink and Oenema, 1998; Huijsmans and De Mol, 1999). Most farmers contract out injection. The heavy equipment used by contractors may be detrimental to the sod and soil quality. The intended improvement of *MS* may in that case carry a price in terms of *SC*. Table 9.6 gives a summary of Dutch research on NH_3 losses in relation to the application technique, indicating that losses decrease with incorporation depth. Placement at

Table 9.6. Ammonia volatilization losses (in % of the ammonia present in manure) from land spreading on arable land and grassland, as affected by the application technique (Huijsmans, 1999).

		Volatilization	
Land use	Technique	Observed range	Average
Arable	Surface application	20–100	68
	Surface application followed by incorporation	1–49	20
	Tine injection	0–40	9
Grassland	Surface application	27–98	68
	Trailing feet injection	8–50	26
	Sod injection	1–25	10
	Tine injection	0–3	1

a high depth, however, does not benefit *SC* values (see Section 9.3.3.1).

This paragraph may have helped to underline that conversion coefficients cannot be changed independently, as strong interactions among farm components do exist. Therefore, evaluations of the impact of separate measures should preferably include the whole farm picture.

9.3.3 From soil-incorporated manure to harvestable crops

9.3.3.1 Fertilizer management strategies

NATURE OF FERTILIZER. First, the value of *SC* is affected by the nature of the fertilizer used. As mineralization from organically bound nutrients usually takes more time than just 1 year, residual effects from manure and other organic inputs may accumulate over time (Magdoff and Amadon, 1980; Sommerfeldt *et al.*, 1988; Dilz *et al.*, 1990). Hence, it takes time before *SC* values associated with organic fertilizers approach the values which are attainable with mineral fertilizers. This also implies that there are long-term consequences of reduced N and P rates, as crops may benefit for many years from soil fertility built up in the past (Wolf *et al.*, 1989; Motavalli *et al.*, 1992; Whitmore and Schröder, 1996). Therefore, *SC* values and hence the nutrient loss and the output per unit input at the whole farm level should be interpreted with great caution when referring to enterprises that have recently adopted a low-input strategy.

In the USA and Europe coarse guidelines are available to help farmers credit the N contribution from manure. However, many advisers only account for the N available in the first season following manuring. The complexity of the matter makes farmers reluctant to refrain from 'insurance' N applications and this may have a negative impact on the environment. Reliable *a priori* estimates based on soil analysis are difficult to make (Jarvis *et al.*, 1996a; Powlson, 1997) and a deliberate postponement of N dressings until after emergence, to better account for the mineralizable N, is unfortunately not always a strategy without risks (Schröder *et al.*, 2000). Even when the long-term N fertilizer value is correctly accounted for, the *SC* value of organic inputs may not be as high as that of mineral fertilizers. The reason is that mineralization may partly occur outside the growing season of crops. Note that the coefficient *SC* integrates more than just one underlying factor; *SC* is the product of the fertilizer value of organic fertilizers relative to mineral fertilizer, the nutrient recovery index of a crop and the nutrient harvest index (Schröder, 2002). In reasonably managed permanent grasslands and arable rotations long-term *SC* values for organic fertilizers of 0.4–0.8 for N and 0.7–1.0 for P appear attainable. Corresponding values for mineral fertilizers can be as high as 0.5–0.9 (N) and 0.8–1.0 (P).

RATE. The utilization of mineral fertilizer and manure can be enhanced by the right time, the right place and certainly the right amount. Incorrect decisions, in particular excessive rates, negatively affect the nutrient recovery and, therefore, *SC* (Greenwood *et al.*, 1989; Schröder *et al.*, 1998; Ten Berge *et al.*, 2000b). Prior knowledge of the composition of manure and an accurate estimate of crop requirements are

Table 9.7. The ratio of nitrogen (N) and phosphorus (P) in crops and manures (based on data in Lammers (1983), Beukeboom (1996) and Beijer and Westhoek (1996)).

	When based on		
		Effectively available N/P	
Product	Total N/P	Short-term	Long-term
Maize grain	4.8		
Wheat grain	5.6		
Maize whole crop silage	7.4		
Grass	8.5		
Legumes	9.7		
Slurry, cattle	6.2	2.1–4.1	2.7–5.3
Farmyard manure, cattle	4.1	1.4–2.1	2.3–3.4
Farmyard manure, goats	3.7	1.4–2.1	2.1–3.4

instrumental to an accurate determination of site-specific application rates and the consequential utilization of manure and fertilizer. Estimates of crop requirements may be too high when recommendations are based on uncritically chosen regression models (Cerrato and Blackmer, 1990; Bullock and Bullock, 1994; Stecker *et al.*, 1995; Schröder *et al.*, 1998) or when inspired by expected yield levels only (Vanotti and Bundy, 1994a,b; Sims *et al.*, 1995; Schlegel *et al.*, 1996; Schröder *et al.*, 1998). Moreover, N and P can be saved and *SC* increased by a realistic instead of pessimistic estimate of the N that is to be released from former organic inputs.

It is generally impossible to meet the N requirements of the crop rotation as a whole with manure only, if excessive applications of P are to be avoided. The reason for this is that the N:P ratio in most crops averages >6, whereas the ratio of effectively available N and P in manures is generally <5 (Table 9.7). Hence, sustainable production systems (i.e. in which P inputs do not exceed P outputs) are always short of N. This relative N deficiency must be met by the presence of legumes in the rotation (grass clover leys, maize–lucerne rotation) or by mineral fertilizer N supplementation. In order to adjust these supplements to the actual mineralization and crop N requirements, indicator-based spoon-feeding strategies can be useful to achieve high *SC* values (Schröder *et al.*, 2000).

TIME. In temperate climates, N should be water-soluble during the growing season to be available to plants, whereas it should be organically bound from late summer to spring in order to avoid losses. Manures differ strongly in the ratio of water-soluble and organically bound N. Hence, manure ideotypes depend on the soil type-imposed time windows for spreading. Manures with a low water-soluble N:organically bound N ratio (farmyard manures) are generally better suited for autumn applications than manures with a high ratio (slurries) (Smith and Chambers, 1993). When applied in spring, farmyard manures may not be able to meet the crop requirements in time, especially not when the associated bedding (e.g. cereal straw) is not yet fully decomposed and hence tends to immobilize soil mineral N (Berry *et al.*, 2002). From late summer-applied slurries, on the other hand, considerable amounts of mineral N can be lost unless this N is timely sequestered with cover crops and successfully transferred to the subsequent spring (Schröder *et al.*, 1997a). Storing manure until spring appears to be the safest way to go in any climate with cool and wet winters, despite a slightly larger $(1 - MS)$ that may result from longer storing. So, when the soil type permits field traffic in early spring, postponement of applications benefits *SC* (Van Dijk, 1985; Unwin *et al.*, 1986; Schröder *et al.*, 1993; Van der Meer and Van der Putten, 1995). Addition of a nitrification inhibitor to autumn-applied slurry may increase *SC* but is not always considered as effective as spring application. The utilization of spring-applied slurry can also be further improved by the addition of a nitrification inhibitor (Schröder *et al.*, 1993; Van der Meer and Van der Putten,

1995). These observations indicate that even spring-applied manure can be exposed to N losses whenever crops take up little N until late spring.

Synchronization may be further improved when the application of manure is partly postponed until after crop emergence (Beauchamp, 1983). Such a deliberate N splitting is common in grassland but not without risks in annual crops such as maize (Schröder *et al.*, 2000). Especially when manure is the source of N, the equipment involved may damage the soil structure and/or crop and compromise *SC*. Moreover, incorporation is needed to avoid NH_3 volatilization as indicated in the previous subsection. In a standing crop, incorporation may damage the root system. Crop damage can be minimized, though, by side-dressing in-between the plant rows. From a plant nutrition perspective, however, this position can be less effective as root length densities between the rows can remain low for a long time, in particular when row distances are wide (Schröder *et al.*, 1996a). Therefore, *SC* and the consequent crop yield do not always benefit from slurry splitting. Benefits may be more evident in the case of mineral fertilizers and at low levels of N availability (Schröder, 1999; Schröder *et al.*, 2000).

So far, this subsection has addressed synchronization problems and associated reductions of *SC* at the start of the growing season. Lack of synchronization between supply and demand can also occur at the end of the season, however. By nature, N mineralization from manure continues beyond the period in which arable crops take up N. That N is redundant and can be lost when the arable crop is followed by a long fallow period after its harvest. The establishment of a catch crop can thus help to save N and transfer it to the next growing season. Unfortunately, crops leaving considerable soil mineral N residues are often harvested relatively late in the season. This implies that the N sequestration potential of catch crops is strongly determined by weather conditions; nevertheless, the utilization of manure-N can be enhanced slightly by catch cropping (Schröder *et al.*, 1996b, 1997a). The role of a catch crop is to recycle residual N and thus improve the apparent recovery of slurry-N in subsequent crops. It must be emphasized that catch crops only improve the utilization of manure if the subsequent crop is in need of N. Hence, N mineralizing from catch crop residues must be correctly accounted for in the fertilizer management of subsequent crops. If not, catch cropping may eventually become a merely cosmetic measure.

PLACE. *SC* is not only determined by the rate and timing of manure and fertilizer applications but also by their place of application. This applies alike to (mobile) N and (immobile) P. Placement has both a vertical and a horizontal component. As for the vertical aspect in the case of N, the reduction of NH_3 losses is a major consideration. This consideration was treated in Section 9.3.2 and led to the recommendation to incorporate manure one way or another, but never too deep.

As for the horizontal aspect of placement, proper attention must be given to spreading techniques. Irregular, patchy spreading patterns increase the heterogeneity of soil fertility. Consequently, some parts of the field become over-fertilized, whereas other parts may become deficient. The application of urine and dung via grazing may conflict with various recommendations presented in the previous sections: rates that may be locally far too high, exposure of the excretions to NH_3 volatilization, and poor timing when grazing is extended into the end of the growing season.

The previous warning for a too patchy distribution of manure must not be seen as a general argument for a uniform distribution. In crops with a wide row distance *SC* of both N and P may benefit from techniques that apply manure or fertilizers close to the anticipated position of the crop roots (Sawyer *et al.*, 1991; Schröder *et al.*, 1997b; Van Dijk and Brouwer, 1998).

SOLID OR LIQUID? In previous paragraphs it has already been stated that solid manure can be a sensible choice when late summer or autumn provide the only time-windows for application. Accurate placement, however, is better served with liquid manures. By nature, the ratio organic matter/N is generally much higher in solid manures than in liquid manures. Solid manures may hence fit well in farming systems that are in need of additional organic matter. However, the N:P ratio in solid manures is generally much lower in solid than in liquid manure, as a consequence of losses associated with the storage and handling of solid manures. One may argue that NH_3 in slurries can be lost easily too, be it on the field instead of the yard. However, relatively simple measures can be taken against NH_3 volatilization from land spreading, as indicated in Section 9.3.2.

On farms of similar intensity, the use of solid manures is associated with (much) smaller yearly applications of directly available N (i.e. ammoniacal N) and larger applications of organically bound N. A larger (equilibrium) supply of soil organic N and more abundant soil life result from this. Consequently, the yearly mineralization is also larger, in accordance with the increased soil organic N supply. These phenomena are sometimes twisted around as if the soil life abundance itself is the cause of the enhanced mineralization. Until now, there is no convincing evidence that the promotion of soil life through the use of farmyard manure contributes to a better recovery of N by crops (Langmeier *et al.*, 2002). Therefore, it seems fair to say that solid manures with their inherent lower N:P ratio need larger N supplementation via either legumes or mineral N fertilizers. The only alternative for that would be to over-apply solid manure in terms of P, but that would not be sustainable.

Preferences for farmyard manure are also inspired by the desire to provide proper bedding for animals. Research directed at the reconciliation of efficient N use and animal welfare is therefore urgently needed.

9.3.3.2 *Crop effects*

The efficiency by which soil nutrients are converted into harvestable crop nutrients (*SC*) is not merely a function of fertilizer management (Section 9.3.3.1), but depends also on crop choice, on crop rotation and on cutting regimes, as crops and associated management differ in their ability to recover nutrients from the soil and allocate them to harvested organs (Prins *et al.*, 1988; Greenwood *et al.*, 1989; Schröder *et al.*, 1996c, 1997a; Vos and Van Der Putten, 2000). According to the present definition, *SC* integrates the nutrient recovery index of a crop and the nutrient harvest index. Hence, 'crop effects' include deciding on which fraction of the crop to harvest and at what growth stage. Just harvesting maize ears instead of the whole crop, for instance, reduces *SC* although it may benefit *FP* and *MS* due to a better digestibility, energy supply in the rumen and consequent utilization of consumed N.

The present definition of *SC* also takes account of harvest losses either by the animal (e.g. trampling damage) or the machine. Avoidance of these losses increases *SC*. Finally, it must be reemphasized that subsystems may strongly interact. For example, crop selection can have a notable impact on N and P uptake by crops or legume fixation of N. This in turn also affects the composition of rations and, therefore, the digestion and metabolism of nutrients (*FP*) by animals.

9.3.4 From harvestable crops to edible feed

The efficiency of converting harvestable crop nutrients into nutrients in feed and bedding material (*CF*) depends on the extent to which losses associated with the senescence of standing crops and processes after cutting, i.e. wilting, conservation and feeding have been avoided. These losses mainly apply to N. Especially crops unnecessarily rich in N may emit N in this section of the ruminant production system. Conservative use of manures and fertilizers during periods of the year when long wilting periods are likely may contribute to a reduction of these losses. Conservation of this kind of crop material in combination with chopped cereal straw may be another measure to increase *CF*. Crop management and favourable weather patterns also reduce harvest N losses.

9.4 The Case of De Marke

Section 9.3 reviewed factors affecting the conversion coefficients, as observed in experiments. It also tried to address some of the positive and negative interactions between coefficients. In Section 9.2, the consequences of the variation of the coefficients at the whole farm level were explored by means of a model. As unexpected interactions may exist between measures taken in the various components of a farm, our fractional knowledge requires further experimental validation, based on real life cases. For exactly that reason the experimental dairy farm De Marke was started in 1989 in the Netherlands (Aarts *et al.*, 1992, 1999a). Progressive farmers and researchers wanting to address societal demands initiated De Marke. The ongoing project is supported by the Dutch ministries of agriculture and environment and the National Dairy Board. The farm is located on a sandy soil where groundwater as well as air quality is a political issue and where reconciliation of economy and ecology is considered difficult.

The experimental farming system was designed to achieve an annual milk production of 11,600 kg/ha. Management of De Marke is directed at improving the conversion of excreted nutrients into harvestable crop nutrients (*MS* and *SC* in Fig. 9.2) and of crop nutrients into milk (*CF* and *FP* in Fig. 9.2). Compensation of the annual nutrient outputs and losses is achieved at De Marke by imported feed rather than mineral fertilizer inputs (IM = 0.26 kg N/kg N and 0.30 kg P/kg P). This incomplete self-sufficiency on feeds made it easier to increase the nutrient efficiency to required levels than when De Marke had been self-sufficient. However, comparable commercial farms have an even higher level of imported feed ($IM = 0.30$ kg N/kg N and 0.43 kg P/kg P).

At De Marke, cattle consume lower quantities of N and P with feed than at comparable commercial farms. This results in a relatively high conversion of feed into milk and meat (*FP* in Fig. 9.1). The improved *FP* value implies that the animals excrete less N. Subsequently, the smaller amounts of manure N and P are efficiently converted into harvestable crop nutrients through a combination of restricted grazing, a better timing of manure spreading and adjustments in the housing of animals and the composition of the rotation (so, *MS* and *SC* in Fig. 9.1) (Aarts *et al.*, 1999a, 2000a,b).

Efficiencies of N conversion at De Marke are approximately 0.68 for *SC*, 0.93 for *CF*, 0.23 for *FP* and 0.91 for *MS*, whereas corresponding figures for commercial farms of similar intensity were approximately 0.63, 0.71, 0.19 and 0.83 (Hilhorst *et al.*, 2001). Taking account of the respective *IM* values, the O/I values on De Marke and the commercial farms were 0.32 and 0.16 kg N/kg N, respectively. The corresponding L/O values were 2.2 and 5.3 kg N/kg N. Efficiencies of P conversion at De Marke are approximately 0.94 for *SC*, 0.94 for *CF*, 0.30 for *FP* and 1.00 for *MS*, whereas corresponding figures for commercial farms of similar intensity were approximately 0.62, 0.91, 0.22 and 1.00 (Aarts *et al.*, 2000a). Taking account of the respective *IM* values, the O/I values on De Marke and the commercial farms were 0.87 and 0.33 kg P/kg P, respectively. The corresponding L/O values were 0.2 and 2.0 kg P/kg P. Paul *et al.* (1998) and Jarvis and Aarts (2000) confirmed the existence of similar positive feedbacks in other projects in which the conversion coefficients of ruminant production systems were simultaneously optimized.

De Marke shows that drastic reductions of the N and P losses are feasible, relative to commercial farms of a similar intensity, especially due to lower inputs of nutrients through fertilizers and feed. Averaged over 1994–1997, the annual N loss of 154 kg N/ha was 63% lower than on commercial farms in the region in the same period, and the P loss of only 3 kg P/ha was 92% lower than on commercial farms (Aarts *et al.*, 1999a, 2000a,b). The price of all measures taken was estimated at almost US$3 per 100 kg milk, so slightly more than US$300/ha (De Haan, 2001).

9.5 Intensity and Nutrient Efficiency

Ruminant production systems are considered to be a major contributor of N and P losses to the environment and hence there is a justified call for control (Neeteson, 2000; Schröder *et al.*, 2003b). The development of control measures is a complicated issue, if only because action (farm management) and response (environmental effect) generally do not coincide in space and time. For instance, the regional water quality, including that in coastal zones, is not only determined by the quality of water directly under and along agricultural land, but also by the discharge of water and nutrients from land use other than agriculture. In addition to this, the initial charge from agriculture is at least temporarily muffled by chemical processes in its course to downstream water systems and groundwater (Oenema *et al.*, 1998). Obviously, the impact of agriculture on the regional water quality becomes more evident if agriculture is the dominant form of land use. The spatial scale at which environmental goals have to be achieved strongly determines to what extent 'dilution' from non-agricultural land use can be taken into account.

Many environmental indicators have the dimensions 'units load per unit area' (Schröder *et al.*, 2003b). Extensification of farming systems is then an option to decrease the environmental impact. Such farming systems are attractive because they provide the opportunity to combine production with other functions such as supplying habitats for wild flora and fauna and improving landscape qualities (e.g. allowing grazing cows instead of keeping them indoors in order to reduce N-leaching risks from too high a number of urine patches). To maintain regional production with such systems, more land may be required. Society, in contrast, may desire that land for other purposes than (multifunctional) agriculture. At a higher spatial scale, the realization of a set of multiple goals may, therefore, be better served by intensive, highly productive, specialized farming systems on a limited area, than by extensive, multifunctional farming systems on a larger area. Note that the multiplications of area and loss per unit area may result in a similar environmental load in both cases. Looking at emissions from this perspective is justified, as environmental effects are often exerted at a national or multinational scale, next to the local effects of emissions. Hence, it is extremely relevant to evaluate farming systems also in terms of their environmental impact per unit output (i.e. per unit food produced) and their utilization of other resources such as land, water, energy and labour (De Wit, 1992). However, intensification and the consequent segregation of functions (rural vs. urban, cultural vs. natural, production intensity vs. environmental quality) may alienate agriculture from society. This may conflict with the desire for rural vitality, the need for transparent and traceable food chains and the human eagerness for fluid but respectable objectives such as proximity, self-sufficiency and 'naturalness'. Too strong a focus on the environmental impact per unit output may also stimulate farmers to fully specialize into either arable or livestock production. Such a development can have a negative effect on the nutrient use efficiency of the society as a whole (Schröder *et al.*, 2003a) or may incur a high consumption of fossil energy due to the high energy demands per unit output, inevitable transregional transport of manures, feedstuffs, bedding materials and farm products including animals (Corré *et al.*, 2003). Moreover, highly intensive and specialized farming systems operating at the slippery edge of environmental requirements probably have higher demands for administration and control, for capital and for knowledge transfer, as suggested by cost–benefit analyses of precision farming (Lowenberg-DeBoer and Boehlje, 1996; De Haan, 2001). The price of these requirements must be accurately weighted against the benefits of low nutrient losses per unit output of such intensive farms.

9.6 Conclusions

Nutrient management involves definitely more than just the operational management on the level of the herd, the yard or the field: processes at one level interact with processes at other levels. At each of these levels measures can be taken to promote an efficient use of nutrients. When measures are evaluated at too small spatial and temporal scales, however, measures may be recommended or rejected incorrectly, as antagonistic or synergistic effects at higher levels of integration are insufficiently accounted for. Evaluations at such higher levels should be common practice in research on nutrient management in addition to the ongoing demand for more detailed disciplinary studies. The complexity of these evaluations may require sophisticated optimization techniques (Ten Berge *et al.*, 2000a).

References

Aarts, H.F.M., Biewinga, E.E. and Van Keulen, H. (1992) Dairy farming systems based on efficient nutrient management. *Netherlands Journal of Agricultural Science* 40, 285–299.

Aarts, H.F.M., Habekotté, B., Hilhorst, G.J., Koskamp, G.J., Van der Schans, F.C. and De Vries, C.K. (1999a) Efficient resource management in dairy farming on sandy soil. *Netherlands Journal of Agricultural Science* 47, 153–167.

Aarts, H.F.M., Habekotté, B. and Van Keulen, H. (1999b) Limits to intensity of milk production in sandy areas in the Netherlands. *Netherlands Journal of Agricultural Science* 47, 263–277.

Aarts, H.F.M., Habekotté, B. and Van Keulen, H. (2000a) Phosphorus (P) management in the 'De Marke' dairy farming system. *Nutrient Cycling in Agroecosystems* 56, 219–229.

Aarts, H.F.M., Habekotté, B. and Van Keulen, H. (2000b) Nitrogen (N) management in the 'De Marke' dairy farming system. *Nutrient Cycling in Agroecosystems* 56, 231–240.

Addiscott, T.M. and Powlson, D.S. (1992) Partitioning losses of nitrogen fertilizer between leaching and denitrification. *Journal of Agricultural Science, Cambridge* 118, 101–107.

Bannink, A. and Hindle, V.A. (2003) Prediction of N intake and N excretion by dairy cows from milk data (in Dutch). Report 03 0008567, Animal Sciences Group, Lelystad, The Netherlands (in press).

Bannink, A. and Van Vuuren, A.M. (1998) *Kation–anion-verschil in melkveerantsoenen als methode oem urine pH en amoniakemissie te beinvloeden* [*Cation Anion Difference in Dairy Cow Rations as a Measure to Influence Urine pH and Ammonia Emission*]. Report 98.032. ID-DLO, Lelystad, The Netherlands, 28 pp.

Bannink, A., Valk, H. and Van Vuren, A.M. (1999) Intake and excretion of sodium, potassium, and nitrogen and the effects on urine production by lactating dairy cows. *Journal of Dairy Science* 82, 1008–1018.

Beauchamp, E.G. (1983) Response of corn to nitrogen in preplant and sidedress applications of liquid dairy cattle manure. *Canadian Journal of Soil Science* 63, 377–386.

Beijer, L. and Westhoek, H. (1996) *Meststoffen voor de rundveehouderij: samenstelling, werking en gebruik.* Publicatie 17. Informatie- en Kennis Centrum Landbouw, Ede, The Netherlands, 109 pp.

Berry, P.M., Sylvester-Bradley, R., Philipps, L., Hatch, D.J., Cuttle, S.P., Rayns, F.W. and Gosling, P. (2002) Is the productivity of organic farms restricted by the supply of available nitrogen? *Soil Use and Management* 18, 248–255.

Beukeboom, J.A. (1996) *Forfaitaire gehalten voor de mineralenboekhouding.* Informatie- en Kennis Centrum Landbouw, Ede, The Netherlands, 22 pp.

Bokhorst, J. and Ten Berg, C. (2001) *Handboek mest & compost: behandelen, beoordelen & toepassen.* Louis Bolk Instituut, Driebergen, The Netherlands, 292 pp.

Bullock, D.G. and Bullock, D.S. (1994) Quadratic and quadratic plus-plateau-models for predicting optimal nitrogen rate of corn: a comparison. *Agronomy Journal* 86, 191–195.

Bussink, D.W. and Oenema, O. (1998) Ammonia volatilization from dairy farming systems in temperate areas: a review. *Nutrient Cycling in Agroecosystems* 51, 19–33.

Canh, T.T., Arnink, A.J.A., Mroz, Z., Jongbloed, A.W., Schrama, J.W. and Verstegen, M.W.A. (1998) Influence of electrolyte balance and acidifying calcium salts in the diet of growing-finishing pigs on urinary pH, slurry pH and ammonia volatilization from slurry. *Livestock Production Science* 56, 1–13.

Carton, O.T. and Jarvis, S.C. (2001) N and P cycles in agriculture. In: De Clercq, P., Gertis, A.C., Hofman, G., Jarvis, S.C., Neeteson, J.J. and Sinabell, F. (eds) *Nutrient Management Legislation in European Countries.* Ghent University, Department of Soil Management and Soil Care, Ghent, Belgium, pp. 4–13.

Cartwright, N., Clark, L. and Bird, P. (1991) The impact of agriculture on water quality. *Outlook on Agriculture* 20, 145–152.

Cerrato, M.E. and Blackmer, A.M. (1990) Comparison of models for describing corn yield response to nitrogen fertilizer. *Agronomy Journal* 82, 138–143.

Chadwick, D.R., Sneath, R.W., Phillips, V.R. and Pain, B.F. (1999) A UK inventory of nitrous oxide emissions from farmed livestock. *Atmospheric Environment* 33, 3345–3354.

Chadwick, D.R., Matthews, R., Nicholson, R.J., Chambers, B. and Boyles, L.O. (2002) Management practices to reduce ammonia emissions from pig and cattle manure stores. In: Venglovský, J. and Gréserová, G. (eds) *10th FAO Ramiran Conference on Recycling of Organic Residues in Agriculture,* 14–18 May 2002, Strbske Pleso. University of Veterinary Medicine, Kosice, Slovak Republic, pp. 219–222.

Corré, W.J., Schröder, J.J. and Verhagen, J. (2003) *Energy Use in Conventional and Organic Farming Systems.* Proceedings No. 511. International Fertiliser Society, York, UK, pp. 1–24.

De Clercq, P., Sinabell, F., Hofman, G., Jarvis, S.C., Neeteson, J.J. and Gertis, A.C. (2001) Discussion and conclusions. In: De Clercq, P., Gertis, A.C., Hofman, G., Jarvis, S.C., Neeteson, J.J. and Sinabell, F. (eds) *Nutrient Management Legislation in European Countries.* Ghent University, Department of Soil Management and Soil Care, Ghent, Belgium, pp. 307–343.

Dijkstra, J., Neal, H.D.St.C., Beever, D.E. and France, J. (1992) Simulation of nutrient digestion, absorption and outflow in the rumen: model description. *Journal of Nutrition* 122, 2239–2256.

De Haan, M.H.A. (2001) Economics of environmental measures on experimental farm 'De Marke'. *Netherlands Journal of Agricultural Science* 49, 179–194.

Dewes, T. (1995) Nitrogen losses from manure heaps. *Biological Agriculture and Horticulture* 11, 309–317.

De Wit, C.T. (1992) Resource use efficiency in agriculture. *Agricultural Systems* 40, 125–151.

Dilz, K., Postmus, J. and Prins, W.H. (1990) Residual effect of long-term applications of farmyard manure to silage maize. A case study to test the Sluijsmans–Kolenbrander model. *Fertilizer Research* 26, 249–252.

Dueck, Th.A., Van Dijk, C.J. and Van der Meer, H.G. (2001) *Neglected Emissions of NOx and NHy.* Note 60. Plant Research International, Wageningen, The Netherlands, 28 pp.

Dunlap, T.F., Kohn, R.A., Dahl, G.E., Varner, M. and Erdman, R.A. (2000) The impact of somatotropin, milking frequency, and photoperiod on dairy farm nutrient flows. *Journal of Dairy Science* 83, 968–976.

Galloway, J.N. (1998) The global nitrogen cycle: changes and consequences. *Journal of Environmental Pollution* 102 (Suppl. 1), 15–24.

Greenwood, D.J., Kubo, K., Burns, I.G. and Draycott, A. (1989) Apparent recovery of fertilizer N by vegetable crops. *Soil Science and Plant Nutrition* 35, 367–381.

Hilhorst, G.J., Oenema, J. and Van Keulen, H. (2001) Nitrogen management on experimental dairy farm 'De Marke'; farming system, objectives and results. *Netherlands Journal of Agricultural Science* 49, 135–151.

Hof, G., Vervoorn, M.D., Lenaers, P.J. and Tamminga, S. (1997) Milk urea nitrogen as a tool to monitor the protein nutrition of dairy cows. *Journal of Dairy Science* 80, 3333–3340.

Huijsmans, J.F.M. (1999) Manure application. In: Steenvoorden, J.H.A.M., Bruins, W.J., Van Eerdt, M.M., Hoogeveen, M.W., Hoogervorst, N., Huijsmans, J.F.M., Leneman, H., Van der Meer, H.G., Monteny, G.J. and De Ruijter, F.J. (eds) *Monitoring of National Ammonia Emissions from Agriculture* (in Dutch). Reeks Milieuplanbureau 6, Alterra, Wageningen, The Netherlands, pp. 65–75.

Huijsmans, J.F.M. and De Mol, R.M. (1999) A model for ammonia volatilization after surface application and subsequent incorporation of manure on arable land. *Journal of Agricultural Engineering Research* 74, 73–82.

Isermann, K. (1993) Territorial, continental and global aspects of C, N, P and S emissions from agricultural ecosystems. In: Wollast, R., Mackenzie, F.T. and Chou, L. (eds) *Interactions of C, N, P and S Biochemical Cycles and Global Change.* Springer-Verlag, Berlin–Heidelberg, Germany, pp. 79–121.

Jarvis, S.C. and Aarts, H.F.M. (2000) Nutrient management from a farming systems perspective. European Grassland Federation. *Grassland Science in Europe* 5, 363–373.

Jarvis, S.C. and Pain, B.F. (1990) *Ammonia Volatilization from Agricultural Land.* Proceedings No. 298. International Fertiliser Society, York, UK, 35 pp.

Jarvis, S.C., Stockdale, E.A., Shepherd, M.A. and Powlson, D.S. (1996a) Nitrogen mineralization in temperate agricultural soil: processes and measurement. *Advances in Agronomy* 57, 187–235.

Jarvis, S.C., Wilkins, R.J. and Pain, B.F. (1996b) Opportunities for reducing the environmental impact of dairy farming management: a systems approach. *Grass and Forage Science* 51, 21–31.

Jonker, J.S., Kohn, R.A. and Erdman, R.A. (1998) Using milk urea nitrogen to predict nitrogen excretion and utilization efficiency in lactating cows. *Journal of Dairy Science* 81, 2681–2692.

Kebreab, E., France, J., Beever, D.E. and Castillo, A.R. (2001) Nitrogen pollution by dairy cows and its mitigation by dietary manipulation. *Nutrient Cycling in Agroecosystems* 60, 275–285.

Kohn, R.A., Dou, Z., Ferguson, J.D. and Boston, R.C. (1997) A sensitivity analysis of nitrogen losses from dairy farms. *Journal of Environmental Management* 50, 417–428.

Lammers, H.W. (1983) *Gevolgen van het gebruik van organische mest op bouwland.* Consulentschap voor Bodemaangelegenheden in de landbouw, Wageningen, The Netherlands, 44 pp.

Langmeier, M., Frossard, E., Kreuzer, M., Mäder, P., Dubois, D. and Oberson, A. (2002) Nitrogen fertilizer value of cattle manure applied on soils originating from organic and conventional farming systems. *Agronomie* 22, 789–800.

Lowenberg-DeBoer, J. and Boehlje, M. (1996) Revolution, evolution or dead-end: economic perspectives on precision agriculture. In: Robert, P.C., Rust, R.H. and Larson, W.E. (eds.) *Precision Agriculture. Proceedings of the Third International Conference.* Minneapolis, USA, pp. 923–944.

Magdoff, F.R. and Amadon, J.F. (1980) Yield trends and soil chemical changes resulting from N and manure application to continuous corn. *Agronomy Journal* 72, 161–164.

McDowell, L.R. (1992) Potassium. In: Cunha, T.J. (ed.) *Minerals in Animal and Human Nutrition.* Academic Press, London, UK, pp. 98–114.

Menzel, R.G. (1991) Long-term field research on water and environmental quality. *Agronomy Journal* 83, 44–49.

Monteny, G.J. and Erisman, J.W. (1998) Ammonia emission from dairy cow buildings: a review of measurement techniques, influencing factors and possibilities for reduction. *Netherlands Journal of Agricultural Science* 46, 225–247.

Motavalli, P.P., Bundy, L.G., Andraski, T.W. and Peterson, A.E. (1992) Residual effects of long-term fertilization on nitrogen availability to corn. *Journal of Production Agriculture* 5, 363–368.

Neeteson, J.J. (2000) Nitrogen and phosphorus management on Dutch dairy farms: legislation and strategies employed to meet the regulations. *Biology and Fertility of Soils* 30, 566–572.

Novotny, V. (1999) Diffuse pollution from agriculture: a worldwide outlook. *Water Science and Technology* 39, 1–13.

Oenema, O., Boers, P.C.M., Van Eerdt, M.M., Fraters, B., Van der Meer, H.G., Roest, C.W.J., Schröder, J.J. and Willems, W.J. (1998) Leaching of nitrate from agriculture to groundwater: the effect of policies and measures in the Netherlands. *Environmental Pollution* 102 (Suppl. 1), 471–478.

Oenema, O., Bannink, A., Sommer, S.G. and Velthof, G.L. (2001) Gaseous nitrogen emission from livestock farming systems. In: Follett, R.F. and Hatfield, J.L. (eds) *Nitrogen in the Environment: Sources, Problems, and Management.* Elsevier Science BV, The Netherlands, pp. 255–289.

Paul, J.W., Dinn, N.E., Kannangara, T. and Fisher, L.J. (1998) Protein content in dairy cattle diets affects ammonia losses and fertilizer nitrogen value. *Journal of Environmental Quality* 27, 528–534.

Pierzynski, G.M., Sims, J.T. and Vance, G.F. (2000) *Soils and Environmental Quality*, 2nd edn. CRC Press LCC, New York, 429 pp.

Powlson, D.S. (1997) *Integrating Agricultural Nutrient Management with Environmental Objectives – Current State and Future Prospects.* Proceedings No. 402. International Fertiliser Society, York, UK, 44 pp.

Prins, W.H., Dilz, K. and Neeteson, J.J. (1988) *Current Recommendations for Nitrogen Fertilisation within the*

E.E.C. in Relation to Nitrate Leaching. Proceedings No. 276. International Fertiliser Society, York, UK, 27 pp.

Sawyer, J.E., Schmitt, M.A., Hoeft, R.G., Siemens, J.C. and Vanderholm, D.H. (1991) Corn production associated with liquid beef manure application methods. *Journal of Production Agriculture* 4, 335–344.

Schepers, A.J. and Meijer R.G.M. (1998) Evaluation of the utilization of dietary nitrogen by dairy cows based on urea concentration in milk. *Journal of Dairy Science* 81, 579–584.

Schiere J.B. and Van Keulen, H. (1999) Rethinking high input systems of livestock production: a case study of nitrogen emissions in Dutch dairy farming. *Tropical Grasslands* 33, 1–10.

Schils R.L.M., Van Der Meer, H.G., Wouters, A.P., Geurink, J.H. and Sikkema, K. (1999) Nitrogen utilization from diluted and undiluted nitric acid treated cattle slurry following surface application to grassland. *Nutrient Cycling in Agroecosystems* 53, 269–280.

Schlegel, A.J., Dhuyvetter, K.C. and Havlin, J.L. (1996) Economic and environmental impacts of long-term nitrogen and phosphorus fertilization. *Journal of Production Agriculture* 9, 114–118.

Schröder, J.J. (1999) Effect of split applications of cattle slurry and mineral fertilizer-N on the yield of silage maize in a slurry-based cropping system. *Nutrient Cycling in Agroecosystems* 53, 209–218.

Schröder, J.J. (2002) Restoring farmers' confidence in manure benefits the environment. In: Venglovsky, J. and Gréserová, G. (eds) *10th FAO Ramiran Conference on Recycling of Organic Residues in Agriculture*, 14–18 May 2002, Strbske Pleso. University of Veterinary Medicine, Kosice, Slovak Republic, pp. 401–415.

Schröder, J.J., Ten Holte, L., Van Keulen, H. and Steenvoorden, J.H.A.M. (1993) Effects of nitrification inhibitors and time and rate of slurry and fertilizer N application on silage maize yield and losses to the environment. *Fertilizer Research* 34, 267–277.

Schröder, J.J., Groenwold, J. and Zaharieva, T. (1996a) Soil mineral nitrogen availability to young maize plants as related to root length density distribution and fertilizer application method. *Netherlands Journal of Agricultural Science* 44, 209–225.

Schröder, J.J., Van Dijk, W. and De Groot, W.J.M. (1996b) Effects of cover crops on the nitrogen fluxes in a silage maize production system. *Netherlands Journal of Agricultural Science* 44, 293–315.

Schröder, J.J., Van Asperen, P., Van Dongen, G.J.M. and Wijnands, F.G. (1996c) Nutrient surpluses on integrated arable farms. *European Journal of Agronomy* 5, 181–191.

Schröder, J.J., Ten Holte, L. and Janssen, B.H. (1997a) Non-overwintering cover crops: a significant source of N. *Netherlands Journal of Agricultural Science* 45, 231–248.

Schröder, J.J., Ten Holte, L. and Brouwer, G. (1997b) Response of silage maize to placement of cattle slurry. *Netherlands Journal of Agricultural Science* 45, 249–261.

Schröder, J.J., Neeteson, J.J., Withagen, J.C.M. and Noij, I.G.A.M. (1998) Effects of N application on agronomic and environmental parameters in silage maize production on sandy soils. *Field Crops Research* 58, 55–67.

Schröder, J.J., Neeteson, J.J., Oenema, O. and Struik, P.C. (2000) Does the crop or the soil indicate how to save nitrogen in maize production? Reviewing the state of the art. *Field Crops Research* 66, 151–164.

Schröder, J.J., Aarts, H.F.M., Ten Berge, H.F.M., Van Keulen, H. and Neeteson, J.J. (2003a) An evaluation of whole-farm nitrogen balances and related indices for efficient nitrogen use. *European Journal of Agronomy* 20, 33–44.

Schröder, J.J., Scholefield, D., Cabral, F. and Hofman, G. (2003b) Sound science and feasible legislation on nutrient management in agriculture: friends or foes? *Environmental Science and Policy* 7, 15–23.

Sharpley, A.N., Smith, S.J. and Naney, J.W. (1987) Environmental impact of agricultural nitrogen and phosphorus use. *Journal of Agriculture and Food Chemistry* 35, 812–817.

Sims, J.T., Vasilas, B.L., Gartley, K.L., Milliken, B. and Green, V. (1995) Evaluation of soil and plant nitrogen tests for maize on manured soils of the Atlantic Coastal Plains. *Agronomy Journal* 87, 213–222.

Smith, K.A. and Chambers, B.J. (1993) Utilizing the nitrogen content of organic manures on farms: problems and practical solutions. *Soil Use and Management* 9, 105–112.

Smits, M.C.J., Valk, H., Elzing, A. and Keen, A. (1995) Effect of protein nutrition on ammonia emission from a cubicle house for dairy cattle. *Livestock Production Science* 44, 147–156.

Smits, M.C.J., Valk, H., Monteny, G.J. and Van Vuuren, A.M. (1997) Effect of protein nutrition on ammonia emission from cow houses. In: Jarvis, S.C. and Pain, B.F. (eds) *Gaseous Nitrogen Emissions from Grasslands*. CAB International, Wallingford, UK, pp. 101–107.

Smits, M.C.J., Van Duinkerken, G. and Monteny, G.J. (2001) *Potential reduction of ammonia emission from dairy farms via animal nutrition* (in Dutch). Note P200236. IMAG, Wageningen, The Netherlands, 37 pp.

Sommerfeldt, T.G., Chang, C. and Entz, T. (1988) Long-term annual manure applications increase soil organic matter and nitrogen, and decrease carbon to nitrogen ratio. *Soil Science Society of America Journal* 52, 1668–1672.

Stecker, J.A., Buchholz, D.D., Hanson, R.G., Wollenhaupt, N.C. and McVay, K.A. (1995) Tillage and rotation effects on corn yield response to fertilizer nitrogen on Aqualf soils. *Agronomy Journal* 87, 409–415.

Ten Berge, H.F.M., Van Ittersum, M.K., Rossing, W.A.H., Van de Ven, G.W.J., Schans, J. and Van de Sanden, P.A.C.M. (2000a) Farming options for the Netherlands explored by multi objective modelling. *European Journal of Agronomy* 13, 263–277.

Ten Berge, H.F.M., Withagen, J.C.M., De Ruijter, F.J., Jansen, M.J.W. and Van der Meer, H.G. (2000b) *Nitrogen Responses in Grass and Selected Field Crops; QUADMOD Parameterisation and Extensions for STONE-Application.* Report 24. Plant Research International, Wageningen, The Netherlands, 44 pp.

Tunney, H., Carton, O.T., Brookes, P.C. and Johnston, A.E. (1997) *Phosphorus Losses from Soil to Water.* CAB International, Wallingford, UK, 467 pp.

Unwin, R.J., Pain, B.F. and Whinham, W.N. (1986) The effect of rate and time of application of nitrogen in cow slurry on grass cut for silage. *Agricultural Wastes* 15, 253–268.

Valk, H. (2002) Nitrogen and phosphorus supply of dairy cows, Stickstof- en fosforvoorziening van melkkoeien. PhD thesis, University of Utrecht, Utrecht, The Netherlands, 204 pp.

Valk, H., Sebek, L.B.J. and Beynen, A.C. (2002) Influence of phosphorus intake on excretion and blood plasma and saliva concentrations of phosphorus in dairy cows. *Journal of Dairy Science* 85, 2642–2649.

Van Bruchem, J., Schiere, H. and Van Keulen, H. (1999a) Dairy farming in the Netherlands in transition towards more efficient nutrient use. *Livestock Production Science* 61, 145–153.

Van Bruchem, J., Verhoeven, F., Brussaard, L. and Tamminga, S. (1999b) Diet and manure characteristics in relation to nitrogen flows in a dairy farming system. In: Van der Heide, D., Huisman, E.A., Kanis, E., Osse, J.W.M. and Verstegen, M.W.A. (eds) *Regulation of Feed Intake.* Proceedings of the Fifth Zodiac Symposium. CAB International, Wallingford, UK, pp. 219–225.

Van der Meer, H.G. and Van der Putten, A.H.J. (1995) Reduction of nutrient emissions from ruminant livestock farms. In: Pollott, G.E. (ed.) *Grassland into the 21st Century: Challenges and Opportunities.* Occasional Symposium 29. British Grassland Society, Reading, UK, pp. 118–134.

Van der Meer, H.G., Thompson, R.B., Snijders, P.J.M. and Geurink, J.H. (1987) Utilization of nitrogen from injected and surface-spread cattle slurry applied to grassland. In: Van der Meer, H.G., Unwin, R.J., Van Dijk, T.A. and G.C. Ennik (eds) *Animal Manure on Grassland and Fodder Crops. Fertilizer or Waste?* Developments in Plant and Soil Sciences. Martinus Nijhoff Publishers, Dordrecht, The Netherlands, pp. 47–71.

Van der Molen J., Van Faassen, H.G., Leclerc, M.Y., Vriesema, R. and Chardon, W.J. (1990) Ammonia volatilization from arable land after application of cattle slurry. I. Field estimates. *Netherlands Journal of Agricultural Science* 38, 145–158.

Van Dijk, T.A. (1985) *Leaching of Plant Nutrients from Arable Land as Affected by Annual Applications of Cattle Slurry* (in Dutch). Rapport 2-85. Institute for Soil Fertility, Haren, The Netherlands. 61 pp.

Van Dijk, W. and Brouwer, G. (1998) Nitrogen recovery and dry matter production of silage maize (*Zea mays* L.) as affected by subsurface band application of mineral nitrogen fertilizer. *Netherlands Journal of Agricultural Science* 46, 139–155.

Van Noordwijk, M. (1999) Nutrient cycling in ecosystems versus nutrient budgets of agricultural systems. In: Smaling, E.M.A., Oenema, O. and Fresco, L.O. (eds) *Nutrient Disequilibria in Agroecosystems; Concepts and Case Studies.* CAB International, Wallingford, UK, pp. 1–26.

Vanotti, M.B. and Bundy, L.G. (1994a) Corn and sorghum. An alternative rationale for corn nitrogen fertilizer recommendations. *Journal of Production Agriculture* 7, 243–249.

Vanotti, M.B. and Bundy, L.G. (1994b) Corn nitrogen recommendations based on yield response data. *Journal of Production Agriculture* 7, 249–256.

Van Vuuren, A.M. and Smits, M.C.J. (1997) Effect of nitrogen and sodium chloride intake on production and composition of urine in dairy cows. In: Jarvis, S.C. and Pain, B.F. (eds) *Gaseous Nitrogen Emissions from Grasslands.* CAB International, Wallingford, UK, pp. 95–99.

Velthof, G.L. and Oenema, O. (1997) Nitrous oxide emission from dairy farming systems in the Netherlands. *Netherlands Journal of Agricultural Science* 45, 347–360.

Velthof, G.L., Van Beusichem, M.L. and Oenema, O. (1998) Mitigation of nitrous oxide emission from dairy farming systems. *Environmental Pollution* 102 (Suppl. 1), 173–178.

Vos, J. and Van Der Putten, P.E.L. (2000) Nutrient cycling in a cropping system with potato, spring wheat, sugar beet, oats and nitrogen catch crops. I. Input and offtake of nitrogen, phosphorus and potassium. *Nutrient Cycling in Agroecosystems* 56, 87–97.

Whitmore, A.P. and Schröder, J.J. (1996) Modelling the change in soil organic C and N and the mineralization of N in response to applications of slurry manure. *Plant and Soil* 184, 185–194.

Wolf, J., De Wit, C.T. and Van Keulen, H. (1989) Modeling long-term crop response to fertilizer and soil nitrogen. I. Model description and application. *Plant and Soil* 120, 11–22.

Index

www.cabi-publishing.org/bookshop

Browse Read and Buy

ANIMAL & VETERINARY SCIENCES
BIODIVERSITY CROP PROTECTION
HUMAN HEALTH NATURAL RESOURCES
ENVIRONMENT PLANT SCIENCES
SOCIAL SCIENCES

CABI *Publishing*
A division of CAB International

Online BOOK SHOP

- Subjects
- Search
- Reading Room
- Bargains
- New Titles
- Forthcoming

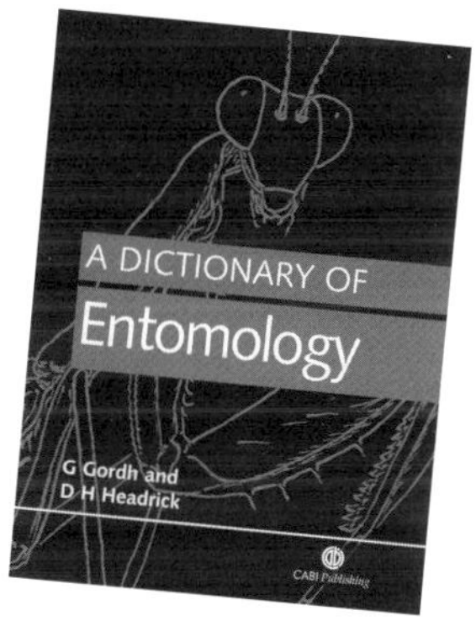

Order & Pay Online!

MasterCard VISA AMERICAN EXPRESS

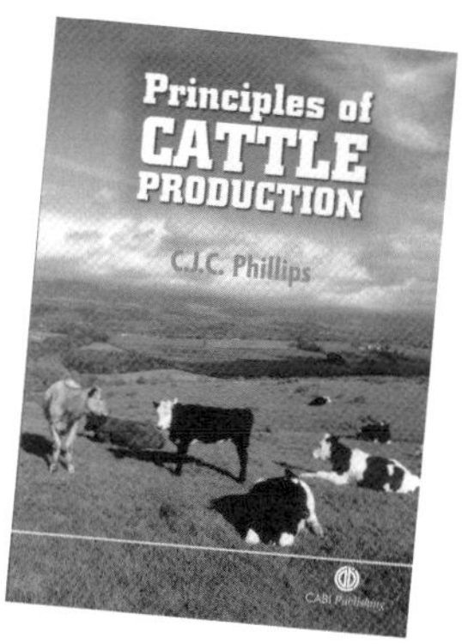

★ FULL DESCRIPTION ★ BUY THIS BOOK ★ BOOK OF THE MONTH

Tel: +44 (0)1491 832111 Fax: +44 (0)1491 829292